Plant Science: Biology and Growth

Plant Science: Biology and Growth

Edited by Austin Balfour

New York

Published by Syrawood Publishing House,
750 Third Avenue, 9th Floor,
New York, NY 10017, USA
www.syrawoodpublishinghouse.com

Plant Science: Biology and Growth
Edited by Austin Balfour

International Standard Book Number: 978-1-68286-585-9 (Hardback)

Cataloging-in-Publication Data

Plant science : biology and growth / edited by Austin Balfour.
p. cm.
Includes bibliographical references and index.
ISBN 978-1-68286-585-9
1. Crop science. 2. Botany. 3. Crop improvement. 4. Growth (Plants). I. Balfour, Austin.
SB91 .P53 2018
631--dc23

TABLE OF CONTENTS

PREFACE

The branch of biology which studies plants is known as plant science. The structure, growth, reproduction, taxonomy and evolution of plants are some of the primary areas studied under plant science. This book provides significant information of this discipline to help develop a good understanding of plant science and related fields. The book, with its detailed analyses and data, will prove immensely beneficial to professionals and students involved in this area at various levels.

This book unites the global concepts and researches in an organized manner for a comprehensive understanding of the subject. It is a ripe text for all researchers, students, scientists or anyone else who is interested in acquiring a better knowledge of this dynamic field.

I extend my sincere thanks to the contributors for such eloquent research chapters. Finally, I thank my family for being a source of support and help.

Editor

1

Differential Expression of Rubisco in Sporophytes and Gametophytes of Some Marine Macroalgae

Chao Wang[1,2], Xiaolei Fan[3], Guangce Wang[1]*, Jianfeng Niu[1], Baicheng Zhou[1]

1 Key Laboratory of Experimental Marine Biology, Institute of Oceanology, Chinese Academy of Sciences, Qingdao, China, 2 Graduate University of the Chinese Academy of Sciences, Beijing, China, 3 Qingdao Institute of Bioenergy and Bioprocess Technology, Chinese Academy of Sciences, Qingdao, China

Abstract

Rubisco (ribulose-1, 5-bisphosphate carboxylase/oxygenase), a key enzyme of photosynthetic CO_2 fixation, is one of the most abundant proteins in both higher plants and algae. In this study, the differential expression of Rubisco in sporophytes and gametophytes of four seaweed species — *Porphyra yezoensis, P. haitanensis, Bangia fuscopurpurea* (Rhodophyte) and *Laminaria japonica* (Phaeophyceae) — was studied in terms of the levels of transcription, translation and enzyme activity. Results indicated that both the Rubisco content and the initial carboxylase activity were notably higher in algal gametophytes than in the sporophytes, which suggested that the Rubisco content and the initial carboxylase activity were related to the ploidy of the generations of the four algal species.

Editor: Sharyn Jane Goldstien, University of Canterbury, New Zealand

Funding: This study was supported by the National Natural Science Foundation of China (No. 30830015), Project for Supporting the National Development (No. 2006BAD09A04) and 863 High Technology projects (No. 2007AA09Z406). The funders had no role in study design, data collection and analysis, decision to publish, or preparation of the manuscript.

Competing Interests: The authors have declared that no competing interests exist.

* E-mail: gcwang@ms.qdio.ac.cn

These authors contributed equally to this work.

Introduction

Algae are an ancient group of unicellular and multicellular organisms. Unlike higher plants, algae have a life history style that is diverse and displays three different styles: species in which the gametophyte predominates, represented by monoecious *Porphyra yezoensis* (Rhodophyte) and dioecious *P. haitanensis* (Rhodophyte); species in which the gametophyte and the sporophyte are of equal dominance, represented by *Bangia fuscopurpurea* (Rhodophyte); and species in which the sporophyte is predominant, represented by *Laminaria japonica* (Phaeophyceae). In contrast to higher plants, both the gametophytes and the sporophytes of algae can photosynthesize and survive independently as autotrophic organisms. Algae are thus ideal organisms for comparative studies on the life histories of different generations of photosynthetic organisms as well as for studies on photosynthesis.

Ribulose-1, 5-biphosphate carboxylase-oxygenase (Rubisco) — a key enzyme of carbon assimilation that is widely distributed in photosynthetic organisms [1] — is crucial for comparative studies among generations of algae. It is also one of the most abundant proteins, accounting for more than 50% of the total soluble protein in C3 plants [2]. In both higher plants and green algae the large subunit (LSU) and small subunit (SSU) of Rubisco are respectively encoded by the chloroplast genome and the nuclear genome [3]. In Rhodophyta and Phaeophyta, however, both the LSU and the SSU are encoded by the chloroplast genome in an operon [4], so the LSU, which provides all the catalytically-essential residues, can directly reflect the activity of the mature Rubisco [5]. A comparative study on Rubisco in different organisms could therefore provide interesting insights into disparities in generations of algae.

In higher plants the abundance of both Rubisco and carboxylase activity are quite different among organs and during the development of leaves and seeds [6–8], while in algae, the activity rates of Rubisco are reported to be conspicuously different among the stipe, meristem and blade of *Laminaria setchellii* [9]. The possible function of Rubisco during the development of organisms has not, however, been investigated. On most occasions, the function of Rubisco has been attributed to the assumed different carbon fixation efficiency. In recent years, it was reported that Rubisco acted in a previously-undescribed metabolic context without the Calvin cycle to increase the efficiency of carbon use during the formation of oil in developing embryos of *Brassica napus* L. (oilseed rape) [10]. Comparative studies on Rubisco activity during algal life cycles, could improve understanding of its function in seaweeds. In this study, we took four species of algae — *Porphyra yezoensis, P. haitanensis, B. fuscopurpurea* (Rhodophyte) and *L. japonica* (Phaeophyceae), which all have dimorphic life cycles, consisting of haploid gametophytes and diploid sporophytes — as typical examples for three types of life history. Comparative research on the levels of mRNA, transcription, enzyme activity and protein expression was then undertaken to investigate differences in Rubisco activity in gametophytes and sporophytes during the process of carbon assimilation.

Materials and Methods

Materials

Filamentous sporophytes of *Porphyra* were cultivated in a 12 hr photoperiod of 36 $\mu mol \cdot m^{-2} \cdot s^{-1}$ fluorescent illumination at 17°C with constant air bubbling. The freshly collected leafy gametophytes were preserved at −20°C after drying in a shady breezy place. The dried blades were resuscitated in 4°C filtrated culture for about a week before use, after which healthy mature blades were selected.

The sporophyte of *L. japonica* was freshly collected and cultivated in a 12 hr photoperiod of 80 $\mu mol{\cdot}m^{-2}{\cdot}s^{-1}$ fluorescent illumination at 9°C for about a month. The gametophytes were cultivated in a 12 hr photoperiod of 20 $\mu mol{\cdot}m^{-2}{\cdot}s^{-1}$ fluorescent illumination at 9°C.

Sporophytes and gametophytes of *B. fuscopurpurea* were both kept in a laboratory culture system. The sporophyte was cultivated in a 12 hr photoperiod of 24 $\mu mol{\cdot}m^{-2}{\cdot}s^{-1}$ fluorescent illumination at 18°C and the gametophyte cultivated in the same photoperiod at 40 $\mu mol{\cdot}m^{-2}{\cdot}s^{-1}$ fluorscent illumination at 13°C.

All the algal culture media was renewed on a weekly basis with bacterial-free seawater containing 0.1 mM KH_2PO_4 and 0.1 mM $NaNO_3$. Before use, the algal material was cleaned three times with pre-cooled filtrated seawater and again (for another three times) with pre-cooled distilled water. Water remaining on materials was wiped away at 4°C in dark conditions. After being weighed, algal material was preserved in liquid nitrogen for later use.

Chloroplast genomic DNA (ctDNA) isolation from leafy gametophytes of *P. yezoensis*

One gram of material was homogenized at 0°C in 10 mL buffer I (containing 5 mM EDTA, 1 mM $MgCl_2$, 1 mM $MnCl_2$, 2 mM $NaNO_3$, 0.5 mM K_2HPO_4 and 0.1 M sucrose). After centrifuging at 350 g at 4°C for 5 min, the supernatant was transferred to another tube and centrifuged at 2,000 g at 4°C for 10 min. The supernatant was discarded and the nuclei pellets were collected. Buffer II (50 mM Tris, 25 mM EDTA, 2% SDS, 50 μg/mL Proteinase K, pH 8.0) was added to resuspend the pellets followed by incubation at 40°C for 3 hr. The pellets were discarded after centrifugation at 12,000 g at 4°C for 10 min. One volume of saturated phenol, saturated phenol-chloroform-isoamyl alcohol (25:24:1) and chloroform-isoamyl alcohol (24:1) were then added, sequentially, to the sample. The aqueous phase was collected and two volumes of absolute ethanol with 0.1 volume (approximately 50 mL) sodium acetate (3 M pH 5.2) were added and gently mixed. The sample was left for at least 20 min at 20°C and then centrifuged at 12,000 g at 4°C for 30 min, after which the supernatant was discarded. After washing with 70% ethanol and drying at room temperature, the pellets were dissolved in 10–50 mL of pure water and preserved at −20°C for use.

Probe synthesis and its hybridizing efficiency determination

The *rbcL* (gene encoding large subunit of Rubisco) of *P. yezoensis* was amplified in a 25 μL PCR reaction system containing 2.5 μL 10× Reaction Buffer (with 15 mM $MgCl_2$), 2.5 μL of dNTPs (2 mM each), 0.5 μL of each primer (10 μM) including rbcl 1 (5′-CTGCAGAAATGGGTTACTGGGATG - 3′), rbcl 2 (5′- CTCGAGGTCTCAACGAAATCAGCT - 3′), 0.3 U of Taq polymerase (Promega), approximately 5–10 ng of template DNA and a corresponding volume of PCR water. The mixtures were subjected to the following conditions: 95°C for 5 min, followed by 35 cycles of 94°C for 45 s, 45°C for 40 s and 72°C for 80 s, and a final extension at 72°C for 10 min. All reactions were conducted on a Mastercycler gradient (Eppendorf). The products were purified for probe synthesis.

Labeling the DNA with Random primer and determination of the hybridizing efficiency were performed according to the protocol of the Dig High Prime DNA Labeling and Detection Starter Kit I (Roche).

Northern blot assay

The sporophytic and gametophytic materials of *P. yezoensis* were separately homogenized in liquid nitrogen; 1 mL Trizol reagent (Invitrogen) was added to 100 mg (F_w) materials and the total RNA was extracted according to the protocol. Northern blot assay was performed according to the protocol of the Dig High Prime DNA Labeling and Detection Starter Kit I (Roche). Prehybridization temperature and hybridization temperature were 50°C, and the incubation temperature with high stringency buffer was 65°C.

Determination of rbcL gene expression in gametophytes and sporophytes by qPCR

Total RNA of four algae, including gametophytes and sporophytes, were extracted with an RNAprep pure plant kit (Tiangen). Reverse transcription of each DNase-treated RNA template was carried out with random primer and M-MLV Reverse Transcriptase (Promega).

The qPCR was performed on a Bio-Rad iQ5 Multicolor Real-time PCR Detection system (Bio-Rad, Hercules, CA, USA) with the total volume of 25 μL reaction mix containing 12.5 μL of 2× SYBR® Premix Ex Taq™ (TaKaRa Biotechnology Co., Ltd.), 2.5 μL of each primer (2 μM), 2.5 μL of diluted cDNA mix and 5 μL of RNase-free water. Thermal cycling conditions were as follows: an initial temperature of 95°C for 3 min, followed by 35 cycles of 95°C for 15 s, annealing for 40 s, and 72°C for 30 s. The confirmation of the specificity of the PCR product was made by analyzing the dissociation curve at the end of each PCR reaction. To maintain consistency, the baseline was set automatically by the software [11]. Besides, the sequences, amplification efficiencies and corresponding annealing temperatures of all the involved primers for each alga were listed in Table 1. To maximize accuracy, both relative and absolute quantification were carried out to determine the different expression level of *rbcL* in gametophytes and sprorophytes in the four algae.

For relative quantification, two house-keeping genes were used as calibrators for each alga in this study to verify successful reverse transcription and to calibrate the cDNA template. The calibrator genes involved in this study were *18S* (gene encoding small subunit ribosomal RNA), *GAPDH* (gene encoding glyceraldehyde-3-phosphate dehydrogenase) and *ACTIN* (gene encoding β-actin), in which *18S* and *GAPDH* were chosen for *P. yezoensis* and *P. haitanensis*, and the *18S* and *ACTIN* were chosen for *B. fuscopurpurea* and *L. japonica*. Three independent reactions together with negative control of each gene were performed and data analysis was carried out using the $2^{-\Delta\Delta Ct}$ method [12].

For absolute quantification, standard curve was generated by plotting the Ct value against the logarithm of the quantity (copy numbers), and three independent serial dilutions of cDNA standard samples were performed at the same time together with unknown samples and negative control reactions [13]. The sample's concentration (ng/μL) was determined by using Qubit fluorometer (invitrogen™). The calculation of the quantities of standard unknown samples and the

Table 1. Primers used in the qPCR assay.

Algae	Gene		Primer sequences (5′-3′)	Size (bp)	Tem (°C)	En (%)
PY	*18S*	F	CGACCGTTTACTGTGAAG	175	58	96.4
		R	GACAATGAAATACGAATGCC			
	rbcL	F	GATGTAGTTCTCAGTTTGGTGGTG	168	58	95.6
		R	ACAAGTTTTAGCAGCGTCCCTC			
	GAPDH	F	CCAACAAGTGGGAGTAAGCG	104	58	95.7
		R	GGACAGAACCGAACAGCGTA			
PH	*18S*	F	CCGTTACTCCTGTGGACCTG	100	56	103
		R	AGGCGAACCTTCAGAGACTTT			
	rbcL	F	AACCATTTATGCGTTGGAGAG	291	56	104
		R	GATTTTTTTGACGAGAGTAAGTAGA			
	GAPDH	F	GGTGCGTCCAAGCATCTGA	113	56	97.3
		R	TGGGTGTAGTCCTGGTCGTTC			
BF	*18S*	F	ACAGGACTTGGGCTCTATTTTG	135	56	100.1
		R	AGATGCTTTCGCAGTGGTTCG			
	rbcL	F	AGCCATTTATGCGTTGGAGAG	234	56	100.3
		R	CATTTTTACGAGCCCAGATTGC			
	ACTIN	F	CCATCTATGAAGGCTACTCGC	193	56	101.4
		R	GCCATCTCCTTGTCGTAATCC			
LJ	*18S*	F	GCCTGAGAAACGGCTACCAC	346	58	105
		R	GACAACCTAATGCCAGCGACAC			
	rbcL	F	CTATTGGTCACCCTGATGGTATTC	174	58	100
		R	CTTTCCATAAATCTAACGCTGCTT			
	ACTIN	F	CCCATCTACGAGGGTTACGCT	249	58	103.8
		R	GTTTCCGTCGGGGAGTTC			

PY: P. Yezoensis; PH: P. Haitanensis; BF: B. Fuscopurpurea; LJ: L. Japonica; F: Forward primer; R: Reverse primer; E_n, the primer's amplification efficiency.

determination of amplification efficiency were performed as follows:

Molecular weight of standard sample (g/mol):

$$W = S \times 330$$

S: base pairs size of single-stranded DNA; 330: the mean molecular weight of single-stranded DNA (Da/bp)

Quantity of standard sample (copies/µL):

$$Q = (C \times 10^{-9})/W \times A$$

C: sample's concentration; A: Avogadro's constant

Linear equation corresponding to standard curve:

$$y = mx + b$$

Quantity of unknown sample (copies/µL):

$$Q' = 10^{(C_t{}' - b)/m}$$

C_t': the C_t value of unknown sample

Amplification efficiency (%):

$$E_n = (10^{-1/m} - 1) \times 100\%$$

Preparation of the total protein and determination of total protein content

Algal material was ground into powder in liquid nitrogen. Buffer III (0.1 M Tris, 1 mM EDTA, 50 M Mascorbic acid., 8 M Urea, 0.1 mM PMSF, 1%Triton X-100, 1% β-mercaptoethano, pH 7.8) was added to the powder material (1 mL/g·F_w) followed by centrifugation at 15,000 g for 30 min. The supernatant was collected as the total protein solution and the total protein content was determined according to the method of Bradford (1976) [14].

Determination of the chlorophyll a content

N, N-dimethylformamide (DMF) was added into the powder obtained from 1 g algal material to extract chlorophyll a in dark conditions at 4°C for 48 hr. Subsequently, centrifugation was carried out at the maximum rotation speed for 10 min and the supernatant was collected and diluted to a final volume of 25 mL. Chlorophyll a content was calculated according to the formulae of Jeffrey & Humphrey (1975) [15].

For Phaeophyta:

$$C(Chla) = (12.65A_{664} - 2.99A_{647} - 0.04A_{625}) \times V_e/(I \times F_w \times 1000)$$

For Rhodophyta:

$$C(Chla) = (11.47A_{664} - 0.40A_{630}) \times V_e/(I \times F_w \times 1000)$$

C (Chla): the content of Chlorophyll a (mg/g·F_w); A: the value of absorbance; V_e: the total volume of the extraction (mL); I: the diameter of the reaction cuvette (cm); F_w: the fresh weight of the algal material (g).

Table 2. The reaction system for determining Rubisco initial carboxylase activity.

Reagents	Volume (mL)
5 mM NADH solution	0.2
50 mM ATP solution	0.2
Rubisco extraction solution	0.1
50 mM Creatine Phosphate solution	0.2
0.2 mM NaHCO3 solution	0.2
Reaction buffer*	1.4
160 U/mL Creatinephosphokinase	0.1
160 U/mL Phosphoglycerate kinase	0.1
160 U/mL glyceraldehyde-3-phosphate Dehydrogease	0.1
dd-water	0.3

Reaction buffer* was 0.1 M Tris-HCl buffer (pH 7.8) containing 12 mM $MgCl_2$ and 0.4 mM EDTA.

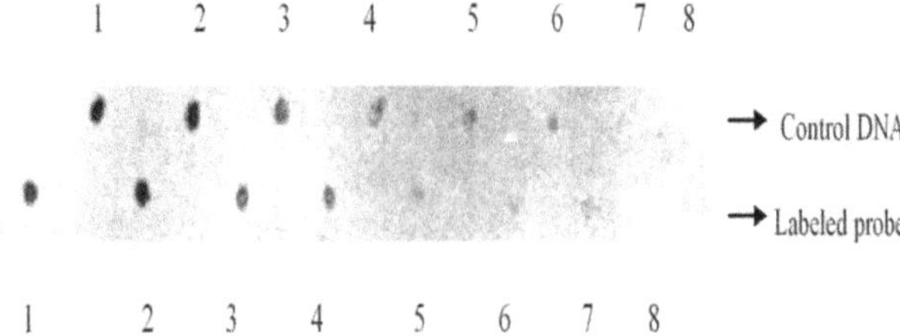

Figure 1. Detection of hybridizing efficiency of the *rbcL* probe from *P. yezoensis*. The Dig High Prime DNA Labeling and Detection Starter Kit I (Roche) was used to detect the hybridizing efficiency of prepared RBCL probe. The probe was labeled with random primer. The control DNA (the first line) which was supplied by the kit was diluted as follows: 1, 1 ng/μL; 2, 10 pg/μL; 3, 3 pg/μL; 4, 1 pg/μL; 5, 0.3 pg/μL; 6, 0.1 pg/μL; 7, 0.03 pg/μL; 8, 0.01 pg/μL; 9, 0 pg/μL. Labeled probe (the second line) was diluted as follows: 1, 1×10^{-2} dilution; 2, 1×10^{-4} dilution; 3, 3.3×10^{-4} dilution; 4, 1×10^{-5} dilution; 5, 3.3×10^{-5} dilution; 6, 1×10^{-6} dilution; 7, 3.3×10^{-6} dilution; 8, 1×10^{-7} dilution; 9, negative control (dd-water).

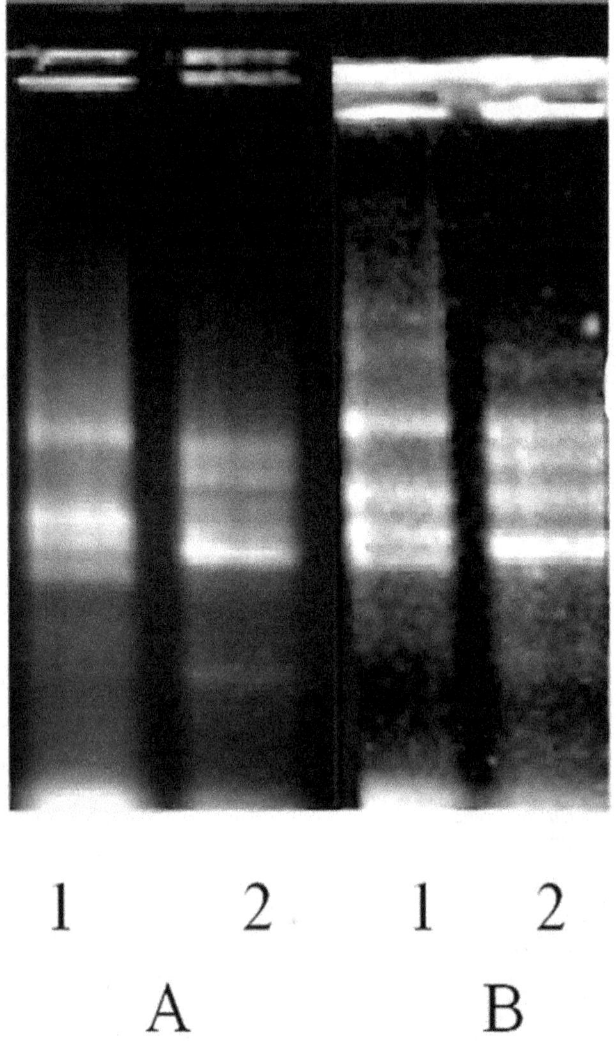

Figure 2. Agarose gel electrophoresis of total RNA of gametophyte and sporophyte of *P. yezoensis*. A, result of native gel electrophoresis. B, result of formaldehyde-agarose gel electrophoresis. 1, Gametophyte. 2, Sporophyte.

SDS-PAGE electrophoresis

SDS-PAGE was carried out to separate protein components according to Laemmli (1970) [16], using 5% condensing gel and 12% separation gel, and using Tris-Gly buffer (0.125 M Tris, 0.96 M Glycine, 0.5%SDS, pH 8.3) as the electrophoresis buffer. One volume of 2× loading buffer (0.25 M Tris-HCl, pH 6.8, 10% SDS (w/v), 0.1% Bromophenol blue (w/v), 0.5% β-Mercatoethanol (v/v)) was added to the total protein solution and heated in boiling water for 5 min, followed by centrifugation at 13,000 g at 4°C for 5 min. Electrophoresis was carried out at a constant voltage of 80 V for 30 min and then 160 V for 90 min. Gels were stained overnight by shaking in the staining solution containing 0.1% Coomassie brilliant Blue R250, 40% Methanol and 1% Acetic acid for protein visualization.

Western blot assay for the LSU of Rubisco of *P. yezoensis*

The Rubisco LSU of *P. yezoensis* was determined by separation on 12% SDS-PAGE followed by electrophoretically transferring to PVDF membrane [17]. After rinsing the membrane with 0.01 M phosphate buffered saline (PBS) three times, for 5 min each, incubation with a 1% (w/v) solution of filtered nonfat dried milk (NFDM) in PBS was carried out for 2 hr with shaking at room temperature. The specific polyclonal antibody against the LSU of Rubisco was diluted with the appropriate ratio of 0.01 M PBS and the incubation occurred during a 12 hr period with shaking at 4°C. The negative control underwent the same treatment at the same time using 1% (w/v) Albumin Bovine V (BSA) instead of the first antibody. Between steps, the membrane was washed three times for 15 min with 0.01 M PBS. The cross-reaction between antibody and protein was detected via a chromogenic reaction in which the anti-rabbit secondary antibody conjugated with horse radish peroxidase (Tiangen).

Determination of Rubisco initial carboxylase activity

Determination of Rubisco initial carboxylase activity was undertaken using the method by Gerard *et al.* (1996) [18].The materials

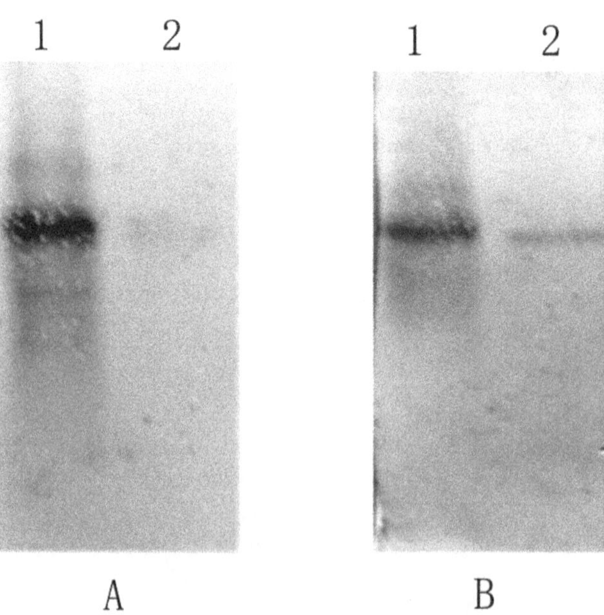

Figure 3. Northern blot results of *P. yezoensis rbcL*. A, comparison result with loading 10 μg total RNA for every lane. B, comparison result with loading total RNA from 25 mg fresh weight material each lane. 1, Gametophyte. 2, Sporophyte.

were ground to a fine powder in liquid nitrogen and homogenized in the pre-cooled Rubisco extraction solution at pH 7.6 (1 mL/g·F_w) containing 40 mM Tris-HCl buffer with 10 mM $MgCl_2$, 0.25 mM EDTA and 5 mM reduced glutathione. The homogenate was centrifuged at 2,000 g for 2 min at 4°C, and the supernatant was collected as the crude solution of Rubisco for the measurement of the enzyme activity. The reaction system was prepared as outlined in Table 2. Distilled water was used as the blank and the OD value at 340 nm of the mixture was recorded as the zero-value. The reaction was initiated by adding 0.1 mL of ribulose-1, 5-bisphosphate (RuBP) into the reaction cuvette and the OD values at 340 nm were recorded every 20 seconds for 3 min. The reduction of the OD value in the first minute was used to calculate the initial carboxylase activity. Since the existence of phosphoglycerate in the extraction solution might affect the determination of Rubisco activity, a control without RuBP was needed. The control system was the same as that of the reaction system except that the crude extraction solution was added at the last step. OD values were determined as described above.The following formula was used to calculate Rubisco initial carboxylase activity:

$$IA = (10N \times \Delta OD)/(6.22 \times 2d\Delta t)$$

IA: the initial carboxylase activity of Rubisco (mmol CO_2/mL·min); ΔOD: the margin of the OD value change in the first minute (OD change of the control was detracted); 6.22: the light density per mmol NADH at 340 nm; N: the dilution factor; d: the diameter of the reaction cuvette (cm); Δt: the time (1 min).

Determination of photosynthetic rate

The experiments of the in vivo chlorophyll fluorescence of PS II (Photosystem II) determination were carried out on Imaging-PAM (Heinz Walz, Effeltrich, Germany), using the automated induction and recovery curve routine provided by the ImagingPam software, with the repetitive application of saturation pulses for assessment of fluorescence. The minimal fluorescence yield (F_0) was determined using 15 min dark-adapted samples. Saturating pulses were applied to obtain the maximum fluorescence yield (F_m).

The following formulas were used to calculate the effective PS II quantum yield (Y II) and the optimum quantum yield (F_v/F_m), which reflect the actual photosynthetic rate and the potential photosynthetic rate respectively [19]:

$$YII = (F_m' - F)/F_m'$$

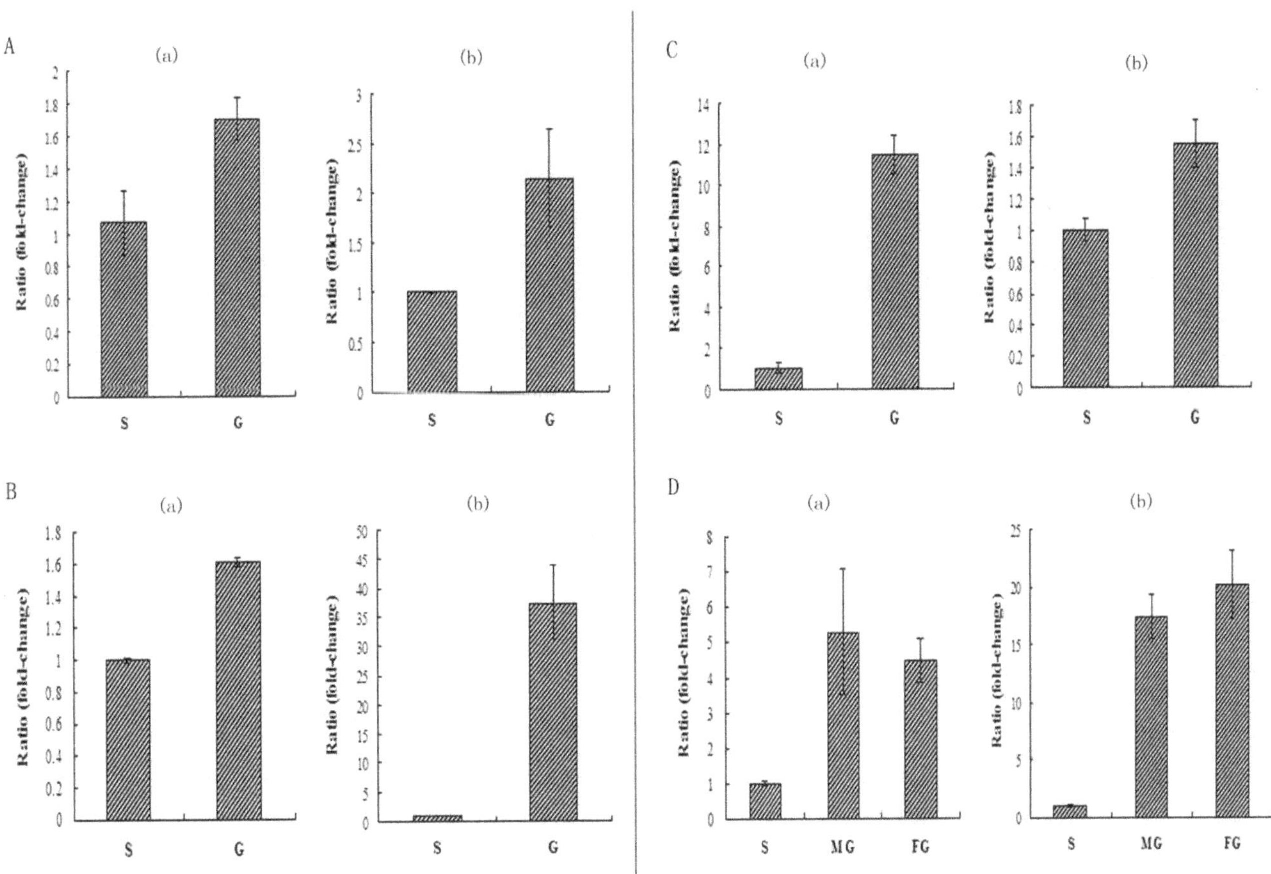

Figure 4. Real-time PCR analysis for the relative expression level (fold change) of rbcL gene in gametophyte and sporophyte of *P. yezoensis, P. haitanensis, B. fuscpurpurea* and *L. japonica.* The cDNA template was amplified and detected by SYBR green using primers of rbcL gene and two reference genes for each alga. A, *P. yezoensis*. B, *P. haitanensis*. C, *B. fuscopurpurea*. D, *L. japonica*. A-(a), B-(a), C-(a) and D-(a) are the results that use *18S* as the reference gene. A-(b) and B-(b) are the results that use *GAPDH* as the reference gene. C-(b) and D-(b) are the results that use *ACTIN* as the calibratorreference gene. S, Sporophyte. G, Gametophyte. MG, Male gametophyte. FG, Female gametophyte. Data are the mean value of three independent experiments (±SD).

$$F_v/F_m = (F_m - F_0)/F_m$$

F_m': the maximum fluorescence yield in illuminated samples

Results

Northern blot assay of *rbcL* mRNA of *P. yezoensis*

In order to obtain the DNA fragment coding for the large subunit of Rubisco, the amplification of *rbcL* from the ctDNA of *P. yezoensis* was carried out. The specificity of the amplified fragment was confirmed by sequence analysis of the PCR products and blasting in the NCBI database. The sequencing result showed 99% similarity with the *rbcL* of *P. yezoensis*. Therefore, it was identified as part gene of *rbcL*, which can be used as the probe (RBCL probe) for northern blot assays.

After labeling, the RBCL probe concentration was estimated to be 100 ng/μL and even a 0.1 pg template produced an obvious hybridizing signal (Fig. 1), indicating that the efficiency which we obtained from the probe was extremely high and sufficient to detect low-abundance mRNAs.

Figure 2 indicates the total RNA from the gametophyte and sporophyte of *P. yezoensis*. It was obvious that whether we used native gel electrophoresis (Fig. 2-A) or formaldehyde-agarose gel electrophoresis (Fig. 2-B), four major bands were visible, which could be the 28S, 26S, 18S and 13S rRNA, top-to-bottom. In Northern blot analysis of total RNA derived from *P. yezoensis*, a single transcript was detected using the RBCL probe. Results are shown in Figure 3, which indicates that, whether based on fresh weight (Fig. 3-B) or the RNA content (Fig. 3-A), the band of Lane 1 (the gametophyte) was brighter than that of Lane 2 (the sporophyte), suggesting that mRNA of *rbcL* was much more abundant in the gametophyte than in the sporophyte.

Analysis of rbcL gene expression in gametophytes and sporophytes of four algae

Relative and absolute quantification of qPCR were carried out to determine differences in the relative (Fig. 4) and absolute expression

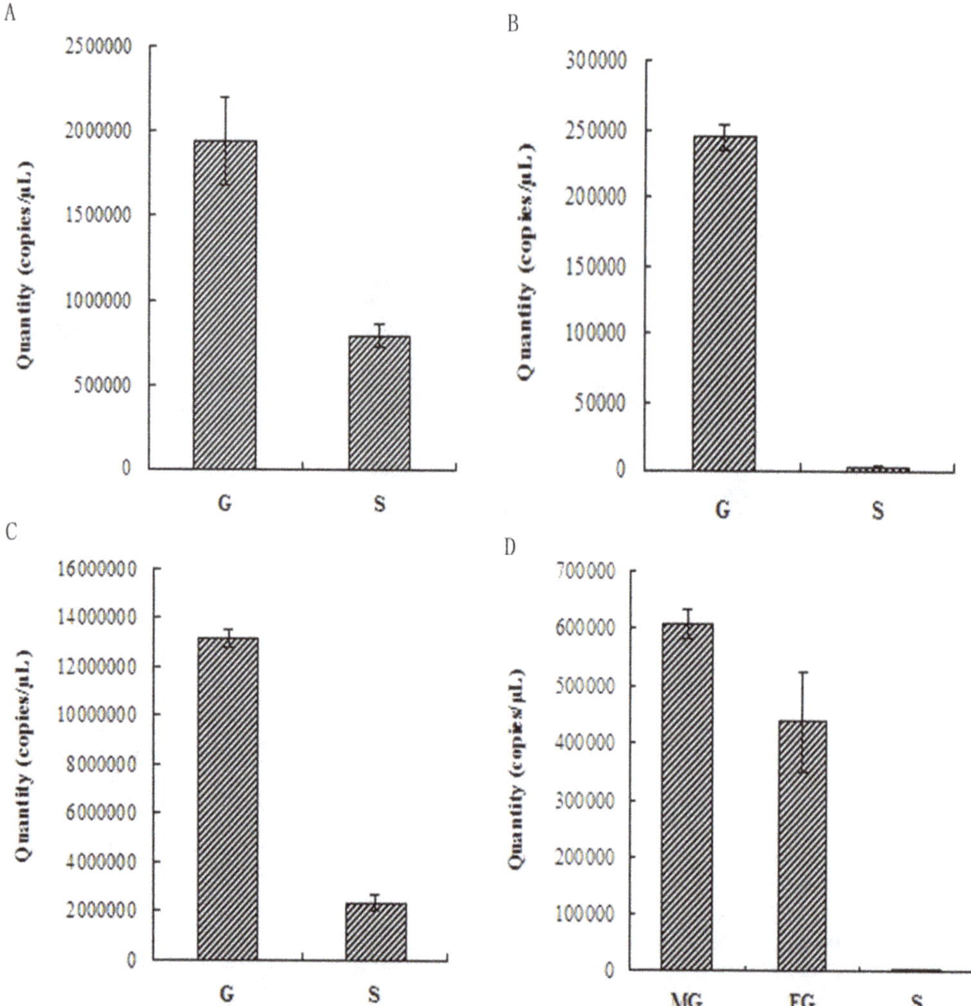

Figure 5. Real-time PCR analysis for the absolute quantification (copy number, copies/μL) of rbcL gene in gametophyte and sporophyte of *P. yezoensis, P. haitanensis, B. fusc-purpurea* and *L. japonica*. Serial diluted cDNAs of each standard samples and unknown samples were amplified and detected by SYBR green and primers of rbcL gene. The quantity (copy number, copies/μL) of rbcL gene of each sample was calculated according to the corresponding standard curve, which was generated by plotting the Ct value against the logarithm of the quantities of the standard samples. A, *P. yezoensis*. B, *P. haitanensis*. C, *B. fuscopurpurea*. D, *L. japonica*. S, Sporophyte. G, Gametophyte. MG, Male gametophyte. FG, Female gametophyte. Data are the mean value of three independent experiments (±SD).

level of the rbcL gene (Fig. 5) respectively between the gametophytes and sporophytes of *P. yezoensis*, *P. haitanensis*, *B. fuscopurpurea* and *L. japonica*. The amplification specificity for the genes of 18S, rbcL, GAPDH and ACTIN were determined by analyzing the dissociation curves of PCR products. Only one peak presented in the dissociation curves for both genes, indicating that the amplifications were specific. From the figures it is obvious that, in the four algae, higher levels of *rbcL* expression were observed in gametophytes with a significant difference both on relative (Fig. 4) and on absolute (Fig. 5) quantification ($P<0.01$), especially for absolute quantification, in which the expression level of *rbcL* in the gametophytes were above ten times more than that in the sporophyte ($P<0.01$) (Fig. 5).

SDS-PAGE assay of the LSU of Rubisco

The total protein of *P. yezoensis* was separated by means of SDS-PAGE, and a 55 kDa distinct band showed up in western blot analysis using Rubisco LSU antibody (Fig. 6-E). Accordingly, the 55 kDa band was identified as the LSU of Rubisco. Since the molecular weights of LSU of Rubisco from *P. yezoensis*, *P. haitanensis*, *B. fuscopurpurea* and *L. japonica* have been reported to be equal, the total protein of *P. yezoensis* was used as the control in this study, to mark the position of Rubisco LSU.

The content of LSU of Rubisco in the algal gametophytes and sporophytes was compared, using SDS-PAGE. In all of the four species (including gametophytes and sporophytes) involved in this study, a protein band at 55 kDa was detected, which represents the LSU of Rubisco (Fig. 6), and similar results were obtained in the four algal species. Loading volumes were determined according to the equal weight of chlorophyll a (Fig. 6-a), the equal weight of total protein (Fig. 6-b), or the equal fresh weight of living materials (Fig. 6-c). The LSU of Rubisco was expressed much more abundantly in the gametophytes than in the sporophytes (Fig. 6).

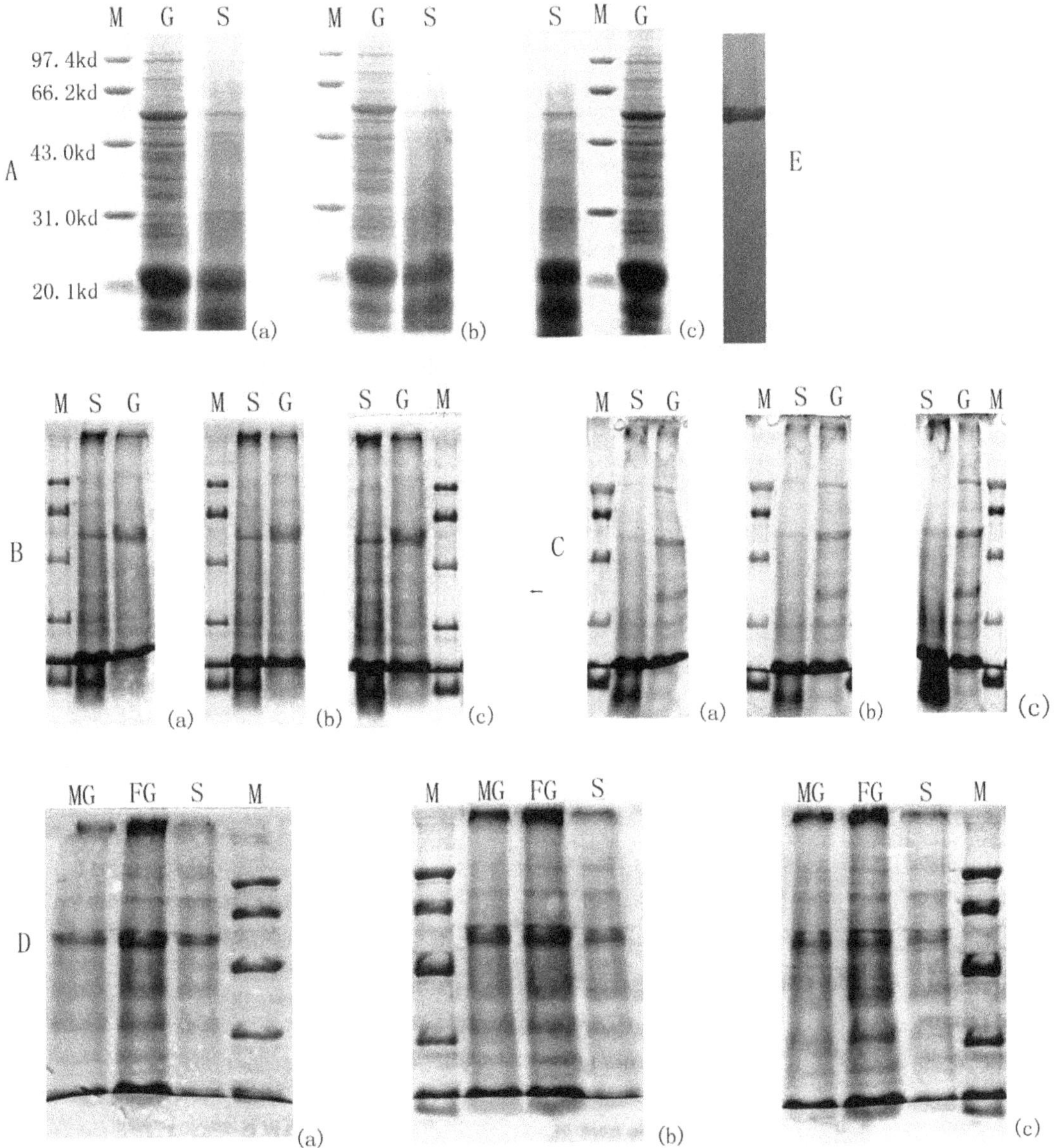

Figure 6. SDS-PAGE electrophoresis of the total soluble protein from gametophytes and sporophyte of *P. yezoensis, P. haitanensis, B. fusc-purpurea* and *L. japonica*. A, *P. yezoensis*. B, *P. haitanensis*. C, *B. fusc-purpurea*. D, *L. japonica*. (a), equal content of chlorophyll a per lane. (b), equal total soluble protein per lane. (c), equal fresh weight of material per lane. E, western blot of Rubisco of *P. yezoensis*. M, Maker. S, Sporophyte. G, Gametophyte. MG, Male gametophyte. FG, Female gametophyte.

Initial carboxylase activity assay of Rubisco

The initial carboxylase activity of Rubisco from gametophytes and sporophytes of *P. yezoensis*, *P. haitanensis*, *B. fuscopurpurea* and *L. japonica* was determined according to the equal fresh weight of living material, the equal weight of total protein, and the equal weight of chlorophyll a, respectively and the detail results were shown in Table 3. From Table 3 it can be seen that, the Rubisco activity rate was about 9-fold greater ($P<0.01$) in the gametophyte than in the sporophyte in *P. yezoensis* — on the basis of equal weight of chlorophyll a and equal fresh weight of living materials. For *P. haitanensis*, the activity of Rubisco was approximately 2.5 times greater ($P<0.01$) in the gametophyte than in the sporophyte on the basis of equal weight of chlorophyll a, equal fresh weight, and equal weight of total protein. As for *B. fuscopurpurea* and *L. japonica*, the ratio of Rubisco activity in the gametophyte and sporophyte reached more than 10 times ($P<0.01$) on the basis of equal weight of chlorophyll a in *B. fuscopurpurea* and equal weight of fresh living material and total protein in *L. japonica*. All the results indicated that the initial carboxylase activities of Rubisco from the gametophytes were much higher than those from the sporophytes.

Assay of photosynthetic rate

The Imaging-PAM analysis was used to determine differences in the photosynthetic rate between the gametophytes and sporophytes of *P. yezoensis*, *P. haitanensis*, *B. fuscopurpurea* and *L. japonica* (Fig. 7 and 8). It was shown in the figures that, for the aglae of *P. yezoensis*, *P. haitanensis* and *B. fuscopurpurea*, the value of effective PS II quantum yield (Y II) of gametophytes was much higher than that of sporophytes ($P<0.01$) (Fig. 7). At the same time, the value of optimum quantum yield (F_v/F_m) of the three algae was also higher in gametophytes than in sporophytes ($P<0.01$) (Fig. 8). On the contrary, for the alga of *L. japonica*, both the value of Y II and F_v/F_m were higher in sporophyte than in gametophyte ($P<0.01$) (Fig. 7, Fig. 8).

Discussion

Generally, three standards are most frequently used in quantifying content and activity of Rubisco in seaweeds. These are: i) per unit of fresh weight of the living material, such as in *Fucus vesiculosus*, *Fucus serratus*, *Cladophora sericea*, *Ulva lactuca*, *Furcellaria fastigiata*, *Chondrus crispus*, *Ceramium rubrum* [20], and *Mastocarpus stellatus* [21] with the shortage of the large deviation caused by the containing water; ii) per unit of content of chlorophyll a, such as in *Enteromorpha clathrat* [22], *Fucus vesiculosus*, *Fucus serratus*, *Cladophora sericea*, *Ulva lactuca*, *Furcellaria fastigiata*, *Chondrus crispus*, *Ceramium rubrum* [20], *Udotea flabellum*, and *Codium decorficatum* [23], with the shortage of the deviation caused by the existence of a large amount of non-chlorophyll photosynthetic pigments; iii) per unit of weight of the total soluble protein, such as in *Ulva lactuca* [24], with the shortage of the deviation caused by the protein in the non-photosynthetic cells.

In order to improve accuracy, in this study the above three standards were all used in quantifying and comparing the carboxylase activity and content of Rubisco from different generations of the four seaweed species — *P. yezoensis*, *P. haitanensis*, *B. fuscopurpurea* (Rhodophyte) and *L. japonica* (Phaeophyceae) — which respectively represent the three main types of life history. Whatever standards we used, the results were highly consistent with each other. In spite of the quantification basis, both the initial carboxylase activity (Tab. 3) and the abundance of Rubisco (Fig. 6) were indeed much higher in the algal haploid gametophytes than in the diploid sporophytes. Moreover, results when using northern blot assay for *P. yezoensis* (Fig. 2) indicated that the mRNA content in the gametophyte was more abundant than that in the sporophyte. When qPCR was used to determine the relative expression level of rbcL gene in gametophytes and sporophytes of the four algae, the results showed that the level of *rbcL* expression in gametophytes was much higher than that in sporophytes ($P<0.01$) (Fig. 4, Fig. 5). All these results indicated that, in terms of levels of transcription and translation, the expression level of Rubisco was much higher in the haploid gametophyte than in the diploid sporophyte.

Although the relationships between protein activity and ploidy are still not certain, the gene dosage effect has been reported in a lot of higher plants: i.e., with the chromosome ploidy increasing, the gene copies and protein content increase accordingly. In higher plants, the relationship between protein activity and ploidy has been studied since 1960s [25–28]. Nevertheless, the gene

Table 3. Initial carboxylase activity of Rubisco from different generations of *P. yezoensis*, *P. haitanensis*, *L. japonica* and *B. fuscopurpurea*.

Algal materials	Initial carboxylase activity of Rubisco		
	mmol CO_2/mg·chla·min	mmol CO_2/mg·pro·min	mmol CO_2/g·F_w·min
L. japonica S	$3.4\times10^{-3}\pm1.30E-4$	$1.2\times10^{-4}\pm7.90E-06$	$4.8\times10^{-4}\pm1.20E-05$
L. japonica M	$4.1\times10^{-3}\pm5.50E-5$	$1.2\times10^{-3}\pm1.90E-04$	$5.3\times10^{-3}\pm3.60E-05$
L. japonica F	$4.0\times10^{-3}\pm1.80E-4$	$4.1\times10^{-3}\pm6.10E-05$	$5.3\times10^{-3}\pm1.40E-04$
P. haitanensis S	$1.0\times10^{-1}\pm8.00E-03$	$2.4\times10^{-2}\pm2.00E-03$	$1.4\times10^{-1}\pm2.20E-02$
P. haitanensis G	$2.7\times10^{-1}\pm5.10E-02$	$5.9\times10^{-2}\pm6.00E-03$	$3.6\times10^{-1}\pm3.00E-02$
B. fuscopurpurea S	$1.7\times10^{-2}\pm9.50E-04$	$5.2\times10^{-3}\pm1.20E-04$	$2.1\times10^{-2}\pm1.40E-03$
B. fuscopurpurea G	$5.7\times10^{-1}\pm1.30E-02$	$3.3\times10^{-2}\pm3.00E-03$	$2.1\times10^{-1}\pm6.00E-03$
P. yezoensis S	$2.8\times10^{-3}\pm1.30E-04$	$2.3\times10^{-4}\pm1.10E-05$	$3.8\times10^{-3}\pm1.60E-04$
P. yezoensis G	$4.5\times10^{-2}\pm2.50E-03$	$9.4\times10^{-4}\pm2.20E-05$	$4.3\times10^{-2}\pm1.90E-03$

Data are the mean value of three independent experiments (±SD).
S: sporophyte;
G: gametophyte;
M: male gametophyte;
F: female gametophyte.

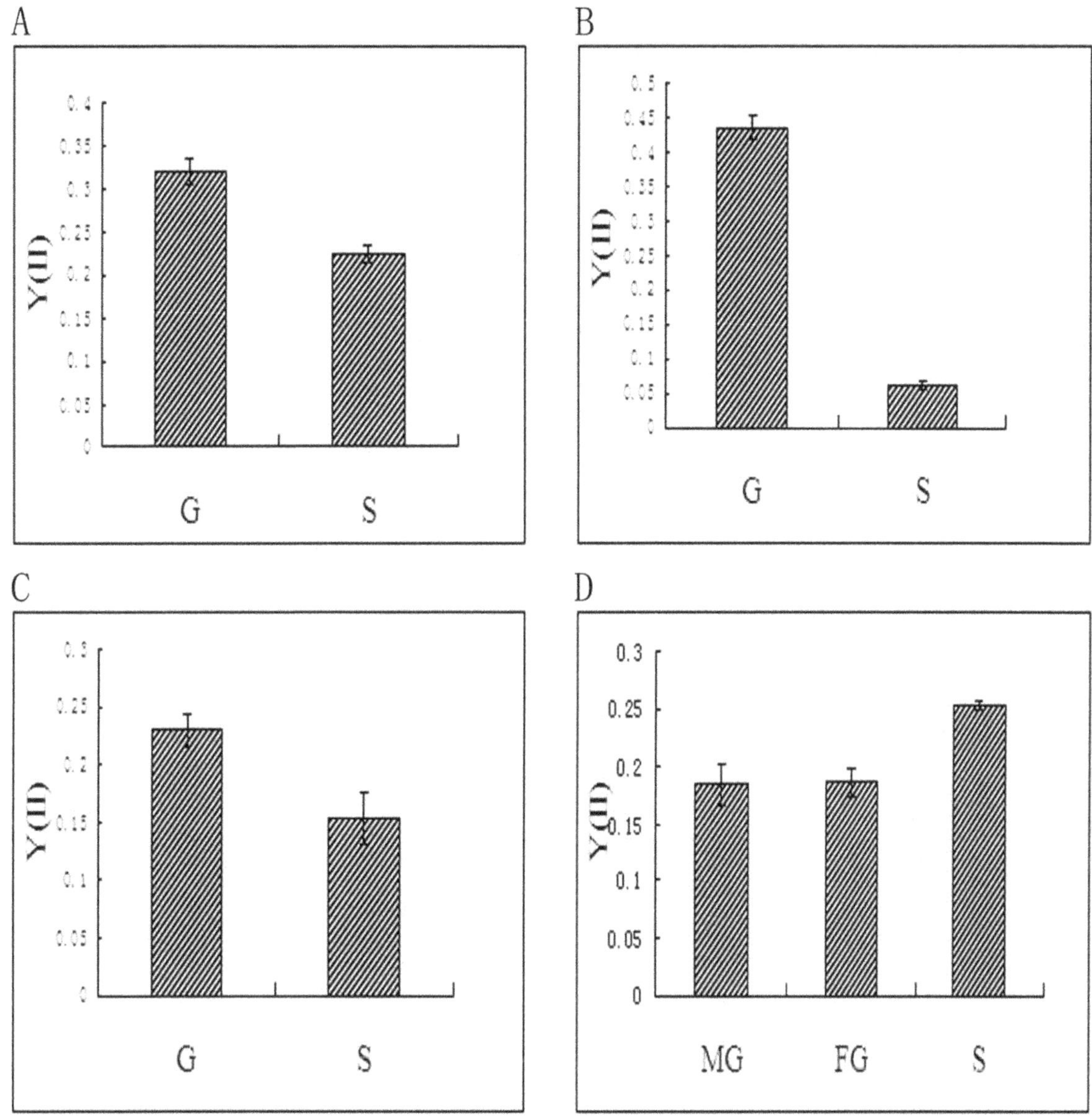

Figure 7. Comparison of the effective PS II quantum yield (Y II) between gametophyte and sporophytes of *P. yezoensis, P. haitanensis, B. fuscopurpurea* and *L. japonica*. A, *P. yezoensis*. B, *P. haitanensis*. C, *B. fuscopurpurea*. D, *L. japonica*. S, Sporophyte. G, Gametophyte. MG, Male gametophyte. FG, Female gametophyte. Data are the mean value of three independent experiments (±SD).

dosage effect is displayed in two completely different ways in higher plants. In one case this effect is conspicuous, as indicated by such as Meyers *et al.* (1982) who demonstrated the gene dosage effect in *Medicago sativa* [29]. In the same year, Dean *et al.* (1982) reported that in *Triticum* species, the content of Rubisco was positively related to chromosome ploidy [30]. Leech *et al.* (1985) also found that the gene dosage effect exists in the *Triticum* species, but he pointed out that this did not apply in two genotypes of hexaploid *Triticum*, which may be due to effect of different genotypes [31]. Warner *et al.* (1989) found that the carboxylase activity of Rubisco also had a positive relationship with chromosome ploidy in the C4 plant *Atriplex confertifolia* [32]. In a different case, however, Bhaskaran *et al.* (1983) found that in a pollen-derived haploid *Saintpaulia ionantha*, Rubisco activity was much higher than that in other multiploids, and in 1987, they again successfully induced a haploid, *Nicotiana tabacum* cv. Burley Ky 14, from the pollen and found that, as with the haploids derived from the same pollen, the specific activity of the Rubisco to bind the CO_2 was much higher than in the diploid parents. They speculated this might be attributed to the mono-genotype [33,34].

According to the results of our study, in which we used four species of algae that had different dominant phases in their lifecycles, there were indications that Rubisco activity (Tab. 3) and content (Fig. 6) were higher in haploid than in diploid stages, which seemed contrary to the gene dosage effect. Results from our study were therefore quite different from the former case of gene dosage effect that existed in most of the higher plants but very similar to the latter one. We therefore speculate that the latter case could be relevant to the distinct encoding character of Rubisco in Rhodophyta and Phaeophyta. In green algae and higher plants, the LSU and SSU of Rubisco are encoded by the chloroplast genome and the nuclear genome respectively [3], and the assemblage and activity of Rubisco are regulated according to

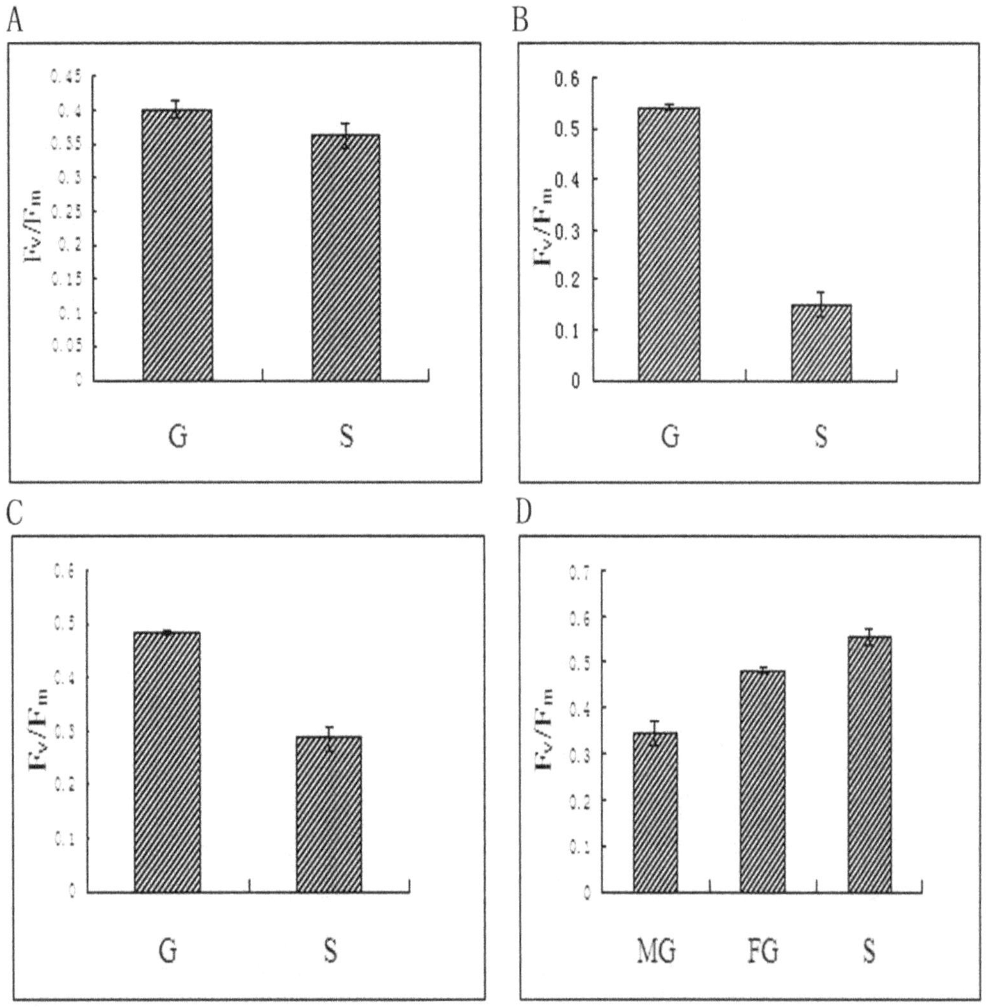

Figure 8. Comparison of the optimum quantum yield (F_v/F_m) between gametophytes and sporophytes of *P. yezoensis, P. haitanensis, B. fuscopurpurea* and *L. japonica*. A, *P. yezoensis*. B, *B. fuscopurpurea*. C, *L. japonica*.A, *P. yezoensis*. B, *P. haitanensis*. C, *B. fuscopurpurea*. D, *L. japonica*. S, Sporophyte. G, Gametophyte. MG, Male gametophyte. FG, Female gametophyte. Data are the mean value of three independent experiments (±SD).

the content of the nuclear-genome-encoded SSU [35,36]. The nucleic gene dosage effect is therefore obvious. In Rhodophyta and Phaeophyta, however, both LSU and SSU of Rubisco are encoded by an operon in the chloroplast genome [4], so nucleic gene dosage effect may only have a minimal effect on the assemblage of Rubisco.

In addition, it was reported in previous studies that gene candidates with differential expression in the sporophytes and gametophytes of some *Porphyra* species have been identified, which significantly contributed to the study of the transition mechanism in different algal generations [37,38]. Nevertheless, as a result of the different habitats of sporophytes and gametophytes, the gene candidates contained a lot of genes that might have been transcribed as an adaptation to the environment. In C3 higher plants, the up-regulated expression of Rubisco was always associated with high rates of photosynthesis, particularly concerning primary production. And the photosynthetic rate assays (Fig. 7, Fig. 8) showed that, for the alage of *P. yezoensis, P. haitanensis* and *B. fuscpurpurea*, the comparison results of photosynthetic rates were corresponding to the comparison results of Rubisco contents (Fig. 6) and activities (Tab. 3) between gametophyte and sporophytes. While for *L. japonica*, the higher Rubisco content (Fig. 6) and activity (Tab. 3) in gametophyte was corresponding to the lower photosynthetic rate (Fig. 7, Fig. 8). Xu *et al.* (1991) had reported that PEPCK (Phosphoenolpyruvate carboxykinase), another important enzyme in carbon fixation in C4 plants, had very high activity in the sporophyte of *L. japonica* [39]. Reiskind *et al.* (1991) also reported that C4-like photosynthetic characteristics in the green alga *Udotea flabellum*, for which the high PEPCK activity with low PEPC (phosphoenolpyruvate carboxylase) activity was a novel characteristic [23]. Fan *et al.* (2007) found a similar result in *P. haitanensis* and speculated that a C4-like carbon fixation pathway may exist in the sporophyte of *P. haitanensis* [37]. Consequently, we assumed that in the sporophyte of these algae, the major carbon fixation pathway may be a C4-like carbon fixation pathway. Since the C4-like pathway could accumulate CO_2 much more efficiently than would be the case for a 'Rubisco pathway', a high abundance of Rubisco would not be necessary. Rubisco could therefore have lower abundance and activity in the sporophytes of these algae.

Comparisons between gametophyte and sporophyte generations, on the level of mRNA content, in higher plants have been carried out since the 1980s. In angiosperms, it is reported that pollen mRNAs of different classes (i.e. the haploid gametophyte generation) were much more abundant than in the corresponding classes of mRNAs in shoots (i.e. the diploid sporophyte generation) of *Tradescantia*/sp. and *Zea mays* [40,41]. This phenomenon was assumed to have resulted from a need to accumulate and store mRNAs prior to rapid translation during pollen germination [42]. Indeed, in most plants the germination and pollen tube growth are relatively rapid events [43], which need corresponding transcripts to be stored and translated during gametophyte germination [42]. Therefore, according to our results, we speculated that the haploid gametophyte of the four algae in our study had a higher abundance of mRNA than that in the diploid sporophyte, the purpose being to facilitate rapid accumulation and storage of transcripts for later differentiation in male and female gametophytes.

To sum up, according to our results, the expression of Rubisco was higher — in terms of the level of transcription, translation and enzymic activity — in haploid gametophytes than in diploid sporophytes. It was speculated that, for the four algae species studied, the nucleic gene dosage had little effect because of the distinct characteristics of Rubisco structure in Rhodophyta and Phaeophyta. Besides, a C4-like carbon fixation pathway may exist in the sporophytes of these algae, so a high abundance of Rubisco would not be necessary for the sporophytes. And the mRNA content in gametophytes was higher than that of sporophytes, which would serve the purpose of accumulating and storing transcripts for later male and female differentiation. The above speculations may explain the reason why Rubisco activity and content in the gametophytes were higher than that in the sporophytes.

Author Contributions

Conceived and designed the experiments: GW BZ. Performed the experiments: CW XF. Analyzed the data: CW XF JN. Contributed reagents/materials/analysis tools: CW XF JN. Wrote the paper: CW XF.

References

1. Zhang G, Wang W, Zou Q (2004) Molecular biology of RuBisCO activase. Plant Physiol Commun 40: 633–637 (in Chinese).
2. Wang WG (1985) Experimental handbook of plant physiology. Shanghai scientific & Technical Publishers. pp 125–128 (in Chinese).
3. Valentin K, Zetsche K (1990) Structure of the Rubisco operon from the unicellular red alga *Cyanidium caldarium*: Evidence for a polyphyletic origin of the plastids. Mol Gen Genet 222: 425–430.
4. Ga YC, Hwan SY, Han GC, Kazuhiro K, Sung MB (2001) Phylogeny of family *Sctyosiphonaceae* (Phaeophyta) from Korea based on sequences of plastid-encoded Rubisco spacer region. Algae 16(2): 145–150.
5. Satoko I, Atsuko M, Seishiro A, Motomi I, Yasuro K, et al. (2009) Molecular adaptation of *rbcL* in the heterophyllous aquatic plant *Potamogeton*. PlosOne 4(2): 1–7.
6. Wanner LA, Gruissem W (1991) Expression dynamics of the tomato rbcS gene family during development. Plant Cell 3(12): 1289–1303.
7. Maayan I, Shaya F, Ratne K, Mani Y, Lavee S, et al. (2008) Photosynthetic activity during olive (Olea europaea) leaf development correlates with plastid biogenesis and Rubisco levels. Physiol Plantrum 134(3): 547–558.
8. Patel M, Berry JO (2008) Rubisco gene expression in C4 plants. J Exp Bot 59(7): 1625–1634.
9. Cabello-Pasini A, S Alberte R (2001) Enzymatic regulation of photosynthetic and light-independent carbon fixation in *Laminaria setchellii* (Phaeophyta), *Ulva lactuca* (Chlorophyta) and *Iridaea cordata* (Rhodophyta). Revista Chilena de Historia Natural 74: 229–236.
10. Schwender J, Goffman F, Ohlrogge JB, Shachar-Hill Y (2004) Rubisco without the Calvin cycle improves the carbon efficiency of developing green seeds. Nature 432: 779–782.
11. Zhang BY, Yang F, Wang GC, Peng G (2009) Cloning and quantitative analysis of the carbonic anhydrase gene from Porphyra yezoensis. J Phycol 45: 290–296.
12. Schmittgen TD, Zakrajsek BA, Mills AG, Corn V, Singer MJ, et al. (2000) Quantitative reverse transcription-ploymerase chain reaction to study mRNA decay: comparison of endpoint and real-time methods. Analytical Biochemistry 285: 194–204.
13. Bustin SA (2000) Absolute quantification of mRNA using real-time reverse transcription polymerase chain reaction assays. Journal of Molecular Endocrinology 25: 169–193.
14. Bradford MM (1976) A rapid and sensitive method for the quantitation of microgram quantities of protein utilizing of protein dye dinding. Anal Biochem 72: 248–254.
15. Jeffrey SW, Humphrey GF (1975) New spectrophotometric equations for determining chlorophylls a, b, c_1 and c_2 in higher plants, algae and natural phytoplankton. *Biochem*. Physiol Pflanzen 167: 191–194.
16. Laemmli UK (1970) Cleavage of structural proteins during the assembly of the head of bacteriophage T4. Nature 227: 680–685.
17. Malcolm MJ, Nguyen NY, Liu TY (1988) Reproducible high yield sequencing of proteins electrophoretically separated and transferred to an inert support. The Journal of Biological Chemistry 263(13): 6005–6008.
18. Gerard VA, Driscoll T (1996) A spectrophotometric assay for RuBisCO activity: application to the kelp *Laminaria saccharina* and implications for radiometric assays. J Phycol 32: 880–884.
19. Lin AP, Wang GC, Yang F, Pan GH (2009) Photosynthetic parameters of sexually different parts of *Porphyra katadai* var. *hemiphylla* (Bangiales, Rhodophyta) during dehydration and re-hydration. Planta 229: 803–810.
20. Beer S, Sand-Jensen K, Vindbaek MT, Nielsen SL (1991) The carboxylase activity of RuBisCO and the photosynthetic performance in aquatic plants. Oecologia 87: 429–434.
21. Dudgeon SR, Davison IR, Vadas RL (1989) Effect of freezing on photosynthesis of intertidal macroalgae: relative tolerance of Chondrus crispus and Mastocarpus stellatus (Rhodophyta). Mar Biol 101: 107–114.
22. He PM, Wu QL, Wu WN, Lu W, Zhang DB, et al. (2004) Pyrenoid ultrastructure and molecular localization of RuBisCO and RuBisCO activase in *Enteromorpha clathrata*. J fish China 28(3): 255–260.
23. Reiskind JB, Bowes G (1991) The role of phosphoenolpyruvate carboxykinase in a marine macroalga with C4-like photosynthetic characteristics. Proc Natl Acad Sci USA 88: 2883–2887.
24. Bischof K, Kräbs G, Wiencke C, Hanelt D (2002) Solar ultraviolet radiation affects the activity of ribulose-1, 5-bisphosphate carboxylase-oxygenaseand the composition of photosynthetic and xanthophyll cycle pigments in the intertidal green alga *Ulva lactuca* L. Planta 215: 502–509.
25. Ciferri O, Sora S, Tiboni O (1969) Effect of gene dosage on tryptophansynthetase activity in *Saccharomyces cerevisiae*. Genetics 61: 567–576.
26. Carlson PS (1972) Locating genetic loci with aneuploids. Mol Gen Genet 114: 273–280.
27. Demaggio AE, Lambrukos J (1974) Polyploidy and gene dosage effects on peroxidase activity in ferns. Biochem. Genet 12: 429–440.
28. Guern M, Gherve L (1980) Polyploidy and aspartate transcarbamylase activity in *Hippocrepis comosa*. Planta 149: 27–33.
29. Meyers SP, Nichols SL, Giani RB, Molin WT, Schrader LE (1982) Ploidy effects in isogenic populations of Alfalfa. 1. Ribulose-1, 5-bisphosphate carboxylase, soluble protein, chlorophyll and DNA of leaves. Plant Physiol 70: 1704–1709.
30. Dean C, Leech RM (1982) Genome expression during normal leaf development. 2. Direct correlation between ribulose bisphosphate carboxylase content and nuclear ploidy in a polyploid series of wheat. Plant Physiol 70: 1605–1608.
31. Leech RM, Leese BM, Jellings AJ (1985) Variation in cellular ribulose-l, 5-bisphosphate carboxylase content in leaves of Triticum genotypes at three levels of ploidy. Planta 166: 259–263.
32. Warner DA, Edwards GE (1989) Effects of Polyploidy on Photosynthetic Rates, Photosynthetic Enzymes, Contents of DNA, Chlorophyll, and Sizes and Numbers of Photosynthetic Cells in the C_4 Dicot *Atriplex confertifolia*. Plant Physiol 91: 1143–1151.
33. Bhaskaran S, Smith RH, Finer JJ (1983) Ribulose Bisphosphate Carboxylase Activity in Anther-Derived Plants of Saintpaulia ionantha Wendl. Shag. Plant Physiol 73: 639–642.
34. Bhaskaran S, Burdick PJ, Smith RH (1987) Crystallization of Ribulose, 1, 5-Bisphosphate Carboxylase of High Specific Activity from Anther-Derived Haploid Plants of *Nicotiana tabacum*. J Exp. Bot 38(2): 270–276.
35. Rodermel S (1999) Subunit control of RuBisCO biosynthesis-a relic of an endosymbiotic past. Photosyn Res 59: 105–123.
36. Miao YG, Li LR (1996) Molecular hybridization of RuBisCO subunits between rice and tobacoo. Acta Phytophysiol Sin 22: 40–44.
37. Fan XL, Fang YJ, Hu SN, Wang GC (2007) Generation and analysis of 5318 express sequence tags (ESTs) from filamentous sporophyte of *Porphyra haitanensis* (Rhodophyte). J Phycol 43: 1287–1294.
38. Asamizu E, Nakajima M, Kitade Y, Saga N, Nakamura Y, et al. (2003) Comparison of RNA expression profiles between the two generations of *Porphyra yezoensis* (Rhodophyta), based on expressed sequence tag frequency analysis. J Phycol 39: 923–930.
39. Xu ZM, Yao NY, Li JZ (1991) Studies on the activity of PEPck in *L. japonica*. Mar Sci 2: 41–45 (in Chinese).

40. Willing RP, Mascarenhas JP (1984) Analysis of the complexity and diversity of mRNA from pollen and shoots of *Tradescantia*. Plant Physiol 75: 865–868.
41. Willing RP, Bashe D, Mascarenhas JP (1988) An analysis of the quantity and diversity of messenger RNAs from pollen and shoots of *Zea mays*. Theor Appl Genet 75: 751–753.
42. Twell D (1994) The diversity and regulation of gene expression in the pathway of male gametophyte development. Molecular and Cellular Aspects of Plant Reproduction, Scott RJ, Stead AD, eds. pp 83–135.
43. Mascarenhas JP (1989) The male gametophyte of flowering plants. The Plant Cell 1: 657–664.

2

Adjusted Light and Dark Cycles Can Optimize Photosynthetic Efficiency in Algae Growing in Photobioreactors

Eleonora Sforza[1], Diana Simionato[2], Giorgio Mario Giacometti[2], Alberto Bertucco[1], Tomas Morosinotto[2]*

1 Dipartimento di Ingegneria Industriale DII, Università di Padova, Padova, Italy, **2** Dipartimento di Biologia, Università di Padova, Padova, Italy

Abstract

Biofuels from algae are highly interesting as renewable energy sources to replace, at least partially, fossil fuels, but great research efforts are still needed to optimize growth parameters to develop competitive large-scale cultivation systems. One factor with a seminal influence on productivity is light availability. Light energy fully supports algal growth, but it leads to oxidative stress if illumination is in excess. In this work, the influence of light intensity on the growth and lipid productivity of *Nannochloropsis salina* was investigated in a flat-bed photobioreactor designed to minimize cells self-shading. The influence of various light intensities was studied with both continuous illumination and alternation of light and dark cycles at various frequencies, which mimic illumination variations in a photobioreactor due to mixing. Results show that *Nannochloropsis* can efficiently exploit even very intense light, provided that dark cycles occur to allow for re-oxidation of the electron transporters of the photosynthetic apparatus. If alternation of light and dark is not optimal, algae undergo radiation damage and photosynthetic productivity is greatly reduced. Our results demonstrate that, in a photobioreactor for the cultivation of algae, optimizing mixing is essential in order to ensure that the algae exploit light energy efficiently.

Editor: Andrew Webber, Arizona State University, United States of America

Funding: TM acknowledges the financial support from Università di Padova (grant CPDA089403) and CUIA. The funders had no role in study design, data collection and analysis, decision to publish, or preparation of the manuscript.

Competing Interests: The authors have declared that no competing interests exist.

* E-mail: tomas.morosinotto@unipd.it

Introduction

Photosynthetic organisms are receiving growing attention, due to their possible exploitation in the production of biofuels for partial replacement of fossil fuels [1–5]. One of the possibilities currently implemented is to produce biodiesel from oil-rich seeds, although several problems are still open, such as limited areal productivity of crops and competition with food production for arable land [6]. One interesting alternative is exploiting some species of algae which are capable of accumulating large amounts of lipids and may thus represent suitable feedstock for biodiesel production. These organisms have also been estimated to have potential oil productivity per area which is ten times higher than that of crops, and they are thus a highly promising source for biomass production in a medium-term perspective. However, intensive research efforts are still needed to exploit this potential to the full in large-scale cultivation systems [4,5,7].

Algae are a group of organisms with very large biological variability [3]: species of the genus *Nannochloropsis* are particularly interesting in this context, because of their ability to accumulate large amounts of lipids which may reach concentrations of up to 65–70% of total dry weight [8–10]. Such massive accumulation of lipids has been shown to be activated in response to stresses such as nitrogen or phosphorus starvation or exposure to excess light [10–14].

In order to exploit these organisms for large-scale biofuel production, it is essential to investigate in detail how various parameters influence productivity. Among the several possibilities, light is a major factor, because it provides all the energy necessary to support metabolism but, if present in excess, may lead to the formation of harmful reactive oxygen species (ROS) and oxidative stress [15].

When cells are exposed to illumination, one component of the photosynthetic apparatus, photosystem II (PSII), is continually damaged and must be continually repaired by re-synthesis of damaged components [16,17]. Photosynthetic organisms exposed to saturating light can also reduce oxidative damage by thermal dissipation of excess energy [15]. Both the repair of damaged photosystems and the dissipation of energy reduce the overall efficiency of light use and should be minimized, if higher productivity is to be achieved.

Algae in photobioreactors are inevitably exposed to variable incident light due to diurnal and seasonal differences in irradiation. *Nannochloropsis* species have been shown to be capable of growing in a large range of illumination intensities, acclimating to changing conditions by optimizing the composition of their photosynthetic machinery to irradiation [8,18,19]. Observed responses to different light intensities include modulation of pigment composition and concentrations of enzymes involved in carbon fixation [19,20].

It should also be noted that algal cultures in photobioreactors have high optical density, which causes highly inhomogeneous light distribution. As a consequence, surface-exposed cells absorb most of the light, leaving only a residual part of the radiation for the cells

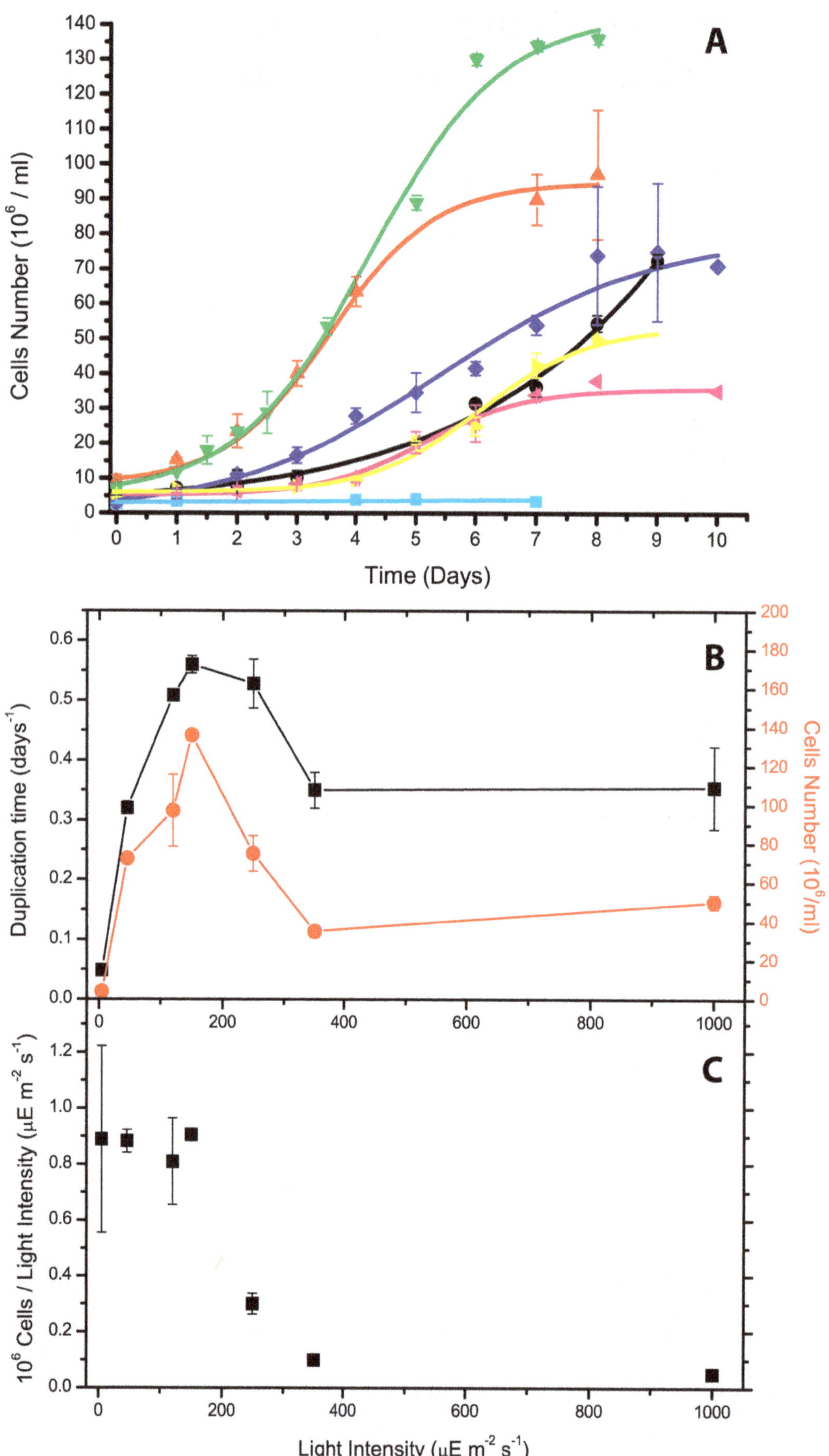
A
Cells Number (10^6 / ml)
Time (Days)
B
Duplication time (days^{-1})
Cells Number (10^6/ml)
C
10^6 Cells / Light Intensity (μE m^{-2} s^{-1})
Light Intensity (μE m^{-2} s^{-1})

Figure 1. *Nannochloropsis salina* growth under different light intensities in a flat-bed photobioreactor. A) Growth kinetics of algae exposed to differing light intensities from 5 to 1000 μE m^{-2} s^{-1}. Data with 5, 50, 120, 150, 250, 350 and 1000 μE m^{-2} s^{-1} shown in light blue, black, red, green, dark blue, pink and yellow, respectively. B) Growth parameters determined from curves in A, specific growth rate (black squares) and cellular concentration after 8 days of growth (red circles). C) Cellular concentrations reported in B normalized to light intensity: this may be used as approximate estimate of biomass production, as no significant deviation of cell size or DW/cell ratio was observed.

underneath, which are thus limited in their growth. Instead, external layers are easily exposed to excess light and they must thermally dissipate up to 80% of their photons in order to avoid radiation damage. This greatly reduces their light use efficiency. Following this idea, it has been shown that the overall efficiency of photobioreactors increases when the light path is diminished, reducing the inhomogeneity of light distribution [21]. Unfortunately, very short light paths are difficult to be implemented in large-scale plants, due to practical and economic reasons.

Another factor to be considered is that cells in photobioreactors are rapidly mixed and move abruptly from darkness to full sunlight [22]. Mixing cycles vary greatly according to cultivation system, and mixing rate on a millisecond time-scale can be achieved in closed tubular reactors or optical fiber-based photobioreactors, thanks to turbulent eddies [22]. Conversely, in raceway ponds, laminar flows often affect efficient mixing [23].

Alternation of light/dark periods has been suggested to be beneficial to photosynthetic efficiency [23–32]. In some cases, the possibility of achieving light integration has been shown, meaning that fluctuating light can be exploited with the same efficiency as continuous light of equal average intensity [31]. However, the experiments reported in the literature focused on very different ranges of flash frequencies, from 5 Hz [31] to 1 kHz [27,33], making a complete comparison of results difficult. Also, while some experiments evaluated flash effects only on photosynthetic oxygen evolution [29,33–35], others evaluated longer-term effects such as growth rates [26,27,32], which again affected the possibility of merging data.

Experiments with photobioreactors also suggest that mixing rates affect photosynthetic productivity and, in particular, that the latter increases with the frequency of light/dark alternation [25,36,37]. However, this conclusion has not always been confirmed, and other reports show that higher mixing rates do not improve photosynthetic efficiency [38,39], clearly indicating that deeper understanding of the influence of light fluctuations on photosynthetic productivity is needed.

In this work, the influence of illumination conditions on *Nannochloropsis salina* growth and lipid productivity was evaluated in a flat-bed photobioreactor, designed to minimize self-shading. Algae were grown under various continuous irradiances but also with dark/light cycles of different intensities and frequencies, simulating changes in illumination occurring in a photobioreactor as a consequence of mixing. Results showed that, if mixing is appropriately optimized, a photobioreactor can exploit even very intense irradiances with high efficiency. Instead, if the alternation of dark and light is not carefully established, pulsed light can inhibit growth.

Results

Growth of *Nannochloropsis salina* in a Flat-bed Panel at Various Illumination Intensities

The influence of light intensity on *Nannochloropsis salina* growth was assessed in a flat-bed photobioreactor, designed to reduce the influence of self-shading on observed growth rates and productivity (Figure S1). All experiments were performed with low optical density cultures, to further reduce the effect of self-shading on growth kinetics. CO_2 and nitrogen (as nitrate) were provided in excess, in order to avoid growth limitation due to these nutrients and to reveal illumination effects only. Figure 1 shows that the light intensity reaching the culture greatly influenced both growth rate and final cellular concentration achieved in the stationary phase. Between 5 and 150 μE m^{-2} s^{-1}, the growth rate increased with light intensity, peaking at 150 μE m^{-2} s^{-1}. Above this limit, any light increase was inhibiting. Under the most intense illumination (350 and 1000 μE m^{-2} s^{-1}), growth curves also showed a detectable lag phase, which did not appear in the other conditions (Figure 1A). However, after a few days, *Nannochloropsis* cells resumed growth even in these conditions.

When biomass productivity, expressed as final cellular concentration, is normalized to light intensity, two distinct regions are clearly identifiable. This ratio is constant up to 150 μE m^{-2} s^{-1}, indicating that irradiation energy is exploited with comparable efficiency in all cultures within this range. Instead, over the 150 μE m^{-2} s^{-1} limit, the ratio decreases drastically, showing that, although the cells are still capable of showing substantial growth, they use light energy with lower efficiency (Figure 1B).

Effect of Pulsed Light on Growth of *Nannochloropsis salina*

As already noted, algae in a photobioreactor are subjected to natural variations in illumination but also to dark/light cycles due to mixing. Accordingly, cells rapidly move from regions where they are fully exposed to sunlight, to others where they are substantially in darkness. In order to understand how algae respond to these conditions, *Nannochloropsis salina* cells were grown in square-wave light/dark cycles to simulate mixing. All experiments were performed providing an average total amount of photons always corresponding to 120 μE m^{-2} s^{-1} of continuous light, a value chosen because it represents the highest intensity at which the cell growth is still light-limited. Once the average intensity (I_a) of the light provided was fixed, the influence of other parameters, such as frequency and intensity of light pulses, could be assessed. As shown in Table 1 and schematized in Figure S2, flashes of two different intensities, 350 and 1200 μE m^{-2} s^{-1}, were used and, in order to provide the same total amount of energy, light was turned on for one-third and one-tenth of the light cycle, respectively (duty cycles (ϕ) of 0.33 and 0.1). In both cases, the influence of different frequencies of light changes, corresponding to different durations of light and dark phases, was explored (see Table 1).

With the strongest flashes (1200 μE m^{-2} s^{-1}), the choice of the frequency of light pulses showed a huge influence on growth performance. At 10 Hz, the growth rate was the same as that of cells exposed to constant moderate light (120 μE m^{-2} s^{-1}, Figure 2A, with nomenclature reported in Table 1). In these conditions, the cells showed complete light integration, meaning that they exploited pulsed light as well as continuous illumination [31]. When light was supplied in pulses at lower frequencies, such as 5 and 1 Hz, growth was greatly inhibited, although the total amount of light and pulse intensity were the same. In the case of 1200-1 and 1200-5 Hz, growth was even slower than under constant, intense light (Figure 2A–C). Biomass productivity, estimated as the number of cells per unit of light intensity

Table 1. Description of pulsed light conditions employed for *Nannochloropsis salina* growth.

Condition	Light Intensity (I_0)	Frequency of light change	Flash time (t_f)	Dark time (t_d)	Integrated light intensity (I_a)	Duty cycle (ϕ)
120	120 μE $m^{-2} s^{-1}$	-	∞	-	120 μE $m^{-2} s^{-1}$	1
1200-10	1200 μE $m^{-2} s^{-1}$	10 Hz	10 ms	90 ms	120 μE $m^{-2} s^{-1}$	0.1
1200-5	1200 μE $m^{-2} s^{-1}$	5 Hz	20 ms	180 ms	120 μE $m^{-2} s^{-1}$	0.1
1200-1	1200 μE $m^{-2} s^{-1}$	1 Hz	100 ms	900 ms	120 μE $m^{-2} s^{-1}$	0.1
350-10	350 μE $m^{-2} s^{-1}$	10 Hz	33.33 ms	66.67 ms	120 μE $m^{-2} s^{-1}$	0.33
350-30	350 μE $m^{-2} s^{-1}$	30 Hz	11 ms	22 ms	120 μE $m^{-2} s^{-1}$	0.33

Alternating cycles of light and dark were all designed to have same integrated light intensity (I_a), corresponding to 120 μE $m^{-2} s^{-1}$ of continuous light. Flashes of two intensities were employed, 1200 and 350 μE $m^{-2} s^{-1}$, with duty cycles of 0.1 and 0.33. Light changes made at different frequencies, 10, 5, 1 Hz with 1200 μE $m^{-2} s^{-1}$ and 10, 30 Hz with 350 μE $m^{-2} s^{-1}$. Pulsed light conditions resulted in precise duration of flashes (**t_f**) and dark (**t_d**). See also Figure S2.

(Figure 2D), showed that cells at 1200-1 and 1200-5 Hz exploit light with greatly reduced efficiency, although the total amount of energy provided is low.

Similar experiments were also performed with 350 μE $m^{-2} s^{-1}$ flashes. In one case, i.e., at 350-30 Hz, growth rate and final cell concentration values were again equivalent to continuous 120 μE $m^{-2} s^{-1}$, showing that cells achieved light integration. In another condition, i.e., 350-10 Hz, growth was inhibited, confirming that the frequency of light changes has an enormous effect on biomass productivity.

Effect of Illumination Conditions on Photosynthetic Apparatus

Fv/Fm is a useful parameter to evaluate photosynthetic efficiency in algae and plants and, in particular, to highlight photoinhibition due to excess illumination [40]. Fv/Fm was monitored in all cultures and cells grown at different levels of continuous light up to 150 μE $m^{-2} s^{-1}$, and all showed similar Fv/Fm values, around 0.62±0.02 (Figure 3). Over this limit, reduced Fv/Fm was correlated with increase of light intensity, indicating that the cells were undergoing photoinhibition.

In pulsed light experiments, Fv/Fm was in the optimal range in both cases when growth was good, 1200 - 10 Hz and 350 - 30 Hz. Instead, in all cases with impaired growth, a reduction in Fv/Fm was also observed, indicating that the cells also underwent photoinhibition, although they were exposed to a low total amount of photons.

Nannochloropsis, like many other algae, responds to different light conditions by modulating the composition of its photosynthetic apparatus, a response called acclimation [19,41]. One regulation commonly observed in photosynthetic organisms exposed to various light intensities is alteration of chlorophyll (Chl) content per cell and the carotenoid (car)/Chl ratio. Under excess illumination, Chl content decreases to reduce light harvesting efficiency, and carotenoids, active in protecting against oxidative stress, are accumulated. As shown in Figure 4, continuous strong light reduces Chl content per cell and increases that of carotenoids.

Instead, clear-cut differences in the light response were observed in cells under pulsed light, which did not accumulate large amounts of carotenoids or even increase their Chl content per cell. The cells thus showed a peculiar response with respect to those exposed to the same total amount of light but provided continuously. This response did not depend particularly on the frequency or duration of light pulses, since similar pigment contents were observed in cells growing well and in others showing light inhibition, with the only exception of the condition 350 - 10 Hz.

Light-stressed Cells Accumulate Lipids in Excess CO_2 Conditions Only with Continuous Illumination

Light intensity has been suggested to influence algal lipid synthesis, with strong illumination inducing their accumulation [1,42]. A transition from control to high light conditions was found to enhance lipid accumulation also in *Nannochloropsis* species [13,14,43].

To assess if this was also the case in the experimental conditions tested here, lipid productivity was monitored in all cultures (Figure 5). In all limiting light conditions up to 150 μE $m^{-2} s^{-1}$, cells at the end of the exponential growth phase had low lipid contents, around 10% DW, corresponding to the constitutive contents of cellular membranes [44]. At irradiances over 150 μE $m^{-2} s^{-1}$, lipid contents increased, reaching a maximum of 70±9% at 350 μE $m^{-2} s^{-1}$. Even considering that gravimetric evaluation of lipid contents has been suggested to present the risk of overestimation [45], these data support the hypothesis that strong illumination stimulates lipid biosynthesis.

It was interesting to observe that, when cells were grown in pulsed light, no induction of lipid accumulation was observed in any of the conditions tested, whether cells were growing well or not.

Discussion

Influence of Light Conditions on Algal Productivity

The effects of light irradiance on *Nannochloropsis* productivity were monitored in a flat-bed reactor designed to minimize cellular self-shading. Since other major factors influencing algal growth (nitrogen and CO_2) were supplied in excess, this system was considered as optimal to reveal the influence of illumination on algal productivity.

One method for verifying whether this assumption is correct is to compare growth rate and final cellular concentration. The former is measured in the first few days of the curve, whereas biomass concentration depends on when cell duplication stops. In the presence of a significant cells shading, high light cultures should have beneficial effects in the late exponential phase, reaching higher concentrations. Instead, Figure 1B shows that growth rates and final cellular concentrations both show a very similar dependence on illumination intensity, supporting the previous assumption that shading effects are minimized in this system.

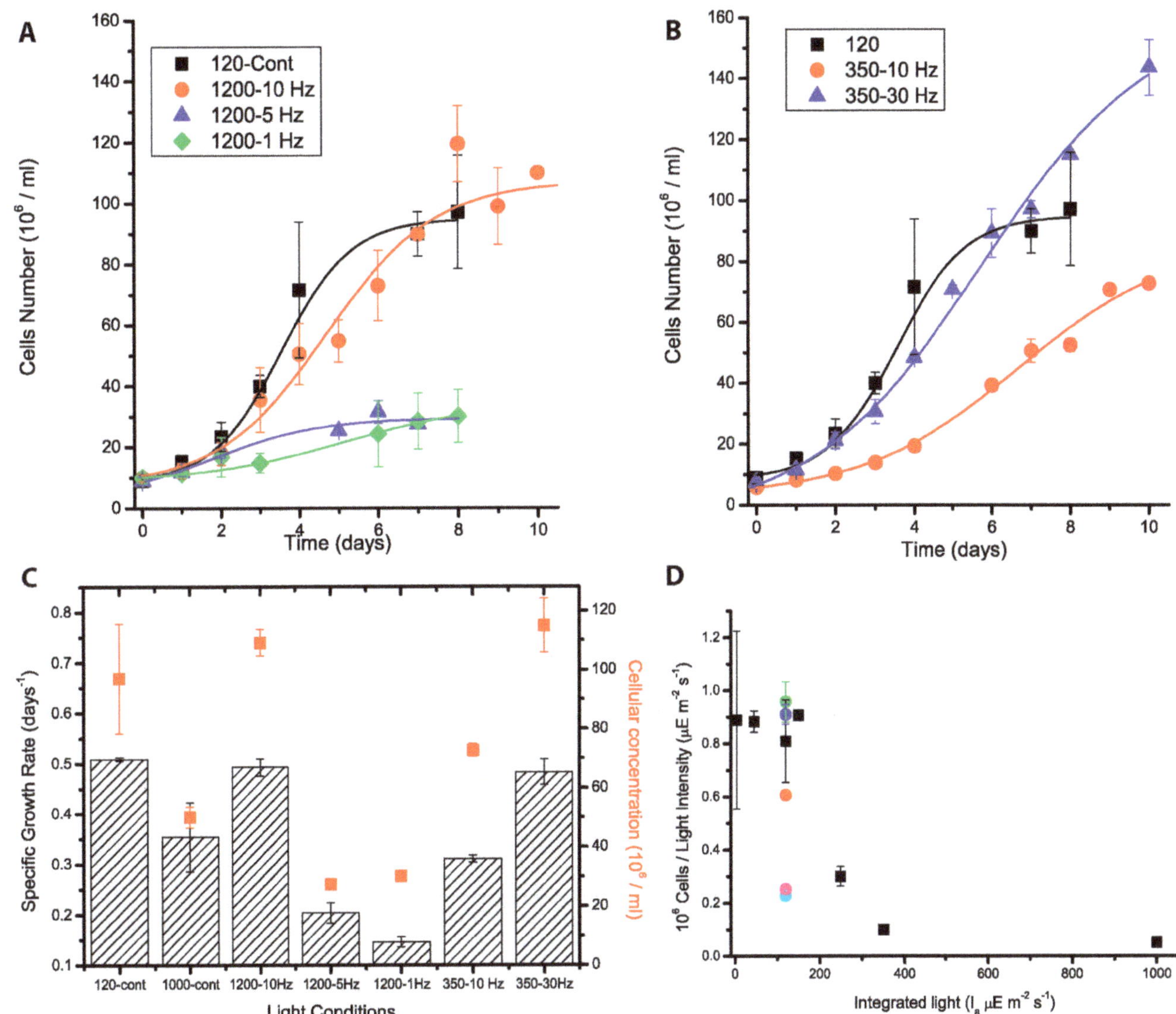

Figure 2. Algal growth kinetics under pulsed light. A–B) *Nannochloropsis* growth curves in pulsed light of differing intensity and frequency, 1200 µE m^{-2} s^{-1} (10, 5, 1 Hz, respectively in red, blue and green, A) and 350 µE m^{-2} s^{-1} (10 and 30 Hz in red and blue, B). Kinetics with 120 µE m^{-2} s^{-1} continuous light reported for comparison (black). C) Growth rate (columns) and cellular concentration after 8 days of growth (red squares) extrapolated from curves in A–B. Values with 120 and 1000 µE m^{-2} s^{-1} constant illumination from Figure 1 reported for comparison. D) Cell concentration in C reported normalized to integrated light intensity. 1200-10, 5, 1 Hz and 350-10, 30 Hz reported in dark blue, pink, light blue, red and green respectively.

Growth dependence on light intensity underwent a first phase during which irradiation was limiting and a second phase in which light had an inhibitory effect (Figure 1B). This was expected, as light may be limiting for growth but, if in excess, leads to oxidative stress [15]. It is worth noting that, at the highest intensity tested (1000 µE m^{-2} s^{-1}), cultures showed growth similar to that at 350 µE m^{-2} s^{-1}, indicating that cells can protect themselves from such strong light excess and yet maintain significant biomass accumulation.

In order to demonstrate the influence of illumination on growth, it was interesting to analyze the final cell concentration values, normalized to illumination intensity (Figure 1C). Since no significant deviations in cell size or weight were observed, this ratio represents an estimate of culture biomass productivity. In this case, in which light is limiting for growth, cells clearly exploit radiation energy with very similar efficiency, probably to the maximum in these conditions. Instead, after the 150 µE m^{-2} s^{-1} limit, productivity falls drastically. The case of 250 µE m^{-2} s^{-1} is interesting to be mentioned, since cells still showed a good growth rate, close to the values found at 120 µE m^{-2} s^{-1}. However, since the light energy provided was more than double, light use efficiency fell by more than 50%. A real large-scale photobioreactor may be imagined as being composed by a overlap of several of these layers: this result indicates when most external cells are exposed to strong illumination, although they are able to cope with the resulting stress, they use available energy with lower efficiency. This drop in productivity of the more external layers, which are also those absorbing most of the light energy, is definitely detrimental for the overall photobioreactor performance.

One strategy against this limitation is suggested by experiments with pulsed light, which showed that even strong light, beyond the saturation point of photosynthesis, can be efficiently exploited for

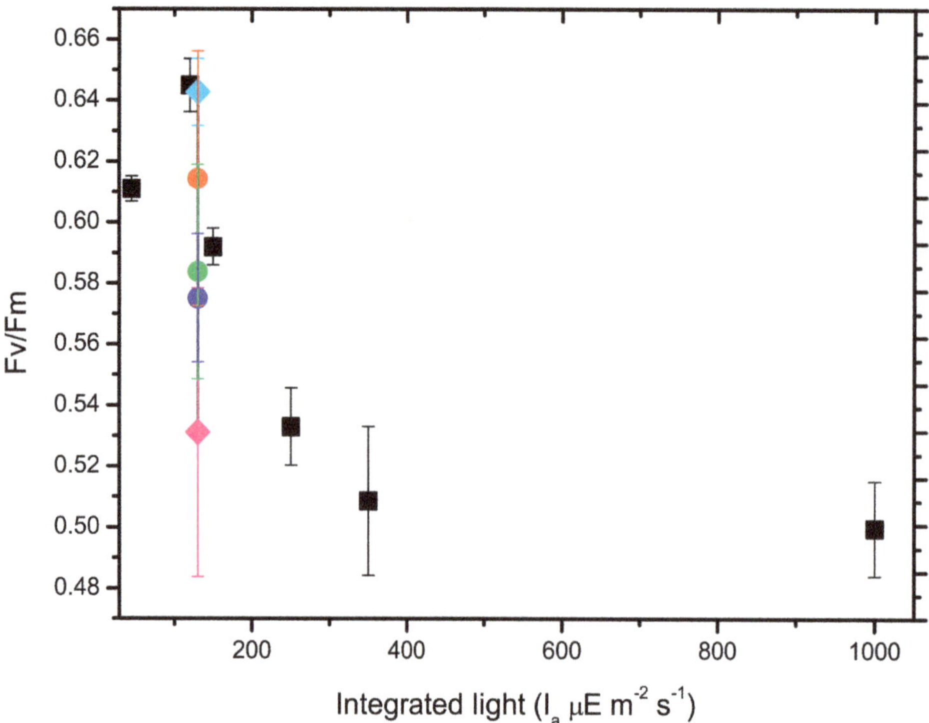

Figure 3. Dependence of photosynthetic efficiency (Fv/Fm) on illumination conditions. Fv/Fm values at end of exponential phase, compared between cells grown under continuous illumination of differing intensity (black squares) and pulsed light of differing intensity and frequency, 1200 µE m^{-2} s^{-1} (10, 5, 1 Hz, red, green and blue circles), 350 µE m^{-2} s^{-1} (10 and 30 Hz, pink and light blue diamonds). Cells at 5 µE m^{-2} s^{-1} were too dilute to provide reliable results.

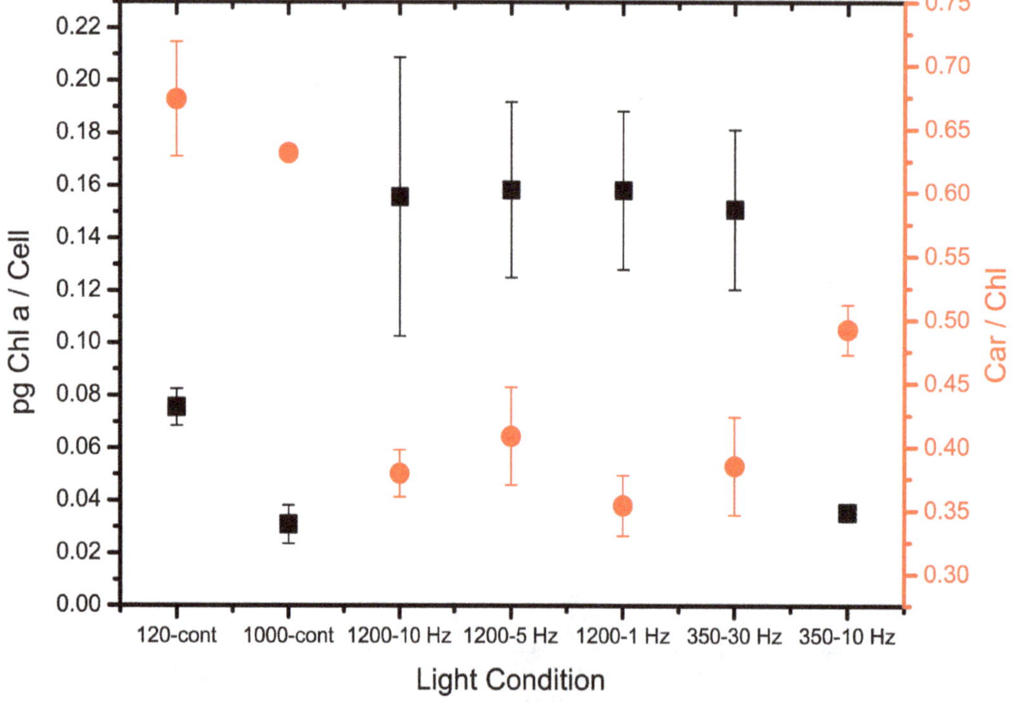

Figure 4. Acclimation response in cells grown in continous vs. pulsed light. Cells grown under differing light intensities, either continuous or pulsed, compared with their Chl content per cell (black) and Chl/Car ratio (red), parameters indicating activation of acclimation response to pulsed light conditions.

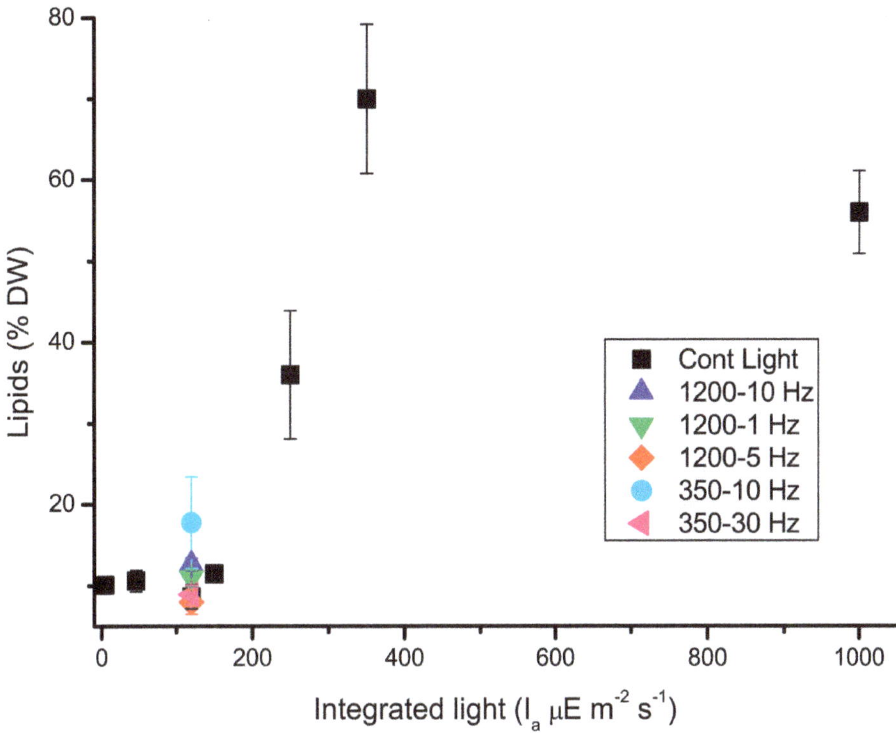

Figure 5. Evaluation of lipid productivity. Dependence of lipid production on illumination intensity. Productivity values with constant light (black squares) compared with the ones with light flashes of 350 and 1200 µE m^{-2} s^{-1} at various frequencies.

growth, as demonstrated by the fact that cells grew equally well at 120 µE m^{-2} s^{-1} and 1200-10 Hz. Note that, at 120 µE m^{-2} s^{-1}, light is still limiting for growth, as shown in Figure 1B–C, implying that, with this optimal alternation of light and dark, cells can use all the energy of pulsed light with the same efficiency they achieve under continuous low illumination [25–27,31]. This is clearly shown by noting that the number of cells per unit of light is the same at 1200-10, 350-30 Hz and all constant low illumination (Figure 2D). This high efficiency is possible because photochemical intermediates produced in a short flash of intense light can be processed further by enzymatic reactions during the following dark period, so that cells perform time integration of the light energy received.

However, similar experiments, at different frequencies, also showed that alternating dark and light may be detrimental, as demonstrated by the growth inhibition observed in the 1200-1 and −5 Hz curves. These cells showed a decrease in Fv/Fm values similar to that observed under constant strong light, indicating that the inhibition of growth was due to photoinhibition. Even more importantly, when the number of cells per unit of light was considered, cells growing at 1200-1 and −5 Hz used light at very low efficiency. For instance, in the 1200-1 Hz experiment, biomass productivity was as low as in 1000 µE m^{-2} s^{-1} continuous light cultures, although the total energy provided was 8.3 times lower. Apparently, in these cases, cells perform worse that those exposed to continuous light, showing that alternating light and dark can also reduce growth efficiency.

In experiments with alternate light, it should be noted that similar results were obtained with light flashes of 1200 and 350 µE m^{-2} s^{-1}, indicating that the intensity of the light pulses does not have much influence (Figure 2C). The data also show that frequency *per se* is not the major parameter determining light influence. On the contrary, the conditions with the highest growth (1200 µE m^{-2} s^{-1} - 10 Hz and 350 µE m^{-2} s^{-1}- 30 Hz) have in common the same length of the illumination phase, which thus appears to have the largest influence on biomass productivity among the parameters considered here.

The optimal duration of light pulses was found to be around 10 ms, which is consistent with the suggested PSII turnover rate in whole cells [4,46]. Accordingly, after photon absorption by the photosystem, 1–15 ms are needed to reset the system, before it is ready to receive another photon [22]. If the illumination is this short, most photons are exploited for photosynthesis and do not lead to the formation of ROS which then causes photoinhibition. These results indicate that even strong light does not cause damage if it only lasts a short time. Conversely, longer exposure allows the generation of ROS and damage and, in this case, the abrupt changes in illumination undergone by the cells are as harmful to the photosynthetic apparatus as constant high light.

A further observation is that a peculiar acclimation response is activated under pulsed light conditions. This response does not depend to any great extent on the frequency or duration of light pulses, since similar pigment contents were observed in cells growing well and in others (1200- 1 and 1200-5 Hz), and also in ones showing light inhibition, with the only exception of the 350-10 Hz level. These results indicate a particular type of acclimation response in pulsed light conditions, which does not depend on the stress perceived by the cells, nor on the total amount of light absorbed. However, this deserves further investigation.

Light Effect on Lipid Production

In the perspective of exploiting algae as feedstock for biodiesel production, biomass growth must be considered together with lipid productivity. The relationship between light intensity and productivity is complex, and many parameters influence the ability of algae to accumulate lipids. The experiments with continuous illumination presented here showed that, under strong continuous light, lipid accumulation is stimulated, as also previously reported [13,14,43]. It is noteworthy that such large lipid production was observed in a medium where nitrogen was provided in excess and was not induced by lack of this nutrient. As shown in Figure S3, lipid accumulation starts in the early phases of culture, when nitrogen cannot be depleted, even if a far higher cell consumption is assumed. This result confirms that excess light can induce lipid accumulation without direct or indirect nitrogen depletion, showing that other signals can induce high lipid accumulation in *N. salina*.

In this context, the seminal influence of carbon dioxide availability should be stressed. Previous results on *Nannochloropsis* cells exposed to various radiation intensities with atmospheric CO_2 showed light stress without any detectable induction of lipids [19]. On the contrary, in the experiments presented here, where carbon dioxide was externally provided, light excess induced lipid accumulation. These results clearly indicate that light stress alone does not induce lipid synthesis, which is instead the result of more complex regulation. With high light and limiting CO_2, the efficiency of the Calvin-Benson cycle is probably limiting, with the consequent accumulation of molecules upstream of carbon dioxide fixation. Instead, with excess CO_2 and strong illumination, molecules downstream of carbon dioxide fixation are more likely to accumulate, eventually triggering triacylglycerol biosynthesis.

This hypothesis is consistent with data obtained with cells under pulsed light. For example, cells exposed to 1200-1 Hz showed light stress and inhibited growth which, however, did not result in any significant lipid accumulation, confirming that light stress alone does not induce lipid biosynthesis, but that its influence is integrated with other metabolic signals.

Materials and Methods

Culture Conditions

Nannochloropsis salina, from SAG, was always grown in sterile filtered F/2 medium [47], with 22 g/l sea salts from SIGMA, 40 mM TRIS HCl pH 8, SIGMA Guillard's (f/2) marine water enrichment solution 1×, modified by the addition of a non-limiting nitrogen concentration ($NaNO_3$ 1.5 g/L). Cultures were maintained and propagated in the same medium, with the addition of 10 g/l of Plant Agar (Duchefa Biochemie). Growth experiments were performed in a flat-bed apparatus 0.8 cm deep (Figure S1). Pre-cultures were grown at 100 $\mu E\ m^{-2}\ s^{-1}$ in the exponential phase, which was diluted to an Optical Density (OD) of 0.45 at 750 nm. The final volume of the photobioreactor was 150 ml. Constant illumination between 5 and 1200 $\mu E\ m^{-2}\ s^{-1}$ was provided with a LED Light Source SL 3500 (Photon Systems Instruments). The light source was also programmed to generate square-wave dark/light cycles at the desired intensities and frequencies (Figure 2). Parameters describing flashes were flash time (t_f), dark time (t_d), duty cycle (ϕ, corresponding to $t_f/(t_f + t_d)$ [26,32]), flash light intensity (I_0) and integrated light intensity (I_a). Temperature was kept at 23±1°C in a growth chamber. CO_2 in excess (mixed at 5% v/v with air) was supplied by bubbling, which also mixed cells. The medium was buffered with 40 mM TRIS HCl, pH 8, to avoid alterations due to excess supply of CO_2. Algal growth was measured by daily changes in optical density OD_{750} (Lambda Bio 40 UV/VIS Spectrometer, Perkin Elmer) and cell numbers were monitored in a Bürker Counting Chamber (HBG, Germany). The specific growth rate was calculated by the slope of the logarithmic phase for number of cells. At least four replicates of all curves/experiments were performed.

In vivo monitoring of photosynthetic parameters. Chlorophyll fluorescence was determined *in vivo* at the end of exponential phase of growth using a Dual PAM 100 from WALZ. Fv/Fm parameter was calculated as (Fm-Fo)/Fo [48] after 20 minutes of dark adaptation.

Pigment Extraction and Analysis

Chlorophyll a and total Car were extracted, at the end of exponential phase of growth, from centrifuged cells of *Nannochloropsis* with 100% N,N'-dimethylformamide for at least 48 hours at 4°C in dark conditions, as in [49]. Pigment concentrations were determined spectrophotometrically with specific extinction coefficients [50,51].

Lipid Analysis

Lipid contents was routinely monitored by measuring fluorescence of Nile Red stained cells, after verification of the linear correlation between the fluorescence signal and the total amount of lipids. $2*10^6$ cells were re-suspended in 1.9 ml of de-ionized sterile water with 2.5 μg/mL NR and incubated for 10 minutes at 37°C [52]. Fluorescence was measured on a spectrofluorometer (OLIS DM45), with excitation wavelength at 488 nm and emission at 580±5 nm [53]. Signals from algal autofluorescence and Nile Red alone were subtracted. For verification of fluorescence signal linearity, total lipids were extracted from dried cells with ethanol-hexane (2.5:1 vol/vol) as solvent in a Soxhlet apparatus for 10 h [54], as reported in the Figure S4. The lipid mass was measured gravimetrically after solvent removal in a rotary evaporator.

Supporting Information

Figure S1 Scheme of the Flat Bed Photobioreactor. The flat-plate photobioreactors were built with transparent materials (polycarbonate) for maximum utilization of light energy. The working volume is 150 ml and the culture is mixed by an air-CO_2 flow from a sparger placed in the bottom of the panel. The amount of CO_2 in air is regulated by two flow meters. The gas flow supplies a non-limiting CO_2 content to the culture. The gas flow for each reactor is regulated using suitable valves.

Figure S2 Pulsed light conditions utilized for *Nannochloropsis salina* growth. Alternated cycles of light and dark were designed to have all the same integrated light intensity (I_a), corresponding to 120 $\mu E\ m^{-2}\ s^{-1}$ of continuous light. Flashes of two different intensities were employed, 350 and 1200 $\mu E\ m^{-2}\ s^{-1}$, with a duty cycle of respectively 0.33 and 0.1. Light changes were performed with different frequencies, respectively 10 and 33 Hz with 350 $\mu E\ m^{-2}\ s^{-1}$ and 10, 5 and 1 with 1200 $\mu E\ m^{-2}\ s^{-1}$. These pulsed light conditions resulted in precise durations of flashes ($\mathbf{t_f}$) and dark ($\mathbf{t_d}$), as reported in Table 1.

Figure S3 Timeline of lipids accumulation in *Nannochloropsis* cells exposed to 350 (black) and 150 (red) $\mu E\ m^{-2}\ s^{-1}$. Lipid content was evaluated each day using Nile Red staining correlated to total lipid concentration quantified gravimetrically (see Figure S4).

Figure S4 Correlation of lipid accumulation in *Nannochloropsis* evaluated by Nile Red staining and gravimetric analysis.

Author Contributions

Conceived and designed the experiments: GMG AB TM. Performed the experiments: ES DS. Analyzed the data: ES DS TM. Contributed reagents/materials/analysis tools: AB GMG. Wrote the paper: TM ES.

References

1. Hu Q, Sommerfeld M, Jarvis E, Ghirardi M, Posewitz M, et al. (2008) Microalgal triacylglycerols as feedstocks for biofuel production: perspectives and advances. Plant J 54: 621–639.
2. Dismukes GC, Carrieri D, Bennette N, Ananyev GM, Posewitz MC (2008) Aquatic phototrophs: efficient alternatives to land-based crops for biofuels. Curr Opin Biotechnol 19: 235–240.
3. Hannon M, Gimpel J, Tran M, Rasala B, Mayfield S (2010) Biofuels from algae: challenges and potential. Biofuels 1: 763–784.
4. Malcata FX (2011) Microalgae and biofuels: a promising partnership? Trends Biotechnol 29: 542–549. S0167-7799(11)00094-1 [pii];10.1016/j.tibtech.2011.05.005 [doi].
5. Chisti Y, Yan JY (2011) Energy from algae: Current status and future trends/ Algal biofuels - A status report. Applied Energy 88: 3277–3279.
6. Singh A, Nigam PS, Murphy JD (2011) Renewable fuels from algae: an answer to debatable land based fuels. Bioresour Technol 102: 10–16.
7. Amaro HM, Guedes AC, Malcata FX (2011) Advances and perspectives in using microalgae to produce biodiesel. Applied Energy 88: 3402–3410.
8. Boussiba S, Vonshak A, Cohen Z, Avissar Y, Richmond A (1987) Lipid and biomass production by the halotolerant microalga *Nannochloropsis salina*. Biomass 12: 37–47.
9. Hodgson PA, Henderson RJ, Sargent JR, Leftley JW (1991) Patterns of variation in the lipid class and fatty-acid composition of *Nannochloropsis oculata* (Eustigmatophyceae) during batch culture.1. The Growth Cycle. Journal of Applied Phycology 3: 169–181.
10. Rodolfi L, Chini ZG, Bassi N, Padovani G, Biondi N, et al. (2009) Microalgae for oil: strain selection, induction of lipid synthesis and outdoor mass cultivation in a low-cost photobioreactor. Biotechnol Bioeng 102: 100–112.
11. Gouveia L, Oliveira AC (2009) Microalgae as a raw material for biofuels production. J Ind Microbiol Biotechnol 36: 269–274.
12. Sforza E, Bertucco A, Morosinotto T, Giacometti GM (2012) Photobioreactors for microalgal growth and oil production with *Nannochloropsis salina:* from lab-scale experiments to large-scale design. Chemical Engineering Research & Design in press.
13. Sukenik A, Carmeli Y, Berner T (1989) Regulation of fatty-acid composition by irradiance level in the eustigmatophyte *Nannochloropsis* sp. Journal of Phycology 25: 686–692.
14. Solovchenko A, Khozin-Goldberg I, Recht L, Boussiba S (2011) Stress-induced changes in optical properties, pigment and fatty acid content of Nannochloropsis sp.: implications for non-destructive assay of total fatty acids. Mar Biotechnol (NY) 13: 527–535. 10.1007/s10126-010-9323-x [doi].
15. Li Z, Wakao S, Fischer BB, Niyogi KK (2009) Sensing and responding to excess light. Annu Rev Plant Biol 60: 239–260.
16. Murata N, Takahashi S, Nishiyama Y, Allakhverdiev SI (2007) Photoinhibition of photosystem II under environmental stress. Biochim Biophys Acta 1767: 414–421.
17. Nixon PJ, Michoux F, Yu J, Boehm M, Komenda J (2010) Recent advances in understanding the assembly and repair of photosystem II. Ann Bot 106: 1–16.
18. Pal D, Khozin-Goldberg I, Cohen Z, Boussiba S (2011) The effect of light, salinity, and nitrogen availability on lipid production by *Nannochloropsis* sp. Appl Microbiol Biotechnol 90: 1429–1441.
19. Simionato D, Sforza E, Corteggiani CE, Bertucco A, Giacometti GM, et al. (2011) Acclimation of *Nannochloropsis gaditana* to different illumination regimes: Effects on lipids accumulation. Bioresour Technol 102: 6026–6032.
20. Fisher T, Minnaard J, Dubinsky Z (1996) Photoacclimation in the marine alga *Nannochloropsis* sp (eustigmatophyte): a kinetic study. Journal of Plankton Research 18: 1797–1818.
21. Richmond A, Cheng-Wu Z, Zarmi Y (2003) Efficient use of strong light for high photosynthetic productivity: interrelationships between the optical path, the optimal population density and cell-growth inhibition. Biomol Eng 20: 229–236.
22. Carvalho AP, Silva SO, Baptista JM, Malcata FX (2011) Light requirements in microalgal photobioreactors: an overview of biophotonic aspects. Appl Microbiol Biotechnol 89: 1275–1288.
23. Grobbelaar JU (2010) Microalgal biomass production: challenges and realities. Photosynth Res 106: 135–144. 10.1007/s11120-010-9573-5 [doi].
24. Xue S, Su Z, Cong W (2011) Growth of *Spirulina platensis* enhanced under intermittent illumination. J Biotechnol 151: 271–277. S0168-1656(10)02063-8 [pii];10.1016/j.jbiotec.2010.12.012 [doi].
25. Gordon JM, Polle JE (2007) Ultrahigh bioproductivity from algae. Appl Microbiol Biotechnol 76: 969–975.
26. Phillips JN, Myers J (1954) Growth rate of *Chlorella* in flashing light. Plant Physiol 29: 152–161.
27. Matthijs HC, Balke H, van Hes UM, Kroon BM, Mur LR, et al. (1996) Application of light-emitting diodes in bioreactors: flashing light effects and energy economy in algal culture (*Chlorella pyrenoidosa*). Biotechnol Bioeng 50: 98–107.
28. Kim ZH, Kim SH, Lee HS, Lee CG (2006) Enhanced production of astaxanthin by flashing light using *Haematococcus pluvialis*. Enzyme and Microbial Technology 39: 414–419.
29. Kok B (1956) Photosynthesis in flashing light. Biochim Biophys Acta 21: 245–258.
30. Nedbal L, Tichy V, Xiong FH, Grobbelaar JU (1996) Microscopic green algae and cyanobacteria in high-frequency intermittent light. Journal of Applied Phycology 8: 325–333.
31. Terry KL (1986) Photosynthesis in modulated light: quantitative dependence of photosynthetic enhancement on flashing rate. Biotechnol Bioeng 28: 988–995. 10.1002/bit.260280709 [doi].
32. Vejrazka C, Janssen M, Streefland M, Wijffels RH (2011) Photosynthetic efficiency of *Chlamydomonas reinhardtii* in flashing light. Biotechnol Bioeng. 10.1002/bit.23270 [doi].
33. Nedbal L, Tichy V, Xiong FH, Grobbelaar JU (1996) Microscopic green algae and cyanobacteria in high-frequency intermittent light. Journal of Applied Phycology 8: 325–333.
34. Brindley C, Fernandez FG, Fernandez-Sevilla JM (2011) Analysis of light regime in continuous light distributions in photobioreactors. Bioresour Technol 102: 3138–3148. S0960-8524(10)01744-X [pii];10.1016/j.biortech.2010.10.088 [doi].
35. Grobbelaar JU, Nedbal L, Tichy V (1996) Influence of high frequency light/ dark fluctuations on photosynthetic characteristics of microalgae photoacclimated to different light intensities and implications for mass algal cultivation. Journal of Applied Phycology 8: 335–343.
36. Meiser A, Schmid-Staiger U, Trosch W (2004) Optimization of eicosapentaenoic acid production by *Phaeodactylum tricornutum* in the flat panel airlift (FPA) reactor. Journal of Applied Phycology 16: 215–225.
37. Qiang H, Richmond A (1996) Productivity and photosynthetic efficiency of *Spirulina platensis* as affected by light intensity, algal density and rate of mixing in a flat plate photobioreactor. Journal of Applied Phycology 8: 139–145.
38. Zijffers JWF, Schippers KJ, Zheng K, Janssen M, Tramper J, et al. (2010) Maximum photosynthetic yield of green microalgae in photobioreactors. Marine Biotechnology 12: 708–718.
39. Kliphuis AMJ, de Winter L, Vejrazka C, Martens DE, Janssen M, et al. (2010) Photosynthetic Efficiency of *Chlorella sorokiniana* in a Turbulently Mixed Short Light-Path Photobioreactor. Biotechnology Progress 26: 687–696.
40. Maxwell K, Johnson GN (2000) Chlorophyll fluorescence - a practical guide. J Exp Bot 51: 659–668.
41. Falkowski PG, LaRoche J (1991) Acclimation to spectral irradiance in algae. J Phycol 27: 8–14.
42. Damiani MC, Popovich CA, Constenla D, Leonardi PI (2010) Lipid analysis in *Haematococcus pluvialis* to assess its potential use as a biodiesel feedstock. Bioresour Technol 101: 3801–3807.
43. Fisher T, Berner T, Iluz D, Dubinsky Z (1998) The kinetics of the photoacclimation response of *Nannochloropsis* sp. (Eustigmatophyceae): a study of changes in ultrastructure and PSU density. Journal of Phycology 34: 818–824.
44. Su CH, Chien LJ, Gomes J, Lin YS, Yu YK, et al. (2011) Factors affecting lipid accumulation by *Nannochloropsis oculata* in a two-stage cultivation process. Journal of Applied Phycology 23: 903–908.
45. Laurens LM, Quinn M, Van WS, Templeton DW, Wolfrum EJ (2012) Accurate and reliable quantification of total microalgal fuel potential as fatty acid methyl esters by in situ transesterification. Anal Bioanal Chem. 10.1007/s00216-012-5814-0 [doi].
46. Dubinsky Z, Falkowski PG, Wyman K (1986) Light harvesting and utilization by phytoplankton. Plant Cell Physiol 27: 1335–1349.
47. Guillard RRL, Ryther JH (1962) Studies of marine planktonic diatoms. I. *Cyclotella nana* Hustedt and *Detonula confervagea* Cleve. Can J Microbiol 8: 229–239.
48. Demmig-Adams B, Adams WW, Barker DH, Logan BA, Bowling DR, et al. (1996) Using chlorophyll fluorescence to assess the fraction of absorbed light allocated to thermal dissipation of excess excitation. Physiologia Plantarum 98: 253–264.
49. Moran R, Porath D (1980) Chlorophyll determination in intact tissues using n,n-dimethylformamide. Plant Physiol 65: 478–479.
50. Porra RJ, Thompson WA, Kriedemann PE (1989) Determination of accurate extinction coefficients and simultaneous equations for assaying chlorophylls a and b extracted with four different solvents: verification of the concentration of chlorophyll standards by atomic absorption spectroscopy. Biochim Biophys Acta 975: 384–394.
51. Wellburn AR (1994) The spectral determination of chlorophyll-A and chlorophyll-B, as well as total carotenoids, using various solvents with spectrophotometers of different resolution. Journal of Plant Physiology 144: 307–313.

52. Chen W, Zhang C, Song L, Sommerfeld M, Hu Q (2009) A high throughput Nile red method for quantitative measurement of neutral lipids in microalgae. J Microbiol Methods 77: 41–47.
53. Greenspan P, Mayer EP, Fowler SD (1985) Nile red: a selective fluorescent stain for intracellular lipid droplets. J Cell Biol 100: 965–973.
54. Molina GE, Robles MA, Gimenez GA, Sanchez PJ, Garcia-Camacho F, et al. (1994) Comparison between extraction of lipids and fatty acids from microalgal biomass. J Am Oil Chem Soc 71: 955–959.

3

Scale-Dependent Effects of Grazing on Plant C: N: P Stoichiometry and Linkages to Ecosystem Functioning in the Inner Mongolia Grassland

Shuxia Zheng*, Haiyan Ren, Wenhuai Li, Zhichun Lan

State Key Laboratory of Vegetation and Environmental Change, Institute of Botany, Chinese Academy of Sciences, Beijing, China

Abstract

Background: Livestock grazing is the most prevalent land use of grasslands worldwide. The effects of grazing on plant C, N, P contents and stoichiometry across hierarchical levels, however, have rarely been studied; particularly whether the effects are mediated by resource availability and the underpinning mechanisms remain largely unclear.

Methodology/Principal Findings: Using a multi-organization-level approach, we examined the effects of grazing on the C, N, and P contents and stoichiometry in plant tissues (leaves and roots) and linkages to ecosystem functioning across three vegetation types (meadow, meadow steppe, and typical steppe) in the Inner Mongolia grassland, China. Our results showed that the effects of grazing on the C, N, and P contents and stoichiometry in leaves and roots differed substantially among vegetation types and across different hierarchical levels (species, functional group, and vegetation type levels). The magnitude of positive effects of grazing on leaf N and P contents increased progressively along the hierarchy of organizational levels in the meadow, whereas its negative effect on leaf N content decreased considerably along hierarchical levels in both the typical and meadow steppes. Grazing increased N and P allocation to aboveground in the meadow, while greater N and P allocation to belowground was found in the typical and meadow steppes. The differences in soil properties, plant trait-based resource use strategies, tolerance or defense strategies to grazing, and shifts in functional group composition are likely to be the key mechanisms for the observed patterns among vegetation types.

Conclusions/Significance: Our findings suggest that the enhanced vegetation-type-level N contents by grazing and species compensatory feedbacks may be insufficient to prevent widespread declines in primary productivity in the Inner Mongolia grassland. Hence, it is essential to reduce the currently high stocking rates and restore the vast degraded steppes for sustainable development of arid and semiarid grasslands.

Editor: Minna-Maarit Kytöviita, Jyväskylä University, Finland

Funding: This project was supported by the State Key Basic Research Development Program of China (2009CB421102) and the National Natural Science Foundation of China (30900193). The funders had no role in study design, data collection and analysis, decision to publish, or preparation of the manuscript.

Competing Interests: The authors have declared that no competing interests exist.

* E-mail: zsx@ibcas.ac.cn

Introduction

Livestock grazing is the most prevalent land-use type of grasslands worldwide, and has the potential to substantially affect community structure and primary productivity [1,2], alter nutrient cycling [3,4,5,6] and carbon (C) and nitrogen (N) pools [7–9] in grasslands. Overgrazing, in particular, has profound effects on multiple ecosystem functions and services, such as N retention, carbon sequestration, biodiversity conservation, and ecosystem stability [10–12]. Previous studies have demonstrated that the impacts of herbivores on ecosystem N cycling can range from positive to negative. The accelerating N cycling hypothesis [4,13–15] predicts that herbivore increases the tissue loss of grazing-tolerant species with high N content and litter quality, and directly deposits urine and feces to enhance soil available N, which stimulates net N mineralization and N utilization by plants, leading to the enhancement in shoot nutrients. The decelerating N cycling hypothesis [9,16,17], on the contrary, argues that grazing depresses the palatable and N-rich species due to herbivore selectivity, but promotes the dominance of those N-poor or defended species with low litter quality, contributing to slow N turnover and utilization, reducing shoot nutrients. Whether nutrient cycling increases or decreases in response to grazing likely depends on many abiotic and biotic factors, such as resource availability (e.g., soil water and nutrients), grazing duration, herbivore species, plant species and functional group compositions, and plant functional traits [9,18–20].

A functional trait-based approach to community ecology has recently emerged as a promising way to understand plant adaptive strategies, plant-herbivore interactions, and how they link to ecosystem functioning [21–23]. The C, N and P contents (%) and stoichiometry in plant tissues (e.g., leaves and roots) are associated with plant growth and ecosystem attributes, which strongly influence ecosystem processes (e.g., N cycling and litter decomposition) and plant responses to various environment variables and disturbance (e.g., precipitation, soil moisture, and grazing pressure) [24]. The N contents of leaves and roots are closely related to plant N uptake and utilization, and indicate N allocation between above- and belowground tissues. Previous studies demonstrated that grazing altered the C and N allocation between plant tissues [9,25,26], and heavy grazing increased N allocation to aboveground tissues [27]. Several studies also proposed that plant N and biomass allocation are largely dependent on resource availability [28–30]. For example, high soil nutrient availability and low light availability would favor plants allocating more N to stems or leaves than roots [28]. Therefore, grazing-induced shifts in resource allocation (e.g., N and P) between above- and belowground tissues may be meditated by resource availability. The stoichiometric ratios of C: N: P in plant leaves and litter have been widely used to determine N versus P limitation on plant growth, primary productivity and litter decomposition [31–34]. In general, plant species can generate positive feedbacks to N cycling, directly through uptake, utilize and loss of nutrients, and indirectly by influencing herbivory selectivity and microbial activity [35,36]. Hence, plant resource-use strategies, above- vs. belowground competitive abilities, and tolerance vs. defense strategies to grazing may be responsible for diverse stoichiometric responses of different species, which are largely dependent on plant functional traits (e.g., physiological and morphological traits) [21,23,37,38].

Leaf and root traits, e.g., N content of leaves and roots, specific leaf area (SLA), specific root length (SRL), net photosynthetic rate (Pn), photosynthetic nutrient use efficiency (PNUE), and water use efficiency (WUE), reflect plant resource-use strategies (acquisitive vs. conservative) and functional advantage in aboveground or light competition versus belowground competition [21,28,38]. The whole plant traits, e.g., plant height, individual biomass, root: shoot ratio, and stem: leaf ratio, reflect plant biomass allocation between tissues and competitive capacity for light [29,39]. In addition, leaf N content and C: N ratio are closely associated with foliar palatability, which further influence herbivore selectivity and plant tolerance or defense strategies to grazing [9,17]. Thus, plant functional traits may provide important insights into the mechanisms underpinning plant responses to grazing. In the long-term, shifts in species or functional group compositions are likely to affect ecosystem-level response and nutrient dynamics [18]. Therefore, a variety of mechanisms may operate at different scales or organizational levels, which may involve positive, neutral, and negative effects. For a given system, the response may dependent upon the balance of these mechanisms. For example, the plant community-level response to grazing is likely determined by species-level responses, species abundance, and environment conditions [40–42]. The counterbalance between positive and negative feedbacks among vegetation types may affect the regional scale response to grazing. Also, the positive or negative effects may accumulate along the hierarchy of organizational levels. Our recent study demonstrated the effects of grazing on leaf traits are scale-dependant and varied with vegetation types [43]. Therefore, it is difficult to identify the mechanisms based on a single scale study, while an integrated research involving multiple organizational levels is critical for elucidating the underlying mechanisms of grazing impacts on ecosystem properties.

The arid and semiarid grasslands on the Mongolia plateau, which include diverse vegetation types and distribute widely across the Eurasian Steppe region, have been historically subjected to continuous grazing by livestock with high stocking rates, leading to widespread degradation in ecosystem function and services in recent decades [12,44]. In the Inner Mongolia grassland, plant growth and primary productivity are co-limited by water and N availability [45], thus grazing impacts on plant functional traits and ecosystem functioning are likely mediated by water availability. Our recent studies suggest plant responses to grazing were greater in mesic than dry systems [43], and for a given ecosystem greater responses were found in wet than dry years [37].

In this study, we examined the effects of grazing on the C, N, and P contents and stoichiometry in plant tissues and soil across three vegetation types (meadow, meadow steppe, and typical steppe) along a soil moisture gradient in the Xilin River Basin of Inner Mongolia grassland. Specifically, we address three questions: first, how does grazing affect the C, N, and P contents and stoichiometry in plant tissues (leaves and roots) at different hierarchical levels (plant species, functional group, and vegetation type level) and soil? Second, how do grazing effects on ecosystem functioning (e.g., above- and belowground standing biomass, C, N and P pools, and nutrient limitation) vary across different vegetation types? Third, what are the major mechanisms underpinning different stoichiometric responses across vegetation types in terms of soil properties, plant functional traits, and shifts in functional group compositions? To explore the underlying mechanisms of how grazing affects plant C: N: P stoichiometry and resource allocation across different levels of hierarchy, we hypothesize that: (1) the effects of grazing on plant C, N, and P contents and stoichiometry may become stronger along the hierarchical levels, i.e., from species to functional group to vegetation type. (2) The grazing effects on plant C: N: P stoichiometry and N and P allocation may strongly depend on resource availability, such as water and N. We expect that grazing may have positive effects on leaf nutrients (e.g., N and P contents) but negative effects on root nutrients in the wet and fertile meadow. In contrast, grazing may have no effects or even negative effects on leaf nutrients but positive effects on root nutrients in the dry and infertile typical steppe. Hence, grazing may elevate N and P allocation to aboveground in the meadow, but it may enhance N and P allocation to belowground in the typical and meadow steppes.

Methods

Study Area

The study was conducted in the Xilin River Basin (43°26′–44°29′N, 115°32′–117°12′E), which is located in the typical steppe region of the Inner Mongolia grassland, northern China, and covers an area of 10 786 km^2, with elevation ranging from 983 to 1469 m. Mean annual temperature is 0.4°C, with the lowest mean monthly temperature −21.4°C in January and the highest 19.0°C in July. Mean annual precipitation is 336.9 mm yr^{-1}, with about 80% occurring in the growing season (May–August). The dominant soil types are typical chestnut and dark chestnut, while meadow soil is a non-zonal soil type in this region [46]. Six pairs of parallel ungrazed and grazed plant communities, i.e., *Carex appendiculata* meadow, *Stipa baicalensis* meadow steppe, *Leymus chinensis* typical steppe, *S. grandis* typical steppe, *Caragana microphylla* typical steppe, and *Artemisia frigida* typical steppe were selected along a soil moisture gradient in the Xilin River Basin. The six pairs of plant communities include one for meadow, one for meadow steppe, and four for typical steppe, with the number of sites being proportional to its area. These communities are subjected to similar climatic conditions, such as temperature and precipitation, but differ in terms of floristic composition and soil properties, particularly soil moisture and nutrients. The ungrazed sites of communities are the permanent field sites of the Inner Mongolia Grassland Ecosystem Research Station (IMGERS), Chinese Academy of Sciences, which have been fenced from grazing for about 20–30 years [42]. In contrast, the grazed sites, located outside the fence of ungrazed sites, have been managed as free grazing pasture (mainly by sheep) since 1950s, thus they have about 60 years of grazing history. More detailed information for the six pairs of communities could be available from Zheng et al. [43].

Field Sampling and Measurements

Field sampling was carried out during 28 July to 14 August, 2007, when the aboveground standing biomass reached its annual peak [42]. At each site, aboveground biomass of plants were sampled by 5–10 quadrats (1 × 1 m each) located randomly within an area of 100 m × 100 m. Ten quadrats were used for meadow steppe and typical steppe, and 5 quadrats were for the more homogeneous meadow community. For the grazed sites, these quadrats were randomly located in the areas that were not subjected to grazing during the current growing season. Within each quadrat, all living biomass and current year dead materials were harvested by clipping to the soil surface, separated to species, and oven dried at 70°C for 24 h to constant mass and weighed. Litter biomass within each quadrat was collected. The aboveground biomass of each species was collected and transported to a laboratory for stem and leaf separation, then they were oven-dried at 70°C for 24 h to constant mass, thus stem biomass, leaf biomass, and plant biomass could be calculated. The total number of species, number of individuals, and aboveground biomass of each species were measured within each quadrat, which was used for estimating species richness, species abundance, and aboveground standing biomass of community at each site. The relative abundance of each species was obtained by calculating the proportion of individual density of each species to the total density. The relative biomass of each species was determined by its biomass ratio to the total community biomass. Belowground biomass was determined in late August, 2007, when it reached the annual peak in the Inner Mongolia grassland. Belowground biomass was sampled by randomly taking two 7-cm diameter soil cores from 0–20 cm depths inside each quadrat, and totally 10–20 samples at each site. Soil was rinsed out from roots under running water over a 1-mm screen, dead materials were picked out, and then root biomass was oven-dried at 65°C and weighed. Within each quadrat, soil samples were collected by taking three 5-cm diameter soil cores from 0–20 cm depths, mixed in situ as one composite sample, and hand-sorted to remove plant materials. Each soil sample was sieved through 2 mm mesh and separated into two parts. One was air-dried and ground to 80-mesh to determine soil total N, P, and organic C contents, the other maintained fresh to measure soil ammonium (NH_4^+-N) and nitrate (NO_3^--N) contents. Soil samples were also taken from 0–20 cm layer with a soil bulk density auger, oven-dried at 105°C for 48 h, and weighed to determine soil bulk density.

After vegetation survey and soil sampling, leaf samples of dominant and common species were collected at each site. For each species, 10–20 fully grown individuals were randomly selected, and 2–3 mature fully expanded leaves were picked, which were further divided into 5 samples. The same number of individuals of each species that were not suffering from grazer bites was also collected at the grazed sites. In this study, 169 shared species at the ungrazed and grazed sites, including 81 genera and 29 families were collected across three vegetation types, that is, 55 species in the meadow, 54 species in the meadow steppe, and 60 species in four communities of the typical steppe. All species were classified into five functional groups based on life forms, i.e., perennial bunchgrasses (PB), perennial rhizome grasses (PR), perennial forbs (PF), annuals and biennials (AB), and shrubs and semi-shrubs (SS). Root samples of 19 shared dominant species were collected at the ungrazed and grazed sites in the meadow (6 species), meadow steppe (6 species) and typical steppe (7 species). For each species, 10 individuals were randomly selected, and root biomass were sampled by taking 7-cm diameter soil cores from 0–20 cm depths after removing aboveground parts, which were further divided into 5 samples. Leaf and root samples were dried at 60°C for 24 h in a forced-draught oven and ground to homogeneity with a ball mill (MM 2000; Retsch GmbH & Co, Haan, Germany) for C, N and P measurements.

The total N, P, and organic C contents (percentage dry mass, %) in soil and plant tissues (leaves and roots) were analyzed using the standard methods [47]. The organic C content was analyzed using $K_2Cr_2O_7$-H_2SO_4 oxidation method. The total N content was determined using the Kjeldahl acid-digestion method with an Alpkem autoanalyzer (Kjektec System 1026 Distilling Unit, Sweden), and total P content was analyzed using the molybdenum blue colorimetric method with a UV/visible spectrophotometer (Beckman Coulter DU 800, USA). Contents of soil NH_4^+-N and NO_3^--N (mg kg^{-1}) were determined using the 2 mol L^{-1} KCl extraction method (solution: sediment = 5:1, 1-h extraction, solid separation by centrifugation followed by Whatman #1 paper filtration) with a flow injection autoanalyzer (FIAstar 5000 Analyzer, Foss Tecator, Denmark). Total available soil N was the sum of ammonium and nitrate contents. The total amounts of C, N and P (g m^{-2}) in aboveground leaf biomass were calculated by multiplying leaf C, N and P contents with leaf biomass of each species within the quadrat. The C, N and P pool of belowground biomass (0–20 cm, g m^{-2}) were calculated from root C, N and P contents multiplied by belowground biomass from 0–20 cm soil depth. Soil C, N and P pool (0–20 cm, g m^{-2}) were obtained from the contents multiplied by soil bulk density [9].

Plant Functional Traits

Plant functional traits, including whole plant traits, e.g., plant height, individual biomass, stem: leaf biomass ratio (SLR), and root: shoot biomass ratio (RSR); leaf traits, e.g., leaf N content, leaf C:N ratio, specific leaf area (SLA), net photosynthetic rate (Pn), photosynthetic nutrient use efficiency (PNUE), and foliar stable carbon isotope composition ($\delta\ ^{13}C$); root traits, e.g., root N content, root C:N ratio, and specific root length (SRL); and reproductive traits, e.g., reproductive allocation (fruit biomass ratio) and seed mass were measured for dominant and common species at the ungrazed sites across three vegetation types. In particular, Pn, PNUE, foliar $\delta\ ^{13}C$, SRL, and reproductive traits were determined for species only in typical steppe. These plant traits were measured on 10–20 individuals for each species in early August. Plant height was measured by the distance from the basal stem to the natural crown of each individual. After the height measurement, aboveground part of each individual was collected and taken back to the laboratory for stem and leaf separation. All leaves of an individual were picked to determine the projected leaf area with a portable leaf area meter (Li-3100C; Li-COR, Lincoln, NE, USA), and the number of leaves were recorded isochronously. Then the stem and leaf samples were oven-dried at 70°C for 24 h to constant mass and weighted. Hence, individual leaf area, dry mass per leaf, stem biomass, leaf biomass, and individual biomass could be calculated, and specific leaf area (SLA, $cm^2\ g^{-1}$) and stem: leaf ratio (SLR) were separately calculated as the ratio of leaf area to dry mass, and ratio of stem biomass to leaf biomass. The net photosynthetic rate (Pn) was measured with a Li-6400 portable photosynthetic system (Li-6400, Li-Cor, Lincoln, NE, USA) at a CO_2 concentration of 400 μmol mol^{-1} (using the built-in Li-Cor 6400 CO_2 controller) and a saturating irradiance of 1500 μmol $m^{-2}\ s^{-1}$ provided by a built-in red LED light source. The photosynthetic nutrient use efficiency (PNUE, μmol CO_2 $mol^{-1}\ s^{-1}$) was defined to equal Pn divided by leaf N content. The foliar

δ ^{13}C value was analyzed with a stable isotope ratio mass spectrometer (MAT-253; Finnigan, San Jose, USA). The root: shoot ratio (RSR) was obtained by calculating the ratio of root biomass to aboveground biomass of each individual. The specific root length (SRL, m g^{-1}) was determined by the ratio of total length of third-order roots to its dry mass [48], and root length was analyzed with a root analysis system (Delta-T Scan, Cambridge, UK). The reproductive allocation was calculated as the ratio of fruit biomass to individual biomass. The mature seeds were collected from 10–20 individuals of each species, and the sampling time was determined based on the mature season of species. Seeds were dried at room temperature for a week. For each species, the mean seed mass (mg $seed^{-1}$) was the average dry mass of 20–100 individual seeds.

Statistical Analysis

Statistical analyses were performed using SAS Version 9.2 (SAS Institute, Cary, North Carolina, USA, 2003). The effects of grazing on C, N, and P contents and C: N: P stoichiometery in soil and plant tissues (leaves and roots) at the species, plant functional group, and vegetation levels were tested with ANOVA, using treatment (grazed and ungrazed), vegetation type (meadow, meadow steppe, and typical steppe), and all interactions as fixed-effects. A total of 169 species present at the paired ungrazed and grazed sites across three vegetation types were classified into three response groups, i.e., decreased, increased, and unchanged, based on their leaf nutrients response to grazing. The effects of grazing on the C, N and P pools of above- and belowground biomass and soil, above- and belowground standing biomass, and relative aboveground biomass of plant functional groups were also examined across three vegetation types. One-way ANOVAs followed by LSD multiple-range tests were performed to test differences in functional traits among four plant functional groups (PB, PR, PF and AB).

Results

Grazing Effects on Leaf Nutrients at Different Levels of Hierarchy

We examined the responses of leaf C, N and P contents and stoichiometry to grazing in 169 shared species at the ungrazed and grazed sites across three vegetation types. At plant species level, grazing significantly increased leaf N and P contents in 62% and 58% of the total species in the meadow, but leaf C content remained unchanged in 55% of species, resulting in reduction in C: N and C: P ratios in most of the species (Fig. 1). Leaf N: P ratio in 33% of species was diminished by grazing, and that in 44% of species was unchanged. In the meadow steppe, leaf C, N, P contents and stoichiometry remained unchanged in more than 50% of species, with other species either increased or decreased in leaf N and P contents and stoichiometry. In the typical steppe, grazing decreased leaf C, N and P contents in 47%, 38%, and 43% of species, respectively; but had no significant effects on them in 38%, 55% and 28% of species. The ratios of C: N, C: P and N: P increased in 32–33% of species, but they remained unchanged in 40–55% of species.

At functional group level, grazing had larger impacts on the meadow than the meadow steppe and typical steppe. The magnitudes of changes in leaf nutrients and stoichiometry were generally greater in perennial forbs (PF), annuals and biennials (AB), and perennial rhizome grasses (PR) than in perennial bunchgrasses (PB) and shrubs and semi-shrubs (SS) (Fig. 2). In the meadow, leaf N and P contents were significantly increased by grazing, while leaf C content remained unchanged in PF, leading to reduction in C: N, C: P and N: P ratios. For AB, leaf N content was significantly increased, leaf C content was slightly reduced, and leaf P content remained unchanged by grazing, resulting in decreased C: N and unchanged C: P and N: P ratios. For PB, grazing slightly increased leaf P content, and leaf C and N contents remained unchanged, with no change in C: N and N: P and reduction in C: P. In the meadow steppe, grazing diminished only leaf P content in PF and had no effects on leaf C and N contents, resulting in no change in C: N but increases in C: P and N: P ratios. In the typical steppe, leaf N content remained unchanged in all functional groups, leaf P content was increased in PR, and leaf C content was decreased in PB and PR, leading to no change in C: N and N: P ratios but reduced C: P ratio.

At vegetation type level, grazing had greater effects on the meadow than the meadow steppe and typical steppe. In the meadow, both leaf N and P contents were increased, while leaf C content was decreased by grazing, causing reduction in C: N and C: P ratios and no change in N: P ratio (Fig. 2). In the meadow steppe, grazing diminished leaf P, and leaf C and N contents remained unchanged, leading to no change in C: N ratio but increases in C: P and N: P ratios. In the typical steppe, grazing only decreased leaf P content, but had no significant effects on C and N contents and C:N:P stoichiometry. When three vegetation types were pooled together, leaf N was significantly enhanced, leaf C was diminished, and leaf P remained unchanged, leading to diminished C: N ratio in the Xilin River Basin (Fig. 2).

Root Nutrients and Stoichiometry

At plant species level, grazing significantly enhanced root N in one of six species in the meadow, diminished root C in three species, resulting in decreased C: N ratio in two species, and increased N: P ratio in one species (Fig. S1). Root P and C: P ratio remained unchanged in all six species. In the meadow steppe, grazing increased root N in two of six species and root P in one species, leading to decreased C: P and N: P ratios of this species. In the typical steppe, grazing significantly increased root N in three and P content in four of seven species, but it decreased root C in three species, contributing to the reduced C: N, C: P and N: P ratios in these species.

At functional group level, grazing generally had greater effects on PF and PR than PB and SS (Fig. 3). In the meadow, grazing decreased root C content and C: N ratio in PF and PB, but had no significant effects on their N and P contents, C: P and N: P ratios. In the meadow steppe, however, grazing only increased root N and P contents in PF, leading to decreased C: P and N: P ratios. In the typical steppe, grazing increased root N and P contents, but decreased C: P and N: P ratios in PR. For PB, however, grazing decreased root C content, while N and P contents remained unchanged, resulting in declines in root C: N and C: P ratios.

At vegetation type level, grazing had greater effects on the typical steppe than the meadow steppe and meadow (Fig. 3). Root C and C: N ratio decreased while N and P contents and C: P and N: P ratios remained unchanged in the meadow. Grazing significantly increased root N and P contents, while root C content remained unchanged in the meadow steppe, leading to reduction in root C: P and N: P ratios. In the typical steppe, root N and P contents were increased, while root C content was slightly decreased by grazing, which leads to decreases in root C: N and C: P ratios and no change in N: P ratio. In the Xilin River Basin, root N content was significantly increased, root C content was decreased, while root P content remained unchanged, resulting in decreased C: N and C: P ratios and unchanged N: P ratio.

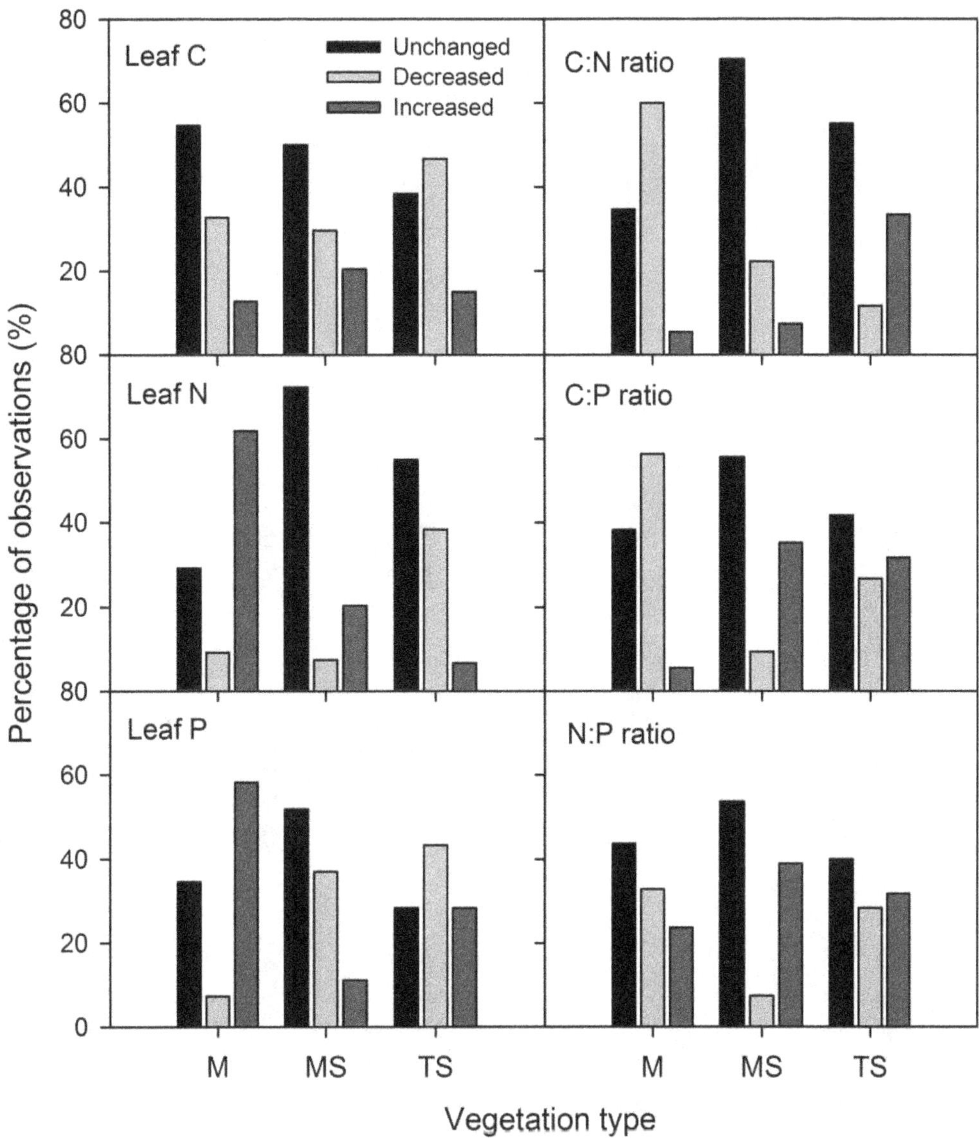

Figure 1. Percentages of species categorized as three groups based on responses of leaf C, N, P contents and stoichiometory to grazing. M, meadow; MS, meadow steppe; and TS, typical steppe. Significant differences ($P<0.05$) between the grazed and ungrazed sites were analyzed for increased, decreased and unchanged response groups.

Soil Nutrients and Stoichiometry

Soil C, N and P contents and total available N (ammonium+nitrate) were much higher in the meadow than the meadow steppe and typical steppe (Fig. S2). Grazing only caused minor shifts in soil nutrients and stoichiometry across the three vegetation types in the Xilin River Basin. Soil P was slightly decreased ($P<0.1$) by grazing in the meadow. Soil C: N was slightly increased ($P<0.1$) in the meadow steppe but decreased in the typical steppe.

C, N and P Pools of Above- and Belowground

In the meadow, grazing increased the N and P pools of leaf biomass by 104–183%, but had no significant effects on C, N and P pools of belowground biomass due to large variations (Fig. 4). In the meadow steppe, however, grazing significantly increased C, N and P pools of belowground biomass by 74–107%, but had no significant effects on those of aboveground biomass. In the typical steppe, grazing decreased the C, N and P pools of both above- (by 46–51%) and belowground biomass (by 17–25%, except for P pool). Grazing had no significant effects on soil C, N and P pools across the three vegetation types, except soil P pool was slightly decreased ($P<0.1$) in the meadow (Fig. 4). In the Xilin River Basin, the C, N and P pools in above- and belowground biomass and soil remained unchanged.

Above- and Belowground Standing Biomass and Functional Group Composition

Grazing significantly decreased the aboveground standing biomass by 34%, but had no significant effect on belowground biomass in the meadow (Fig. 5). In the meadow steppe, however, belowground biomass was enhanced by 71%, but aboveground biomass remained unchanged. Grazing diminished the aboveground biomass by 48% and belowground biomass by 23% in the typical steppe. On average, both the above- and belowground standing biomass were unchanged in the Xilin River Basin. In the

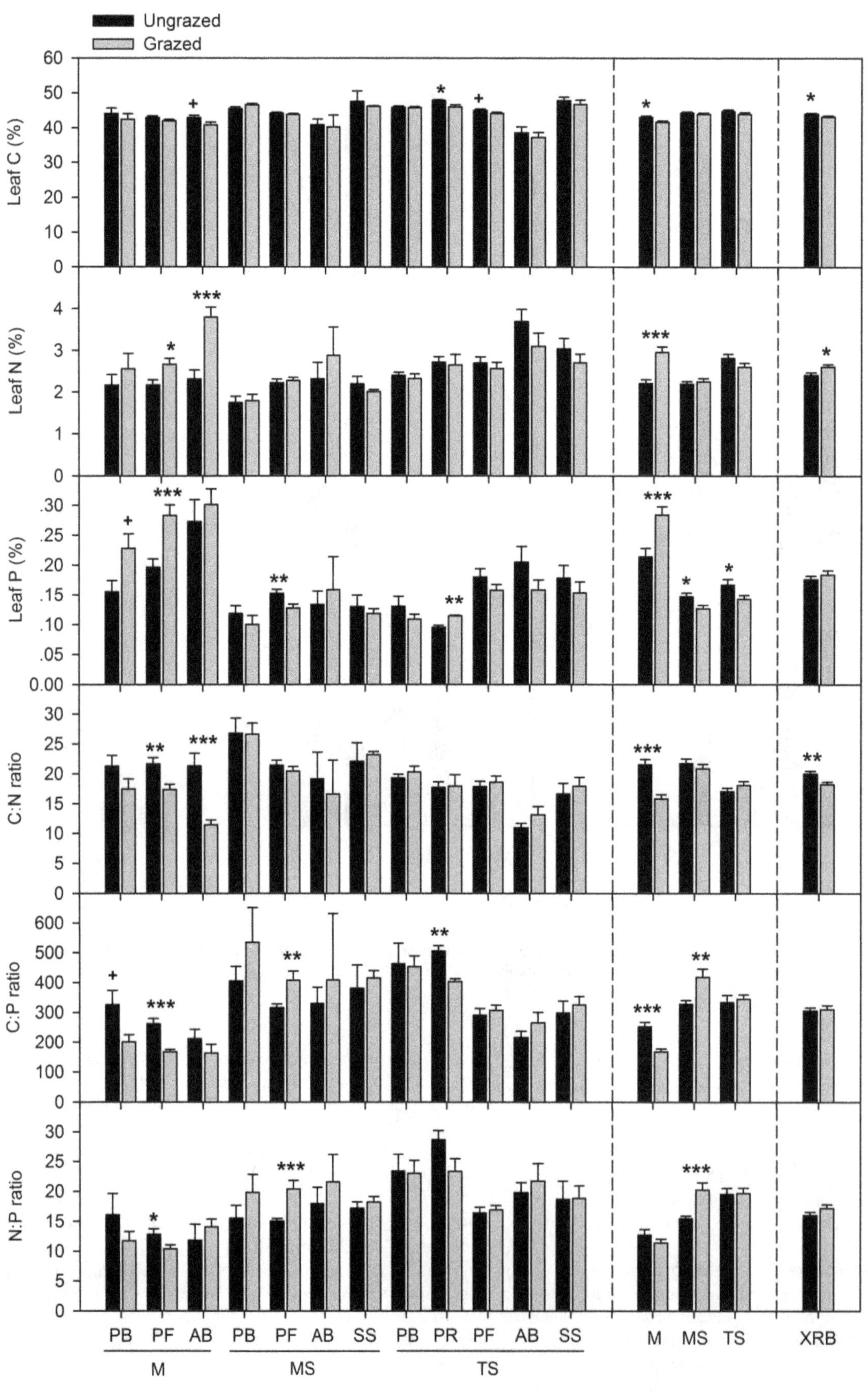
Ungrazed
Grazed
Leaf C (%)
Leaf N (%)
Leaf P (%)
C:N ratio
C:P ratio
N:P ratio
PB PF AB PB PF AB SS PB PR PF AB SS
M MS TS
M MS TS
XRB
Hierarchical level

Figure 2. Effects of grazing on leaf C, N, P contents and stoichiometory at plant functional group and vegetation type levels. The error bars are mean+SE. PB, perennial bunchgrasses; PR, perennial rhizomatous grasses; PF, perennial forbs; AB, annuals and biennials; SS, shrubs and semi-shrubs; M, meadow; MS, meadow steppe; TS, typical steppe; and XRB, Xilin River Basin. Significant differences between the grazed and ungrazed sites are reported from ANOVA as +, $0.05<P<0.1$; *, $P<0.05$; **, $P<0.01$; ***, $P<0.001$.

meadow, the relative biomass of PF was increased by 95%, while that of PB was decreased by 65% (Fig. 5). In the meadow steppe, grazing increased the relative biomass of PR by 99%, but decreased that of PB by 33%. In the typical steppe, grazing increased the relative biomass of PB by 98%, but decreased PR by 66% and PF by 45%. The relative biomass of AB and SS, in contrast, remained unchanged across three vegetation types.

Plant Functional Traits

Multi-trait comparison showed that different functional groups differed substantially ($P<0.05$) in 9 out of 15 plant functional traits, including whole plant traits, leaf traits, root traits, and reproductive traits (Table 1). The PB exhibited greater plant height, individual biomass, root: shoot ratio, stem: leaf ratio, leaf C: N ratio, and root N content, but lower leaf N content, specific leaf area (SLA), root C: N ratio, and reproductive allocation. In contrast, AB showed lower plant height, individual biomass, root: shoot ratio, stem: leaf ratio, leaf C: N ratio, and root N content, but higher leaf N content, SLA, root C: N ratio, and reproductive allocation. Moreover, AB generally had higher net photosynthetic rate (Pn), photosynthetic nutrient use efficiency (PNUE), and water use efficiency indicated by $\delta\ ^{13}C$ value. For PF and PR (i.e., *Leymus chinensis*), the values of functional traits were intermediate between PB and AB. For example, these species showed relatively lower leaf N content, SLA, Pn, and reproductive allocation, but higher plant height, individual biomass, leaf C: N ratio, and root N content than AB. Compared to PB and PF, PR and AB usually exhibited higher specific root length (SRL) and smaller seed mass.

Discussion

Hierarchical Plant C: N: P Stoichiometric Responses to Grazing

Our findings suggest that the effects of grazing on plant C, N, and P contents and stoichiometry increased progressively along the hierarchy of organizational levels, i.e., from plant species to functional group to vegetation type. Grazing enhanced vegetation-type-level N contents characterized by the increased N contents and decreased C: N ratios of plant tissues (leaves and roots) in the Xilin River Basin. However, grazing has contrasting effects on plant stoichiometry and linkage to resource allocation across the three vegetation types, as indicated by the changes in N and P contents, C: N and C: P ratios of leaves and roots at different hierarchical levels (species, functional group, and vegetation type level). In the meadow, grazing enhanced leaf N and P contents but reduced C: N and C: P ratios at three levels of hierarchy. The magnitudes of positive effects of grazing on leaf N and P contents and negative effects on C: N and C: P ratios increased along the hierarchy of organizational levels. In contrast, the magnitude of reduction in leaf N content and increase in C: N ratio diminished considerably along the hierarchy of organizational levels in the typical and meadow steppes. Compared to the leaf stoichiometry, root C: N: P stoichiometry exhibited opposite response patterns to grazing. In the typical and meadow steppes, the magnitudes of positive effects of grazing on root N and P contents and negative effects on C: N and C: P ratios increased along hierarchical levels. In the meadow, however, the magnitude of root stoichiometric responses to grazing decreased at higher hierarchical levels. These findings support our original hypothesis and further suggest that plant stoichiometric responses to grazing are largely mediated by site conditions and are often scale-dependent. Greater and positive leaf responses (i.e., leaf N and P contents increased) but weaker root responses were found in wetter habitats (meadow). On the contrary, stronger and positive root responses (i.e., root N and P contents increased) but weaker and negative leaf responses were found in drier habitats (typical and meadow steppes). This indicates that grazing increases resource allocation to aboveground tissues in the meadow, but it enhances N and P allocation to belowground tissues in the typical and meadow steppes. These results are consistent with the prediction of our second hypothesis, and are also corroborated by previous studies [9,43]. The minor shifts in soil C, N, and P contents and stoichiometry across the three vegetation types indicate that soil responses to grazing may be subject to a time-lag effect compared to the strong vegetation responses [49,50].

Our results showed the alterations of C, N and P pools in above- and belowground biomass were caused by changes both in absolute biomass and tissue nutrient contents (leaves and roots). For example, grazing diminished the C, N and P pools of aboveground biomass in the typical steppe, which is mainly attributable to sharply decrease in leaf biomass (by 47%). Grazing enhanced N and P pools of belowground biomass, caused by increases in root N and P contents. The increments in aboveground N and P pools are also mainly attributable to the increased leaf N and P contents in the meadow. In the meadow steppe, however, the increments in belowground C, N and P pools are due to both the increased root biomass (by 71%) and N and P contents. This indicates that the shifts in resource allocation between above- and belowground biomass in response to grazing may change with vegetation type or site conditions. The C, N and P pools of topsoil (0–20 cm) were less affected by grazing across the three vegetation types, because both soil bulk density and nutrient contents remained relatively unchanged. This is consistent with the results from a 7-yr field experiment conducted in the N-limited Minnesota oak savanna [9]. Compared to the meadow steppe and typical steppe, the meadow had much higher above- and belowground standing biomass, and C, N and P pools of above- and belowground biomass and soil. The minor shifts in these plant and soil properties in the Xilin River Basin are attributable to the counterbalance between the positive and negative responses to grazing among vegetation types. This implies that ecosystem resilience to grazing is relatively high in these grasslands, although they have been subjected to free-grazing for 20–30 years. Our findings further suggest that grazing effects on C: N: P stoichiometry are closely linked to ecosystem functions, such as resource allocation and nutrient pools.

Mechanisms Underpinning Grazing-induced Changes in Plant C: N: P Stoichiometry

Several mechanisms are likely to be responsible for the differential effects of grazing on plant C: N: P stoichiometry across different vegetation types. First, soil properties, particularly soil water and nutrient availability, are likely to be responsible for the contrasting effects of grazing on plant stoichiometry between the meadow and typical steppe. The field holding capacity in the meadow (52%) is two times higher than that in the typical steppe

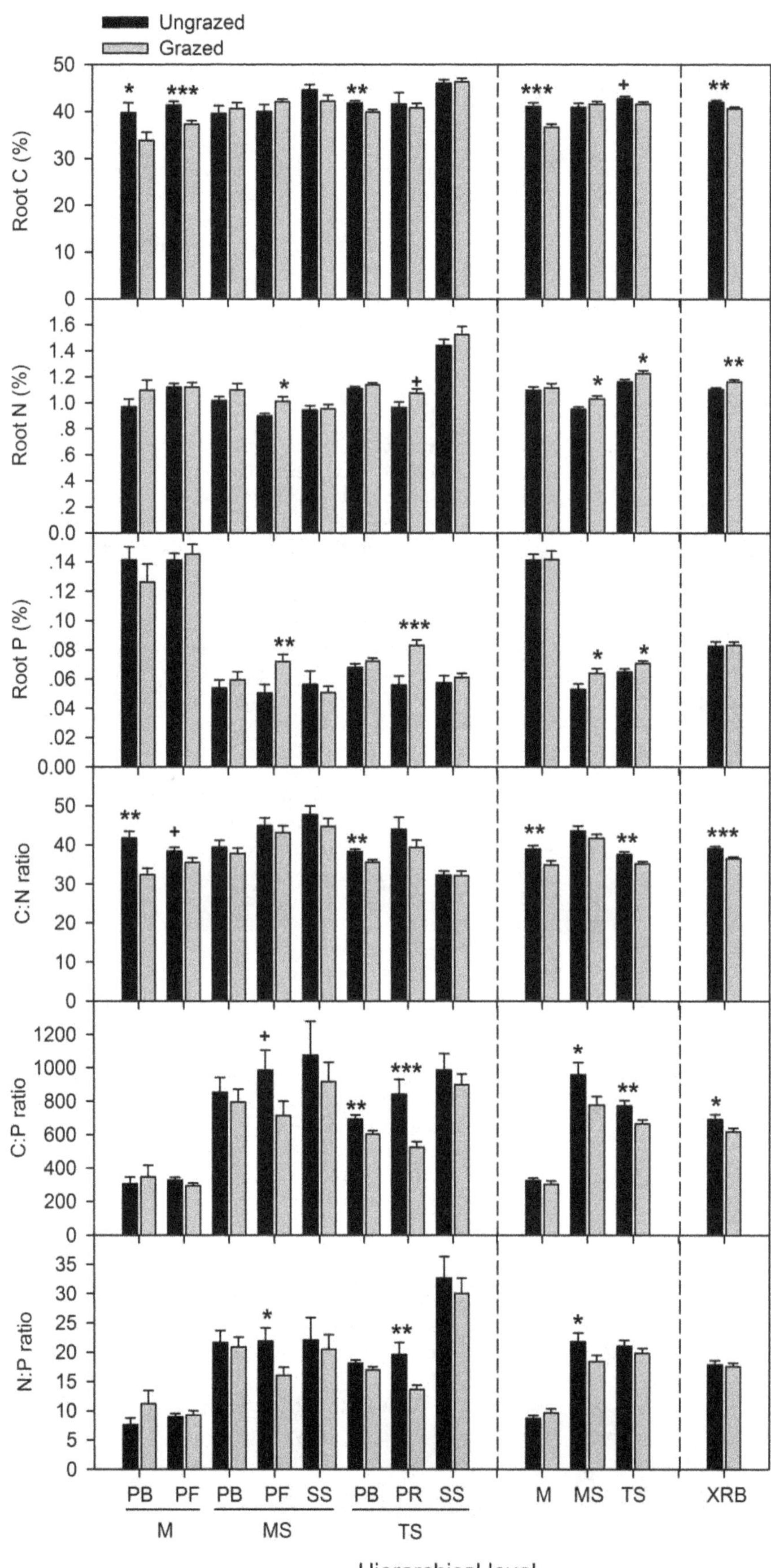
Ungrazed
Grazed
Root C (%)
Root N (%)
Root P (%)
C:N ratio
C:P ratio
N:P ratio
PB PF PB PF SS PB PR SS M MS TS XRB
M MS TS
Hierarchical level

Figure 3. Effects of grazing on root C, N, P contents and stoichiometory at plant functional group and vegetation type levels. The error bars are mean+SE. Significant levels and all symbols are derived as in Fig. 2.

(26%), which may facilitate the utilization of urine and dung deposited by herbivores and improve the soil available N directly, thereby promote plant nutrient absorption [3,19]. In the typical steppe, however, the low soil moisture and high temperature in the growing season might inhibit the process of N mineralization and nutrient utilization by plants [51]. This is because water availability is a key limiting factor in arid and semiarid grasslands, which is tightly coupled with soil N availability to affect plant growth and primary productivity [42,45]. In addition, soil carbon and nutrients (N and P) and cation exchange capacity (CEC) are much higher, but soil pH and bulk density are lower in the meadow than the typical steppe [43]. These factors favor tolerant species in the meadow to take up nutrients quickly and accelerate tissue regrowth to compensate for biomass loss by herbivores [9,35]. Young tissues usually have higher nutrient contents than mature or senescent ones. In the meadow steppe, soil water and nutrient availability are intermediate between the meadow and typical steppe, resulting in a moderate response of plant stoichiometry to grazing. Our findings suggest that the effects of grazing on plant C, N, and P contents and stoichiometry are likely mediated by resource availability. These results are consistent with our original predictions that grazing generally has positive effects on leaf nutrients (e.g., N and P contents) in wet and fertile habitats but had no effects or even negative effects on leaf nutrients in dry and infertile habitats.

Second, plant functional traits, which are closely linked to plant resource-use strategies, above- vs. belowground competitive abilities, and tolerance vs. defense strategies to grazing, are responsible for the diverse stoichiometric responses of different functional groups. The annuals and biennials exhibit acquisitive resource-use strategy in fertile habitats, such as high leaf N content and specific leaf area (SLA), high specific root length (SRL) and root C: N ratio, high net photosynthetic rate (Pn), photosynthetic nutrient use efficiency (PNUE), and water use efficiency (δ ^{13}C), but low root N content, root: shoot ratio, and leaf C: N ratio, corresponding to great aboveground/light competitive ability.

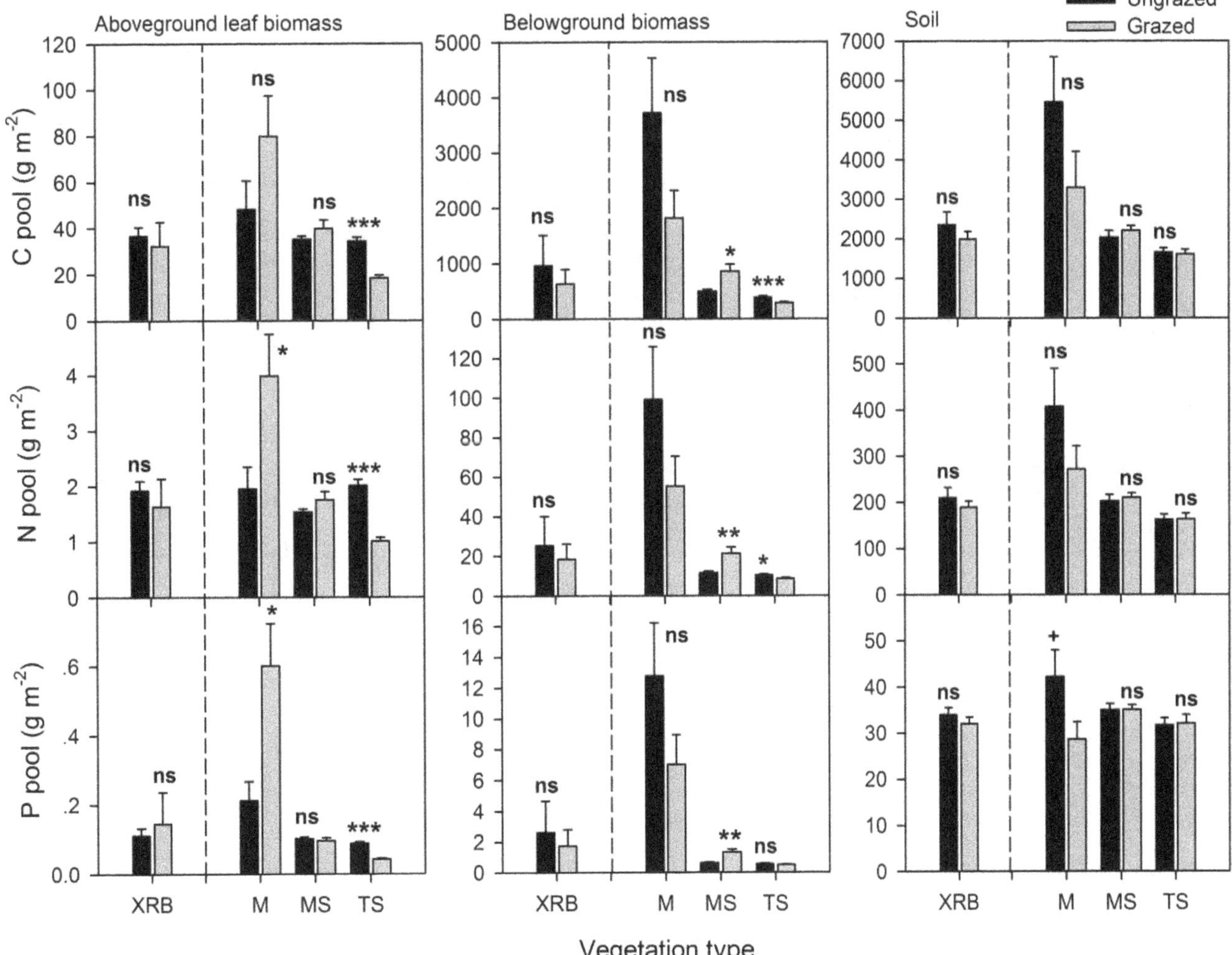

Figure 4. Effects of grazing on C, N, and P pools in above- and belowground biomass and soil (0–20 cm). The error bars are mean+SE. Significant differences between the grazed and ungrazed sites are reported from ANOVA as ns, $P>0.1$; +, $0.05<P<0.1$; *, $P<0.05$; **, $P<0.01$; ***, $P<0.001$. All symbols are derived as in Fig. 2.

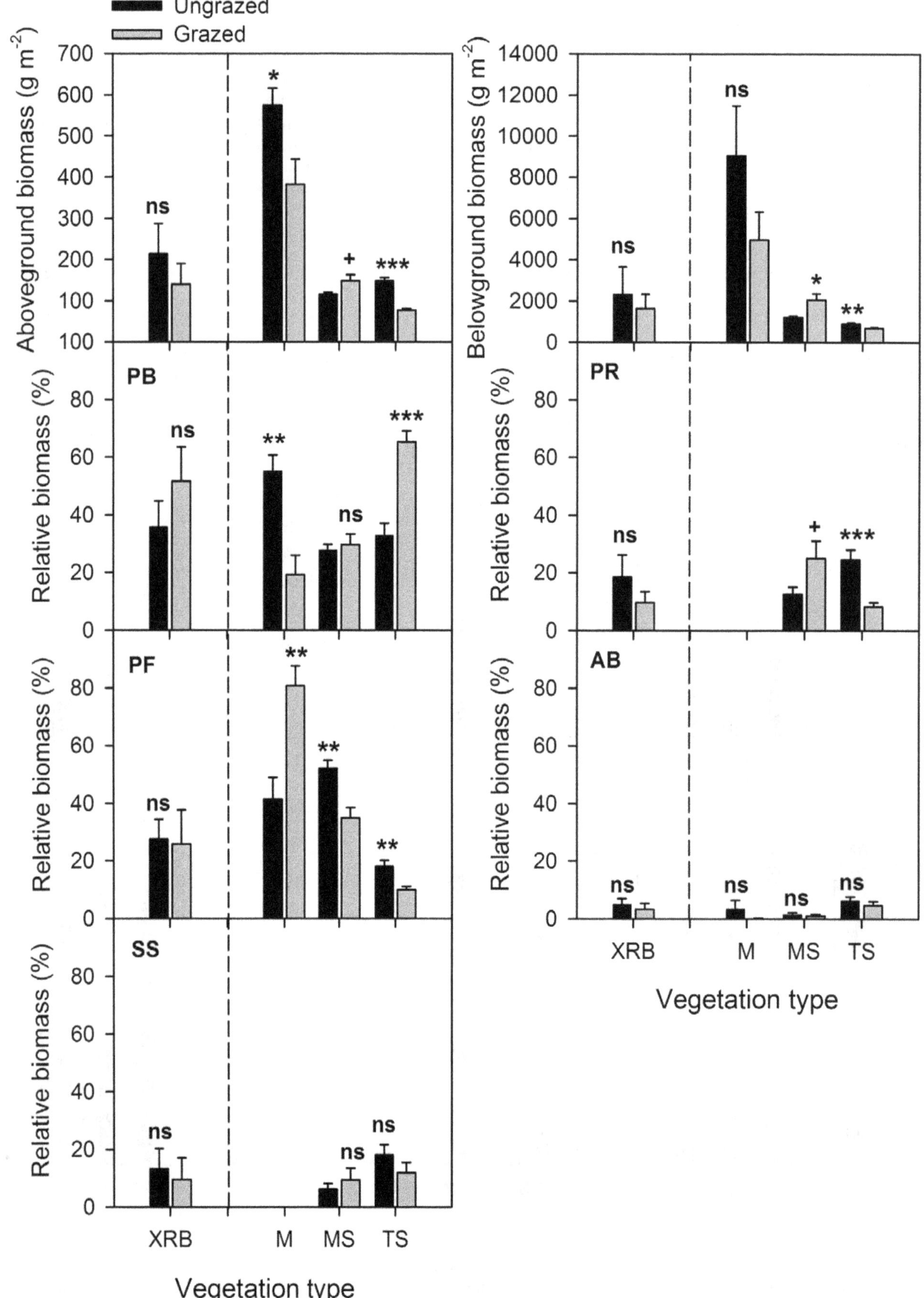
Ungrazed
Grazed
Aboveground biomass (g m-2)
Belowground biomass (g m-2)
PB
PR
PF
AB
SS
Relative biomass (%)
ns
*
+
**
***
XRB
M
MS
TS
Vegetation type

Figure 5. Effects of grazing on above- and belowground standing biomass and plant functional group composition. The error bars are mean+SE. PB, perennial bunchgrasses; PR, perennial rhizomatous grasses; PF, perennial forbs; AB, annuals and biennials; SS, shrubs and semi-shrubs; M, meadow; MS, meadow steppe; TS, typical steppe; and XRB, Xilin River Basin. Significant levels are as in Fig. 4.

These species also adopt tolerance strategy to grazing, reflecting by high foliar palatability (e.g., high leaf N content, but low leaf C: N ratio) to improve herbivore selectivity, high biomass allocation to leaves (e.g., high leaf N content and SLA, but low stem: leaf ratio, root: shoot ratio, and root N content) to accelerate shoot regrowth following defoliation. Thus annuals and biennials generally allocate more resource to shoots and increase leaf nutrients in response to grazing. In addition, these annuals and biennials have high recruitment capacity, reflected by reproductive traits of small seed, high reproductive output, and short life history. Hence, these species have the competitive advantage in resource exploitation and they maintain relatively stable species compositions across three vegetation types as indicated by their relative biomass.

In contrast, most perennial bunchgrasses are xerophyte which adopt conservative resource-use strategy in dry and nutrient-poor habitats. These species generally exhibit traits characterized by high root N content, root: shoot ratio, and leaf C: N ratio, but low leaf N content, SLA, and SRL, corresponding to great belowground competitive ability. The greater plant height, individual biomass and stem: leaf ratio of these species suggest the functional advantage in light interception, thus they can maintain relatively higher light, nitrogen and water use efficiencies, as reflected by the high Pn, PNUE and $\delta^{13}C$. In addition, perennial bunchgrasses adopt defense strategy to grazing, indicated by low foliar palatability (e.g., low leaf N content and SLA, but high leaf C: N ratio) to decrease herbivore selectivity and thereby restrain compensatory growth. Therefore, the perennial bunchgrass dominated plant communities have the advantage in belowground resource competition, leading to weak leaf stoichiometric responses to grazing.

Perennial rhizomatous grasses (i.e., *Leymus chinensis*) are meso-xerophyte which generally adopt both acquisitive and conservative resource-use strategies characterized by relatively higher plant height, leaf N content, and photosynthetic rate (Pn) for above-ground/light competition, and higher root N content and specific root length (SRL) for belowground competition. In addition, these species have high foliar palatability and allocate more biomass to leaves thus exhibiting more tolerance following grazing. Moreover, *L. chinensis* has well-developed laterally spread rhizomes to reproduce vegetative tillers, of which rhizome biomass accounted for 30% of the total plant biomass. The perennial forbs are dominant species in the meadow and meadow steppe and are also widely distributed in the typical steppe. The traits of perennial forbs associated with resource-use strategy, above- vs. below-ground competitive ability, and grazing-tolerance are intermediate between the annuals and biennials and perennial bunchgrasses, leading to a moderate response to grazing. The differences in functional traits among different functional groups indicate the fundamental trade-offs between productivity and persistence of species, and further reflect the contrasting species-specific tolerance and defense strategies to grazing. In the Inner Mongolia grassland, the meadow is mainly dominated by perennial forbs and annuals and biennials, with more acquisitive resource-use strategies, great aboveground competitive ability, and fast regrowth following defoliation, suggesting a potential for increasing resource allocation to aboveground tissue and leaf nutrient contents as response to grazing. The typical steppe, however, is dominated by perennial bunchgrasses, with more conservative resource-use strategies, strong belowground competitive ability, and low compensatory growth, indicating a shift towards increasing belowground allocation and root nutrient contents in response to grazing. In the meadow steppe, perennial forbs and perennial bunchgrasses are two dominant life forms, which drive an intermediate response to grazing between the meadow and typical steppe. These results are consistent with the prediction of our second hypothesis that grazing enhances N and P allocation to aboveground in the meadow, but it increases N and P allocation to belowground in the typical and meadow steppes.

Third, the shifts in plant functional group compositions are the major mechanisms driving vegetation-type-level stoichiometric responses and nutrient dynamics. In the meadow, the shift in dominance from perennial bunchgrasses to perennial forbs leads to increase in leaf nutrients (N and P) but no change in root nutrients in response to grazing. On the contrary, the shift in dominance from perennial rhizomatous grasses to perennial bunchgrasses results in the increases in root nutrients (N and P) but no change or even decreases in leaf nutrients in response to grazing. In the meadow steppe, perennial forbs were replaced by perennial rhizomatous grasses and perennial bunchgrasses, leading to an intermediate response between the meadow and typical steppe.

N Versus P Limitation Across Vegetation Types

Our results suggest that grazing potentially increases P limitation in the semiarid grassland of the Xilin River Basin, as evidenced by the enhanced N content and N: P ratio. The N: P ratio was, on average, 17.2 at the grazed sites, which is higher than the threshold of 16:1 for P limitation [31]. In addition, the average C: P ratio was 311 at grazed sites, which is higher than the hypothesized threshold of 250:1 required for efficient growth of P-rich herbivores feeding on comparably C-rich plants, indicating potential P deficiency in these habitats [32]. However, the effects of grazing on nutrient limitation differed across the three vegetation types. Neither N nor P limitation was observed in the meadow, N and P co-limitation in typical steppe, and more P limitation in the meadow steppe. In the meadow, soil nutrients, N and P contents of leaves and roots, and community biomass of above- and belowground were generally higher than those in the typical steppe and meadow steppe. Grazing decreased C: P ratio from 253 to 168 (lower than 250:1), while N: P ratio (11.4) remained unchanged, which suggests the community is not subject to either N or P limitation. In the typical steppe, both the N: P (19.7) and C: P ratios (345) were relatively less affected by grazing. These factors together with the lowest soil N availability, above- and belowground biomass indicate that the typical steppe is co-limited by N and P. In the meadow steppe, however, grazing enhanced N: P (20.3) and C: P (418) ratios, indicating a shift towards P limitation. In this study, the mean N: P ratio of meadow (12.7 for ungrazed vs. 11.4 for grazed) is lower than previously documented threshold of 14:1 for N limitation, and N: P ratio of typical steppe (19.6 for ungrazed vs. 19.7 for grazed) is out of the range between 14 and 16 for N and P co-limitation [31], indicating that using foliar N: P ratio to determine nutrient limitation needs more careful interpretation. The limitation threshold may be ecosystem-specific or dependent on site conditions.

In the meadow, grazing enhanced leaf N and P contents, but had little effects on root N and P contents across different

Table 1. Plant functional traits of four functional groups across three vegetation types in the Xilin River Basin of Inner Mongolia grassland.

PFGs	Whole plant traits				Leaf traits						Root traits			Reproductive traits	
	Plant height (cm)	Individual biomass (g)	RSR ($g \cdot g^{-1}$)	SLR ($g \cdot g^{-1}$)	Leaf N (%)	Leaf C:N ratio ($g \cdot g^{-1}$)	SLA ($cm^2 \cdot g^{-1}$)	Pn (μmol $CO_2 \cdot m^{-2} \cdot s^{-1}$)	PNUE (μmol $CO_2 \cdot mol^{-1} \cdot s^{-1}$)	$\delta^{13}C$ (‰)	Root N (%)	Root C:N ratio ($g \cdot g^{-1}$)	SRL ($m \cdot g^{-1}$)	Seed mass ($mg \cdot seed^{-1}$)	RA (%)
PB	42.42a	1.99a	0.66ab	2.50a	2.26b	20.96a	93.62b	14.42ab	120.96	−24.40	1.11a	38.17b	97.63	3.57	6.30b
PR	34.28ab	0.56ab	0.28b	0.87b	2.59ab	18.63ab	59.59b	16.50ab	104.04	−26.41	1.01a	45.20b	122.55	1.89	0.25c*
PF	32.72b	1.81a	1.52a	1.33b	2.32b	20.63a	114.21ab	9.80b	68.02	−26.45	1.12a	40.90b	95.54	3.51	12.07b
AB	18.91c	0.48b	0.49b	1.21b	2.79a	17.49b	133.64a	20.35a	139.74	−23.63	0.75b	67.62a	104.10	0.70	24.70a
P value	<0.001	0.044	0.018	0.025	0.021	0.092	0.009	0.112	0.182	0.19	<0.001	<0.001	0.846	0.333	0.005

Abbreviations: PFGs, plant functional groups; PB, perennial bunchgrasses; PR, perennial rhizomatous grasses; PF, perennial forbs; AB, annuals and biennials; RSR, root: shoot biomass ratio; SLR, stem: leaf biomass ratio; SLA, specific leaf area; Pn, net photosynthetic rate; PNUE, photosynthetic nutrient use efficiency; $\delta^{13}C$, Carbon isotope ratio; SRL, specific root length; RA, reproductive allocation. *, The dominant dispersal mode for perennial rhizomatous grass (i.e., *Leymus chinensis*) is rhizome, with rhizome biomass accounting for 30% of the total plant biomass. *P* values following one-way ANOVAs indicate differences in plant functional traits among four functional groups. Different lowercases represent significant differences among plant functional groups (LSD multiple-range tests, $P<0.05$). The sample sizes of functional traits (plant height, individual biomass, SLR, SLA, leaf N and C: N ratio, and root N and C: N ratio) are 27, 5, 104 and 23 for PB, PR, PF and AB, respectively across meadow, meadow steppe and typical steppes. The sample sizes of functional traits (RSR, Pn, PNUE, $\delta^{13}C$, SRL, seed mass and RA) are 5, 3, 10–21 and 8 for PB, PR, PF and AB, respectively in typical steppe.

hierarchical levels. This indicates more nutrients are reallocated to aboveground for supporting shoot regrowth, and P availability maybe sufficient to support stoichiometrically balanced growth for plants. Our results are corroborated by a previous study [52], which demonstrated that when soil nutrients are abundant, plant nutrient contents (especially P) may be unchanged or even increase following defoliation. It is likely because plants need to allocate sufficient P to recover from defoliation in order to produce ribosomal RNA to replace the N-rich photosynthetic proteins lost to grazers [32,53]. On the contrary, the enhanced root N and P contents but the unchanged leaf N and declined leaf P content observed in the typical steppe and meadow steppe, suggesting more nutrients are invested to belowground. Previous studies also found grazing-tolerant species exhibit positive feedbacks to herbivory by promoting root exudation of carbon to stimulate rhizospheric microbial populations for soil N mineralization [26,36]. In our study, we found the dominant perennial bunchgrasses, such as *S. grandis*, *A. cristatum*, *Achnatherum sibiricum*, *C. squarrosa*, and *L. chinensis* in typical steppe generally have roots with rhizosheath, which favors nutrient absorption and water reservation in resource-poor habitats [54]. Tian et al. [55] reported most of dominant and subdominant plant species in typical steppe are colonized by arbuscular mycorrhizal fungi, which could improve plant P uptake, thereby lessen the defoliation-induced P limitation [56]. Nevertheless, these compensatory feedbacks may be insufficient to prevent widespread N or P limitation across a broad geographic region, especially in arid and semiarid grasslands.

Conclusions

Using a multi-organization-level approach, our study demonstrates the effects of grazing on the C, N and P contents and stoichiometry of plant tissues (leaves and roots) are scale-dependant, and they may change with vegetation type or site conditions. However, soil nutrients exhibit weak responses. Grazing-induced shifts in resource allocation between above- and belowground and nutrient limitation as indicated by leaf stoichiometry differed across vegetation types. Grazing increases N and P allocation to aboveground in the meadow, but it enhances N and P allocation to belowground in the typical and meadow steppes. Neither N nor P limitation was found in the meadow, with N and P co-limitation in the typical steppe and P limitation in the meadow steppe. Our findings suggest that the enhanced vegetation-type-level N contents by grazing and species compensatory feedbacks may be insufficient to prevent widespread declines in primary productivity and nutrient pools in the Inner Mongolia grassland, although they could moderately mitigate the impacts of nutrient limitation. Hence, it is essential to reduce the currently high stocking rates and restore the vast degraded steppes for sustainable development of arid and semiarid grasslands.

Supporting Information

Figure S1 Effects of grazing on root C, N, P contents and stoichiometory of dominant species across three vegetation types. The error bars are mean+SE. Significant differences between the grazed and ungrazed sites are reported from ANOVA as +, $0.05<P<0.1$; *, $P<0.05$; **, $P<0.01$; ***, $P<0.001$. Abbreviations: Ag, *Agrostis gigantea*; Ca, *Carex appendiculata*; Br, *Blysmus rufus*; Pan, *Potentilla anserina*; Ib, *Inula britanica*; Ss, *Sium suave*; Sb, *Stipa baicalensis*; Ac, *Agropyron cristatum*; Sc, *Serratula centauroides*; As, *Allium senescens*; Pac, *Potentilla acaulis*; Af, *Artemisia frigida*; Lc, *Leymus chinensis*; Sg, *Stipa grandis*; Cs, *Cleistogenes squarrosa*;

Ks, *Koeleria cristata*; Cm, *Caragana microphylla*; M, meadow; MS, meadow steppe; TS, typical steppe.

Figure S2 Effects of grazing on soil C, N, P contents, inorganic N (NH_4^+-N and NO_3^--N) and stoichiometory across three vegetation types. The error bars are mean+SE. M, meadow; MS, meadow steppe; TS, typical steppe; and XRB, Xilin River Basin. Significant differences between the grazed and ungrazed sites are reported from ANOVA as +, $0.05<P<0.1$; *, $P<0.05$; **, $P<0.01$.

Acknowledgments

We appreciate Valerie T. Eviner for helpful comments on an early version of this manuscript. We thank Xinrui Chen, Chunmei Pan, Li Qiang, and many others at the Inner Mongolia Grassland Ecosystem Research Station (IMGERS), Chinese Academy of Sciences for their help with fieldwork.

Author Contributions

Conceived and designed the experiments: SZ HR. Performed the experiments: SZ HR WL ZL. Analyzed the data: SZ HR. Wrote the paper: SZ HR.

References

1. Sasaki T, Okayasu T, Jamsran U, Takeuchi K (2008) Threshold changes in vegetation along a grazing gradient in Mongolian rangelands. Journal of Ecology 96: 145–154.
2. Frank DA (2005) The interactive effects of grazing ungulates and aboveground production on grassland diversity. Oecologia 143: 629–634.
3. Frank DA, Evans RD (1997) Effects of native grazers on grassland N cycling in Yellowstone National Park. Ecology 78: 2238–2248.
4. McNaughton SJ, Banyikwa FF, McNaughton MM (1997) Promotion of the cycling of diet-enhancing nutrients by African grazers. Science 278: 1798–1800.
5. Semmartin M, Aguiar MR, Distel RA, Moretto AS, Ghersa CM (2004) Litter quality and nutrient cycling affected by grazing-induced species replacements along a precipitation gradient. Oikos 107: 148–160.
6. Pastor J, Cohen Y, Hobbs T (2006) The role of large herbivores in ecosystem nutrient cycles. In: Danell K, Bergström R, Duncan P, Pastor J, eds. Large mammalian herbivores, ecosystem dynamics, and conservation: Cambridge University Press. 289–325.
7. Franzluebbers AJ, Stuedemann JA, Schomberg HH, Wilkinson SR (2000) Soil organic C and N pools under long-term pasture management in the Southern Piedmont USA. Soil Biology and Biochemistry 32: 469–478.
8. Golluscio RA, Austin AT, Martínez GCG, Gonzalez-Polo M, Sala OE, et al. (2009) Sheep grazing decreases organic carbon and nitrogen pools in the Patagonian steppe: Combination of direct and indirect effects. Ecosystems 12: 686–697.
9. Ritchie ME, Tilman D, Knops JMH (1998) Herbivore effects on plant and nitrogen dynamics in oak savanna. Ecology 79: 165–177.
10. Bagchi S, Ritchie ME (2010) Introduced grazers can restrict potential soil carbon sequestration through impacts on plant community composition. Ecology Letters 13: 959–968.
11. Wu HH, Dannenmann M, Fanselow N, Wolf B, Yao ZS, et al. (2011) Feedback of grazing on gross rates of N mineralization and inorganic N partitioning in steppe soils of Inner Mongolia. Plant and Soil 340: 127–139.
12. White R, Murray S, Rohweder M (2000) Pilot analysis of global ecosystems: grassland ecosystems. Washington D.C.: World Resources Institute.
13. Holland EA, Detling JK (1990) Plant-response to herbivory and belowground nitrogen cycling. Ecology 71: 1040–1049.
14. Frank DA, Groffman PM (1998) Ungulate vs. landscape control of soil C and N processes in grasslands of Yellowstone National Park. Ecology 79: 2229–2241.
15. McNaughton SJ (1985) Ecology of a grazing ecosystem: the Serengeti. Ecological Monographs 55: 259–294.
16. Bryant JP, Provenza FD, Pastor J, Reichardt PB, Clausen TP, et al. (1991) Interactions between woody-plants and browsing mammals mediated by secondary metabolites. Annual Review of Ecology and Systematics 22: 431–446.
17. Pastor J, Naiman RJ (1992) Selective foraging and ecosystem processes in boreal forests. American Naturalist 139: 690–705.
18. Bardgett RD, Wardle DA (2003) Herbivore-mediated linkages between aboveground and belowground communities. Ecology 84: 2258–2268.
19. Augustine DJ, McNaughton SJ (2006) Interactive effects of ungulate herbivores, soil fertility, and variable rainfall on ecosystem processes in a semi-arid savanna. Ecosystems 9: 1242–1256.
20. Wardle DA, Bardgett RD, Klironomos JN, Setala H, van der Putten WH, et al. (2004) Ecological linkages between aboveground and belowground biota. Science 304: 1629–1633.
21. Díaz S, Hodgson JG, Thompson K, Cabido M, Cornelissen JHC, et al. (2004) The plant traits that drive ecosystems: Evidence from three continents. Journal of Vegetation Science 15: 295–304.
22. Díaz S, Lavorel S, de Bello F, Quetier F, Grigulis K, et al. (2007) Incorporating plant functional diversity effects in ecosystem service assessments. Proceedings of the National Academy of Sciences of the United States of America 104: 20684–20689.
23. Díaz S, Lavorel S, McIntyre S, Falczuk V, Casanoves F, et al. (2007) Plant trait responses to grazing - a global synthesis. Global Change Biology 13: 313–341.
24. Elser JJ, Fagan WF, Kerkhoff AJ, Swenson NG, Enquist BJ (2010) Biological stoichiometry of plant production: metabolism, scaling and ecological response to global change. New Phytologist 186: 593–608.
25. Holland JN, Cheng WX, Crossley DA (1996) Herbivore-induced changes in plant carbon allocation: assessment of below-ground C fluxes using carbon-14. Oecologia 107: 87–94.
26. Bardgett RD, Wardle DA, Yeates GW (1998) Linking above-ground and below-ground interactions: how plant responses to foliar herbivory influence soil organisms. Soil Biology and Biochemistry 30: 1867–1878.
27. Shan YM, Chen DM, Guan XX, Zheng SX, Chen HJ, et al. (2011) Seasonally dependent impacts of grazing on soil nitrogen mineralization and linkages to ecosystem functioning in Inner Mongolia grassland. Soil Biology and Biochemistry 43: 1943–1954.
28. Tilman D (1988) Plant strategies and the dynamics and structure of plant communities. Princeton, New Jersey: Princeton University Press.
29. Poorter H, Nagel O (2000) The role of biomass allocation in the growth response of plants to different levels of light, CO_2, nutrients and water: a quantitative review. Australian Journal of Plant Physiology 27: 595–607.
30. Pan Q, Bai Y, Wu J, Han X (2011) Hierarchical plant responses and diversity loss after nitrogen addition: Testing three functionally-based hypotheses in the Inner Mongolia grassland. PLoS ONE 6: e20078.
31. Koerselman W, Meuleman AFM (1996) The vegetation N:P ratio: a new tool to detect the nature of nutrient limitation. Journal of Applied Ecology 33: 1441–1450.
32. Elser JJ, Fagan WF, Denno RF, Dobberfuhl DR, Folarin A, et al. (2000) Nutritional constraints in terrestrial and freshwater food webs. Nature 408: 578–580.
33. Gusewell S (2004) N : P ratios in terrestrial plants: variation and functional significance. New Phytologist 164: 243–266.
34. Aerts R (1997) Climate, leaf litter chemistry and leaf litter decomposition in terrestrial ecosystems: a triangular relationship. Oikos 79: 439–449.
35. Hobbie SE (1992) Effects of plant-species on nutrient cycling. Trends in Ecology and Evolution 7: 336–339.
36. Hamilton EW III, Frank DA (2001) Can plants stimulate soil microbes and their own nutrient supply? Evidence from a grazing tolerant grass. Ecology 82: 2397–2402.
37. Zheng SX, Lan ZC, Li WH, Shao RX, Shan YM, et al. (2011) Differential responses of plant functional trait to grazing between two contrasting dominant C_3 and C_4 species in a typical steppe of Inner Mongolia, China. Plant and Soil 340: 141–155.
38. Craine JM, Froehle J, Tilman DG, Wedin DA, Chapin FS (2001) The relationships among root and leaf traits of 76 grassland species and relative abundance along fertility and disturbance gradients. Oikos 93: 274–285.
39. Cornelissen JHC, Lavorel S, Garnier E, Díaz S, Buchmann N, et al. (2003) A handbook of protocols for standardised and easy measurement of plant functional traits worldwide. Australian Journal of Botany 51: 335–380.
40. Naeem S (1998) Species redundancy and ecosystem reliability. Conservation Biology 12: 39–45.
41. Loreau M, Hector A (2001) Partitioning selection and complementarity in biodiversity experiments. Nature 412: 72–76.
42. Bai YF, Han XG, Wu JG, Chen ZZ, Li LH (2004) Ecosystem stability and compensatory effects in the Inner Mongolia grassland. Nature 431: 181–184.
43. Zheng SX, Ren HY, Lan ZC, Li WH, Wang KB, et al. (2010) Effects of grazing on leaf traits and ecosystem functioning in Inner Mongolia grasslands: scaling from species to community. Biogeosciences 7: 1117–1132.
44. Jiang GM, Han XG, Wu JG (2006) Restoration and management of the Inner Mongolia Grassland require a sustainable strategy. AMBIO 35: 269–270.
45. Bai YF, Wu JG, Xing Q, Pan QM, Huang JH, et al. (2008) Primary production and rain use efficiency across a precipitation gradient on the Mongolia plateau. Ecology 89: 2140–2153.
46. Bai YF, Li LH, Wang QB, Zhang LX, Zhang Y, et al. (2000) Changes in plant species diversity and productivity along gradients of precipitation and elevation in the Xilin River Basin, Inner Mongolia. Acta Phytoecologica Sinica 24: 667–673.
47. Sparks DL, Page AL, Helmke PA, Loeppert RH, Soltanpour PN, et al. (1996) Methods of soil analysis Part 3: Chemical methods. Madison, WI, USA: Soil Science Society of America, Inc. and American Society of Agronomy, Inc.

48. Pregitzer KS, DeForest JL, Burton AJ, Allen MF, Ruess RW, et al. (2002) Fine root architecture of nine North American trees. Ecological Monographs 72: 293–309.
49. Milchunas DG, Lauenroth WK (1993) Quantitative effects of grazing on vegetation and soils over a global range of environments. Ecological Monographs 63: 327–366.
50. Zhou ZY, Li FR, Chen SK, Zhang HR, Li GD (2011) Dynamics of vegetation and soil carbon and nitrogen accumulation over 26 years under controlled grazing in a desert shrubland. Plant and Soil 341: 257–268.
51. Shan YM, Chen DM, Guan XX, Zheng SX, Chen HJ, et al. (2011) Seasonally dependent impacts of grazing on soil nitrogen mineralization and linkages to ecosystem functioning in Inner Mongolia grassland. Soil Biology and Biochemistry 43: 1943–1954.
52. McNaughton SJ, Chapin FS III (1985) Effects of phosphorus nutrition and defoliation on C_4 graminoids from the serengeti plains. Ecology 66: 1617–1629.
53. Matzek V, Vitousek PM (2009) N : P stoichiometry and protein : RNA ratios in vascular plants: an evaluation of the growth-rate hypothesis. Ecology Letters 12: 765–771.
54. McCully ME (1999) Roots in soil: unearthing the complexities of roots and their rhizospheres. Annual Review of Plant Physiology and Plant Molecular Biology 50: 695–718.
55. Tian H, Gai JP, Zhang JL, Christie P, Li XL (2009) Arbuscular mycorrhizal fungi associated with wild forage plants in typical steppe of eastern Inner Mongolia. European Journal of Soil Biology 45: 321–327.
56. van der Heijden MGA, Klironomos JN, Ursic M, Moutoglis P, Streitwolf-Engel R, et al. (1998) Mycorrhizal fungal diversity determines plant biodiversity, ecosystem variability and productivity. Nature 396: 69–72.

4

Fundamental Limits on Wavelength, Efficiency and Yield of the Charge Separation Triad

Alexander Punnoose[1,2]*, Liza McConnell[2], Wei Liu[2], Andrew C. Mutter[2], Ronald Koder[2]*

1 Instituto de Física Teórica, Universidade Estadual Paulista, São Paulo, Brazil, **2** Department of Physics, City College of the City University of New York, New York, New York, United States of America

Abstract

In an attempt to optimize a high yield, high efficiency artificial photosynthetic protein we have discovered unique energy and spatial architecture limits which apply to all light-activated photosynthetic systems. We have generated an analytical solution for the time behavior of the core three cofactor charge separation element in photosynthesis, the photosynthetic cofactor triad, and explored the functional consequences of its makeup including its architecture, the reduction potentials of its components, and the absorption energy of the light absorbing primary-donor cofactor. Our primary findings are two: First, that a high efficiency, high yield triad will have an absorption frequency more than twice the reorganization energy of the first electron transfer, and second, that the relative distance of the acceptor and the donor from the primary-donor plays an important role in determining the yields, with the highest efficiency, highest yield architecture having the light absorbing cofactor closest to the acceptor. Surprisingly, despite the increased complexity found in natural solar energy conversion proteins, we find that the construction of this central triad in natural systems matches these predictions. Our analysis thus not only suggests explanations for some aspects of the makeup of natural photosynthetic systems, it also provides specific design criteria necessary to create high efficiency, high yield artificial protein-based triads.

Editor: Carl J. Bernacchi, University of Illinois, United States of America

Funding: RLK gratefully acknowledges support by the following grants: FA9550-10-1-0350 from the Air Force Office of Scientific Research and the NIH National Center for Research Resources to CCNY (NIH 5G12 RR03060). ACM gratefully acknowledges support from the Center for Exploitation of Nanostructures in Sensor and Energy Systems (CENSES) under NSF Cooperative Agreement Award Number 0833180. These funders had no role in study design, data collection and analysis, decision to publish, or preparation of the manuscript.

Competing Interests: The authors have declared that no competing interests exist.

* E-mail: alexander.punnoose@gmail.com (AP); koder@sci.ccny.cuny.edu (RK)

Introduction

Solar energy conversion machines found in nature utilize a number of small molecules, called cofactors, which serve as discrete sites for the binding of a single electron [1]. Charge separation in these proteins is effected via a cascade of several individual electron transfer (ET) events initiated by the absorption of a photon at a central cofactor termed the primary-donor [2]. These protein machines typically contain numerous cofactors arranged so as to enable the movement of the electron and the oxidizing equivalent away from the primary-donor in opposite directions [3,4]. The resultant potential energy is then coupled to some chemical reaction or reactions which create a storable, diffusable form of chemical energy.

Chemists have made an intensive effort over the past forty years to recreate the charge separation capability of these devices in synthetic systems [5–11]. The minimal construct that can achieve long-lived charge separation contains a primary-donor along with two other cofactors to facilitate the separation and prevent the fast relaxation of the electron back to the groundstate of the primary-donor (see Figure 1A). This has been termed the photosynthetic cofactor triad (PCT) [7,11–13]. Research efforts have aimed at engineering protein-based PCTs, either through the reengineering of natural proteins [14,15] or de novo design of new artificial proteins [16,17]. An optimal PCT construct will maximize the yield of the charge separated state and minimize energy loss while maintaining the state for as long as necessary before decaying to the groundstate. These performance metrics are intimately related to the microscopic ET rates which themselves are a function of the reduction potentials and the spatial arrangement of the three cofactors. Given the large expense and long time scale of these design efforts [2,18–20], it is important to understand the optimal structure and properties of this molecule from the beginning of the design process. The key is to identify the set of microscopic parameters which when manipulated can effect maximum benefit during the design process.

Clearly, numerical simulations of the rate equations to map out the optimal set of ET rates for the entire construct involve a large parameter space [21]. For this reason there has been little theoretical analysis of the optimal structure and properties of the cofactor triad and its many sequential ETs. There are several semiclassical equations which predict ET rates that are well validated, in particular the semiclassical Marcus expression [22–26]. These are all complicated functions with a number of terms. The challenge in a complex system such as the PCT is to select the formalism which will give a meaningful analytical expression for its behavior. For example, Cho and Silby, in 1995, derived the time-dependent behavior of a molecular dyad structure composed of two cofactors and three states in the limit of a very large reduction potential difference between the excited state primary donor and the acceptor site [27].

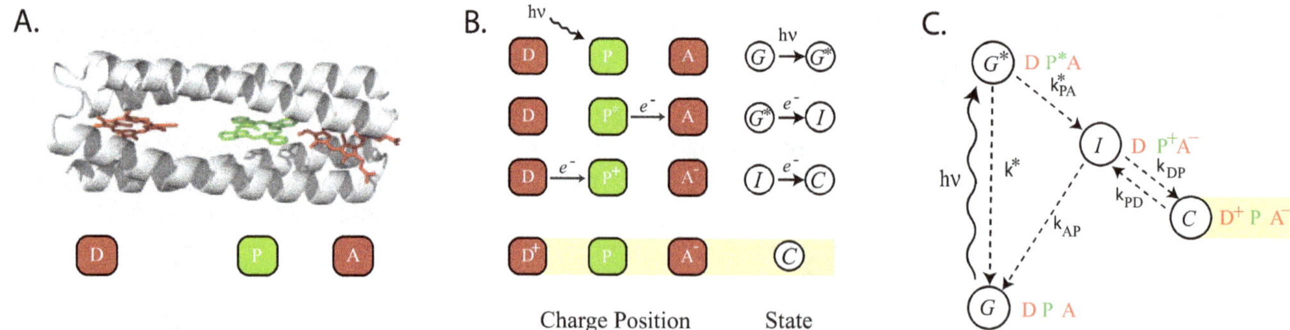

Figure 1. Structure and function of the photosynthetic triad. (A) Molecular detail of an idealized artificial charge separation construct, a self-assembling de novo designed protein. (B) Discrete steps in the formation of the charge separated state: The primary-donor molecule P in the ground state configuration $G \equiv$ DPA absorbs a photon of the correct frequency to form $G^* \equiv$ DP*A, where P* is the photoexcited state of P. The excited electron transfers to the acceptor cofactor, A, forming the intermediate state $I \equiv$ DP$^+$A$^-$. The donor cofactor, D, then transfers an electron into P, resulting in the charge separated state $C \equiv$ D$^+$PA$^-$. (C) Energy level diagram of the states in B. The k-variables denote the corresponding microscopic single-electron ET rates. In this scheme, the direct long range tunneling between D and A (i.e., $C \rightarrow G$) and the 'thermal back reaction' [33] between P and A (i.e., $I \rightarrow G^*$) are not considered. As explained in the main text, their magnitudes can be significantly suppressed without affecting the efficiency and yield.

In this work, we solve the rate equations analytically for a generic molecular triad with four states and obtain closed-form expressions relating the lifetime and yield of the charge separated state to the ET rates. The equations allow us to isolate the relevant ratios of rate constants that control the yield and the lifetime. These conditions are used to set bounds on the physical distances and potentials that makeup the PCT using a standard semiclassical model which incorporates Marcus theory for the ΔG dependence and an exponential drop-off of the ET rate with distance [28] as parameterized in the Moser-Dutton ruler [29,30]. We report two major findings: first, that the highest yield occurs when the primary-donor cofactor is closest to the acceptor cofactor and second, that the highest yield and efficiency occurs when the absorption frequency of the primary-donor is more than twice the reorganization energy of the first electron transfer. Interestingly, we demonstrate that natural systems seem to obey these rules despite their much higher degree of complexity.

Methods

The basic PCT arrangement for long lived charge separation is depicted schematically in Figure 1A [7], and the microscopic steps leading to charge separation are shown explicitly in Figure 1B and energetically in Figure 1C: upon photoexcitation of the site of charge separation or primary-donor (P) to P*, the excited electron transfers to an acceptor molecule (A). A donor molecule (D) then transfers an electron to P, thus blocking the unproductive charge recombination via back electron transfer, to create a fully charge separated state, $C \equiv$ D$^+$PA$^-$.

The state C principally relaxes back to the ground state, $G \equiv$ DPA, by one of two mechanisms: direct long ET between A$^-$ and D$^+$, or a two step recombination process via the intermediate state, $I \equiv$ DP$^+$A$^-$, followed by electron transfer from A$^-$ to P$^+$, i.e, from state $I \rightarrow$ G. When the molecules are arranged linearly, as in Figure 1A, the first short-circuit reaction mechanism is considerably suppressed, and is therefore neglected in our model. The ET rates for the two step process are k_{PD} and k_{AP}, respectively. The reverse transition back to the excited state P* from A is also suppressed; below we demonstrate that the corresponding ET rate is exponentially suppressed for energy differences larger than 60–100 meV between P* A and P$^+$A$^-$, which we show is much less than 10% of the output energy and therefore does not affect our general conclusions. Similarly, since the energy difference for the ET from P to A is in the eV range, thermal excitation from the groundstate to the acceptor is not considered at room temperatures.

The master equations describing the transitions between the states corresponding to the scheme in Figure 1C are:

$$\frac{dG^*}{dt} = -(k^* + k^*_{PA})G^* \tag{1a}$$

$$\frac{dI}{dt} = k^*_{PA}G^* - (k_{DP} + k_{AP})I + k_{PD}C \tag{1b}$$

$$\frac{dC}{dt} = k_{DP}I - k_{PD}C \tag{1c}$$

$$\frac{dG}{dt} = k^*G^* + k_{AP}I \tag{1d}$$

The transition rates between different configurations is governed by the *microscopic* ET rates. The specific ET involved is encoded in the subscript, for example, k_{DP}, denotes the ET rate for the $D \rightarrow P$ transition. The complete list of transitions are: $G^* \xrightarrow{k^*_{PA}} I \xrightarrow{k_{DP}} C \xrightarrow{k_{PD}} I \xrightarrow{k_{AP}} G$ and $G^* \xrightarrow{k^*} G$. As explained earlier, the $C \rightarrow G$ short-circuit and the reverse $I \rightarrow G^*$ transitions are suppressed in our scheme. The rate k^* is the combined direct relaxation rate, fluorescent and otherwise, from the photoexcited state P* to its groundstate P.

Setting either of the two rates k_{AP} or k_{PD} to zero in Equation 1 prevents the state C (the charge separated state) from decaying into the ground state G creating a steadystate at long times. A finite k_{AP} and/or k_{PD} will, on the other hand, force C to decay in a finite time, which we call the lifetime of the charge separated state. To study this decay and determine the population (yield) of state C, it is convenient to solve for the evolution of $C(t)$ analytically. The solution is presented in the next section.

Results

Analytical solution of the PCT

Our goal in this section is two-fold: to obtain the conditions under which a charge separated state can be maintained in a quasi-steadystate (QSS) for a desired length of time, determined, for example, by the optimal throughput rate, and to derive simple explicit formulas for the lifetime of the QSS and the maximum yield of C. To this end, we first analytically solve Equation 1 for $C(t)$. For the initial conditions, we note that the equations being homogeneous, the solutions scale with the initial population $G^*(0)$, which is determined by the efficiency of the photoexcitation process. Hence, given a non-zero "source" $G^*(0)$ at $t=0$, we assume that $I(0)=C(0)=G(0)=0$, i.e., they are initially unpopulated.

We first note that Equation 1a can be integrated to give $G^*(t)=G^*(0)e^{-(k^*+k^*_{PA})t}$. Substituting $G^*(t)$ into Equation 1c for C and after taking a second time derivative to eliminate I, we arrive at the following second order equation for $C(t)$:

$$\frac{d^2C}{dt^2}+k_1\frac{dC}{dt}+k_2C=Se^{-(k^*+k^*_{PA})t} \tag{2}$$

where

$$k_1=(k_{AP}+k_{DP}+k_{PD}) \text{ and } k_2=k_{AP}k_{PD} \tag{3}$$

The source term on the right.

$$S=G^*(0)k_{DP}k^*_{PA} \tag{4}$$

The boundary condition $dC(t)/dt|_{t=0}=0$ is obtained by setting $I(0)=C(0)=0$ in Equation 1c. Since the coefficients appearing in Equation 2 are real, it can be written as:

$$(\frac{d}{dt}+k_+)(\frac{d}{dt}+k_-)C=Se^{-(k^*+k^*_{PA})t} \tag{5}$$

where the *macroscopic* rate constants

$$k_\pm=\frac{1}{2}k_1\pm\frac{1}{2}\sqrt{k_1^2-4k_2} \tag{6}$$

are the roots of the algebraic equation: $D^2+k_1D+k_2=0$. (Negative roots are used to keep the rates $k_\pm$ positive.) Using the following identity, where k is a constant,

$$(\frac{d}{dt}+k)f(t)=e^{-kt}\frac{d}{dt}(e^{kt}f(t)) \tag{7}$$

we rewrite Equation 5 as

$$e^{-k_+t}\frac{d}{dt}\left[e^{(k_+-k_-)t}\frac{d}{dt}(e^{k_-t}C)\right]=Se^{-(k^*+k^*_{PA})t} \tag{8}$$

Equation 8 is now easily integrated to obtain the solution for $C(t)$. Separating the constant source term as $C(t)=Sc(t)$, the time-dependent part $c(t)$ equals

$$c(t)=\frac{e^{-(k^*+k^*_{PA})t}}{(k^*+k^*_{PA}-k_+)(k^*+k^*_{PA}-k_-)}+\frac{e^{-k_+t}}{(k_+-k_-)(k_+-k^*-k^*_{PA})}+\frac{e^{-k_-t}}{(k_--k_+)(k_--k^*-k^*_{PA})} \tag{9}$$

This completes our derivation of $C(t)$.

A typical time evolution of $C(t)$ is depicted in Figure 2. This graph exhibits several aspects characteristic of charge separation in a PCT [12]: an exponential buildup of state C with a rate constant k_+, followed by a QSS plateau region that decays exponentially at the rate k_-. The length of time for which the QSS persists is termed the charge separation lifetime (τ_C), and the population of C at the plateau stage is termed the yield of charge separation (Y_C). The rate $(k^*+k^*_{PA})$ is the rate at which the state I is initially populated by G^* (see Equation 1b); I then subsequently populates C at the rate k_+. Physically, any system designed to spatially separate charges has to be able to transfer an electron from the photoexcited primary-donor $\mathbf{P}^*$ to the acceptor A (which is controlled by the rate k^*_{PA}) before it decays back to the ground state at the rate k^*. We thus restrict our analysis to

$$k^*\ll k^*_{PA} \tag{10}$$

[We use the strong inequality ($\ll$) to emphasize at least an order-of-magnitude smallness.]

The determinants of QSS lifetime and yield

In the previous section, we observed that a QSS is reached at intermediate times provided

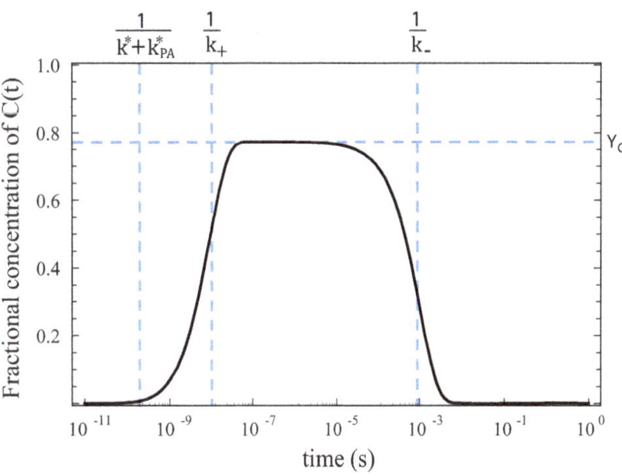

Figure 2. The evolution of the charge separated state $C(t)$ derived in eqn:ct. $C(t)$ is normalized by $G^*(0)$, which we take to be unity. Rate constants are chosen as $k^*=10^9s^{-1}$, $k^*_{PA}=5\times10^9s^{-1}$, $k_{AP}=5\times10^6s^{-1}$, $k_{DP}=5\times10^7s^{-1}$, and $k_{PD}=1.2\times10^4\ s^{-1}$. Relevant timescales are labeled on the upper axis and are marked by vertical lines (see eqn:kpm for definitions of $k_\pm$). A central quasi-steadystate (QSS) plateau region is formed when these timescales are well separated. We define the decay time of the QSS, $\tau_C=1/k_-$, as the lifetime of the charge separated state. The horizontal line marks the yield, Y_C, defined as the value of C in QSS. Analytical expressions for τ_C and Y_C are derived in Equations 14 and 17, respectively.

$$k_- \ll k_+ \text{ and } k^*_{PA} \tag{11}$$

When k_- is significantly different from k_+ and k^*_{PA}, the lifetime of the QSS, and thus that of the charge separated state, can be defined as $\tau_C = 1/k_-$. Our key observation is that k_- in Equation 6 can be made as small as we require by arranging either or both k_{PD} and k_{AP} to be sufficiently small. More precisely, we find that the constraints on the macroscopic rate constants in Equation 11 are satisfied if the microscopic rates obey:

$$k_{PD} \ll k_{DP} \text{ and } k_{AP} \lesssim k^*_{PA} \tag{12}$$

We recognize that k_{AP} is a downhill transfer that can be fast or slow depending on the driving force of the ET determined by where it lies in the Marcus curve. k_{PD}, on the other hand, involves an energetically uphill electron transfer which is always slower than its corresponding downhill transfer (i.e., $k_{PD} < k_{DP}$). We therefore only demand a strong constraint for k_{PD} compared to that for k_{AP} in Equation 12.

To prove that the conditions in Equation 12 are sufficient to establish a QSS, we first show that *independent* of the magnitude of k_{AP} the term under the square-root in eqn:kpm, besides being positive, satisfies the stronger constraint

$$k_1^2 \gg 4k_2 \text{ when } k_{PD} \ll k_{DP} \tag{13}$$

We show this by expanding the square-root in eqn:kpm and analyzing the behavior of $k_\pm$ for small and large k_{AP}. For small $k_{AP} \ll k_{DP}$, we see that $k_- \approx k_{AP}(k_{PD}/k_{DP})$ and $k_+ \approx k_{DP}$, while for large $k_{AP} \gg k_{DP}$, they reduce to $k_- \approx k_{PD}$ and $k_+ \approx k_{AP}$. It is immediately clear that assuming $k_{PD} \ll k_{DP}$ is sufficient to satisfy Equation 13 for all values of k_{AP}. Note that, since $k_+ \approx k_{AP}$ for large k_{AP}, the second condition in Equation 11, namely $k_- \ll k^*_{PA}$, is automatically satisfied if we restrict $k_{AP} \lesssim k^*_{PA}$. Hence, the conditions on the macroscopic rate constants in Equation 11 for a QSS to exist are met when the microscopic rate constants obey the constraints in Equation 12.

The importance of the observation that Equation 13 is satisfied for all values of k_{AP} is that it allows us to expand the square-root in Equation 6 to derive simple closed-form expressions for τ_C and Y_C. They can be analyzed to identify the key optimization parameters controlling the lifetime and yield of the charge separated state. Thus an almost exact expression for the lifetime τ_C is obtained after expanding the square-root for the leading non-zero value of k_-

$$\frac{1}{\tau_C} = k_- = \frac{k_{PD} \times (k_{AP}/k_{DP})}{1 + (k_{AP}/k_{DP}) + (k_{PD}/k_{DP})} \tag{14}$$

Similarly, to find the yield Y_C, we first note in Equation 9 that the QSS behavior of $C(t)$ for times $t \gg 1/k_+$ and $1/k^*_{PA}$ is well approximated by the surviving third term denoted below as $C_Q(t)$.

$$C_Q(t) = \frac{G^*_f e^{-k_- t}}{1 + (k_{PD}/k_{DP}) + (k_{AP}/k_{DP})} \tag{15}$$

$$G^*_f = \frac{G^*(0)}{1 + (k^*/k^*_{PA})} \tag{16}$$

To obtain the above expressions we used Equation 11 to justify keeping only the leading order terms in the expansion of the square-root in Equation 6, namely, $k_+ = k_1$ and $k_- = 0$. G^*_f denotes the fraction of the initial population of the photoexcited state $G^*(0)$ that remains after direct transition to the grounstate (predominantly fluorescence). Since $C_Q(0)$ is the maximum value that $C(t)$ attains, namely, its value at the plateau (see Figure 2), before decaying to the groundstate, we define the yield, Y_C, as:

$$Y_C = C_Q(0) = \frac{G^*_f}{1 + (k_{PD}/k_{DP}) + (k_{AP}/k_{DP})} \tag{17}$$

The expressions for τ_C and Y_C derived in Equations 14 and 17 are the main results of this section. They are compared in Figure 2 with the exact solution for $C(t)$ (Equation 9); the agreement is excellent. When combined with the conditions in Equation 12 for a QSS to exist, they provide all the necessary information for the design of highly optimized PCTs.

Maximizing the QSS yield and lifetime: microscopic constraints

The advantage of having formulas Equations 14 and 17 for τ_C and Y_C is that they enable us to identify the primary control parameters that have the largest affect on the performance of the PCT. From Equations 10–17 we conclude that the relevant ratios of the five microscopic ET rates $\{k^*, k^*_{PA}, k_{AP}, k_{DP}, k_{PD}\}$ are

$$\alpha = \frac{k_{PD}}{k_{DP}},\ \alpha^* = \frac{k_{AP}}{k^*_{PA}}\ and\ \beta = \frac{k_{AP}}{k_{DP}},\ \beta^* = \frac{k^*}{k^*_{PA}} \tag{18}$$

They control the formation, yield and lifetime of the charge separated state. To make the dependence explicit, we rewrite Equations 14 and 17 as functions of the dimensionless ratios here:

$$k_{PD}\tau_C = \frac{1 + \alpha + \beta}{\beta} \tag{19a}$$

$$y_c = \frac{Y_C}{G^*(0)} = \frac{1}{(1 + \beta^*)(1 + \alpha + \beta)} \tag{19b}$$

where $k_{PD}\tau_C$ and y_c are the normalized lifetime and yield, respectively. [Note that all the individual rate constants can be expressed in terms of an appropriate combination of the dimensionless ratios and k_{PD}.]

In terms of these ratios, the conditions for the formation of a QSS in Equation 12 translates to

$$\alpha \ll 1 \text{ and } \alpha^* \lesssim 1 \tag{20}$$

Although no fundamental restrictions on β and β^* exist, it follows from Equation 19b that the yield is substantially suppressed when they are $\gtrsim 1$. Hence, to maximize the yield, we demand

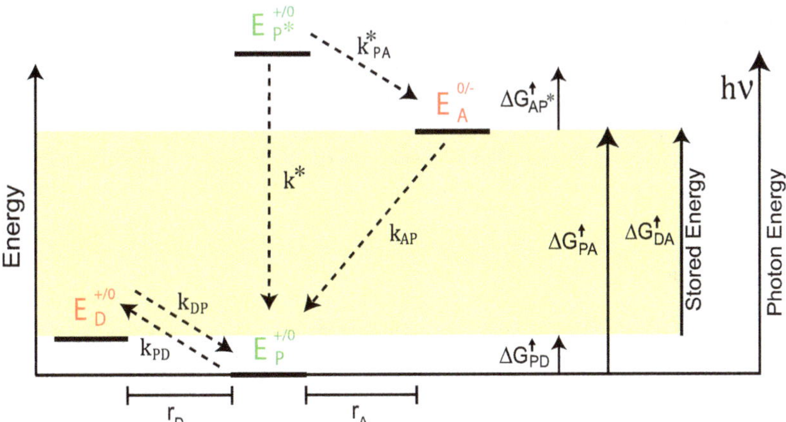

Figure 3. The physical characteristics of a PCT corresponding to the scheme in Figure 1C shown with the distances and reduction potentials marked explicitly. The edge-to-edge separations of the D-P pair and the P-A pair are labeled as r_D and r_A, respectively. The vertical axis is in the direction of increasing energy. The respective reduction potentials are defined in terms of the half-cell potentials, $E^{+/0}=E^0-E^+$ and $E^{0/-}=E^--E^0$ (final minus the initial state). The driving force, $\Delta G^{\uparrow}$, for an uphill electron transfer, say, $D^+P\rightarrow DP^+$, is defined as $\Delta G^{\uparrow}_{PD}=E_D^{+/0}-E_P^{+/0}>0$. The corresponding driving force for the downhill transfer $\Delta G^{\downarrow}=-\Delta G^{\uparrow}$. Given the driving forces and the distances, the rate constants are derived using the Moser-Dutton ruler. Note that the rate k^*, which we take to be the combined relaxation rate, fluorescent and otherwise, of the photoexcited state P^* to its groundstate P, does not follow the Moser-Dutton ruler and must therefore be given. An incoming photon with the correct frequency hv is absorbed by P to create the photoexcited state P^*. We assume that the ET rates involving P^* can be expressed in terms of the reduction potential of the state P, i.e., $E_{P^*}^{+/0}=E_P^{+/0}+hv$, where $E^0_{P^*}=E^0_P+hv$ and $E^+_{P^*}=E^+_P$. Hence the input energy $hv=\Delta G^{\uparrow}_{PA}+\Delta G^{\uparrow}_{AP^*}$. The output energy is the stored energy $\Delta G^{\uparrow}_{DA}$ in the charge separated state.

$$\beta \ll 1 \text{ and } \beta^* \ll 1 \tag{21}$$

The condition $\beta^* \ll 1$ justifies the arguments leading to Equation 10 and therefore no new condition is obtained. Note that since $\tau_C \sim 1/\beta$, a $\beta \ll 1$ also implies long life-times. We wish to emphasize that while restricting α and α^* to $\ll 1$ is necessary for a QSS to form, the conditions on β and β^* ensure a high QSS concentration or yield and a long lifetime.

This completes our analysis of the fundamental constraints on the microscopic rate constants derived to maximize the yield and the lifetime of charge separated states in the QSS regime. It is model-free in the sense that we have not utilized any particular equation to calculate the ET rates and we have not determined any specifics in terms of spatial constraints or electron affinities. We have only derived the limits of optimal values for the rate constants themselves. We now discuss in detail the physical constraints that Equations 20 and 21 impose on the energetics and architecture of the cofactor triads.

Engineering guidelines for optimal PCTs

The physical characteristics of a PCT involve the differences in reduction potentials and distances between the cofactors. Such a construct consistent with the scheme in Figure 1C is shown in Figure 3 where the energies and the distances are marked explicitly. Two more relevant metrics, geometric (overall size) and energetic (efficiency), are introduced below.

Separation distance

We believe that a linear construct is better because it maximizes the distance between the acceptor and donor, thus preventing relaxation by short circuiting direct electron transfer between these sites. We note, however, that this ideal is not found in all natural systems. The maximum distance of charge separation is thus:

$$R_C = r_A + r_D \tag{22}$$

where r_A and r_D are the edge-to-edge separations of the P-A pair and D-P pair, respectively. Since the ET rates are determined primarily by the edge-to-edge distance [29], the width of P does not play a part in any of the microscopic rate constants delineated in our scheme in Figs. 1B and 3. We do not therefore include the actual width of the primary-donor site P, which even further helps to eliminate the short circuiting A→D electron transfer. Instead, we introduce a second distance parameter:

$$\Delta r = r_A - r_D \tag{23}$$

that can be varied, keeping R_C fixed, to optimize the output.

Charge-separation efficiency

An optimal light-activated charge separation construct should also maximize the available useful energy stored in the charge separated state $C=D^+PA^-$ (see Figure 3). The energy stored in C can be expressed in terms of the driving force as $\Delta G^{\uparrow}_{DA}=E_A^{0/-}-E_D^{+/0}=\Delta G^{\uparrow}_{PA}-\Delta G^{\uparrow}_{PD}$ (see Figure 3). Defining the charge separation efficiency, η, as the ratio of the stored energy $\Delta G^{\uparrow}_{DA}$ to the input photon energy, hv, we get:

$$\eta = \frac{\Delta G^{\uparrow}_{DA}}{hv} = \frac{1}{hv}(\Delta G^{\uparrow}_{PA} - \Delta G^{\uparrow}_{PD}) \tag{24}$$

Thus, consideration of the reduction potentials of each cofactor in the PCT adds a third performance metric η, to Y_C and τ_C, to optimize.

The Moser-Dutton ruler

In Equations 20 and 21 we identified and derived constraints on the ratios of the ET rate constants $\{\alpha,\alpha^*,\beta,\beta^*\}$ for optimal charge transfer in a PCT construct. These rates are determined by the individual values of the reduction potentials and the spatial separations of the cofactors. For this we need explicit equations that relate the rate constants to these variables. To this end, we use the Moser-Dutton ruler, a set of empirical equations that is widely used to simply and accurately predict ET rates in proteins [19,29,31]. The ruler predicts a rate constant, $k_{et}^{\downarrow}$, for a downhill electron transfer at room temperature, i.e., when the driving force $\Delta G^{\downarrow}<0$ as

$$\log k_{et}^{\downarrow}=13-0.6(r-3.6)-3.1\frac{(\Delta G^{\downarrow}+\lambda)^2}{\lambda} \tag{25}$$

λ here is the reorganization energy in eV and the term in which it appears is the Marcus term which depicts the hyperbolic dependence of the ET rate on the driving force for electron transfer [23]. Reverse or uphill electron transfer is modified by a Boltzmann term to give:

$$\log k_{et}^{\uparrow}=13-0.6(r-3.6)-3.1\frac{(\Delta G^{\uparrow}+\lambda)^2}{\lambda}+\frac{\Delta G^{\uparrow}}{0.06} \tag{26}$$

The transfer rates are predicted in units of s^{-1}. The energies $\Delta G^{\uparrow}=-\Delta G^{\downarrow}>0$ are measured in eV, and r is the edge-to-edge distance between the cofactors in Å. All logs are to base 10. We note that the use of the Moser-Dutton ruler restricts our analysis from here on only to protein-based PCTs. Other PCTs can be analyzed in a similar way provided the appropriate expressions for the ET rates are used.

Using the Moser-Dutton ruler to express the microscopic variables $\{\alpha,\alpha^*,\beta,\beta^*\}$ in terms of the physical variables $\{\Delta G_{PD}^{\uparrow},\Delta G_{PA}^{\uparrow},R_c,\Delta r\}$ of the PCT, we get:

$$\log\alpha = -29\Delta G_{PD}^{\uparrow}\approx-\frac{\Delta G_{PD}^{\uparrow}}{0.035} \tag{27a}$$

$$\log\alpha^* = -3.1\times4\left(\frac{hv}{2\lambda}-1\right)\left(\Delta G_{PA}^{\uparrow}-\frac{hv}{2}\right) \tag{27b}$$

$$\log\beta = -\frac{3.1}{\lambda}(\Delta G_{PA}^{\uparrow}-\Delta G_{PD}^{\uparrow})(\Delta G_{PA}^{\uparrow}+\Delta G_{PD}^{\uparrow}-2\lambda) -0.6\Delta r \tag{27c}$$

$$\log k_{PD} = 13-3.1\frac{(\Delta G_{PD}^{\uparrow}+\lambda)^2}{\lambda}+\frac{\Delta G_{PD}^{\uparrow}}{0.06}- 0.3(R_C-7.2)+0.3\Delta r \tag{27d}$$

$$\beta^* = \left(\frac{k^*}{k_{PD}}\right)\left(\frac{\alpha\alpha^*}{\beta}\right) \tag{27e}$$

(Refer to Equations 22 and 23 for the definitions of R_C and Δr and $hv=\Delta G_{PA}^{\uparrow}+\Delta G_{AP^*}^{\uparrow}$.) Since k_{PD} is the only dimensionful quantity we need, its form is given explicitly in Equation 27d. The last parameter β^* depends on k^*, which because it involves the combined relaxation rate, fluorescent and otherwise, of the photoexcited state P* to its groundstate P, it cannot be estimated using the Moser-Dutton ruler. It is assumed to be a given quantity in our analysis. And finally, we have assumed that the reorganization energy λ is the same for the entire construct. Only minor modifications to Equation 27c are necessary if this last assumption is relaxed. The qualitative features of our conclusions are robust, although certain quantitative predictions will have to be reworked.

Our final goal is to use the relations in Equation 27 to set general bounds on the physical makeup of a generic protein-based PCT to optimize its performance. We do this by adjusting the physical parameters $\{\Delta G_{PD}^{\uparrow},\Delta G_{PA}^{\uparrow},\Delta r\}$ (for a fixed size R_C of the construct) to satisfy the constraints in Equations 20 and 21 on the rates. This way, we are guaranteed that the performance metrics $\{Y_C,\tau_C\}$ are optimized. The efficiency metric η (defined in Equation 24) is then determined for a given τ_C and Y_C. Clearly, configurations with a large difference in $\Delta G_{PA}^{\uparrow}$ and $\Delta G_{PD}^{\uparrow}$ will ensure a high efficiency. It is therefore desirable to arrive at an independent set of constraints for the energies and the distances. We argue that this is mostly possible, primarily because the α variables depend only on the driving forces and not on the separation distances (see Equations 27a and 27b). Hence any condition on the α variables translates to conditions on ΔG, independently of the distance. These considerations are analyzed in detail next.

Maximizing the QSS yield and lifetime: physical constraints

In Equations 20 and 21 we identified two sets of conditions necessary for the creation of a QSS with a high charge separation yield and lifetime. We can now see what effects these constraints impose on the physical makeup of a PCT. In particular, our aim is to arrive at a set of independent constraints for the energies and distances.

Energy constraints - fundamental limits on the absorption wavelength of the primary-donor

From Equation 27a we note that the ratio $\alpha=k_{PD}/k_{DP}$ is a factor 10 smaller for approximately every 35 meV difference in the reduction potentials $\Delta G_{PD}^{\uparrow}$. We conclude that the first constraint for the existence of a QSS, $\alpha\ll1$ stated in Equation 20, is satisfied provided:

$$\Delta G_{PD}^{\uparrow}\gtrsim 35\text{ mV} \tag{28}$$

Since α depends exponentially on $\Delta G_{PD}^{\uparrow}$, the strong inequality on α translates to a weak inequality on $\Delta G_{PD}^{\uparrow}$. This simple observation is the key to the viability of our general mathematical analysis of a PCT. Small adjustments to the physical parameters can drive the system into the QSS regime and affect large changes in the performance metrics thus allowing us to arrive at a set of practically realizable bounds on the physical parameters.

From Equation 27b we see that the second condition in Equation 20, $\alpha^*\ll1$, is satisfied if $\Delta G_{PA}^{\uparrow}\gtrsim hv/2$ for photon energies

$$hv\gtrsim 2\lambda \tag{29}$$

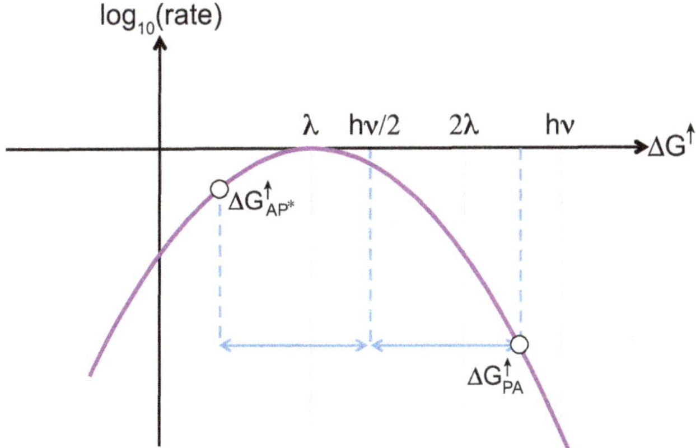

Figure 4. The optimal range for $\Delta G^{\uparrow}_{PA}$ and $\Delta G^{\uparrow}_{AP^*}$ are shown on the Marcus curve. A high yield, high efficiency QSS formation in a triad requires that back electron transfer from A to P be so downhill as to be well into the Marcus inverted region. To see this, we substitute $h\nu = \Delta G^{\uparrow}_{PA} + \Delta G^{\uparrow}_{AP^*}k$ in Equation 27b so that $\log \alpha^* \approx -((\Delta G^{\uparrow}_{PA} + \Delta G^{\uparrow}_{AP^*})/2 - \lambda)(\Delta G^{\uparrow}_{PA} - \Delta G^{\uparrow}_{AP^*})$, from which it immediately follows that the condition $\alpha^* \ll 1$ is satisfied if the mean value $(\Delta G^{\uparrow}_{PA} + \Delta G^{\uparrow}_{AP^*})/2 = h\nu/2 > \lambda$. The condition $\alpha^* \ll 1$ was derived in Equation 20 to be a necessary condition for the formation of a QSS.

The low energy range $h\nu \lesssim 2\lambda$ results in a small $\Delta G^{\uparrow}_{PA} \lesssim h\nu/2$ which reduces the efficiency as seen from Equation24, and is therefore not a useful range for our purpose.

Equation 21 presents the constraints necessary to maximize the yield Y_C. The energy dependent term for β in Equation 27c is formed out of the product of the sum and difference of the driving forces. The difference term is proportional to the efficiency η and is therefore always positive. To maintain $\beta \ll 1$ for all η, the sum must satisfy $\Delta G^{\uparrow}_{PA} + \Delta G^{\uparrow}_{PD} \gtrsim 2\lambda$. Since a large $\Delta G^{\uparrow}_{PA} > \Delta G^{\uparrow}_{PD}$ increases the efficiency from Equation 24, for practical purposes it is sufficient to ensure that $\Delta G^{\uparrow}_{PA} > 2\lambda$. The two conditions for $\Delta G^{\uparrow}_{PA}$ can be combined as:

$$\max\left[\frac{h\nu}{2}, 2\lambda\right] \lesssim \Delta G^{\uparrow}_{PA} \lesssim h\nu \tag{30}$$

where max$[a,b]$ implies the larger of the two variables a and b. Thus, a $\Delta G^{\uparrow}_{PA} > 2\lambda$ satisfies conditions for both high Y_C and η, and a $\Delta G^{\uparrow}_{PA} < 2\lambda$ will either result in a loss of efficiency or yield. We thus predict that high yield, high efficiency QSS formation in a triad requires that back electron transfer from A to P be so downhill as to be well into the Marcus inverted region (see Figure 4.) This greatly slows the rate of this ET, allowing the donor molecule time to re-reduce the primary-donor molecule.

The upper limit is necessary to facilitate the electron transfer from the photoexcited primary-donor $P^* \to A$. This, however, has certain limitations: as $\Delta G^{\uparrow}_{PA}$ approaches the photoexcitation energy $h\nu$, the back ET governed by the uphill rate $k^{*\uparrow}_{AP}$ (which is set to zero in our scheme) will become relevant - see Figure 3. This will provide yet another route for the charge separated state to decay to the groundstate thus reducing the performance. Hence, we put an upper cut-off on $\Delta G^{\uparrow}_{PA}$. To obtain this upper cut-off, we study the behavior of the ratio of the uphill vs downhill rates using the Moser-Dutton ruler and find that (similar to the ratio α in Equation 27a)

$$\log\left(\frac{k^{*\uparrow}_{AP}}{k^{*}_{PA}}\right) = -\frac{\Delta G^{\uparrow}_{AP^*}}{0.035} = -\frac{(h\nu - \Delta G^{\uparrow}_{PA})}{0.035} \tag{31}$$

This clearly suggests that as long as the uphill driving force $\Delta G^{\uparrow}_{AP^*} = h\nu - \Delta G^{\uparrow}_{PA} > 35$ meV, the back reaction is exponentially suppressed. By direct simulation (not shown here), we find that our results obtained by setting $k^{*\uparrow}_{AP} = 0$ are unaffected if a difference of the order of $60 - 100$ meV or higher is maintained. Although larger values will reduce the efficiency, a difference of 100 meV affects the efficiency by less than 10%.

Finally, we analyze β^* in Equation 27e. Since it can be expressed in terms of the ratios α, α^* and β, no new energy constraints can be obtained. Instead, we show below that requiring $\beta^* \ll 1$ (Equation 21) provides useful insight into the geometrical construction of the PCT.

Distance constraints - optimal placement of the three cofactors

Since $\beta^* \sim 1/\beta$ (see Equation 27e), maintaining $\beta \ll 1$ can only be done at the expense of increasing β^* which is counterproductive as we require both β and $\beta^* \ll 1$ for maximum yield (Equation 21). An optimal compromise can be reached by adjusting the distance $\Delta r = r_A - r_D$. To see this, we write the Δr dependence explicitly in Equation 27e by combining it with Equations 33 and 34 to give $\log \beta^* \sim 0.3\Delta r + (\text{energy} - \text{dependent terms})$. Hence, once the energy terms are optimized for a fixed total length R_C, we predict that arranging for

$$\Delta r < 0 \tag{32}$$

will significantly reduce β^*. Provided $\beta \ll 1$, an order-of-magnitude suppression of β^* is achieved if $0.3\Delta r \sim -1$, i.e., $\Delta r \sim -3$ A. Hence, Δr can be fine-tuned to increase the yield.

There is a relatively simple physical explanation for our prediction. This is due in part to the fact that while electron transfer rates in proteins are strongly distance dependent, the equilibria are not. As noted in Equation 19b, there are three pairs of rate constants which determine quantum yield, of which the α parameter involves electron transfers that are the forward and reverse of each other between P and D. Since ratios involving forward and backward rates for any particular triad are not dependent on distance, the α parameter is distance independent. On the other hand, the remaining two pairs, β^* and β, are distance dependent: the first is the initial ET between A and P which competes with the relaxation processes encompassed by $k^\star$. In this case the actual rate, and therefore the distance, plays a direct role in the final yield. This constrains the primary-donor and acceptor cofactors to be close enough to out-compete $k^\star$. The second is the competing pair of electron transfers which can occur from state I: reverse electron transfer from A to P forming G vs. the formation of C by electron transfer from D to P. This second pair is a weaker constraint given the fact that electron transfer from A to P is well into the Marcus inverted region. Viewed in this light it is not surprising that optimal arrangements move the primary-donor and acceptor closer together (see Figure 4 for a demonstration of this using commonly observed parameters). The full dependence of Y_C and η on Δr are discussed further in the discussion.

Finally, regarding the total size $R_C = r_A + r_D$: the R_C dependence of τ_C and Y_C in Equations 19a and 19b are fully governed by the rate constant k_{PD} defined in Equation 27d. It follows that

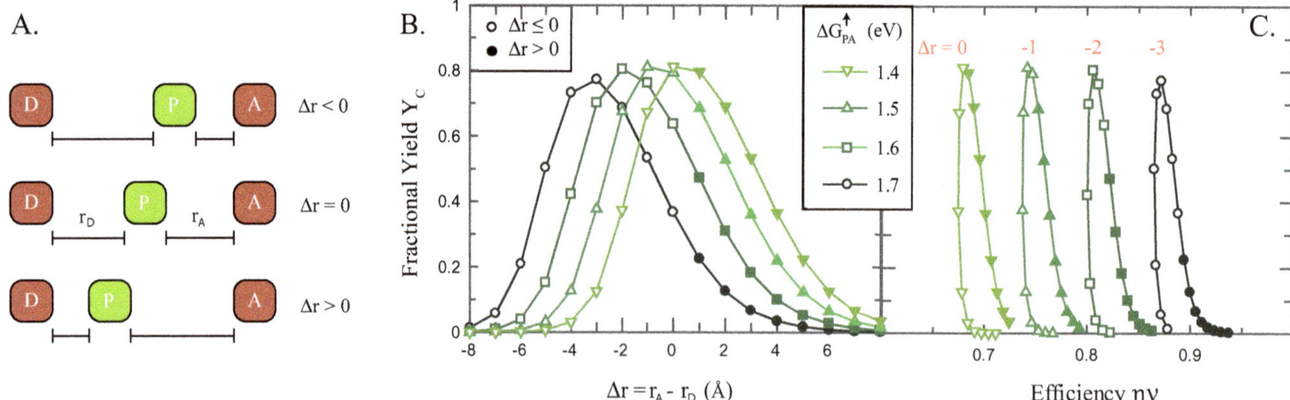

Figure 5. Sensitivity of the yield and efficiency of a typical PCT to $\Delta G^{\dagger}_{PA}$ and Δr. The following parameters are fixed: The light frequency $h\nu = 1.8$ eV, the reorganization energy $\lambda = 0.7$ eV, the size $R_C = 16$A, and the relaxation rate at $1/k^* = 1$ ns. The driving force $\Delta G^{\dagger}_{PD}$ for each choice of $\Delta G^{\dagger}_{PA}$ and Δr is obtained by solving Equation 19a setting $\tau_C = 1$ ms. All the relations necessary to invert Equation 19a for $\Delta G^{\dagger}_{PD}$ can be found in Equation 27. (A) Illustration of possible changes in Δr made while keeping R_C fixed. (B) Predictions for Y_C made using differing values of $\Delta G^{\dagger}_{PA}$ plotted as a function of Δr. Open symbols are used to indicate $\Delta r \leq 0$ and solid symbols for $\Delta r > 0$. Note that in each case, a maximum Y_C of ≈ 0.81 is achieved at some optimal $\Delta r < 0$. (C) Re-plot of the same data explicitly showing the variation in yield and efficiency as $\Delta G^{\dagger}_{PA}$ and Δr are varied. η is defined in Equation 24. Legends mark different Δr values varied in 1A increments evaluated at the same points as in (B). The Δr value at the maximum are labeled explicitly.

while the lifetime grows with R_C as $\tau_C \approx 10^{0.3R_C}$, the yield is suppressed as $Y_C \approx 1/(1 + \text{constant} \times 10^{0.3R_C})$.

This completes our analysis of the fundamental constraints on the physical parameters for an optimized PCT. The guidelines listed in Equations 28–32 are relevant to any protein based PCTs where the Moser-Dutton ruler is applicable.

Discussion

In Equations 28–32, we arrived at a set of constraints on the physical makeup of a high performance PCT capable of creating and maintaining a high yield charge separated state in a QSS for a significant length of time. We now apply these results to study the efficiency of such PCTs.

While analysis of many PCT constructs focus on the charge separation lifetime, τ_C [6], it is clear that the arrangement which gives the longest possible lifetime will oftentimes make a less efficient solar energy conversion component. The charge separated state must only last as long as the mechanism for extracting this potential energy requires. After this condition is met, factors which maximize the yield and efficiency of QSS formation (Y_C and η) are paramount, as these determine the eventual power output. Thus, in the following we fix the lifetime $\tau_C = 1$ ms in our analysis.

In Figure 5, we start with a set of parameters that are typical of photonic energy transduction in proteins: The reorganization energy λ varies in the range of 0.7 to 1.4 eV for cofactors bound within typical native proteins, with λ taking higher values with decreasing hydrophobicity in the local cofactor environment [19]. Light frequencies are in the near infrared and higher. We use $\lambda = 0.7$ eV and $h\nu = 1.8$ eV (690 nm) as a starting point. Note that $h\nu > 2\lambda$ is satisfied consistent with Equation 29. To satisfy the energy constraint in Equation 30, we choose the range $\Delta G^{\dagger}_{PA} = 1.4 - 1.7$ eV. [Note that the highest value for $\Delta G^{\dagger}_{PA}$ is 100 meV less than $h\nu$ for the reasons described following Equation 31.] Electron tunneling distances in biology range from 4–14 Å, with the shorter limit that of Van der Waals contact and the longer setting a millisecond time limit on electron transfer rates [29,31]. Hence, the sum of the distances between the cofactors are typically in the range $8\text{A} \leq R_C \leq 28\text{A}$; we use $R_C = 16\text{A}$. Instead of specifying the final parameter, $\Delta G^{\dagger}_{PD}$, we specify the QSS lifetime $\tau_C = 1$ ms and solve Equation 19a for $\Delta G^{\dagger}_{PD}$ for different values of $\Delta G^{\dagger}_{PA}$ and Δr (for a fixed R_C). For self-consistency, we check that the $\Delta G^{\dagger}_{PD}$ values obtained using these parameters all satisfy Equation 28 in the optimal range.

Several things are immediately apparent upon inspection of the data in Figure 5. First, the yield Y_C in Figure 5B is strongly dependent on Δr, with a maximum value in each case being reached at a configuration where the primary-donor P is closer to the acceptor site than the donor site as predicted in Equation 32. Second, the efficiency η in Figure 5C is considerably enhanced as $\Delta G^{\dagger}_{PA}$ is increased closer to the maximum value $h\nu$.

To gain further insight on the dependence of the metrics Y_C and η on the size R_C and self-relaxation (fluorescence) rate k^*, we study the variation of the optimized PCT metrics, i.e., the metrics obtained after adjusting Δr for maximum yield (i.e, we track the location of maximum yield in Figue 5 as R_C and k^* are varied). We find that once optimized for Δr, the two metrics Y_C and η are mostly orthogonal in terms of their determinants. This is demonstrated by Figure 6A where the maximum yield is seen to be strongly suppressed with increasing R_c while the efficiency at maximum yield is robust. The former is due to the decrease in β^* caused by the increase in distance, resulting in a smaller partitioning factor $G^*_f = G^*(0)/(1 + \beta^*)$ (Equation 16). As Figure 6B demonstrates, this loss can be alleviated by decreasing the rate of self-relaxation, k^*, of the excited primary-donor P^*.

Conclusions

We have generated an analytical solution for the time behavior of the PCT and explored its dependence on the architecture, the reduction potentials of its components, and the absorption frequency of the primary-donor cofactor. Our primary findings are two: First, that a high efficiency, high yield PCT will have an absorption frequency more than twice the reorganization energy

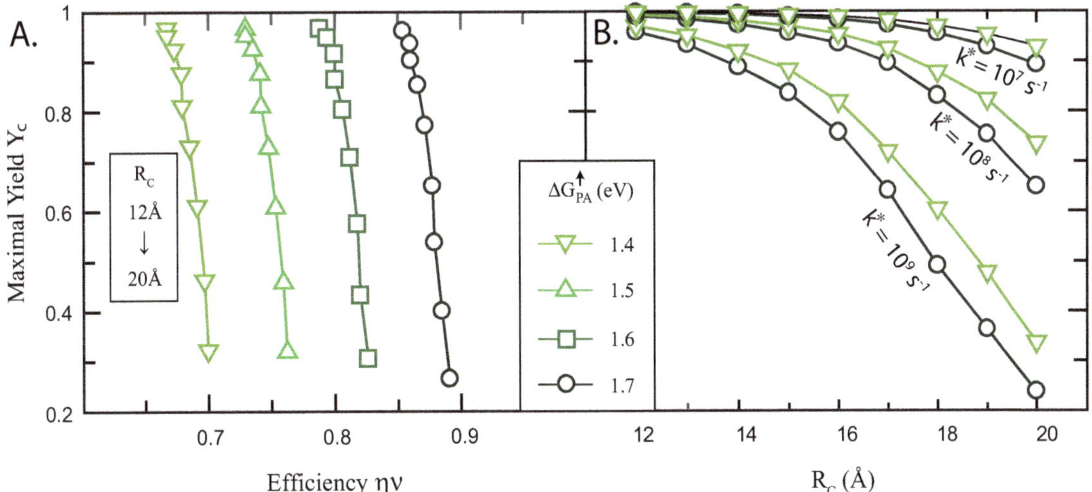

Figure 6. (A) Orthogonality of the yield, Y_C, and the energy storage efficiency, η, of QSS formation by the PCT. For each point, $\Delta G_{PD}^{\uparrow}$ and Δr are set to the values that maximizes Y_C within the limits set by hv,k^* and τ_C as in Figure 5. Y_C is strongly sensitive to the separation distance, R_C, and η is primarily sensitive to $\Delta G_{PA}^{\uparrow}$. (B) The decrease in the maximal values of Y_C with increasing R_C plotted for different values of k^* and $\Delta G_{PA}^{\uparrow}$. At large values of R_C the optimized yield is primarily dependent on k^*.

of the first electron transfer, and second, that the distance metric Δr (the relative distance of the acceptor and the donor from the primary-donor) plays an important role in the determination of the yields.

We remark that our use of the Moser-Dutton ruler clearly does not capture all the subtle details of protein ET reactions. For example, the assumption that ET rates drop-off exponentially with distance ignores possible effects of the intervening medium when present [32]. Secondly, some experimental results point to an asymmetric Marcus curve [33], that is known to be relevant when certain high-frequency intramolecular vibrations are active, are not accounted for. It is a simple matter to include these effects into Equations 20 and 21, which as we noted earlier are model-free, they provide fundamental constraints on the microscopic rate constants derived to maximize the yield and the lifetime of the charge separated states in the QSS regime. Further work is needed to study the quantitative effects these corrections will have on our conclusions. We show below, however, that our analysis incorporating the simple Moser-Dutton ruler is able to successfully explain a number of remarkable features observed in Nature.

Implications for natural systems

The first implication sets a long-wavelength limit or red-edge [34] for efficient solar energy conversion. It is estimated that the reorganization energies scale from 0.7 eV to 1.4 eV for typical proteins and cofactors bound in local environments varying from less to more hydrophobic [19]. These values predict that the longest effective wavelength for solar energy conversion is about 890 nm, correlating to the lower value. Longer wavelengths are possible, but this would necessitate a loss in either yield or energy. Our analysis is primarily limited by the fact that we include only three discrete sites for electron localization. Natural photosynthetic proteins have additional acceptor molecules, which enable the stepwise diffusion of the electron further away from the primary-donor. Their effect on the behavior of the PCT is unclear.

However, we do note in Figure 7 that at present the observed wavelength limits for oxygenic photosynthesis, an energetically demanding process in that it must create oxidizing potentials high enough to oxidize water [2], are within the values [35] we predict. Furthermore the limits observed for charge separation in any natural organism, those from bacterial photosynthesis [34], are within 150 nm, or 160 meV, of that predicted by our model, as shown in Figure 7. This suggests that efforts to re-engineer natural systems to utilize longer wavelengths of light, and thus garner a greater fraction of the solar emission spectrum [36] will result in considerable losses of either yield or conversion efficiency to do so.

The other prediction is that yields are maximized by placing the primary-donor closer to the acceptor than the donor cofactor. This

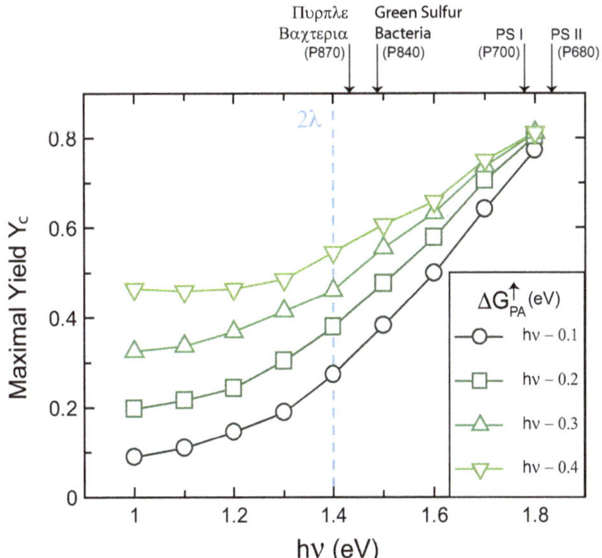

Figure 7. Predicted long-wavelength limit or red-edge for efficient solar energy conversion. Photon energies smaller than 2λ cause a loss in either yield or energy storage efficiciency. For each point, the value of Δr used maximizes Y_C within the constraints set by hv,k^* and τ_C as in Figure 5. The $\Delta G_{PA}^{\uparrow}$ values are calculated as $hv-x$ where x=0.1,0.2,.0.3 and 0.4 eV. Wavelength limits of natural systems depicted above the axis are taken from [34].

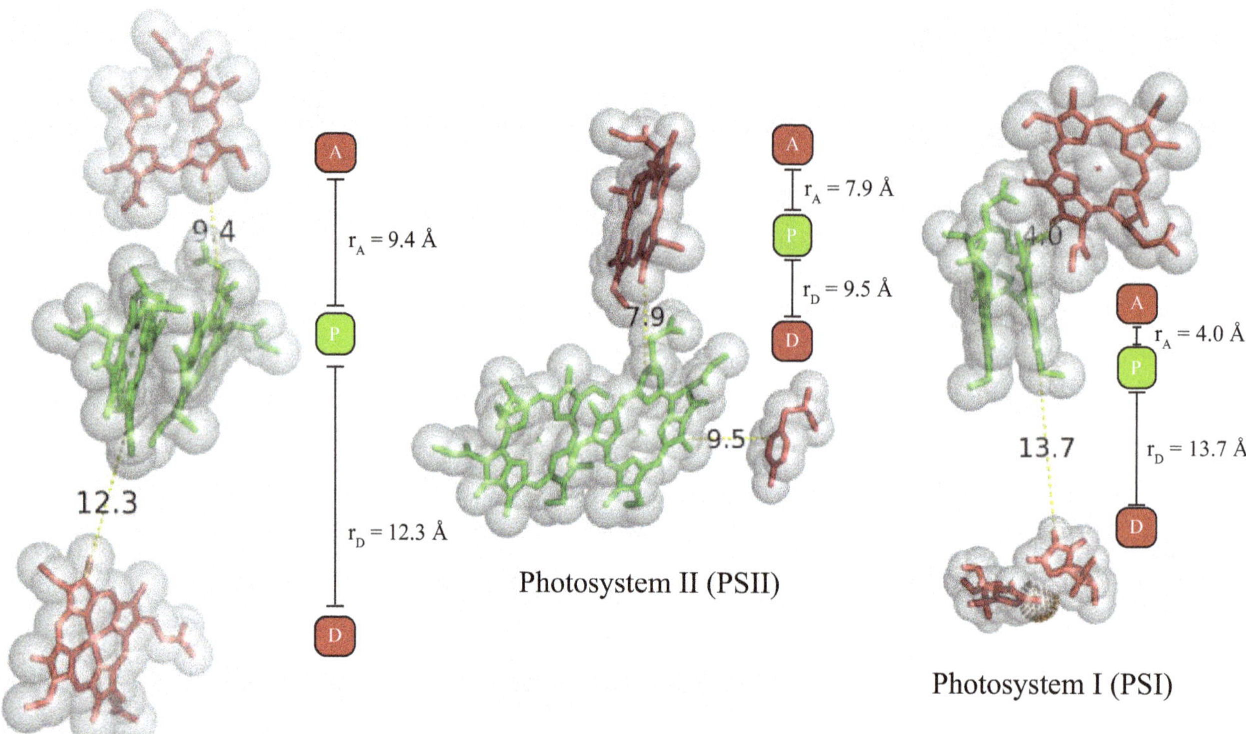

Figure 8. Representative structures of natural photosynthetic cofactor triads. Primary donors P are colored green with the donor D and acceptor A cofactors colored red in each structure. Distances are measured edge-to-edge. (left) Reaction Center complex from *Blastochloris viridis* (PDB ID 2X5U) [37], (center) Photosystem II from *Thermosynechococcus elongates* (PDB ID 3BZ1) [38], and (right) is Photosystem I Plastcyanin complex from *Prochlorothrix hollandica* created by computational docking [39]. Images and distances were created using Pymol.

again may be altered when further discrete electron binding sites are added to the construct, but we again note that for the limited subset of photosynthetic proteins which have structures which include the donor cofactor, the primary-donors are indeed positioned in this manner (Figure 8).

Engineering parameters for artificial charge separation constructs

This analysis sets out the optimal physical composition of an artificial protein-based charge separation construct. It demonstrates that efficient, high yield charge separation can be engineered with ΔG values that are both feasible to engineer and within the ranges observed in natural systems. It further identifies the molecular properties which are important targets for engineering improved PCTs. Principle among these is the control of the reorganization energy, λ. A smaller value of λ will enable the utilization of longer wavelengths of light, enabling the possible utilization of a larger fraction of the solar emission spectrum. There have been very few experimental determinations of λ values within a protein, and even less work on manipulating or optimizing its magnitude. However, it is apparent that it will be important to be able to manipulate this parameter effectively.

While we have identified Δr as a critical parameter for high yield constructs, at smaller cofactor separation distances the tolerances for Δr are very small. The large changes engendered by even a 1 A change in Δr make high yield small constructs difficult to create. Larger constructs have broader Δr maxima, but in this case yields are reduced due to unproductive primary-donor relaxation rates, or k^* (see Equation 16). Consequently the creation of primary-donor cofactors with longer excited state lifetimes is paramount. As Figure 6B demonstrates, longer lifetime cofactors will enable significantly larger constructs, and thus eases the optimization of Δr. We further note that while our analysis uses the protein-specific Moser-Dutton ruler, which models coupling as an exponential drop-off in electron transfer rate with distance, the model-free portion of this analysis leading to Equations 20 and 21 are applicable to synthetic constructs as well. The distance dependence in these systems depends strongly on the nature of the bridging elements which connect the triad cofactors, and the analysis presented here predicts that the coupling must in general be as strong as possible between the primary-donor and acceptor. In a protein this means putting them close together since the "bridge" is always the same. In a bridged system this means choosing a bridge that maximizes the coupling, but it doesn't necessarily mean bringing them closer together.

Acknowledgments

RLK would like to thank Art Van der Est, Brock University Chemistry department, Thomas Haines, CCNY Chemistry department, and Marilyn Gunner, CCNY Physics department, for helpful suggestions and discussion pertaining to this manuscript. WL and AP would like to thank the Department of Cell & Molecular Biology and the Department of Physics and Engineering Physics at Tulane University for their kind hospitality.

Author Contributions

Conceived and designed the experiments: AP LM RLK. Performed the experiments: AP LM WL ACM RLK. Analyzed the data: AP LM WL ACM RLK. Contributed reagents/materials/analysis tools: AP LM WL ACM RLK. Wrote the paper: AP LM RLK.

References

1. Williamson A, Conlan B, Hillier W, Wydrzynski T (2011) The evolution of photosystem II: insights into the past and future. Photosynthesis Research 107: 71–86.
2. McConnell I, Li GH, Brudvig GW (2010) Energy conversion in natural and artificial photosynthesis. Chemistry & Biology 17: 434–447.
3. Moser CC, Page CC, Dutton PL (2005) Tunneling in PSII. Photochemical & Photobiological Sciences 4: 933–939.
4. Moser CC, Page CC, Dutton PL (2006) Darwin at the molecular scale: selection and variance in electron tunnelling proteins including cytochrome c oxidase. Philosophical Transactions Of The Royal Society B-Biological Sciences 361: 1295–1305.
5. Meyer TJ (1989) Chemical approaches to artificial photosynthesis. Accounts Of Chemical Research 22: 163–170.
6. Imahor H, Guldi DM, Tamaki K, Yoshida Y, Luo C, et al. (2001) Charge separation in a novel artifical photosynthetic reaction center lives 380 ms. J Am Chem Soc 123: 6617–6628.
7. Gust D, Moore TA, Moore AL (2001) Mimicking photosynthetic solar energy transduction. Accounts Of Chemical Research 34: 40–48.
8. Alstrum-Acevedo JH, Brennaman MK, Meyer TJ (2005) Chemical approaches to artificial photo-synthesis. 2. Inorganic Chemistry 44: 6802–6827.
9. Malak RA, Gao ZN, Wishart JF, Isied SS (2004) Long-range electron transfer across peptide bridges: The transition from electron superexchange to hopping. Journal of the American Chemical Society 126: 13888–13889.
10. Kodis G, Terazono Y, Liddell PA, Andreasson J, Garg V, et al. (2006) Energy and photoinduced electron transfer in a wheel-shaped artificial photosynthetic antenna-reaction center complex. J Am Chem Soc 128: 1818–1827.
11. Moore GF, Hambourger M, Poluektov MGOG, Rajh T, Gust D, et al. (2008) A bioinspired con- struct that mimics the proton coupled electron transfer between P680(center dot)+ and the Tyr(z)-His190 pair of photosystem II. Journal of the American Chemical Society 130: 10466.
12. Gust D, Moore TA, Makings LR, Liddell PA, Nemeth GA, et al. (1986) Photodriven electron-transfer in triad molecules - a 2-step charge recombination reaction. Journal of the American Chemical Society 108: 8028–8031.
13. Gust D, Moore TA, Moore AL (2009) Solar fuels via artificial photosynthesis. Accounts of Chemical Research 42: 1890–1898.
14. Hay S, Wallace BB, Smith TA, Ghiggino KP, Wydrzynski T (2004) Protein engineering of cytochrome b(562) for quinone binding and light-induced electrons transfer. Proceedings of the National Academy of Sciences of the United States of America 101: 17675–17680.
15. Conlan B, Cox N, Su JH, Hillier W, Messinger J, et al. (2009) Photo-catalytic oxidation of a di-nuclear manganese centre in an engineered bacterioferritin 'reaction centre'. Biochimica Et Biophysica Acta-Bioenergetics 1787: 1112–1121.
16. Fry HC, Lehmann A, Saven JG, DeGrado WF, Therien MJ (2010) Computational design and elaboration of a de novo heterotetrameric alpha-helical protein that selectively binds an emissive abiological (porphinato) zinc chromophore. Journal of the American Chemical Society 132: 3997–4005.
17. Braun P, Goldberg E, Negron C, von Jan M, Xu F, et al. (2011) Design principles for chlorophyll-binding sites in helical proteins. Proteins-Structure Function And Bioinformatics 79: 463–476.
18. Koder RL, Dutton PL (2006) Intelligent design: the de novo engineering of proteins with specified functions. Dalton Transactions 25: 3045–3051.
19. Moser CC, Anderson JLR, Dutton PL (2010) Guidelines for tunneling in enzymes. Biochmicia et Bophysica Acta-Bioenergetics 1797: 1537–1586.
20. Nanda V, Koder RL (2010) Designing artificial enzymes by intuition and computation. Nature Chemistry 2: 15–24.
21. Zusman LD, Beratan DN (1999) Electron transfer in three-center chemical systems. Journal Of Chemical Physics 110: 10468–10481.
22. Marcus RA (1956) Theory of oxidation-reduction reactions involving electron transfer. 1. Journal Of Chemical Physics 24: 966–978.
23. Marcus RA, Sutin N (1985) Electron transfers in chemistry and biology. Biochim Biophys Acta 811: 265–322.
24. Marcus RA (1993) Electron transfer reactions in chemistry. Theory and experiment. Reviews of Modern Physics 65: 599–610.
25. Hopfield JJ (1974) Electron-transfer between biological molecules by thermally activated tunneling. Proc Natl Acad Sci USA 71: 3640–3644.
26. Redi M, Hopfield JJ (1980) Theory of thermal and photoassisted electron-tunneling. Journal Of Chemical Physics 72: 6651–6660.
27. Cho M, Silbey RJ (1995) Nonequilibrium photoinduced electron transfer. Journal Of Chemical Physics 103: 595–606.
28. DeVault D, Parkes JH, Chance B (1967) Electron tunelling in cytochromes. Nature 215: 642–644.
29. Page CC, Moser CC, Chen X, Dutton PL (1999) Natural engineering principles of electron tunnelling in biological oxiation-reduction. Nature 402: 47–51.
30. Moser CC, Page CC, Chen X, Dutton PL (2000) Electron Transfer in Natural Proteins: Theory and Design. In: Scrutton NS, editor, Enzyme-Catalyzed Electron and Radical Transfer, Plenum/Kluwer Press, The Netherlands, volume 3. pp 1–30.
31. Moser CC, Keske JM, Warncke K, Farid RS, Dutton PL (1992) Nature of biological electron-transfer. Nature 402: 796–802.
32. Gray HB, Winkler JR (2003) Electron tunneling through proteins. Q Rev Biophys 36: 341–372.
33. Xu Q, Gunner MR (2000) Temperature dependence of the free energy, enthalpy and entropy of P+Q(A)(-) charge recombination in Rhodobacter sphaeroides R-26 reaction centers. Journal of Physical Chemistry B 104: 8035–8043.
34. Kiang NY, Siefert J, Govindjee, Blankenship RE (2007) Spectral signatures of photsynthesis. I. Review of earth organsims. Astrobiology 7: 222–251.
35. Chen M, Schliep M, Willows RD, Cai ZL, Neilan BA, et al. (2010) A red-shifted chlorophyll. Science 329: 1318–1319.
36. Blankenship RE, Tiede DM, Barber J, Brudvig GW, Fleming G, et al. (2011) Comparing photo- synthetic and photovoltaic efficiencies and recognizing the potential for imporvement. Science 332: 805–809.
37. Wohri AB, Katona G, Johansson LC, Fritz E, Malmerberg E, et al. (2010) Light-induced structural changes in a photosynthetic reaction center caught by laue diffraction. Science 328: 630–633.
38. Guskov A, Kern J, Gabdulkhakov A, Broser M, Zouni A, et al. (2009) Cyanobacterial photosystem II at 2.9-Å resolution and the role of quinones, lipids, channels and chloride. Nature Structural & Molecular Biology 16: 334–342.
39. Myshkin E, Leontis NB, Bullerjahn GS (2002) Computational simulation of the docking of prochlothrix hollandica plastocyanin to photosystem 1: Modeling the electron transfer complex. Biophysical Journal 82: 3305–3313.

5

Interactive Effect of Herbivory and Competition on the Invasive Plant *Mikania micrantha*

Junmin Li[1,2,3], Tao Xiao[1], Qiong Zhang[1], Ming Dong[1,3]*

1 College of Life and Environmental Sciences, Hangzhou Normal University, Hangzhou, China, **2** Institute of Ecology, Taizhou University, Linhai, China, **3** State Key Laboratory of Vegetation and Environmental Change, Institute of Botany, Chinese Academy of Sciences, Beijing, China

Abstract

A considerable number of host-specific biological control agents fail to control invasive plants in the field, and exploring the mechanism underlying this phenomenon is important and helpful for the management of invasive plants. Herbivory and competition are two of the most common biotic stressors encountered by invasive plants in their recipient communities. We predicted that the antagonistic interactive effect between herbivory and competition would weaken the effect of herbivory on invasive plants and result in the failure of herbivory to control invasive plants. To examine this prediction, thus, we conducted an experiment in which both invasive *Mikania micrantha* and native *Coix lacryma-jobi* were grown together and subjected to herbivory-mimicking defoliation. Both defoliation and competition had significantly negative effects on the growth of the invader. However, the negative effect of 75% respective defoliation on the above- and below-ground biomass of *Mikania micrantha* was alleviated by presence of *Coix lacryma-jobi*. The negative effect of competition on the above- and below-ground biomass was equally compensated at 25%, 50% and 100% defoliation and overcompensated at 75% defoliation. The interactive effect was antagonistic and dependent on the defoliation intensity, with the maximum effect at 75% defoliation. The antagonistic interaction between defoliation and competition appears to be able to release the invader from competition, thus facilitating the invasiveness of *Mikania*, a situation that might make herbivory fail to inhibit the growth of invasive *Mikania* in the invaded community.

Editor: Harald Auge, Helmholtz Centre for Environmental Research – UFZ, Germany

Funding: Support for the project was provided through funding from National Natural Science Foundation of China (30800133 and 39825106), China Postdoctoral Science Foundation (20080440557), National Natural Science Foundation of Zhejiang Province of China (Y5110227) and the Innovative R & D Projects of Hangzhou Normal University. The funders had no role in the study design, data collection and analysis,decision to publish, or preparation of the manuscript.

Competing Interests: The authors have declared that no competing interests exist.

* E-mail: dongming@hznu.edu.cn

Introduction

Invasive plants pose severe threats to biological diversity and ecosystems [1], and many methods have been used to control invasive plants. Biological control, i.e., using natural enemies to control invasion success, has received much attention [2,3] and has been highly successfully used to control noxious weeds, such as *Senecio jacobaea* [4] and *Ageratina riparia* [5]. Biological control, being effective and having a low cost and relatively high environmental safety, has been widely accepted [6]. However, many natural enemies have recently been verified as being inefficient in biologically controlling invasive plants in the invaded communities [7,8], even though the host-specific agents were efficient in pot experiments. Thus, exploring the mechanism underlying this phenomenon would be important and useful in developing future biological controls of invasive species.

It has been noted that the failure of biocontrol might be due to the focus on simple predator-prey relationships and the disregard of more complex interactions in the invaded community [8]. In a natural ecosystem, herbivory and competition are two of the most common biotic stressors that plants encounter [9,10], and both play important roles in shaping the structure and dynamics of the community [11]; this is true for both the invasive plants and the invaded community [11]. It is well known that both herbivory and competition from native competitors in the invaded community can negatively affect invasive plants and reduce their growth and fitness [12,13]. Inter-specific competition and herbivory can have synergistic effects on the performance of the attacked invasive host plant [14–16] and, as a result, release native neighbours from competition [17], thus limiting invasive success in the invaded community and facilitating the restoration of the native community [18]. However, only few studies have revealed the independent [19] and antagonistic [10,20] interactive effects of herbivory and competition on invasive plants. We predicted that the antagonistic interactive effect between herbivory and competition could induce the compensatory growth of invasive plants and weaken the effect of herbivory on invasive plants, which would release invasive plants from the naeighbouring competitors and result in the failure of herbivory to control invasive plants. Obviously, an understanding of the interactive effect of herbivory and competition on the performance of invasive plants and the structure and dynamics of the invaded community is important to predict the effectiveness of biological agents on the invasive plants in an invaded community.

Mikania (Asteraceae) (hereafter referred to as *Mikania*), a perennial weed native to Central and South America, was introduced into China in ca. 1919 and subsequently became an invader. *Mikania* has caused serious and extensive damage to many Chinese ecosystems, particularly in recent decades [21]. *Mikania* rarely behaves as a weed in its native range because it encounters

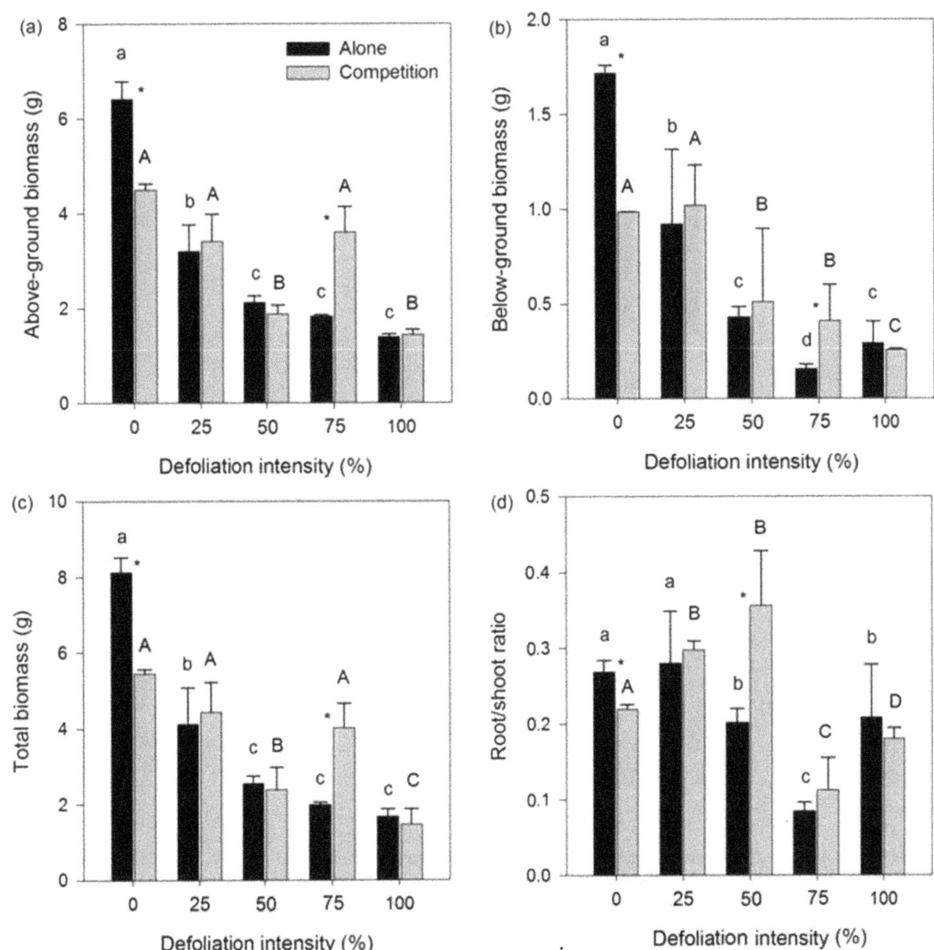

Figure 1. Effect of defoliation and competition on the above-ground (a), below-ground (b), total biomass (c) and root/shoot ratio (d) of invasive *Mikania micrantha.* Values are means ±standard deviation. The different lowercase and uppercase letters indicate significant differences ($p<0.05$) among the defoliation intensities of invasive *Mikania micrantha* growing with and without a native competitor, respectively. * indicates a significant difference ($p<0.05$) between the treatments with or without the competitor.

strong natural enemies in its habitats [22]. Since 1989, herbivores, such as *Liothrips mikaniae*, were introduced to Malaysia, India and China but failed in the biological control of *Mikania* [23]; however, the main reason for the failure is still unknown.

Coix lacryma-jobi (Poaceae) (hereafter referred to as *Coix*) is a native annual grass, commonly occurring in the communities that are subject to invasion by *Mikania*. We conducted an experiment in which invasive *Mikania* was growing with native *Coix* and was treated with defoliation-mimicking herbivory to examine the interactive effect between herbivory and competition on invasive *Mikania*. We predicted that an antagonistic interaction between herbivory and competition from native species would enhance the performance of the invasive *Mikania* and release it from competition. In particular, we addressed the following questions: 1) Can competition from the native neighbouring *Coix* affect the response of the invasive *Mikania* to defoliation? 2) Can defoliation affect the impact of competition on the invasive *Mikania* and release it from competition? 3) Is the interaction between defoliation and competition antagonistic?

Moreover, the extent to which plants respond to herbivory might be dependent on the intensity of herbivory [24]. Puettmann and Saunders found that the compensatory growth of *Pinus strobes* seedlings varied with the competitive conditions and clipping intensity [24]. Accordingly, we also aimed to address the following question: 4) Does the intensity of defoliation affect the interactive effect? In this study, some physiological traits of invasive *Mikania* were also measured to explore the mechanical responses to the interaction between defoliation and competition.

Actinote thalia pyrrha (Fabricius), a natural enemy in the native range of *Mikania*, is currently being introduced to India [25] and China [26,27] to control *Mikania*. *A. thalia pyrrha* is verified as a potential agent of biological control, as the insect consumes all of the young leaves and stems of *Mikania* [26]. The results of our research could provide information for the management of invasive *Mikania* and also for the application of natural enemies to control invasive *Mikania*.

Materials and Methods

Study Site

We conducted our pot experiment in the village of Dengshuiling, southeast of Dongguan City (E 113°31′ −114°15′; N 22°39′−23°09′), Guangdong Province, China. The area has a marine subtropical climate, with a mean annual precipitation of 1819.9 mm, mean annual temperature of 23.1°C and mean annual sunshine time of 1873.7 hr. The zonal vegetation is

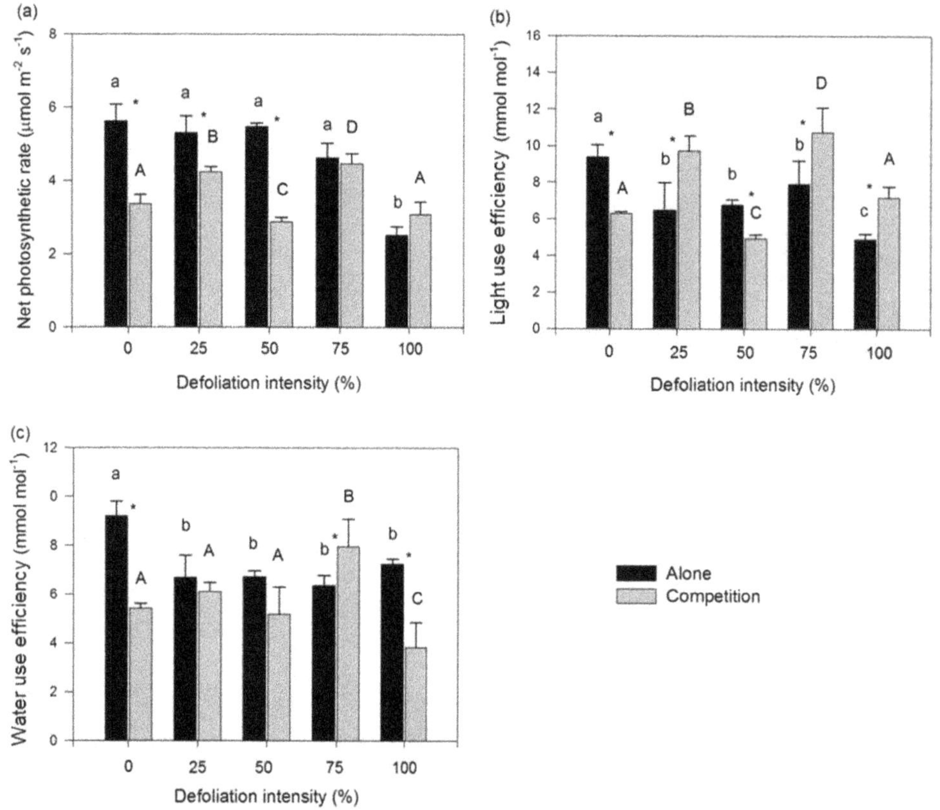

Figure 2. Effect of defoliation and competition on the net photosynthetic rate (a), light use efficiency (b) and water use efficiency (c) of invasive *Mikania micrantha*. Values are means ±standard deviation. The different lowercase and uppercase letters indicate significant differences ($p<0.05$) among the different intensities of defoliation of invasive *Mikania* micrantha growing with and without native competitor, respectively. * indicates significant differences ($p<0.05$) between the treatments with or without competitor.

subtropical evergreen broadleaved forest codominated by *Dactyloctenium aegyptium, Paederia scandens* and *Pharbitis nil. Mikania* began to invade this area in the early 1990 s and spread extensively in shrublands and old fields.

Experimental Design and Measurements

Invasive *Mikania* was collected from the fields surrounding Dengshuiling and then propagated using cuttings. The site is located in an open and abandoned field, and no specific permits were required for the described field studies. Native *Coix* was germinated from seeds that were purchased from Shandong Heze Chinese Medicine Institute. We filled our experimental pots (3 L) with field-collected red clay soil mixed with sand (3:1).

Artificial defoliation has been employed extensively as a method of simulating herbivore attack [12,28–30] and has recently been used to simulate biological agents to control invasive plants [20,31,32]. Although artificial defoliation does not always elicit the same results as true herbivory, it can allow researchers to control the amount of defoliation precisely [20]. We used defoliation to mimic the herbivory that plants are likely to encounter in nature. A factorial combination of defoliation intensities (0%, 25%, 50%,

Table 1. *F* values of the two-way ANOVAs for testing the effects of defoliation (different intensities) and competition (with or without) on the growth and physiological traits of *Mikania* micrantha.

Traits	Competition	Defoliation	Competition × Defoliation
Above-ground biomass	0.033	**133.707*** **	**23.486*** **
Below-ground biomass	0.868	**38.135*** **	**6.510* **
Total biomass	0.287	**103.219*** **	**18.381*** **
Root/shoot ratio	3.142	**25.187** **	**6.724** **
Net photosynthetic rate	**103.013** **	**52.913*** **	**29.446*** **
Water use efficiency	**37.650** **	**7.529*** **	**16.473*** **
Light use efficiency	**4.807*** **	**19.137*** **	**16.161*** **

Figures in bold are significant at $p<0.05$; Significance levels: *$p<0.05$, **$p<0.01$, ***$p<0.001$.

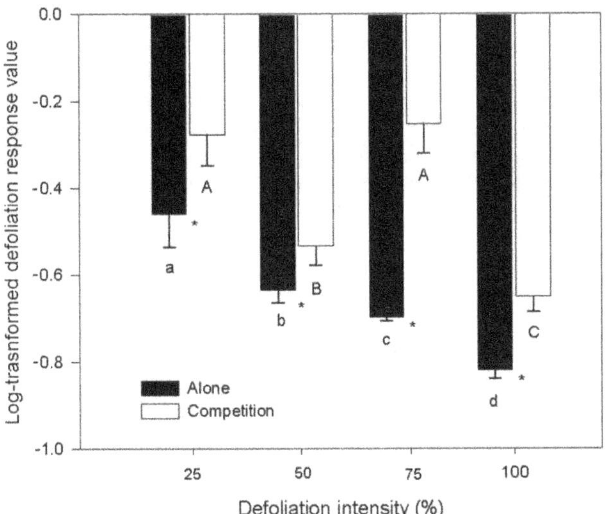

Figure 3. Log-transformed response values of *Mikania micrantha* with and without competition to defoliation intensities. Values are means±standard deviation. The different lowercase and uppercase letters indicate significant differences ($p<0.05$) among the defoliation intensities of invasive *Mikania* micrantha growing with and without native competitor, respectively. * indicates significant differences ($p<0.05$) between the treatments with or without competitor.

75% or 100%) and competition (with or without) were applied to treat invasive *Mikania*. A total of 10 treatments were used in this experiment, and 5 replicates were used for each treatment, amounting to 50 pots. For the experiment without competition, an individual *Mikania* plant was transplanted into each pot; for the competition treatment, an individual *Mikania* plant and one *Coix* plant of similar size were transplanted together into each pot with a distance of 15 cm between them. The pots were irrigated with tap water twice daily and fertilised with 50% Hoagland's nutrient solution once per week [33]. Bamboo sticks (1 m long) were inserted into the soil near *Mikania* to allow the plant to climb. Three weeks after transplantation, *Mikania* plants of similar size were chosen for defoliation. Herbivory by *A. thulia pyrrha* on *Mikania* can remove all of the leaves [26]. To simulate a realistic intensity of herbivory, five intensities were included in this experiment: (1) 0% defoliation, (2) 25% defoliation, (3) 50% defoliation, (4) 75% defoliation, and (5) 100% defoliation. These treatments constituted removing 0%~45% of the total aboveground biomass at the time of clipping to simulate zero to moderate aboveground herbivory [34]. The defoliation of *Mikania* was performed by removing each leaf with scissors, leaving the petiole attached to the stem.

After four weeks from the date of the first defoliation, a second defoliation at different levels was conducted on the newly emerging leaves. The physiological responses of plants to defoliation have received considerable attention and are considered a potential mechanism of the compensatory growth response to defoliation [35]. After three weeks from the date of the second defoliation, the net photosynthetic rate (P_n), transpiration rate (E) and leaf photosynthetically active radiation (PAR) of the *Mikania* plants were measured using a portable photosynthesis and transpiration system (LCA-4, Analytical Development Co. Ltd, Hoddesdon, UK) on the terminal leaflet of the third mature leaf from the top of the plant. The measurements were performed between 9:00 and 11:00 am under light intensity of 1400 μmol m^{-2} s^{-1}, leaf temperature of 30°C, CO_2 concentration of

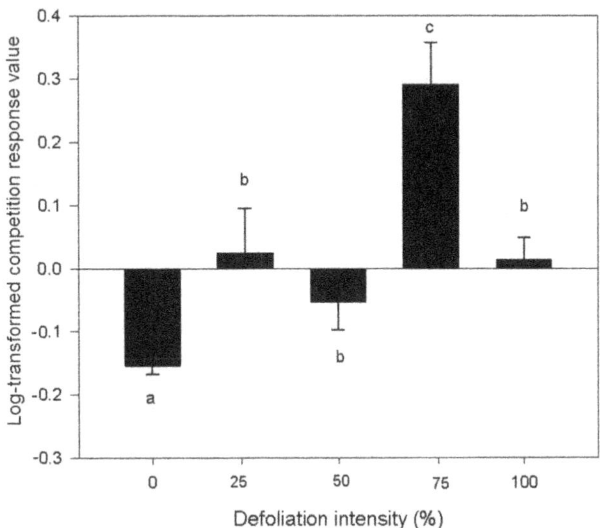

Figure 4. Log-transformed response values of *Mikania micrantha* to competition at different intensities of defoliation. Values are means±standard deviation. Different lower case letters indicate significant differences between the defoliation intensities at $p<0.05$.

350 ppm and relative moisture of 55%. The light use efficiency was calculated as P_n/PAR [36], and the water use efficiency as P_n/E [37]. The harvested plants were then separated into shoots and roots and dried for 48 h at 80°C to determine the final total biomass.

Data Analyses

To investigate the effects of herbivory and competition on the growth of *Mikania* in more detail, we calculated four response indices for the above-ground biomass of *Mikania*: defoliation responses (DR = with defoliation/without defoliation) and competition responses (CR = with competition/without competition [10]. This calculation is based on a null model that competition and

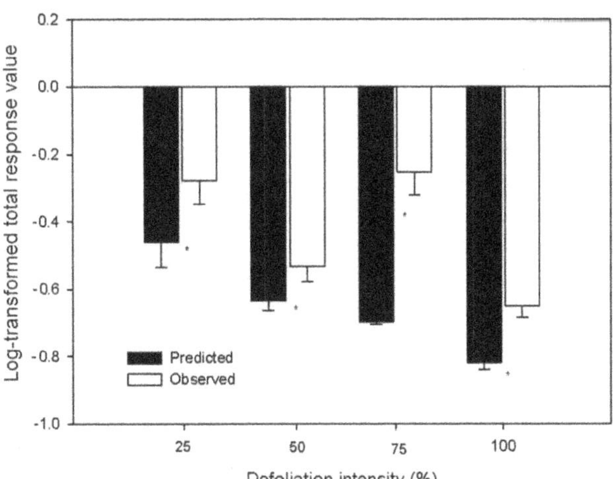

Figure 5. Log-transformed total predicted and observed response values to defoliation and competition of invasive *Mikania micrantha* defoliated at different intensities. Values are means±standard deviation. The different lowercase letters indicate a significant difference between the defoliation intensities at $p<0.05$. * indicates significant differences ($p<0.05$) between the treatments with or without competitor.

herbivory do not interact and respond multiplicatively on a linear scale. If DR or CR = 1, there would be no effect of competition or herbivory on plant growth. If DR or CR <1, there would be a negative effect; If DR or CR>1, there would be a positive effect. We also calculated TR_{pred} (DR × CR) to indicate the simple multiplicative effects of competition and herbivory together on plant growth and TR_{true} (with defoliation and competition/ without defoliation and competition) to indicate the observed combined effect of both competition and herbivory [10]. If $TR_{pred} > TR_{true}$, there would be a synergistic interaction between competition and herbivory; If $TR_{pred} < TR_{true}$, there would be an antagonistic interaction. If $TR_{pred} = TR_{true}$, there would be no interaction.

Two-way analysis of variances (ANOVAs) were used to analyse the factorial effect of the defoliation intensities (0%, 25%, 50%, 75% or 100%) and competition (with or without) on the growth of the invasive plant, with defoliation and competition as the main factors. One-way ANOVAs were used to analyse the effect of defoliation on the growth of the invasive plant with or without the competitor. In the ANOVAs, the response indices were log-transformed and the other indices were log-transformed only when the assumption of homoscedasticity of the indices was not met. The homogeneity of the variance was evaluated using Levene's test. Statistical significance was taken at $p<0.05$.

Results

Effect of Competition on *Mikania* Responses to Defoliation

Defoliation significantly decreased the above- and below-ground and total biomass of *Mikania* growing alone, whereas 25% and 75% defoliation had no significant effect on the growth of *Mikania* growing with native *Coix* except for the below-ground biomass (Fig. 1). Defoliation intensities from 50% to 100% significantly reduced the root/shoot ratio of *Mikania* when growing alone, whereas 25% and 50% defoliation significantly increased the ratio of *Mikania* growing with native *Coix* (Fig. 1). Defoliation significantly decreased the light use efficiency and water use efficiency, yet only 100% defoliation significantly decreased the net photosynthetic rate of *Mikania* growing alone (Fig. 2). Defoliation intensities of 25% and 75% significantly increased the net photosynthetic rate and light use efficiency while 75% defoliation significantly increased the water use efficiency of *Mikania* growing with native *Coix* (Fig. 2). The two-way ANOVAs results showed that defoliation had a significant effect on all of the growth and physiological traits of *Mikania* (Table 1).

In terms of the above-ground biomass, the defoliation response values of *Mikania* were all less than 0 and decreased with increasing defoliation intensities (Fig. 3), indicating a negative effect of defoliation on *Mikania*, regardless of the presence of competition: the more leaves that were removed, the more the above-ground biomass was decreased. However, the response values to the defoliation intensity of *Mikania* growing with native *Coix* were all significantly higher than those of *Mikania* growing alone, particularly at 75% defoliation (Fig. 3), indicating a compensatory growth of *Mikania* to defoliation was induced by the growth of native *Coix*.

Effect of Defoliation on *Mikania* Responses to Competition

Competition significantly decreased the above- and below-ground and total biomass of *Mikania* at 0% defoliation (Fig. 1). When *Mikania* was treated with 25%, 50% and 100% defoliation, competition had no effect on its growth; in contrast, competition significantly increased growth when *Mikania* was treated with 75% defoliation (Fig. 1). Competition significantly decreased the root/ shoot ratio at 0% defoliation and significantly increased the root/ shoot ratio at 50% defoliation but had no effect at 25%, 75% and 100% defoliation (Fig. 1). Competition significantly decreased the net photosynthetic rate, light use efficiency and water use efficiency at 0% defoliation, whereas 75% defoliation resulted in a similar net photosynthetic rate and a greater water use efficiency; 25%, 75% and 100% defoliation increased the light use efficiency, with a statistical significance at 100% defoliation (Fig. 2). The two-way ANOVAs results showed that competition had a significant effect on the root/shoot ratio, net photosynthetic rate, light use efficiency and water use efficiency (Table 1).

Based on the above-ground biomass, the competition response values of *Mikania* at 0% and 50% defoliation were less than 0, whereas those at 25%, 75% and 100% defoliation were more than 0, indicating that competition had a negative effect on the growth of *Mikania* at 0% and 50% defoliation but had a positive effect on the growth of *Mikania* at 25%, 75% and 100% defoliation. The competition response values of *Mikania* at different defoliation intensities were higher than those without defoliation, particularly at 75% (Fig. 4), indicating that defoliation could alleviate the negative effect of competition on the growth of *Mikania.*

Interactive Effect of Competition and Defoliation on *Mikania*

Both competition and defoliation significantly reduced the growth of *Mikania* compared to the plants of the species grown without competition and defoliation (Fig. 1). The TP_{true} values were all significantly higher than the TP_{pred} values, regardless of the intensity with which *Mikania* was defoliated, indicating an antagonistic interactive effect between competition and defoliation on the growth of *Mikania* (Fig. 5). The two-way ANOVAs results showed that the defoliation × competition interaction had a significant effect on all of the growth and physiological traits of *Mikania* (Table 1).

Discussion

Both competition and herbivory by native species could affect the invasiveness of introduced species and often limit the success of invasive species in a recipient community [18]. However, the compensatory growth responses of plants after herbivory damage can alleviate the potential deleterious effects of herbivory and can have a positive impact on the fitness of plants [38], intensifying the negative impact on the native neighbour and releasing the invasive species from competition [39]. Walling and Zabinski have found that the competitive ability of invasive *Centaurea maculosa* to outgrow native plants was intensified by the compensatory growth produced by defoliation, which resulted in a greater capture of resources [31]. In our study, just as we predicted, the effect of the interaction between competition and defoliation on the growth of *Mikania* was less than their individual effects, indicating an antagonism. Similar antagonistic effects have also been found in invasive *Centaurea melitensis* [20], *Centaurea solstitialis* [34] and *Poa annua* [10]. In the present study, the antagonistic interactive effect of defoliation and competition from native *Coix* on invasive *Mikania* and the consequent compensatory growth of *Mikania* might be one of the possible mechanisms why host-specific biological control agents could not successfully control invasive plants in an invaded community.

It has been commonly verified that plants may compensate for tissue losses due to defoliation, resulting in increased growth relative to non-defoliated plants [30,40]. Different from these conclusions, in this study, defoliation had a negative effect on the growth of invasive *Mikania* growing alone: growth declined with increasing defoliation intensities. However, the negative effect of defoliation may be modified by competition. The response values to different defoliation intensities tested on *Mikania* growing with native *Coix* were all significantly higher than those of *Mikania* growing alone, indicating a compensatory growth of *Mikania* induced by competition in response to defoliation, particularly at 75% defoliation. This result indicates that native *Coix* could help invasive *Mikania* be more vigorous after defoliation.

Although the mechanism underling the compensatory growth of *Mikania* that is induced by the competition is unknown, the underground network between the roots of invasive *Mikania* and native *Coix* mediated by mycorrhizae might be a possible mechanism. Although it is still unknown why defoliation can induce a potential transfer of nutrients between a plant and a neighbouring plant, evidence using stable isotopes verified that defoliation could change the underground nitrogen flow [41] and that carbon could be transferred via mycorrhizae from native neighbouring plants to the invasive plant [42]. Native *Coix* is a mycorrhizal plant [43], and the soil in the *Mikania* community is rich in fungi [44]. It has also been verified that native neighbours are capable of enhancing compensatory growth of invasive plants to defoliation in the presence of soil fungi [20,34]. Further atention should be paid to the underground mechanism.

The successfully invasive plants are always strong competitors of the native plant species, however, native plants has been verified as a major force in the resistance of exotic invasions [3,45]. In this study, competition from native *Coix* did significantly decrease the growth of invasive *Mikania* because of the limited resources. However, the negative effect of competition on the growth of *Mikania* may be modified by defoliation. The response values of *Mikania* to competition increased at each defoliation intensity, indicating a release from native competitor *Coix* induced by defoliation, particularly at 75% defoliation. The release of *Mikania* from competition that can be induced by defoliation could increase the number of invasive plants and allow the domination of niche spaces to the detriment of native species [46], perhaps facilitating the invasiveness of *Mikania* and helping to shape the structure and dynamics of the invaded communities.

Plants have the ability to (at least partially) compensate for herbivory only above a certain threshold level of damage [29], and this threshold can differ among plant species. Yu et al. found that invasive *Alternanthera philoxeroides* can only rapidly recover from 50% defoliation [47]. Similarly, in the present study, when the native *Coix* was present, 75% defoliation induced the compensatory growth of invasive *Mikania*. Many morphological and physiological mechanisms have been proposed to explain the compensatory growth that follows herbivory or defoliation [30], such as the increased allocation of substrates from the roots to shoots [48] and the increased photosynthetic rate of the regrowing tissue [49]. In our study, 75% defoliation decreased the root/shoot ratio and significantly increased net photosynthetic rate, light use efficiency and water use efficiency. The resources stored in the roots were shifted to the shoots, significantly reducing the root/shoot ratio [50]. Barton found that *Plantago lanceolata* (Plantaginaceae) seedlings were plastic in their resource allocation between the shoots and roots, resulting in compensatory growth [50]. This type of strong compensatory growth due to phenotypic plasticity and the physiological acclimation of invasive *Mikania* was maximised at 75% defoliation.

Although artificial defoliation has been widely used to mimic the effect of truly herbivory on plants [12,28–30,51], there are undeniably significant differences between defoliation and herbivory [52]. Artificial defoliation can only mimic the effect of the loss of leaf area which decreased the ability of plants to intercept light [53] but not the effect in responding to the physiological and chemical interactions (e.g., due to nutrient supply) between herbivores and plants. In spite of some pitfalls, artificial defoliation has been used more often in herbivory research than real herbivores for easily and precisely controlling, targeted effect and efficient experimental designs [53]. And there were only a few cases (as low as 3%) with the outcomes where artificial and natural damage had opposite effects on plants. The biological control agent of *Mikania* are found to consume all of the young leaves and stems of *Mikania* [26], so the defoliation can at least partially mimic the effect of the loss of leaf area caused by the biological control agent.

In conclusion, our results suggest that natural herbivory might not necessarily be safely used as a potential agent to control invasive *Mikania* in the field because of the induced compensatory growth of *Mikania* by native *Coix*. Further studies should consider the interactions at the intertrophic and multitrophic levels in invaded communities as well as among more factors including, e.g., nutrient supply which seems difficult to investigate with simulated herbivore, whereby the ecological risk of the releasing of the biological control agents can be comprehensively evaluated.

Acknowledgments

We thank T. Suwa for the useful comments on the paper and E. Leila and K. Bill for the English editing.

Author Contributions

Conceived and designed the experiments: JL MD. Performed the experiments: JL. Analyzed the data: JL MD. Contributed reagents/materials/analysis tools: JL. Wrote the paper: JL MD.

References

1. Mack RN, Simberloff D, Lonsdale WM, Evans H, Clout M, et al. (2001) Biotic invasions: causes, epidemiology, global consequences, and control. Ecological Applications 10: 689–710.
2. Vilà M, Weiner J (2004) Are invasive plant species better competitors than native plant species? -evidence from pair-wise experiments. Oikos 105: 229–238.
3. Keane RM, Crawley MJ (2002) Exotic plant invasions and the enemy release hypothesis. Trends in Ecology and Evolution 17: 164–170.
4. McEvoy P, Cox C, Coombs E (1991) Successful biological control of ragwort, *Senecio jacobaea*, by introduce insects in Oregon. Ecology Applications 1: 430–442.
5. Barton JE, Fowler SV, Gianotti AF, Winks CJ, de Beurs M, et al. (2007) Successful biological control of mist flower (*Ageratina riparia*) in New Zealand: Agent establishment, impact and benefits to the native flora. Biological Control 40: 370–385.
6. Müller-Schärer H, Schaffner U, Steinger T (2004) Evolution in invsaive plants: implications for biological control. Trends in Ecology and Evolution 19: 417–422.
7. Dray JR FA, Center TD, Wheeler GS (2001) Lessons from unsuccessful attempts to establish *Spodoptera pectinicornis* (Lepidoptera: Noctuidae), a biological control agent of waterlettuce. Biocontrol Science and Technology 11: 301–316.
8. Pearson DE, Callaway RM (2003) Indirect effects of host-specific biological control agents. Trends in Ecology and Evolution 18: 456–461.
9. Boege K (2010) Induced responses to competition and herbivory: natural selection on multi-trait phenotypic plasticity. Ecology, 91: 2628–2637.
10. Schädler M, Brandl R, Haase J (2007) Antagonistic interactions between plant competition and insect herbivory. Ecology 88: 1490–1498.
11. Doyle R, Grodowitz M, Smart M, Owens C (2007) Separate and interactive effects of competition and herbivory on the growth, expansion, and tuber formation of *Hydrilla verticillata*. Biological Control 41: 327–338.
12. Ferrero-Serrano Á, Collier TR, Hild AL, Mealor BA, Smith T (2008) Combined impacts of native grass competition and introduced weevil herbivory on Canada thistle (*Cirsium arvense*). Rangeland Ecology & Management 61: 529–534.

13. Sheppard AW, Smyth MJ, Swirepik A (2001) The impact of a root-crown weevil and pasture competition on the winter annual *Echium plantagineum*. Journal of Applied Ecology 38: 291–300.
14. Turner PJ, Morin L, Williams DG, Kriticos DJ (2010) Inte ractions between a leafhopper and rust fungus on the invasive plant *Asparagus asparagoides* in Australia: A case of two agents being better than one for biological control. Biological Control 54: 322–330.
15. Sciegienka JK, Keren EN, Menalled FD (2011) Interactions between two biological control agents and an herbicide for Canada thistle (*Cirsium arvense*) suppression. Invasive Plant Science & Management 4: 151–158.
16. Crawley MJ (1997) Plant-herbivory dynamics. *Seeds: the ecology of regeneration in plant communities* (eds M. Fenner M), 401–474. CAB International, Wallingford.
17. Newingham BA, Callaway RM (2006) Shoot herbivory on the invasive plant, *Centaura maculosa*, does not reduce its competitive effects on conspecific and natives. Oikos 114: 397–406.
18. Levine JM, Adler PB, Yelenik SG (2004) A meta-analysis of biotic resistance to exotic plant invasions. Ecological Letters 7: 975–989.
19. Suwa T, Louda SM, Russell FL (2010) No interaction between competition and herbivory in limiting introduced *Cirsium vulgare* rosette growth and reproduction. Oecologia 162: 91–102.
20. Callaway RM, Newingham B, Zabinski CA, Mahall BE (2001) Compensatory growth and competitive ability of an invasive weed are enhanced by soil fungi and native neighbours. Ecology Letters. 4: 429–433.
21. Zhang LY, Ye WH, Cao HL, Feng HL (2004) *Mikania micrantha* H.B.K. in China—an overview. Weed Research 44: 42–49.
22. Cock MJW (1982) The biology and host specificity of *Liothrips mikaniae* (Priesner) (Thysanoptera: Phlaeothripidae), a potential biological control agent of *Mikania micrantha* (Compositae). Bulletin of Entomological Research 72: 523–533.
23. Waterhouse DF (1994) Biological control of weeds: Southeast Asia prospects. Canberra, ACIAR.
24. Puettmann KJ, Saunders MR (2001) Patterns of growth compensation in eastern white pine (*Pinus strobes* L.): the influence of herbivory intensity and competitive environments. Oecologia 129: 376–384.
25. Desmier de chenon R (2003) Feeding preference tests of two Nymphalid butterflies, *Acinote thalia pyrrha* and *Actinote anteas* from South America for the biocontorl of *Mikania micrantha* (Asteraceae) in South East Asia. *Exotic pest and their control* (eds R.J. Zhnag, C.Q. Zhou, H, Pang), 201, Sun Yat Sen University Press, Guangzhou.
26. Li ZG, Han SC, Guo MF, Li LY (2003) Rearing *Actinote thalia pyrrha* and *Actinote anteas* on potted *Mikania micrantha*. Entomolog Knowledge 40: 561–564.
27. Li ZG, Han SC, Guo MF (2004) Biology and host specificity of *Actinote anteas*, a biocontrol agent for contorlling *Mikania micrantha*. Chinese Journal of Biological Control 20: 170–173.
28. Richards JH (1984) Root growth response to defoliation in two *Agropyron* bunchgrasses with an improved root periscope. Oecologia 64: 21–25.
29. Ruiz R, Ward D, Saltz D (2008) Leaf compensatory growth as a tolerance strategy to resist herbivory in *Pancratium sickenbergeri*. Plant Ecology 198: 19–26.
30. Ballina-Gómez HS, Iriarte-Vivar S, Orellana R, Santiago LS (2010) Compensatory growth responses to defoliation and light availability in two native Mexican woody plant species. Journal of. Tropical Ecology 26: 163–171.
31. Walling SZ, Zabinski CA (2006) Defoliation effects on arbuscular mycorrhizae and plant growth of two native bunchgrasses and an invasive forb. Applied Soil Ecology 32: 111–117.
32. Watt MS, Whitehead D, Kriticos DJ, Gous SF, Richardson B (2007) Using a process-based model to analyse compensatory growth in response to defoliation: simulating herbivory by a biological control a biological control agent. Biological Control 43: 119–129.
33. Bacilio-Jiménez M, Aguilar-Flores S, del Valle MV, Pérez A, Zepeda A, et al. (2001) Endophytic bacteria in rice seeds inhibit early colonization of roots. Soil Biology & Biochemistry 33: 167–172.
34. Callaway RM, Kim J, Mahall BE (2006) Defoliation of *Centaurea solstitialis* stimulates compensatory growth and intensifies negative effects on neighbors. Biological Invasions 8: 1389–1397.
35. Vanderklein DW, Reich PB (1999) The effect of defoliation intensity and history on photosynthesis, growth and carbon reserves of two conifers with contrasting leaf lifespans and growth habits. New Phytologist, 144: 121–132.
36. Long SP, Baker NR, Rains CA (1993) Analyzing the responses of photosynthetic CO_2 assimilation to long-term elevation of atmospheric CO_2 concentration. Vegetation 104: 33–45.
37. Hamid MA, Agata W, Kawamitsu Y (1990) Photosynthesis, transpiration and water use efficiency in four cultivars of mungbean, *Vigna radiate* (L.) Wilczek. Photosynthetica 24: 96–101.
38. McNaughton SJ (1983) Compensatory plant growth as a response to herbivory. Oikos 40: 329–336.
39. Callaway RM, Deluca TH, Belliveau WM (1999) Biological-control herbivores may increase competitive ability of the noxious weed *Centaurea maculosa*. Ecology, 80: 1196–1201.
40. McNaughton SJ (1979) Grazing as an optimization process–grass ungulate relationships in the Serengeti. American Naturalist 113: 691–703.
41. Ayres E, Dromph KM, Cook R, Ostle N, Bardgett RD (2007) The influence of below-ground herbivory and defoliation of a legume on nitrogen transfer to neighbouring plants. Functional Ecology 21: 256–263.
42. Carey EV, Marler MJ, Callaway RM (2004) Mycorrhizae transfer carbon from a native grass to an invasive weed: evidence from stable isotopes and physiology. Plant Ecology 172: 133–141.
43. Charoenpakdee S, Phosri C, Dell B, Choonluechanon S, Lumyong S (2010) Compatible arbuscular mycorrhizal fungi of *Jatropha curcas* and spore multiplication using cereal crops. Mycosphere 1: 195–204.
44. Li WH, Zhang CB, Jiang HB, Xing GR, Yang ZY (2006) Changes in soil microbial community associated with invasion of the exotic weed, *Mikania micrantha* H.B.K. Plant and Soil 281: 309–324.
45. Mitchell CE, Agrawal AA, Bever JD, Gilbert GS, Hufbauer RA, et al. (2006) Biotic interactions and plants invasions. Ecology Letters 9: 726–740.
46. Tilman D (2004) Niche tradeoffs, neutrality, and community structure: a stochastic theory of resource competition, invasion, and community assembly. Proceedings of the National Academy of Sciences USA 101: 10854–10861.
47. Yu LF, Yu D, Liu CH, Xie D (2010) Flooding effects on rapid responses of the invasive plant *Alternanthera philoxeroides* to defoliation. Flora 205: 449–453.
48. Dyer MI, Acra MA, Wang GM, Coleman DC, Freckman DW, et al. (1991) Source-sink carbon relations in two *Panicum colratum* ecotypes in response to herbivory. Ecology 72: 1472–1483.
49. Delting JK, Dyer MI, Winn DT (1979) Net photosynthesis, root respiration, an regrowth of *Bouteloua gracilis* following simulated grazing. Oecologia 41: 127–134.
50. Barton KE (2008) Phenotypic plasticity in seedling defense strategies: compensatory growth and chemical induction. Oikos, 117: 917–925.
51. Johnson MTJ (2011) Evolutionary ecology of plant defense against herbivores. Functional Ecology 25: 305–311.
52. Lehtila K, Boalt E (2004) The use and usefulness of artificial herbivory in plant-herbivore studies. In Weisser WW and Siemann E (Eds.) Ecological Studies. Vol 173: Insects and Ecosystem Function, 2004: Springer-Verlag Berlin Heidelberg. 257–275.
53. Trumble JT, Kolodny-Hirsch DM, Ting IP (1993) Plant compensation for anthropod herbivory. Annual Reviews of Entomology 38: 93–119.

6

Enhanced Lipid Productivity and Photosynthesis Efficiency in a *Desmodesmus* sp. Mutant Induced by Heavy Carbon Ions

Guangrong Hu[1], Yong Fan[1], Lei Zhang[1], Cheng Yuan[1], Jufang Wang[2], Wenjian Li[2], Qiang Hu[3], Fuli Li[1]*

1 Shandong Provincial Key Laboratory of Energy Genetics, Qingdao Institute of Bioenergy and Bioprocess Technology, Chinese Academy of Sciences, Qingdao, PR China, **2** Institute of Modern Physics, Chinese Academy of Sciences, Lanzhou, PR China, **3** Laboratory for Algae Research and Biotechnology (LARB), College of Technology and Innovation, Arizona State University, Mesa, Arizona, United States of America

Abstract

The unicellular green microalga *Desmodesmus* sp. S1 can produce more than 50% total lipid of cell dry weight under high light and nitrogen-limitation conditions. After irradiation by heavy $^{12}C^{6+}$ ion beam of 10, 30, 60, 90 or 120 Gy, followed by screening of resulting mutants on 24-well microplates, more than 500 mutants were obtained. One of those, named D90G-19, exhibited lipid productivity of 0.298 g $L^{-1}\cdot d^{-1}$, 20.6% higher than wild type, likely owing to an improved maximum quantum efficiency (Fv/Fm) of photosynthesis under stress. This work demonstrated that heavy-ion irradiation combined with high-throughput screening is an effective means for trait improvement. The resulting mutant D90G-19 may be used for enhanced lipid production.

Editor: Paul Jaak Janssen, Belgian Nuclear Research Centre SCK/CEN, Belgium

Funding: This study was supported by the International cooperative project of Chinese Academy of Sciences (31010103907), and carried out in the framework of the Joint Research Laboratory for Sustainable Aviation Biofuels Collaboration Agreement, and also supported by the "Western Light" Talents Co-Scholar Program of Chinese Academy of Sciences (No. Y106140XBL). The funders had no role in study design, data collection and analysis, decision to publish, or preparation of the manuscript.

Competing Interests: The authors have declared that no competing interests exist.

* E-mail: lifl@qibebt.ac.cn

Introduction

Under suitable environmental conditions, microalgae synthesize fatty acids mainly for the production of membrane glycerolipids, such as glycolipids and phospholipids. However, under unfavorable growth conditions, many microalgae change their lipid biosynthetic pathways to produce large amounts of neutral lipid (20–50% of cell dry weight), mostly in the form of triacylglycerol (TAG), which mainly are stored in cytosolic lipid bodies [1]. Recently, oleaginous microalgae have been regarded as potential next-generation feedstocks for biofuels (e.g., biodiesel and jet fuel) because they exhibit higher photosynthetic efficiency and a greater lipid production rate than terrestrial oil crops [2–4]. Microalgae do not compete for precious arable land with grain crops and many species can be cultured in wastewater or salt water [2]. However, naturally occurring microalgae that have been used so far produce much lower amounts of neutral lipid than the theoretical maximum [1]. Many methods such as physical or chemical mutagenesis and genetic engineering tools may be applied to the production strains for improving lipid production [5–7].

In recent years, some genetic engineering approaches have been applied to several algal species aiming at improving lipid production, including *Chlamydomonas reinhardtii* [8,9], *Phaeodactylum tricornutum* [10], *Cyclotella cryptica* [11], and *Nannochloropsis* sp. [5]. However, only moderate success was achieved, mainly because the lipid metabolism, particularly the functions of key genes and enzymes involved in lipid synthesis and accumulation in these organisms, is not well understood. Furthermore, some transcription factors involved in the regulation of plant lipid accumulation were suggested as the second-generation targets to improve the lipid contents [12–14].

On the other hand, various physical mutagens have been successfully applied to crop breeding and genetic manipulation of microorganisms for trait improvement. As a physical irradiation source, heavy-ion beam induces a broad range of mutations i.e. base substitutions, small and large insertions/deletions, translocation and inversions in the genomes of microalgae. The use of a heavy-ion beam as mutagen has the following advantages: (1) It has a high linear energy transfer (LET) with energetic heavy ions that can produce dense ionization along their trajectories, and cause complex and irreparable damages of DNA. Hence mutations induced by heavy-ion may show a broad spectrum and a high frequency; (2) it causes higher relative biological effectiveness (RBE) compared with low-LET γ-rays and x-rays [15]; (3) it can be controlled so as to deposit high energy at precise positions, in contrast to low-LET irradiation that may cause large deletions, translocations or rearrangements in the genome of a given organism [16]. Like other random mutagenesis methods, heavy-ion irradiation can generate thousands of mutants, but it is a very laborious work to screen for desirable mutants. Establishment of a high-throughput screening method will be helpful for microalgal trait improvements.

The green microalga *Desmodesmus* sp. (Sphaeropleales, Scenedesmaceae) S1 strain was oval, 5–10 μm in length. It had a thin, rim-like, pyrenoid-containing chloroplast that enlarged with age. It was once identified as *Pseudochlorococcum* sp. [17]. However, it should be a strain in the genus *Desmodesmus* according to its phylogenetic relationship with 22 Chlorophyta taxa based on their ITS sequences [18]. *Desmodesmus* sp. S1 exhibited rapid growth and reproduction while possessing high lipid content (up to 55% of dry weight) in BG-11 culture medium. The preliminary experiments showed that a high biomass productivity of 0.5 g $L^{-1}\cdot d^{-1}$ can be achieved in *Desmodesmus* sp. S1 cultures under stress conditions, which was superior to many microalgal strains used for biofuel production [1,2].

In this paper, we introduced a heavy-ion irradiation method to induce mutagenesis of *Desmodesmus* sp. A large number of *Desmodesmus* mutants were screened by coupling a chlorophyll fluorometer with a Nile red staining method. We demonstrated that a mutant D90G-19 exhibited a higher photosynthetic efficiency and higher lipid content, thus resulting in higher lipid productivities than WT.

Materials and Methods

Strains and Culture Conditions

Desmodesmus sp. S1 strain was isolated in Arizona, USA and provided by Professor Qiang Hu at Arizona State University, ASU. The wild type (WT) and mutant seed cultures were grown in BG-11 culture medium in a column photobioreactor (60 cm high, diameter 4 cm, 300 mL culture volume) under continuous illumination of a low light intensity of 50 μmol photons $m^{-2}\cdot s^{-1}$. Culture mixing was provided by aeration with compressed air containing 2% CO_2. This treatment was referred to as LL+N. After 4 days, the seed culture were collected by centrifugation (3200 ×g, 5 min) and transferred into 300 mL or 10 L nitrogen-limited BG-11 medium (4.25 mM $NaNO_3$) in column or panel photobioreactors, respectively. The initial OD_{750} of algal culture was about 0.2. WT and mutant cultures were grown under continuous illumination of 400 μmol photons $m^{-2}\cdot s^{-1}$ with aeration containing 2% CO_2 (which was referred to as HL-N) for two weeks. All strains were cultured in 3 replicates at room temperature (25±2°C).

Pre-weighed Whatman GF filter paper (Whatman International Ltd., Maidstone, UK) was recorded as W1. A 10 mL culture sample was filtered through the pre-weighed filter paper which was then dried at 105°C for at least 8 hours. Final weight of the filter paper was recorded as W2. The difference between W1 and W2 was the dry weight of algal biomass.

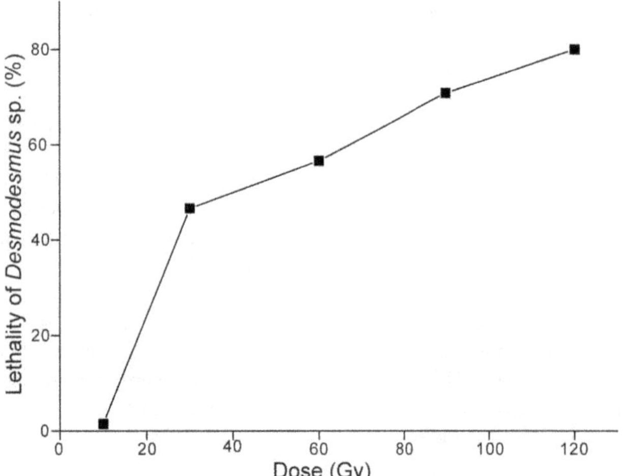

Figure 1. Effects of irradiation on the cells viability of *Desmodesmus* sp. S1.

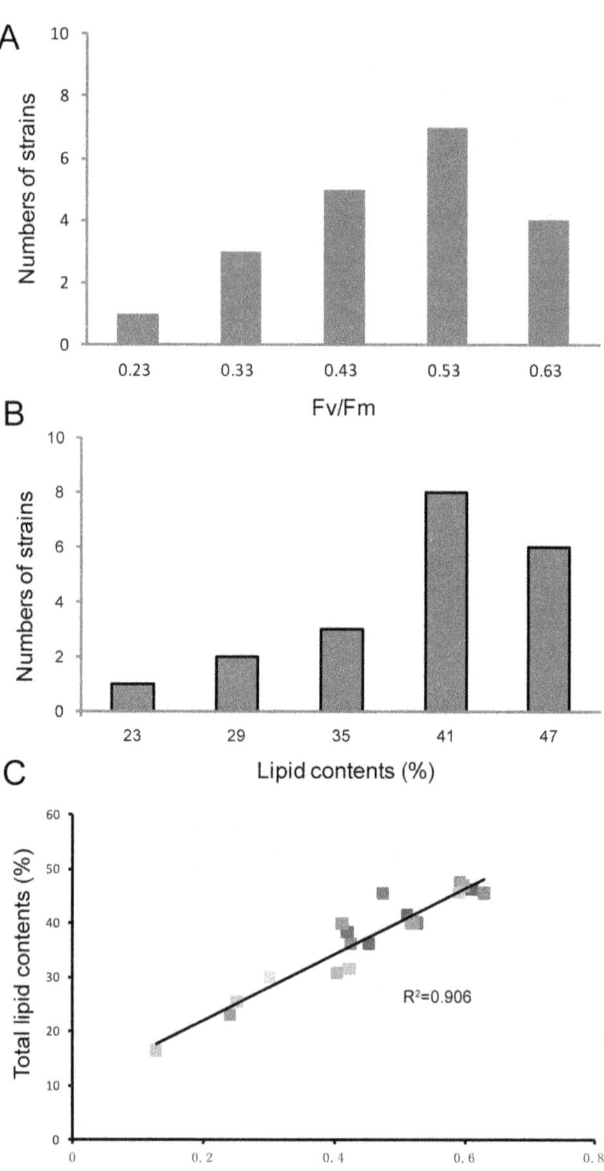

Figure 2. Distribution histogram of *Desmodesmus* sp. mutant phenotypes. (A) Distribution histogram of photosynthesis efficiency mutants (PEMs), (B) distribution histogram of lipid-over-production mutants (LOMs), and (C) the relationship between lipid contents and high light-adapted photosystem II efficiency under stress conditions in the PEMs and LOMs of *Desmodesmus* sp. S1. The distribution histograms of LOMs and PEMs were obtained using SPSS 10.0.

Microalgal suspension was passed through a filter paper (0.22 μm) to remove cells, and nitrate in the filtrate was analyzed according to the method of Collos [19].

Heavy-ion Irradiation

Desmodesmus sp. S1 strain was maintained in BG-11 medium in 100 mL Erlenmeyer flask under low light illumination of 100 μmol photons $m^{-2}\cdot s^{-1}$ at room temperature for 3 days. Algal cells at the exponential growth phase were collected by centrifugation (3200 ×g, 3 min) and washed with sterile water,

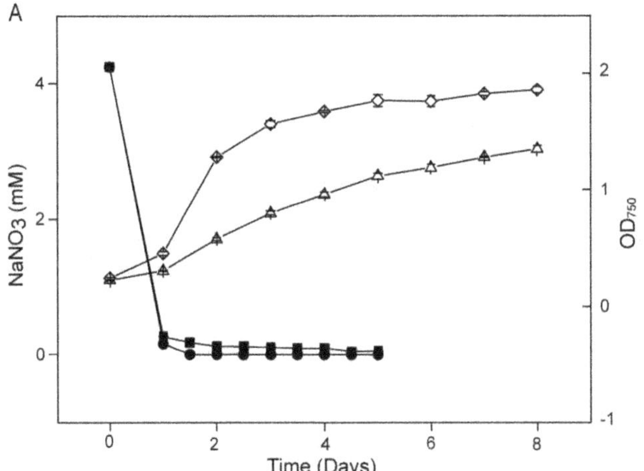

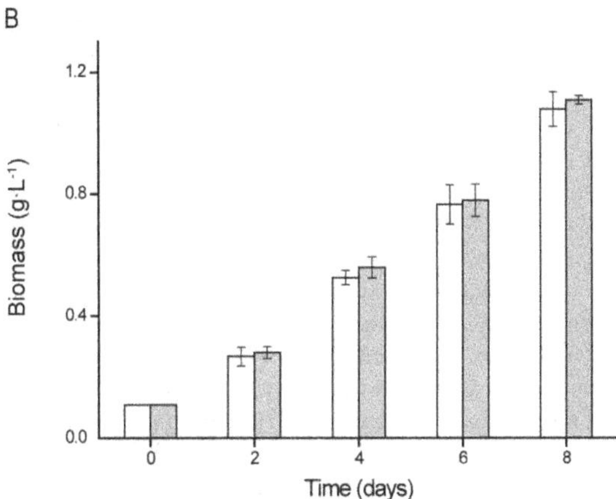

Figure 3. The consumption of $NaNO_3$ and growth kinetics of *Desmodesmus* sp. (A) The consumption of $NaNO_3$ in the nitrogen-limited (4.25 mM) BG-11 medium (closed symbols) and growth kinetics (open symbols) of *Desmodesmus* sp. S1 under low light (closed square, open triangle) and high light (closed circle, open diamond), and (B) biomass concentration of *Desmodesmus* sp. S1 wild type (WT, open square) and mutant D90G-19 (closed square) cultivated in BG-11 medium under LL+N conditions (50 μmol photons $m^{-2}\cdot s^{-1}$, BG-11 medium with 17 mM $NaNO_3$).

then resuspended in fresh BG-11 medium. The cell concentration was adjusted to 1×10^6 cells·mL^{-1}, and exposed to $^{12}C^{6+}$ ion beam provided by the Heavy Ion Research Facility at Lanzhou (HIRFL), Institute of Modern Physics of Chinese Academy of Sciences (CAS). Irradiation treatments were conducted at dosages of 10, 30, 60, 90 and 120 Gy, calculated from particle fluencies and LET, and there were at least three algae samples for every dose treatment.

Mutant Isolation

After irradiation, algae cells were plated on BG-11 agar plates in triplicate and cultured at 25°C under low light (50 μmol photons $m^{-2}\cdot s^{-1}$) until algal colonies occurred on the plates. The colonies derived from the irradiated cells were selected and transferred to BG-11 agar plates several times to obtain purified monoclonal strains, which were regarded as putative mutants and constituted the mutant library.

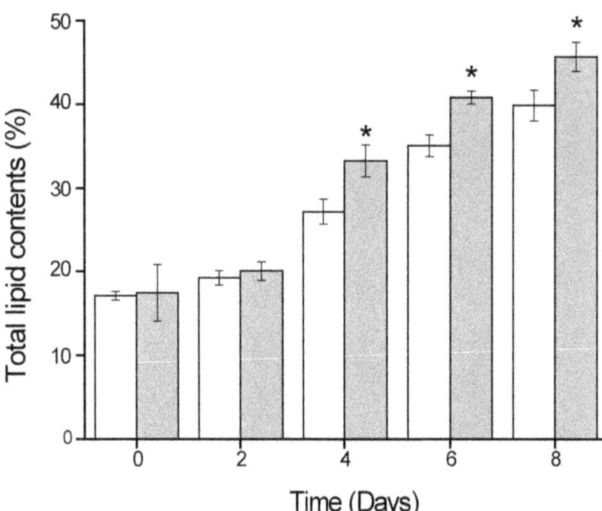

Figure 4. Total lipid contents of *Desmodesmus* sp. WT and D90G-19 were grown under HL-N conditions (300–400 μmol photons $m^{-2}\cdot s^{-1}$, nitrogen-limited BG-11 medium with 4.25 mM $NaNO_3$). The total lipid of the mutant D90G-19 was significantly higher than WT at day 4 ($P=0.0051$), day 6 ($P=0.006$) and day 8 ($P=0.0039$). D90G-19: closed square; WT: open square.

Analysis of Photosynthesis Efficiency and Oxygen Evolution Rate

Algal samples (cells suspension or colonies on plate) were placed in a chamber of the Imaging-PAM (pulse-amplitude) Chlorophyll Fluorometer (Walz, Effeltrich, Germany), and the Fv/Fm values of samples were determined. Three replicates for each mutant were analyzed.

Two milliliter of algal suspension was placed in the chamber of Chlorolab 2 (Hansatech Instruments Ltd., King's Lynn, UK), and the oxygen evolution rate was determined according to the manufacturer's instructions.

Screening of Lipid and Photosynthesis Mutants

The putative mutants and wild type were inoculated into 24-well microplates where each well contained 2.5 mL BG-11 culture medium, and cultured under low light (<100 μmol photons $m^{-2}\cdot s^{-1}$) at 25°C for 6 days. At the sixth day, OD_{750} of the cultures were measured with a spectrophotometer (UV-2600, Unico Instruments Co., Ltd., Shanghai, China). The 24-well microplates were subjected to a high light intensity of 300 μmol photons $m^{-2}\cdot s^{-1}$ for 6 more days. The neutral lipid content of

Table 1. Composition of total lipid in the *Desmodesmus* sp.

Lipid contents[a]	WT	D90G-19
neutral lipid (%)	78.14±1.52	86.97±1.28
Glycolipid (%)	14.85±0.97	10.39±0.88
phospholipid (%)	6.86±0.83	2.64±0.40
G:P[b]	2.18±0.15	3.96±0.26

[a]Microalgae cells were harvested after grown under HL-N (300–400 μmol photons $m^{-2}\cdot s^{-1}$, nitrogen-depleted BG-11 medium with 4.25 mM $NaNO_3$) for 8 days. Mean ± SE with three replicates.
[b]G:P: ratio of glycolipid to phospholipid.

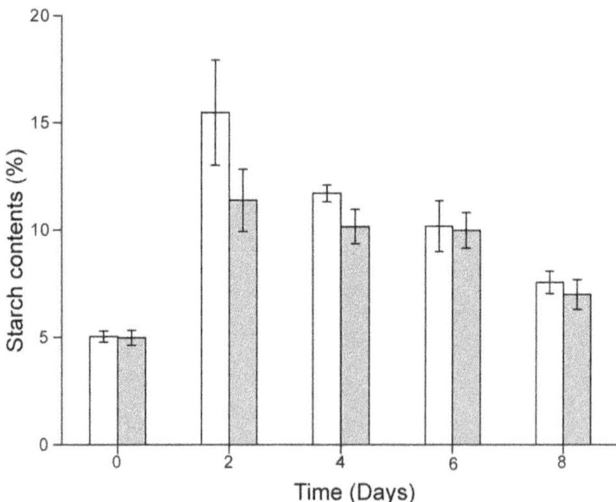

Figure 5. Starch contents of *Desmodesmus* sp. WT and D90G-19. The initial starch content was measured in algae cells grown under LL+N for 4 days. Afterwards, the cells were grown under HL-N (300–400 μmol photons $m^{-2}\cdot s^{-1}$, nitrogen-limited BG-11 medium with 4.25 mM $NaNO_3$). D90G-19: closed square; WT: open square.

mutants and wild type was determined by a modified protocol of Chen [20]. Ten microliter Nile red solution (125 $\mu g\cdot mL^{-1}$ in acetone) was added into individual wells containing 2.0 mL of cultures. The 24-well microplates were vortexed for 1 min and incubated at 40°C for 10 min. After the algal cells were stained, fluorescence emissions were recorded with Synergy HT (Biotek Instruments Inc., Winooski, VT) with the excitation and emission wavelengths of 490 nm and 580 nm, respectively. Three replicates for each mutant were analyzed. As the Nile red staining method was microalgae species and physiological state sensitive [20], the linear regression equation ($y = (x-21.98)\times4.45$) and its correlation coefficient of R^2 (0.972) between Nile red fluorescence intensity (x) and lipid concentration (y: $\mu g\cdot mL^{-1}$) of *Desmodesmus* sp. S1 had been determined before using the Nile red fluorescence method. Then the data of lipid concentration were converted into a percentage of dry weight of algal biomass.

The putative mutants were also inoculated into BG-11 plates and cultured under light (200 μmol photons $m^{-2}\cdot s^{-1}$) at 25°C. After 6 days, the Fv/Fm values of algal mutants were analyzed with an Imaging-PAM Chlorophyll Fluorometer (Walz, Effeltrich, Germany). Three replicates for each mutant were analyzed.

Lipid Extraction, Quantification and Separation

Total lipids were extracted and quantified according to the Bigogno's method [21] with minor modifications.

Total lipids were further separated by column chromatograph using a silica gel (60–200 mesh, Sigma-Aldrich, St. Louis, MO, USA) according to the following procedures: 100 mL chloroform, acetone and methanol were used as the eluent to collect the neutral lipid class, glycolipids and phospholipids, respectively. After evaporating the solvents using a rotary evaporator (RE-52AA, Yarong Inc., Shanghai, China), the remaining individual lipid fractions were dissolved in chloroform and transferred into pre-weight vials. Chloroform was removed using nitrogen evaporator and the residuals were freeze-dried for at least 12 h and weighed. The difference between the final weight and the weight before freeze-drying was the weight of lipid fractions.

Total lipid, neutral lipid, glycolipid or phospholipid were transmethylated with 2% H_2SO_4 in methanol at 85°C for 2.5 h. Heptadecanoic acid (C17:0, 3 $mg\cdot mL^{-1}$) was used as an internal standard. Gas chromatograph analysis was performed with a GC system (7890A, Agilent technologies, Inc. CA).

Pigments and Starch Analysis

The algal pellets were ground with quartz sands in liquid nitrogen until algal cells were broken completely. Then the pellets were ground with 2–3 mL ice-cold acetone and repeated for 2–3 times. All solutions were collected. After centrifugation at 3200×g for 5 minutes, the supernatant was transferred to a new tube. The acetone was evaporated under nitrogen gas and the residuals were lyophilized for 5 hours. Afterwards, 5 mL 80% acetone was added, and the solution absorbance at 663 nm, 646 nm and 470 nm were measured with a spectrophotometer (UV-2600, Unico Instruments Co., Ltd., Shanghai, China, resolution range: 4 nm). The contents of chlorophyll a, chlorophyll b and carotenoid were calculated based on the equation of Wellburn [22].

The starch content was analyzed with the starch assay kit (SA20, Sigma-Aldrich, St. Louis, MO). About 0.1–0.5 gram microalgal sample was ground into powder in liquid nitrogen, then transferred into a flask. After adding 25 mL deionized water, the solution was adjusted to pH of 5–7, and boiled for 3 minutes with gentle stirring. Afterwards, the solution was autoclaved at 135°C for 1 hour. When the temperature of the solution was reduced to about 60°C, deionized water was added to make a total volume of 100 mL. The extracted starch was hydrolyzed to glucose by amyloglucosidase. After subsequent phosphorylation by hexokinase and ATP, the glucose-6-phosphate was oxidised to 6-phosphogluconate by glucose-6-phosphate dehydrogenase in the presence of nicotinamide adenine (NAD). During this reaction, an equimolar amount of NAD was reduced to NADH, which resulted in the consequent increase in absorbance at 340 nm that was directly proportional to the glucose concentration. The starch concentration ($\mu g\cdot mL^{-1}$) was determined according to the formula described in the kit, and was converted into a percentage of dry weight of biomass.

Statistical Analysis

A two-tailed paired t-test was applied to ascertain significant differences using SPSS (Statistical Product and Service Solutions) version 10.0 (SPSS Inc., Shanghai, China) and the level of statistical significance was $P<0.05$.

Results and Discussion

Lethality, Mutation Frequency of the Microalgal Cells Irradiated by $^{12}C^{6+}$ Ion Beam, and Mutants Screening

The lethality of *Desmodesmus* sp. S1 after high-LET heavy $^{12}C^{6+}$ ion beam irradiations was shown in Fig. 1. The relationship between the lethality of *Desmodesmus* sp. S1 and the irradiation dose of $^{12}C^{6+}$ beam was fitted to a logistic curve equation ($y = -22.20+21.09\times\ln(x-6.91)$, $R^2 = 0.98$), which indicated the death rate of cells increased with increasing the radiation dosage from 10 to 120 Gy, with the highest lethality occurring between 90 to 120 Gy. The LD50 value was 37.57 Gy and the lethality of algal cells reached 80% at 120 Gy.

After heavy-ion beam treatment, the algae colonies that appeared on agar plates were considered putative mutants. A preliminary screening of the putative mutants by light microscopy showed that the morphological characteristics (e.g. colony appearance, cell shape and size etc.) of the putative mutants were indistinguishable from the wild type cells. Then photosynthetic characteristics (e.g., Fv/Fm, quantum efficiency of PSII) were used

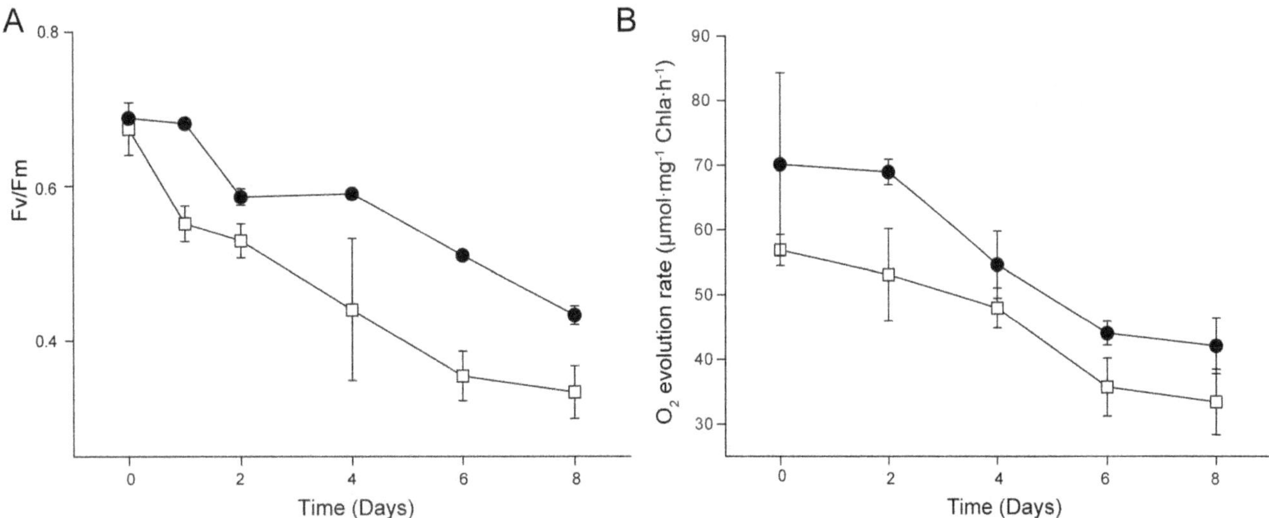

Figure 6. Photosynthesis efficiency of *Desmodesmus* sp. (A) Potential maximum quantum efficiency (Fv/Fm) and (B) oxygen evolution rate. WT and D90G-19 were grown under HL-N (300–400 μmol photons $m^{-2} \cdot s^{-1}$, nitrogen-limited BG-11 medium with 4.25 mM $NaNO_3$). D90G-19: closed circle; WT: open square.

as alternative parameters to further characterize the putative mutants by using a chlorophyll fluorescence technique [23]. Previously, the chlorophyll fluorescence has been used as a sensitive, quantitative parameter to analyze the photosynthetic characteristics of cultivars [24–26] as well as to characterize microalgal mutants [27,28].

Here, we applied a chlorophyll fluorometer Imaging-PAM to our mutant screening effort. The Imaging-PAM can measure several photosynthetic parameters (e.g., Fv/Fm) of 96 algal samples in a 96-well microplate within 10 minutes. Under favorable conditions (e.g., nutrient replete and low light), the Fv/Fm value of *Desmodesmus* sp. WT cells was 0.6–0.7. The colonies with significantly different Fv/Fm values from WT were identified as possible photosynthesis efficiency mutants (PEMs). The frequency of PEMs induced by 30, 60, and 90 Gy of $^{12}C^{6+}$ beam were 14.5%, 25.8% and 28.5%, respectively. The dose of 90 Gy represented a trade off between cell survival rate and mutation frequency. As a result, about 500 mutants were obtained by this treatment.

Previous studies with terrestrial plant materials such as seeds, leaves and other organs indicated that the mutation rate ranged from 8.4% to 17.8% [15,16,29] when irradiated with heavy-ion beam, which was higher than that induced by traditional mutagenesis, such as x-ray, γ-ray or EMS [15]. The mutation rate of up to 28.5% in *Desmodesmus* sp.S1 induced by heavy-ion beam was considerably higher than those of plant materials, which implied that heavy-ion beam can be an effective method for mutagenesis of microalgae.

All colonies were also subjected to screening for lipid-overproduction mutants (LOMs) using a modified Nile red method in conjunction with the determination of growth rate, as indicated by optical density of algal culture measured at 750 nm.

The mutants obtained through the first round screening with Imaging-PAM and Nile red fluorescence method would be subjected to the next rounds of screening. Among these mutants, a wide range of phenotypic distribution of PEMs was observed and approximately in line with the normal distribution (Fig. 2A). A similar normal distribution of phenotypes also occurred among the LOMs (Fig. 2B). In the following screening procedures, the lipid contents and photosynthetic efficiency of mutants were assayed when they were cultured in a column bioreactor. Lipid contents of many PEMs was significantly different with WT under stress conditions, thus were also identified as LOMs (Table S1), which indicated that heavy-ion irradiation induced mutations of unicellular microalgae had a wide spectrum and a high frequency as occurred to plants and mammalian cells [16,30,31]. A positive correlation ($R^2 = 0.906$) existed between the lipid content and high light-adapted PSII efficiency (Fv/Fm) under stress conditions

Table 2. Chlorophyll a, chlorophyll b and carotene contents in *Desmodesmus* sp.

Days	Chla (μg/mg)[a]		Chlb (μg/mg)		Carotenoids (μg/mg)		Chla/Chlb		(Chla+Chlb)/C[b]	
	WT	D90G-19	WT	D90G-19	WT	D90G-19	WT	D90G-19	WT	D90G-19
2	2.57±0.22	2.54±0.63	0.73±0.06	0.66±0.12	1.23±0.11	1.2±0.19	3.50	3.81	2.69	2.67
4	2.22±0.87	2.18±0.15	0.60±0.02	0.63±0.08	1.25±0.38	1.31±0.11	3.70	3.66	2.26	2.07
6	1.28±0.08	1.24±0.07	0.36±0.04	0.38±0.08	0.82±0.04	0.84±0.03	3.53	3.27	2.00	1.93
8	1.15±0.14	1±0.09	0.49±0.01	0.30±0.06	1.05±0.3	0.81±0.04	2.35	3.29	1.86	1.6

[a]Microalgae cells were grown under HL-N (300–400 μmol photons $m^{-2} \cdot s^{-1}$, nitrogen-depleted BG-11 medium with 4.25 mM $NaNO_3$). Mean ± SE with three replicates.
[b]Carotenoids.

Table 3. Fatty acid compositions of individual lipid classes in *Desmodesmus* sp.[d].

FA	Total Lipid		Neutral lipid		Glycolipid		Phospholipid	
	WT	D90G-19	WT	D90G-19	WT	D90G-19	WT	D90G-19
C16:0	17.08±0.65	15.03±0.43	17.03±0.44	15.15±0.60	16.65±1.20	13.55±1.05	16.94±2.92	15.42±1.48
C16:1	3.35±0.67	3.62±0.54	3.45±1.21	4.02±0.82	4.74±2.23	5.50±1.09	3.77±0.13	3.54±0.34
C16:2	2.47±0.37	3.08±0.27	2.29±0.70	3.05±0.40	3.34±1.01	4.20±0.33	7.02±2.99	9.42±1.21
C16:3	3.09±0.69	3.82±0.44	2.82±0.06	3.21±0.46	3.91±0.46	4.13±0.64	7.03±1.30	7.58±1.21
C16:4	1.47±0.65	1.53±0.28	1.11±0.21	1.14±0.22	2.26±0.88	1.96±0.42	6.45±3.53	6.03±1.20
C18:0	9.73±2.68	9.60±1.23	10.62±1.93	10.74±0.20	9.15±0.71	7.08±0.57	5.79±1.24	4.27±1.35
C18:1	44.81±1.42	45.05±1.21	46.23±2.04	46.86±0.66	39.32±2.67	40.88±1.29	17.42±2.52	17.06±0.92
C18:2	7.99±1.04	7.91±1.14	6.64±0.90	6.48±0.73	9.16±0.67	10.30±1.43	13.88±1.01	13.37±0.87
C18:3	9.38±1.78	9.58±0.98	7.74±0.21	8.28±0.67	11.11±1.39	11.93±1.50	22.95±4.14	24.67±3.95
C20:0	0.85±0.24	0.59±0.08	0.88±0.13	0.65±0.03	ND[c]	ND	ND	ND
C20:1	0.90±0.13	0.70±0.09	0.84±0.38	0.60±0.01	ND	ND	ND	ND
C16:C18	0.38	0.38	0.37	0.37	0.45	0.42	0.69	0.71
unsatd[a]	73.45	75.29	71.14	73.63	73.84	78.89	78.52	81.68
∇/mol[b]	1.13	1.18	1.05	1.10	1.23	1.31	1.79	1.87

[a]percentage of unsaturated fatty acids in total fatty acids.
[b]the degree of fatty acid unsaturation [41].
[c]not detected.
[d]Microalgae were harvested after grown under HL-N (300–400 μmol photons $m^{-2}{\cdot}s^{-1}$, nitrogen-depleted BG-11 medium with 4.25 mM $NaNO_3$) for 8 days. Mean ± SE with three replicates.

(Fig. 2C). Similar phenomena were also observed in naturally-occurring microalgae strains [32] and crop cultivars [33] because all carbohydrates including lipids in the photoautotroph were ultimately derived from photosynthesis. On the contrary, not every mutant with high Fv/Fm can produce high yields because the yields were also affected by other metabolic process such as respiration. We noted some mutants with higher Fv/Fm value than wild type that did not produce more biomass or lipids (data unpublished).

Current procedures for the screening of microalgae mutants that have a high yield of biomass and high lipid productivity suffer from two obstacles. Firstly, in unicellular microalgae morphological differences between wild type and most mutants are difficult to discern. Secondly, the quantification of the biomass and lipid contents of microalgae cultivated in conventional culture systems (e.g., flasks, carboys, bags, glass columns) is laborious, although some rapid detection methods such as the Nile red fluorescence method have become available. Therefore, we developed a new strategy to accelerate the screening of microalgae mutants with high yields of biomass and lipids based on the determination of the Fv/Fm value. All mutants were firstly subjected to the analysis of Fv/Fm value under different light intensities by Imaging-PAM. Afterwards, the biomass and lipid yields of those mutants with high Fv/Fm value were quantified in 24-well microplates. Only those mutants that have higher photosynthetic efficiency and lipid yields at least 10% greater than wild type were selected for further study, whereas the other mutants were discarded.

It would otherwise take at least 12 days to get these parameters analyzed for a mutant if it were cultured in a traditional culture system. Assuming that a maximum of 20 mutants could be tested every time, it will take at least 28 days to completely analyze 48 mutants. However, by applying the new strategy, analysis of 48 mutants can be achieved on six 24-well microplates within 8 days. Therefore, the efficiency of the new screening method was higher than that of traditional methods.

Biomass, Total Lipid and Starch Contents of Wild Type and Mutant D90G-19 under HL-N

After 3–4 rounds of the screening process mentioned above, a mutant named D90G-19 with higher photosynthetic efficiency and lipid contents than WT was obtained. In a batch culture mode, once inoculated in a BG-11 medium (4.25 mM Nitrate) and exposed to light illumination, *Desmodesmus* sp. S1 grew rapidly for 4–5 days after a temporary lag phase (0–24 hours), and reached a stationary phase thereafter (Fig. 3A). During the first 2 days, almost all nitrates in the medium were consumed by algal cells. When WT and D90G-19 were cultivated under high light and nitrogen limitation (HL-N) conditions, the specific growth rates of WT and mutant D90G-19 were 0.598 and 0.630 $g{\cdot}L^{-1}{\cdot}d^{-1}$ in 8 days, respectively, and the maximum biomass (dry weight) concentration of 4.95 and 5.23 $g{\cdot}L^{-1}$ were obtained in the WT and D90G-19 cultures (Fig. S1), but the differences were not statistically significant. Under low light conditions, the mutant D90G-19 also showed the same biomass profile as wild type (WT) (Fig. 3B).

While little difference in the lipid content was observed in WT and mutant D90G-19 grown under low light (LL+N), significantly higher amounts of total lipid were obtained in D90G-19 than in the WT when exposed to high light and under nitrogen limitation (HL-N) conditions (Fig. 4). As lipid production is a function of the lipid content and biomass concentration, the total lipid productivity of mutant D90G-19 was 0.298 $g{\cdot}L^{-1}{\cdot}d^{-1}$ during 8 days, which was 20.6% greater than that produced by WT (0.247 $g{\cdot}L^{-1}{\cdot}d^{-1}$).

In order to determine the stability of the phenotypic traits of D90G-19, it was cultivated in a 15 L panel photobioreactor under HL-N. It was shown that D90G-19 accumulated more lipids than

WT (Fig. S2B), although the biomass of D90G-19 was comparable with that of WT (Fig. S2A). During 14 days, the total lipid productivity of D90G-19 was 0.105 $g{\cdot}L^{-1}{\cdot}d^{-1}$, and 17.1% greater than that of the WT (0.089 $g{\cdot}L^{-1}{\cdot}d^{-1}$), showing that the mutant D90G-19 had the potential for large scale biofuel production.

Total lipid includes neutral lipids, glycolipids and phospholipids along with small amounts of sterol esters and pigments. Neutral lipids are non-membrane lipids mainly in the form of triacylglycerols (TAGs) stored in oil droplets in the cytosol of the cells [34]. In *Desmodesmus* sp. WT, the concentrations of neutral lipid, glycolipid, and phospholipid on a per total lipid basis were 78.1%, 14.9% and 6.9%, respectively, when cultivated under HL-N for 8 days. Under the same conditions, the neutral lipid, glycolipid and phospholipid contents in D90G-19 were 87.0%, 10.4% and 2.6% respectively (Table 1). The neutral lipid content of D90G-19 was significantly greater than that of WT ($P = 0.0093$). On the contrary, the glycolipid and phospholipid contents of D90G-19 were lower than that of WT ($P = 0.02$ and $P = 0.0052$ respectively). In addition, the glycolipids contents were greater than phospholipid in the total lipid of WT and mutant D90G-19 under HL-N, which indicated the glycolipids were more prevalent in cell membrane of the mutant D90G-19 ($P = 0.03$) and might help the mutant to acclimate the phosphate-limiting growth conditions by providing more alternatives for phospholipids [35].

Besides neutral lipid, starch is another major storage compound in many microalgae. A study showed that inhibition of starch synthesis resulted in a ten-fold increase of TAG in a starchless mutant of *C. reinhardtii* [8]. The total lipid content of the mutant D90G-19 was shown to be larger than that of the WT. In order to determine whether excess lipids accumulated in D90G-19 might be partially at the expense of starch, the starch contents of D90G-19 and WT were measured quantitatively. When grown under LL+N conditions for 4 days, WT and D90G-19 accumulated a basal amount of starch (about 5% of cells dry weight). After a shift to HL-N condition, WT and D90G-19 increased their starch contents by respectively 15.5% and 11.4% on day 2. Afterwards, the starch contents of WT and D90G-19 decreased to the same level (ca. 7% of dry weight) (Fig. 5) from the second day to the eighth day, but the lipid content of D90G-19 increased and was significantly higher than that of the WT at day 4, 6, and 8 (Fig. 4). Therefore, the additional neutral lipids in the D90G-19 should be synthesized via *de novo* fatty acid synthesis. Considering that the starch synthesis pathway in the D90G-19 was not blocked because the mutant D90G-19 can synthesize the same amounts of starch as the WT, it was possible that the conversion of starch to lipids may also occur in the cells [17].

Characterization of Photosynthesis Efficiency of *Desmodesmus* sp

Microalgae absorb light energy and convert it into chemical energy stored in the form of biochemical compounds such as carbohydrates, proteins and lipids. In order to elucidate the mechanism responsible for the higher lipid production by D90G-19 under stress conditions, the photosynthetic efficiency of the mutant was measured. The value of Fv/Fm reflected the potential maximum quantum efficiency of PSII. The Fv/Fm value was about 0.6–0.7 when WT and mutant D90G-19 were grown under optimal culture conditions (e.g. LL+N). Under stress (HL-N), the Fv/Fm value of WT decreased immediately whereas the decline of Fv/Fm in D90G-19 somewhat lagged behind, especially in the first day and from day 2 to day 4 (Fig. 6A). The results indicated that the mutant D90G-19 might tolerate higher light intensity and have the potential to use quanta more effectively at PSII reaction centers. Likewise, the O_2 evolution rate of D90G-19 was faster than that of WT (Fig. 6B), which was consisted with the trend of Fv/Fm. The contents of chlorophyll a, chlorophyll b and carotenoids in WT and mutant D90G-19 were tested (Table 2). There was no significant difference in chlorophyll and carotenoid contents between the WT and D90G-19. The increase in potential quantum efficiency of PSII in D90G-19 may be the result from the changes in peripheral antenna complexes associated with PSII. Alternatively, the electron transport chain between PSI and PSII may be affected.

It was reported that the partitioning of energy at PSII was disrupted by the decrease of starch synthesis in *Nicotina sylvestris* [36]. Blocking the competing starch synthesis pathway may facilitate carbon flux partitioning into lipid synthesis [8], but it may also lead to decrease of photosynthetic efficiency and growth impairment [37]. Our results suggested that the mutant exhibited higher photosynthetic efficiency under stress conditions, which resulted in more photosynthetically fixed carbon that could be redirected into neutral lipid biosynthesis.

Fatty Acid Profiles of the Different Classes of Glycerolipids in the *Desmodesmus* sp. WT and Mutant under HL-N

The fatty acid acyl profiles of total lipid, neutral lipid, glycolipid and phospholipid in the mutant D90G-19 were similar to those of the WT (Table 3). As for the unsaturation of fatty acid acyl in lipids, there was no significant difference between the WT and mutant D90G-19, although the lipid unsaturation of D90G-19 was slightly higher. Among the three major classes of glycerolipids, phospholipids contained the highest amounts of unsaturated acyl moieties contributing likely to the fluidity of cell membranes. However, the majority of membrane glycerolipids are glycolipid rather than phospholipid, MGDG and DGDG are the predominant species of chloroplast membrane under HL-N. There are two pathways for the assembly of lipid precursors in plant and microalgae: a *de novo* lipid assembly pathway in the plastid and a pathway located at the endoplasmic reticulum (ER) [38]. Glycerolipids synthesized by the ER pathway have a different molecular species composition (18-carbon fatty acids in the *sn*-2 position of the glycerol backbone) from those produced by the plastid pathway (16-carbon fatty acids in *sn*-2) [39]. The results (table 3) showed that 18-carbon fatty acids acyl chains were predominant in three classes of glycerolipids, especially in neutral lipids, implying that the ER was the major site for the synthesis of storage neutral lipid in *Desmodesmus*, which was similar to many other plants and algae [7,34,38,40].

Conclusions

This work represented the first attempt to use heavy carbon ions ($^{12}C^{6+}$) to induce mutagenesis of oleaginous microalgae to obtain mutants with enhanced lipid production potential. The mutation rate was about 20–30% when *Desmodesmus* cells were treated with 60–120 Gy heavy-ion beam, and a wide spectrum of phenotypes in lipid contents and photosynthetic efficiency was observed. Of numerous genuine mutants obtained, D90G-19 exhibited 20.6% higher lipid productivity than WT when they were cultivated in the column bioreactors, likely owing to an improved quantum efficiency of photosynthesis under stress conditions.

Supporting Information

Figure S1 Biomass concentration of *Desmodesmus* sp. WT (open squares) and D90G-19 (closed square) when cultivated in nitrogen-limited medium with 4.25 mM

$NaNO_3$ and high light illumination (300–400 μmol photons $m^{-2}\cdot s^{-1}$) in a column photobioreactor.

Figure S2 Biomass (A) and total lipid content (B) of *Desmodesmus* sp. WT (open square) and D90G-19 (closed square) when cultivated in nitrogen-limited medium with 4.25 mM $NaNO_3$ and high light illumination (300–400 μmol photons $m^{-2}\cdot s^{-1}$) in a 15 L panel photobioreactor. The total lipid contents of mutant D90G-19 was significantly higher than WT at day 10 (P = 0.026) and day 12 (P = 0.036).

Table S1 Fv/Fm value and lipid contents of WT and 20 mutants. Mean ± SE with three replicates.

Acknowledgments

We thank Dr. Dongyuan Zhang for technical assistance and discussion.

Author Contributions

Conceived and designed the experiments: GH FL. Performed the experiments: YF LZ CY. Analyzed the data: GH. Contributed reagents/materials/analysis tools: JW WL. Wrote the paper: GH FL QH.

References

1. Hu Q, Sommerfeld M, Jarvis E, Ghirardi M, Posewitz M, et al. (2008) Microalgal triacylglycerols as feedstocks for biofuel production: perspectives and advances. Plant J 54: 621–639.
2. Chisti Y (2007) Biodiesel from microalgae. Biotechnol Adv 25: 294–306.
3. Wijffels RH, Barbosa MJ (2010) An Outlook on Microalgal Biofuels. Science 329: 796–799.
4. Stephens E, Ross IL, King Z, Mussgnug JH, Kruse O, et al. (2010) An economic and technical evaluation of microalgal biofuels. Nat Biotech 28: 126–128.
5. Kilian O, Benemann CSE, Niyogi KK, Vick B (2011) High-efficiency homologous recombination in the oil-producing alga Nannochloropsis sp. Proc Natl Acad Sci USA.
6. Eichler-Stahlberg A, Weisheit W, Ruecker O, Heitzer M (2009) Strategies to facilitate transgene expression in Chlamydomonas reinhardtii. Planta 229: 873–883.
7. Radakovits R, Jinkerson RE, Darzins A, Posewitz MC (2010) Genetic engineering of algae for enhanced biofuel production. Eukaryot Cell 9: 486–501.
8. Li Y, Han D, Hu G, Dauvillee D, Sommerfeld M, et al. (2010) Chlamydomonas starchless mutant defective in ADP-glucose pyrophosphorylase hyper-accumulates triacylglycerol. Metab Eng 12: 387–391.
9. Li Y, Zhang X, Hu Q, Sommerfeld M (2011) A type-2 acyl-coa:diacylglycerol acyltransferase gene is essential for endoplasmic reticulum-based triacylglycerol synthesis in *Chlamydomonas reinhardtii*. J Phycol 47: S59-S59.
10. Zaslavskaia LA, Lippmeier JC, Kroth PG, Grossman AR, Apt KE (2000) Transformation of the diatom Phaeodactylum tricornutum (Bacillariophyceae) with a variety of selectable marker and reporter genes. J Phycol 36: 379–386.
11. Dunahay TG, Jarvis EE, Dais SS, Roessler PG (1996) Manipulation of microalgal lipid production using genetic engineering. Biotechnol Appl Biochem 57–8: 223–231.
12. Cernac A, Benning C (2004) WRINKLED1 encodes an AP2/EREB domain protein involved in the control of storage compound biosynthesis in Arabidopsis. Plant J 40: 575–585.
13. Ohto M-a, Fischer RL, Goldberg RB, Nakamura K, Harada JJ (2005) Control of seed mass by APETALA2. Proc Natl Acad Sci USA 102: 3123–3128.
14. Butelli E, Titta L, Giorgio M, Mock H-P, Matros A, et al. (2008) Enrichment of tomato fruit with health-promoting anthocyanins by expression of select transcription factors. Nat Biotech 26: 1301–1308.
15. Mei M, Deng H, Lu Y, Zhuang C, Liu Z, et al. (1994) Mutagenic effects of heavy ion radiation in plants. Adv Space Res 14: 363–372.
16. Abe T, Matsuyama T, Sekido S, Yamaguchi I, Yoshida S, et al. (2002) Chlorophyll-deficient Mutants of Rice Demonstrated the Deletion of a DNA Fragment by Heavy-ion Irradiation. J Radiat Res 43: S157-S161.
17. Li Y, Han D, Sommerfeld M, Hu Q (2011) Photosynthetic carbon partitioning and lipid production in the oleaginous microalga Pseudochlorococcum sp. (Chlorophyceae) under nitrogen-limited conditions. Bioresource Technology 102: 123–129.
18. Hu Q, Han D, Summerfeld M (2010) Novel Pseudochlorococcum species and uses therefor. U. S. Patent US 2010/0267085 A1, filed June. 29, 2007, and issued Oct. 21, 2010.
19. Collos Y, Mornet F, Sciandra A, Waser N, Larson A, et al. (1999) An optical method for the rapid measurement of micromolar concentrations of nitrate in marine phytoplankton cultures. J Appl Phycol 11: 179–184.
20. Chen W, Zhang C, Song L, Sommerfeld M, Hu Q (2009) A high throughput Nile red method for quantitative measurement of neutral lipids in microalgae. Journal of Microbiological Methods 77: 41–47.
21. Bigogno C, Khozin-Goldberg I, Boussiba S, Vonshak A, Cohen Z (2002) Lipid and fatty acid composition of the green oleaginous alga Parietochloris incisa, the richest plant source of arachidonic acid. Phytochemistry 60: 497–503.
22. Wellburn AR (1994) The Spectral Determination of Chlorophyll-a and Chlorophhyll-B, as Well as Total Carotenoids, Using Various Solvents with Spectrophotometers of Different Resolution. J Plant Physiol 144: 307–313.
23. Maxwell K, Johnson GN (2000) Chlorophyll fluorescence–a practical guide. J Exp Bot 51: 659–668.
24. O'Neill PM, Shanahan JF, Schepers JS (2006) Use of Chlorophyll Fluorescence Assessments to Differentiate Corn Hybrid Response To Variable Water Conditions Crop Sci 46: 681–687.
25. Wang QA, Lu CM, Zhang QD (2005) Midday photoinhibition of two newly developed super-rice hybrids. Photosynthetica 43: 277–281.
26. Tang YL, Wen XG, Lu CM (2005) Differential changes in degradation of chlorophyll-protein complexes of photosystem I and photosystem II during flag leaf senescence of rice. Plant Physiology and Biochemistry 43: 193–201.
27. Förster B, Osmond CB, Boynton JE, Gillham NW (1999) Mutants of Chlamydomonas reinhardtii resistant to very high light. Journal of Photochemistry and Photobiology B: Biology 48: 127–135.
28. Ossenbühl F, Göhre V, Meurer J, Krieger-Liszkay A, Rochaix J-D, et al. (2004) Efficient Assembly of Photosystem II in Chlamydomonas reinhardtii Requires Alb3.1p, a Homolog of Arabidopsis ALBINO3. The Plant Cell Online 16: 1790–1800.
29. Miyazaki K, Suzuki K-i, Iwaki K, Kusumi T, Abe T, et al. (2006) Flower pigment mutations induced by heavy ion beam irradiation in an interspecific hybrid of Torenia. Plant Biotechnol 23: 163–167.
30. Kagawa Y, Yatagai F, Suzuki M, Kase Y, Kobayashi A, et al. (1995) Analysis of mutations in the human HPRT gene induced by accelerated heavy-ion irradiation. J Radiat Res 36: 185–195.
31. Morimoto S, Honma M, Yatagai F (2002) Sensitive Detection of LOH Events in a Human Cell Line after C-ion Beam Exposure. J Radiat Res 43: S163-S167.
32. Pan Y-Y, Wang S-T, Chuang L-T, Chang Y-W, Chen C-NN (2011) Isolation of thermo-tolerant and high lipid content green microalgae: Oil accumulation is predominantly controlled by photosystem efficiency during stress treatments in Desmodesmus. Bioresour Technol 102: 10510–10517.
33. Murchie EH, Chen YZ, Hubbart S, Peng SB, Horton P (1999) Interactions between senescence and leaf orientation determine in situ patterns of photosynthesis and photoinhibition in field-grown rice. Plant Physiology 119: 553–563.
34. Ohlrogge J, Browse J (1995) Lipid biosynthesis. Plant Cell 7: 957.
35. Benning C, Beatty JT, Prince RC, Somerville CR (1993) The sulfolipid sulfoquinovosyldiacylglycerol is not required for photosynthetic electron-transport in *Rhodobacter sphaeroids* but enhances growth under phosphate limitation. Proc Natl Acad Sci USA 90: 1561–1565.
36. Peterson RB, Hanson KR (1991) Changes in photochemical and fluorescence yields in leaf tissure from normal and starchless Nicotiana sylvestris with increasing irradiance. Plant Sci 76: 143–151.
37. Li Y, Han D, Hu G, Sommerfeld M, Hu Q (2010) Inhibition of starch synthesis results in overproduction of lipids in Chlamydomonas reinhardtii. Biotechnol Bioeng 107: 258–268.
38. Ohlrogge JB, Jaworski JG (1997) Regulation of fatty acid synthesis. Annual Review of Plant Physiology and Plant Molecular Biology 48: 109–136.
39. Heinz E, Roughan PG (1983) Similarities and Differences in Lipid-Metabolism of Chloroplasts Isolated from 18–3 and 16–3 Plants. Plant Physiology 72: 273–279.
40. Benning C (2009) Mechanisms of Lipid Transport Involved in Organelle Biogenesis in Plant Cells. Annual Review of Cell and Developmental Biology 25: 71–91.
41. Chen GQ, Jiang Y, Chen F (2008) Salt-Induced Alterations in Lipid Composition of Diatom Nitzschia Laevis (Bacillariophyceae) under Heterotrophic Culture Condition. J Phycol 44: 1309–1314.

7

Comparison of Photoacclimation in Twelve Freshwater Photoautotrophs (Chlorophyte, Bacillaryophyte, Cryptophyte and Cyanophyte) Isolated from a Natural Community

Charles P. Deblois, Axelle Marchand, Philippe Juneau*

Department of Biological Sciences, TOXEN, Ecotoxicology of Aquatic Microorganisms Laboratory, Université du Québec à Montréal, Montréal, Québec, Canada

Abstract

Different representative of algae and cyanobacteria were isolated from a freshwater habitat and cultivated in laboratory to compare their photoacclimation capacity when exposed to a wide range of light intensity and to understand if this factor may modify natural community dominance. All species successfully acclimated to all light intensities and the response of phytoplankton to increased light intensity was similar and included a decrease of most photosynthetic pigments accompanied by an increase in photoprotective pigment content relative to Chl *a*. Most species also decreased their light absorption efficiency on a biovolume basis. This decrease not only resulted in a lower fraction of energy absorbed by the cell, but also to a lower transfer of energy to PSII and PSI. Furthermore, energy funnelled to PSII or PSI was also rearranged in favour of PSII. High light acclimated organisms also corresponded to high non-photochemical quenching and photosynthetic electron transport reduction state and to a low Φ'_M. Thus photoacclimation processes work toward reducing the excitation pressure in high light environment through a reduction of light absorption efficiency, but also by lowering conversion efficiency. Interestingly, all species of our study followed that tendency despite being of different functional groups (colonial, flagellated, different sizes) and of different phylogeny demonstrating the great plasticity and adaptation ability of freshwater phytoplankton to their light environment. These adjustments may explain the decoupling between growth rate and photosynthesis observed above photosynthesis light saturation point for all species. Even if some species did reach higher growth rate in our conditions and thus, should dominate in natural environment with respect to light intensity, we cannot exclude that other environmental factors also influence the population dynamic and make the outcome harder to predict.

Editor: Brett Neilan, University of New South Wales, Australia

Funding: NSCERC Discovery Grant #145590074 (http://www.nserc-crsng.gc.ca/index_eng.asp), FQRNT-Action concertée #2009-CY-130520 (http://www.fqrnt.gouv.qc.ca/). The funders had no role in study design, data collection and analysis, decision to publish, or preparation of the manuscript.

Competing Interests: The authors have declared that no competing interests exist.

* E-mail: juneau.philippe@uqam.ca

Introduction

In aquatic environment, success of microalgae and cyanobacteria depends on their individual capacity to convert light into biochemical energy through photosynthetic light reactions and to transform carbon and nutrients into biomass. Because of the physical properties of water and the presence of suspended particles, available light intensity for photosynthesis is highly variable in freshwater habitat [1,2]. During sunny days, light intensity at the surface of waterbodies can be high enough to induce photoinhibition and cellular damage to exposed photosynthetic organisms [3,4,5]. Simultaneously, only few meters below the surface (sometime less), light intensity become limiting for photosynthesis and may represent less than 1% of surface irradiance [2,6]. In order to cope with such variability, photosynthetic organisms have developed an array of phenotypic adjustments including photoacclimation processes [2,7,8,9]. Photoacclimation to low or high light environments involves mid to long term adjustments of the photosynthetic apparatus and includes down regulation and *de novo* synthesis of cell constituents such has photosynthetic and non-photosynthetic pigments, photosystems I and II (PSI and PSII), RUBISCO, as well as changes in cell ultrastructure [8,10,11,12,13]. When exposed to high light environments, photoacclimation responses of most algae and cyanobacteria include a decrease of photosynthetic pigments (chlorophylls and phycobiliproteins) combined with an increase in photoprotective carotenoids [14,15,16]. The decrease in chlorophyll *a* (Chl *a*) is normally associated to a decrease in the number of photosystems, while a decrease in accessory pigments (Chl *b*, *c*, *d*, and phycobiliproteins) is associated to a decrease in the size of the light harvesting complexes (LHC) [9]. Combined to an increase in carotenoid (Car) content relative to Chl *a*, these adjustments allow a reduction of the excitation pressure on the photosynthetic apparatus and protect the organism against light induced reactive oxygen species damages [17,18,19]. On the other hand, under light limiting conditions, algae and cyanobacteria adjust their cellular constituents to increase light absorption efficiency [7,9,20].

Table 1. Group, species names and relevant morphological and physiological characteristics including pigments for each of the 12 species selected for this study.

Groups	Species names	Codes	Characteristics	Major pigments
Chlorophyte	*Ankistrodesmus falcatus*	CHL1	Colonial (2–6 cells) non-motile elongate cells,	Chl *a, b, c*
	Pandorina morum	CHL2	Colonial (16 cells), flagelate, big size	Chl *a, b, c*
	Oocystis lacustris	CHL3	Single non-motile ovoid cell or small colony (2–3 cells).	Chl *a, b, c*
	Pediastrum boryanum	CHL4	Colonial, planar, non-motil	Chl *a, b, c*
	Chlamydomonas snowii	CHL5	Unicellular, flagelates	Chl *a, b, c*
Bacillariophyte	*Aulacoseira granulata* var. *angustissima*	BAC2	Elongate curvated filament of 2–3 cells, non-motil, very low pigmentation	Chl *a, b*
	Fragilaria crotonensis	BAC3	Colonial, non-motil	Chl *a, b*
Cryptophyte	*Cryptomonas obovata*	CRY1	Unicellular, flagelates	PE/Chl *a, d*
Cyanophyte	*Phormidium mucicola*	CYA1	Small 0.8 μm non-buoyant rod-like colony (up to 4 cells), toxic.	PC/APC/Chl *a*
	Microcystis flos-aquae	CYA2	Colony (non-mucilaginous), small spherical cells 2 μm diameter, buoyant, toxic.	PC/APC/Chl *a*
	Aphanizomenon flos-aquae	CYA3	Association of numerous buoyant filamentous colony, toxic.	PC/APC/Chl *a*
	Anabaena spiroïdes	CYA4	Filamentous colony, buoyant, toxic.	PC/APC/Chl *a*

The code is the abbreviation associated to the group and used in the figures.

The wide diversity observed in phytoplankton is impressive, and may influence their light utilization efficiency. Interspecific variation of size between individual cells can reach many orders of magnitude from sub-micrometric picoplankton up to micro-planktonic species [21,22]. While many species remain single cells (e.g. *Chlamydomonas* sp., *Cryptomonas* sp., *Navicula* sp.), others grow into structured colonies (e.g. *Pediastrum* sp., *Pandorina* sp., *Volvox* sp., *Merismopedia* sp.), filaments (e.g. *Aphanizomenon* sp., *Anabaena* sp.) or more or less defined clusters of cells (e.g. *Microcystis* sp., *Sphaerocystis* sp., *Ankistrodesmus* sp.) [23,24]. Colonial organization provides some benefits such as protection against grazing, but this characteristic also comes with inconvenience such as increased density inducing stronger sinking rate and increased self-shading leading to lower light availability [21,25,26]. Some of these species may also have flagella or vacuoles permitting them to move in the water column in order to optimize light harvesting [2]. This broad diversity influencing their light utilization is not limited to the morphological properties of phytoplankton, but can also be seen at the photosynthetic, biochemical or physiological levels such as distinct pigmentation [27]. Chlorophyll *a* has a crucial role in the photosystem reaction center (RC) core and in the light harvesting complexes of oxygenic phytoplankton, thus, this pigment is common to all species. Nevertheless, light harvesting capacity also differs between species because of variability in composition of pigments such as chlorophyll *b*, *c*, *d*, carotenoïds and phycobiliproteins [20,28]. Thus the great diversity of phytoplankton characteristics influencing photosynthesis and light harvesting may influence the efficiency of photoacclimation processes.

Many lakes from the eastern Townships in Québec (Canada) are impacted on a periodic basis by cyanobacterial bloom apparitions [29]. This phenomenon is a visible consequence of changes in algal community equilibrium, but the factors influencing this dynamic are not fully understood. Since light is a factor that can modify algal community [1,30,31], comparing photoacclimation responses of various species may help to estimate if this factor contribute to the periodic community imbalance observed in these aquatic ecosystems. In this study, we isolated 12 species of phytoplankton belonging to different algal groups from a single algal assemblage of a temperate dimictic eutrophic lake. Species were selected for their different sizes and strategies to harvest light (pigments, movement) in order to compare their photoacclimation responses and to determine their active light range and photoacclimation capacity. We showed that the general photoacclimation processes among the studied species were similar, but the extent of the responses varies providing possible selective advantages to some species.

Materials and Methods

Sampling and cell culture

In mid-July of 2008, water from the euphotic zone of the Réservoir Choinière, (Eastern Townships, Québec, Canada) was collected and inoculated into bold basal medium (BBM) enriched with carbonate (25 mg L^{-1}) and silicate (80 mg L^{-1}) (BBMsi). Species that successfully grew in that medium were isolated and cultivated in laboratory. From the species initially isolated, 12 species were selected in order to have a diversity of algal groups (Chlorophyte, Bacillariophyte, Cryptophyte and Cyanophyte) and traits: colonial, unicellular, flagellates, buoyant and different cell sizes (see Table 1 for details). Throughout the experiment all species were grown in 125 ml of fresh BBMsi in 250 ml flasks and periodically (frequency depending on growth rate) transferred into fresh medium to maintain the cells in exponential growth phase which provides reproducible physiological characteristics. Periodic inoculum transfers also minimized dead cell accumulation which in any cases remained negligible as confirmed by monitoring, on a daily basis, the stability of the maximum PSII quantum yield (F_V/F_M) and by microscopic observations.

Each species was acclimated for several weeks in an environmental growth chamber (MTR30, Conviron, Manitoba, Canada) with a light:dark cycle of 16: 8 at 21°C to seven light intensities: 14, 43, 76, 191, 341, 583 and 1079 μmol photons (PAR) m^{-2} s^{-1} (measured with a US-SQS/L Micro quantum sensor, Heinz Walz GmbH, Effeltrich, Germany, in the center of the culture flask containing 125 mL BBMsi). Both fluorescent (cool white fluorescent tubes Philips F72T8/TL841/HO) and incandescent bulbs (Philips 60 W) were used and in our conditions (one growth chamber containing all cultures simultaneously) light quality was similar for all light intensities (confirmed by spectroradiometric measurements; HR2000 UV+Vis, Ocean Optics Inc, USA). All measurements were done on independent triplicates of fresh and healthy cultures. For each trial, a new culture was prepared from the previous one and was allowed to grow until maximum PSII quantum yield and similar cell conditions (F_0, F_M and F_V/F_M) were attained. Moreover, using fixed signal gain and dark acclimation period during chlorophyll fluorescence measurements allowed to use the F_0 as a proxy of cell concentration from day to day [32] and this method was used to estimate the growth rate for each trial.

Chlorophyll fluorescence measurement

Induction curves (IC) were measured using WATER-Pulse-Amplitude-Modulated fluorometer (WATER-PAM) (Heinz Walz GmbH, Effeltrich, Germany) on 15 min dark adapted algal suspensions (3 mL) using standard IC protocol with the actinic light carefully matched to growth photon flux density (PFD). The PSII maximum and operational quantum yields (F_V/F_M and Φ'_M), the non-photochemical quenching (NPQ) and the unquenched fluorescence level (UQF_{rel}) were calculated from each IC [33,34]. For cyanobacteria, F_M was estimated at the end of each IC using 50 µM Diuron (DCMU) in presence of actinic light [35].

Pigments determination

At each sampling, known volumes of culture were filtered under dime green light on GF/F filters (Whatman, USA) and kept frozen at −80°C until pigment and phycobiliprotein extraction and determination. Chlorophylls (Chls) and carotenoids (Car) were extracted 5 min in 4 mL of boiling methanol, rigorously vortexed 1 min and kept at −80°C for overnight extraction. Prior to measurement, the extract was filtered on GF/F and the optical density was read between 350 and 800 nm with a Cary 300 WinUV spectrophotometer (Varian, USA). The average OD from 750 to 800 nm was used to correct for sample turbidity. The concentrations of Chl *a*, *b*, *c* and *d* were estimated according to [36], while carotenoid concentration was estimated following [37]. Phycobilioproteins: phycocyanin (PC), allophycocyanin (APC), and phycoerythrine (PE), were extracted using 4 freeze-thaw cycles in 0.1 M potassium phosphate buffer (pH 6.8), sonicated on ice between the second and third cycle (2 watts for 1 min., sonic dismembrator model 100-Fisher Scientific, USA), and finally centrifuged at 5000× *g* for 15 min. The absorbance spectra of the supernatant was recorded between 500 and 700 nm using a Cary 300 WinUV spectrophotometer (Varian, USA) and pigment concentrations were calculated according to the equation given in [38].

Cell division rate

The cell concentration and size were measured using a Multisizer III Coulter counter (Beckman Coulter Inc, Fullerton, USA) when cell morphology allowed it, while for the other species (colony, filament or non-spherical cells) a sample was fixed with Lugol solution for measurement and counting under a microscope. Species specific growth rate (μ_d) was calculated from these data and fitted to growing light intensity (PFD) with a classic 4 parameter PE curve model with photoinhibition [39]:

$$\mu_d = \mu_M \cdot (1 - e^{\frac{-\alpha \cdot PFD}{\mu_M}}) \cdot e^{\frac{-\beta \cdot PFD}{\mu_M}} \quad (1)$$

where α represents the initial slope of the curve, β is a light dependent inhibition constant and μ_M is the theoretical maximum growth rate. From these coefficients, the secondary parameters: achieved maximal growth rate (μ_{MAX}) and its corresponding light intensity ($E_M{}^{\mu d}$) were calculated [40].

Biooptical measurements

All measurements were carried under dime green light. Sample were concentrated (5x) by gentle filtration on polycarbonate membrane filter 0.8 µm pore size (Millipore, USA) and resuspended in 3 mL of BBMsi directly in the measuring quartz cuvette. Cell lost on the filter was negligible. The *in vivo* light absorption spectrum of the concentrated solution (O.D. m^{-1}) was measured (between 350 and 800 nm) with a Cary winUV spectrophotometer (Varian, USA) using the integrating sphere attachment. Immediately after this measurement, 10 µL of DCMU was added to the sample at a final concentration of 50 µM and the sample was maintained 1 minute under white light (500 µmol photons m^{-2} s^{-1}) to eliminate any variable fluorescence. Following this treatment, the sample was immediately transferred in a CaryEclipse spectrofluorometer (Varian, USA) and the *in vivo* fluorescence excitation spectrum (400–700 nm) of the cell suspension was monitored at 730 nm. To avoid light scattering from the apparatus and cell sample, a long pass glass filter (RG695, Schott, AG, Mainz, Germany) was placed in front of the emission beam [20].

Chl a specific absorption coefficient

The *in vivo* light absorption spectrum (O.D. m^{-1}) was corrected for sample turbidity (average OD_{750_800}) and converted to Chl *a* specific absorption coefficient $a^*_\varphi(\lambda)$ (m^2 mg Chl a^{-1}) according to eq. 2:

$$a^*_\varphi(\lambda) = \frac{2.3 \cdot (O.D.)}{(d \cdot Chl)} \quad (2)$$

where 2.3 is the Log to Ln conversion factor, *d* is the cuvette path length (m^{-1}) and *Chl* is the corresponding Chl *a* concentration (mg m^{-3}). The fluorescence excitation spectra was quantum corrected using the dye Basic Blue 3 (4.1 g L^{-1}) which corrects for instrument specific wavelength variation of the excitation beam intensity [41]. The corrected spectra was then scaled to a^*_φ (λ) using the no-overshoot procedure to obtained the PSII Chl *a* specific absorption coefficient: a^*_{PSII} (λ) [20]. We averaged a^*_φ (λ) and a^*_{PSII} (λ) between 400 and 700 nm or in the red band, between 670 and 680 nm, to obtain the light absorption coefficients specific to the whole cell (a^*_φ), to PSII and associated LHCII (a^*_{PSII}), to Chl *a* in whole cell (a^*_φ (red)) or to Chl *a* associated to PSII (a^*_{PSII} (λ)). High wavelength absorption efficiency (high $a^*_\varphi(\lambda)$ or $a^*_{PSII}(\lambda)$) is only significant when photons of that energy are available, hence, both coefficients were spectrally weighted according to the light spectrum *E*(λ) of the growth chamber. This correction was done by normalizing the *E*(λ) area to unity and by multiplying this dimensionless spectrum with $a^*_\varphi(\lambda)$ or $a^*_{PSII}(\lambda)$ which yield the spectrally weighted ($\bar{a}^*_\varphi(\lambda)$

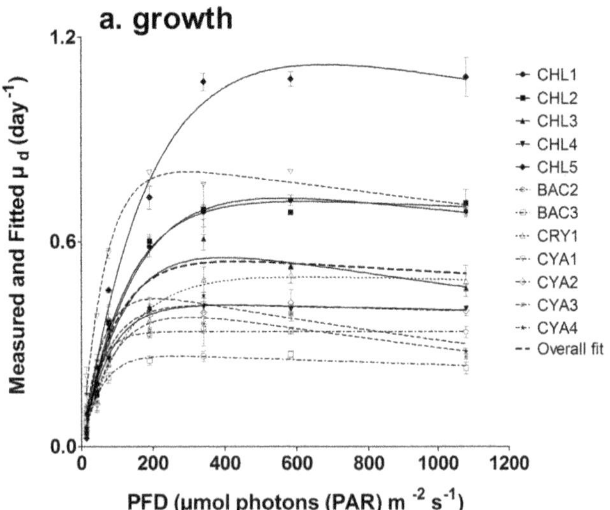

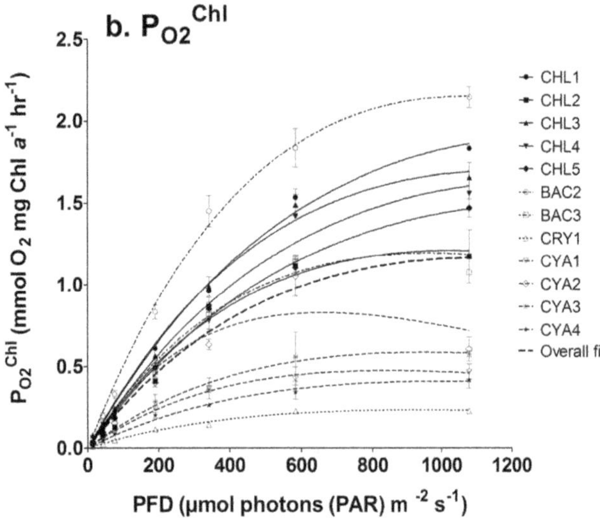

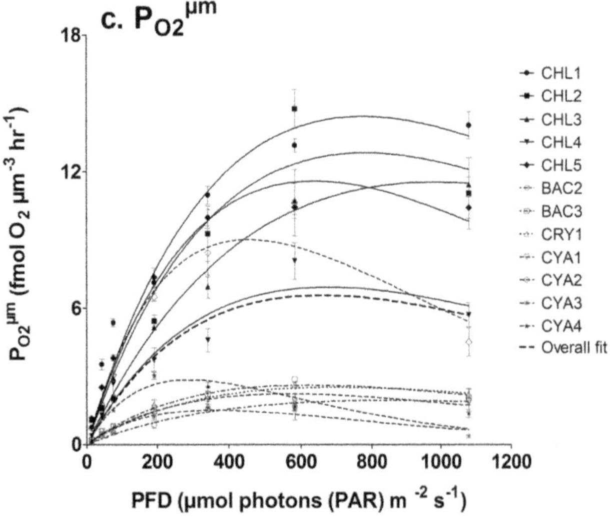

Figure 1. Responses of phytoplankton to light intensities; a. Cell division rate (μ_d). b. Oxygen production normalized to Chl *a* ($P_{O2}{}^{Chl}$) or c. to biovolume ($P_{O2}{}^{\mu m}$) obtained at each growing light intensities of photoacclimated phytoplankton (see Table 1 for the species list). The corresponding fits for growth or photosynthesis versus irradiance curve (PE curve) were obtained using eq. 1 (for μ_d) or eq. 4 (for P_{O2}). Overall fit represents the result obtained for the whole data set.

and $\bar{a}^*_{PSII}(\lambda)$) coefficient to be utilized in oxygen production estimates (see below).

Oxygen production estimate

The best method to estimate oxygen production rate per chlorophyll unit ($P_{O2}{}^{Chl}$) using biooptical approach and chlorophyll fluorescence measurement was showed to be the method relying on $\bar{a}^*_{PSII}$ to calculate light available to PSII photochemistry [42]. Therefore, we calculated the $P_{O2}{}^{Chl}$ data accordingly:

$$P_{O2}Chl = \Phi'_M \cdot PFD \cdot \Gamma \cdot \bar{a}^*_{PSII} \cdot 3.6 \quad (3)$$

where $\bar{a}^*_{PSII}$ represent light absorption specific to PSII (m^2 mg chl a^{-1}) spectrally weighted to available light intensity and spectrum over the PAR range, PFD is the light intensity in the growth chamber (μmol photons (PAR) m^{-2} s^{-1}), Φ'_M is the PSII operational quantum yield, Γ is the minimum theoretical quantum requirement of PSII in order to evolve one O_2 molecule (0.25 O_2 per electron) [43], and 3.6 convert second to hour (3600) and μmol to mmol giving the final dimension for $P_{O2}{}^{Chl}$ of mmol O_2 mg chl a^{-1} hr^{-1}. For some analysis, $P_{O2}{}^{Chl}$ was converted to oxygen production rate per biovolume unit ($P_{O2}{}^{\mu m}$) express in fmol O_2 μm^{-3} hr^{-1}. Each rate obtained for individual species at their specific growth light intensity was plotted against acclimation PFD (PE curve) and fitted to the waiting in-line function using eq. 4 [44]:

$$P_{O2} = A \cdot PFD \cdot Kw \cdot e^{-Kw \cdot PFD} \quad (4)$$

where, P_{O2} is oxygen production rate normalized to Chl *a* ($P_{O2}{}^{Chl}$) or biovolume ($P_{O2}{}^{\mu m}$), *A* and *Kw* are scaling factors for the height of the curve and X-axis respectively and PFD was the light intensity (μmol photons (PAR) m^{-2} s^{-1}) in the growth chamber. From this function, we estimated the saturation ($P_{SAT}{}^{Chl}/P_{SAT}{}^{\mu m}$) and maximum ($P_M{}^{Chl}/P_M{}^{\mu m}$) rates of oxygen production and their corresponding light intensity ($E_K{}^{Chl}/E_K{}^{\mu m}$ and $E_M{}^{Chl}/E_M{}^{\mu m}$) following the equation presented in [44].

Statistical analysis

All analysis were made in JMP 6.0 (SAS institute, USA) or GraphPad Prism software version 5.00 for Windows (GraphPad Software, San Diego California USA). The confidence interval (CI) at 95% was calculated for each coefficient in eq. 1 and eq. 4 using matrice inversion [44]. These coefficients value ± CI were compared by ANOVA and post Hoc Tukey Kramer mean comparison tests. Comparison between light limited and light saturated conditions was done for each species independently with student t-test, while two-way ANOVA was used to compare light limitation and light saturation responses between phylogenic groups [45]. Achieved maximal growth rate (μ_{MAX}) obtained for individual species was compared to the fit obtained for all species grouped (All species) using Dunnett's test ($p<0.05$) [46]. Subsequent comparison was done with Tukey Kramer test ($p<0.05$) to rank species in each subgroup (higher, equal or lower than All species μ_{MAX}).

Results

Cell division rate and primary production

Our results showed that all species successfully acclimated to all growth light conditions and that their specific cell division rates (μ_d day^{-1}) varied between 0.024 and 1.12 (Fig. 1a). Specific maximal growth rates (μ_{MAX}) of *Ankistrodesmus falcatus, Pandorina morum, Chlamydomonas snowii* and *Phormidium mucicola* were significantly higher compared to the overall averaged μ_{MAX} of 0.54($\pm$0.05) using Dunnett's mean comparison (Table 2). In this group of fast growing species, mean comparison (Tukey HSD) showed that *C. snowii* reached the highest μ_{MAX}, while there was no significant difference between *P. mucicola, A. falcatus* and *P. morum.* Above growth light saturation (between 200 and 400 µmol photons PAR m^{-2} s^{-1}: data not shown), these fast growing species attained μ_d values that allowed population to double in a day or less (μ_d>0.693). Such level was not reached for any light conditions in other species of the present study (Fig. 1a). It is worth to notice that although *C. snowii* has the highest μ_d values under saturating irradiance, this species also has the lowest μ_d (0.024$\pm$0.002) when grown at 14 µmol photons (PAR) m^{-2} s^{-1}, indicating that this light intensity was very close to its compensation point. Among the remaining species, *Aulacoseira granulata* var. *angustissima, Fragilaria crotonensis* and *Aphanizomenon flos-aquae* had significantly lower μ_{MAX} compared to the overall value and formed the group of slow growing species (Table 2). In this group, *F. crotonensis* had the lowest μ_{MAX} and therefore the slowest growing species of this study.

Light intensity required to reach μ_{MAX} ($E_M{}^{\mu d}$) varied between 187($\pm$6) and 605($\pm$39) µmol photons m^{-2} s^{-1} for all species with an average of 401($\pm$29) µmol photons m^{-2} s^{-1} (Table 2). The species showing the lowest $E_M{}^{\mu d}$ were cyanophytes (except *Microcystis flos-aquae*) and the bacillariophyte *F. crotonensis*. Although they reached μ_{MAX} at similar light intensity, *F. crotonensis* was able to achieve constant growth rate above this point, while up to 35% growth inhibition was observed for the cyanophytes (except *M. flos-aquae*) (Fig. 1a). Growth inhibition was also observed for the cholorophyte *Oocystis lacustris* despite its high $E_M{}^{\mu d}$. From all species, the cyanophyte *P. mucicola* was the best low light adapted organism since it achieved the highest μ_d at PFD below 191 µmol photons (PAR) m^{-2} s^{-1} (Fig. 1a). On the other hand, the best high light adapted organism was *C. snowii* because of its high μ_{MAX} and μ_d at light above 191 µmol photons (PAR) m^{-2} s^{-1}.

Oxygen production ($P_{O2}{}^{Chl}$) estimates for all species varied between 0.008 and 2.239 mmol O_2 mg Chl *a*$^{-1}$ hr^{-1} and the resulting PE curves depicted typical increases of photosynthetic activity in function of acclimation light intensity (Fig. 1b). When comparing the maximal oxygen production rate ($P_M{}^{Chl}$), our data showed significant differences between algal groups (Table 3). It was higher for the bacillariophytes (1.67$\pm$SD of 0.54 mmol O_2 mg chl *a*$^{-1}$ hr^{-1}) and chlorophytes (1.59$\pm$SD of 0.27 mmol O_2 mg chl *a*$^{-1}$ hr^{-1}), while it was lower for the cyanophytes with 0.58 ($\pm$0.19) mmol O_2 mg chl *a*$^{-1}$ hr^{-1} and the cryptophyte with 0.23 ($\pm$0.02) mmol O_2 mg chl *a*$^{-1}$ hr^{-1} (Table 3). Oxygen production was also normalized to biovolume ($P_{O2}{}^{\mu m}$ unit: fmol O_2 µm^{-3} hr^{-1}) and allowed to compare species with respect to biomass, provided that cellular volume (µm^3) was a good proxy of species biomass [47]. Fitting $P_{O2}{}^{\mu m}$ with eq. 4, yielded different parameter estimates ($P_M{}^{\mu m}$ and $E_M{}^{\mu m}$) compared to the result obtained from $P_{O2}{}^{chl}$ data (Fig. 1b and 1c). When comparing the maximal oxygen production rate per biovolume ($P_M{}^{\mu m}$), our data showed that chlorophytes reached the highest value with an average of 11.46($\pm$2.82) fmol O_2 µm^{-3} hr^{-1} (Table 3). Comparatively, it was up to 10 times lower for bacillariophytes, cryptophytes and cyanophytes (except *M. flos-aquae*) and varied between 1.50 and 2.84 fmol O_2 µm^{-3} hr^{-1} (Table 3). The light intensities at which $P_M{}^{Chl}$ and $P_M{}^{\mu m}$ were achieved ($E_M{}^{Chl}$ and $E_M{}^{\mu m}$) were also different and our data showed that $E_M{}^{\mu m}$ was significantly lower for all species (Table 3). For $E_M{}^{Chl}$, it varied

Table 2. Achieved maximum growth rate (μ_{MAX}) and light intensity required to reach that rate ($E_M{}^{\mu d}$) estimated from growth versus irradiance fit (R^2 of each fit are presented) using eq. 1.

Species	$E_M{}^{\mu d}$ (µmol photons m^{-2} s^{-1})	Error	μ_{MAX} (day^{-1})	Error	R^2 of Fit	p-value Dunnett's (for μ_{MAX})	Tukey HSD (for μ_{MAX})
C. snowii (CHL5)	585	25	1.12	0.14	0.98	<.0001	a
P. mucicola (CYA1)	279	5	0.80	0.03	0.98	0.0002	b
A. falcatus (CHL1)	503	30	0.73	0.08	0.96	0.0095	b
P. morum (CHL2)	588	39	0.72	0.08	0.95	0.0151	b
O. lacustris (CHL3)	340	2	0.55	0.06	0.91	1	a
C. obovata (CRY1)	605	39	0.50	0.05	0.94	0.9688	ab
A. spiroïdes (CYA4)	187	6	0.43	0.03	0.87	0.2497	ab
M. flos-aquae (CYA2)	369	64	0.41	0.06	0.73	0.1189	b
P. boryanum (CHL4)	403	19	0.41	0.02	0.97	0.1189	b
A. flos-aquae (CYA3)	252	6	0.38	0.05	0.81	0.0257	a
A. granulata (BAC2)	375	18	0.34	0.01	0.95	0.0032	a
F. crotonensis (BAC3)	255	20	0.26	0.02	0.90	<.0001	b
All species	401	29	0.54	0.05	0.48	1	-

The achieved maximal growth rate (μ_{MAX}) obtained for individual species was compared to the fit obtained for all species grouped (All species) using Dunnett's test ($p<0.05$). Subsequent comparisons with Tukey test ($p<0.05$) were done to rank species in each subgroup (higher, equal or lower than All species μ_{MAX}). Presented error corresponds to the 95% confidence interval.

Table 3. Calculated parameters and associated errors (95% interval) of PE curves fitted (see also Fig. 1.1b and c) using waiting in line function (eq. 4) for each species or combined all data (All species).

	P_M^{Chl} (mmol O^2 mg Chl a^{-1} hr^{-1})	error	E_M^{Chl} (μmol photons m^{-2} s^{-1})	error	$P_M^{\mu m}$ (fmol O^2 μm^{-3} hr^{-1})	error	$E_M^{\mu m}$ (μmol photons m^{-2} s^{-1})	error
A. falcatus (CHL1)	1.91	0.09	1376	112	14.44	1.29	774	74
P. morum (CHL2)	1.21	0.16	1022	174	12.82	2.02	780	133
O. lacustris (CHL3)	1.69	0.13	1169	131	11.54	0.94	979	101
P. boryanum (CHL4)	1.64	0.17	1335	225	6.92	1.15	678	113
C. snowii (CHL5)	1.50	0.09	1361	131	11.58	0.81	639	43
A. granulata (BAC2)	2.15	0.15	1061	98	1.95	0.42	851	211
F. crotonensis (BAC3)	1.19	0.12	961	152	2.61	0.27	622	62
C. obovata (CRY1)	0.23	0.02	926	77	2.52	0.32	697	89
P. mucicola (CYA1)	0.48	0.07	843	148	2.22	0.32	570	76
M. flos-aquae (CYA2)	0.83	0.15	660	119	9.95	2.14	480	89
A. flos-aquae (CYA3)	0.59	0.12	951	232	1.50	0.29	363	57
A. spiroïdes (CYA4)	0.41	0.06	1002	176	2.84	0.42	287	33
All species	1.17	0.18	1129	243	6.56	1.24	660	123

Presented error corresponds to the 95% confidence interval.

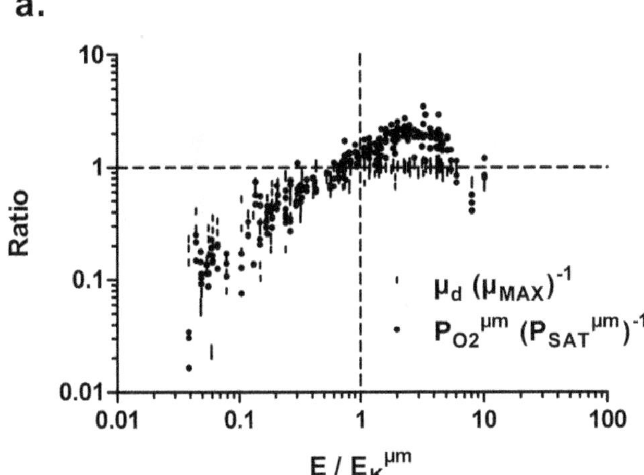

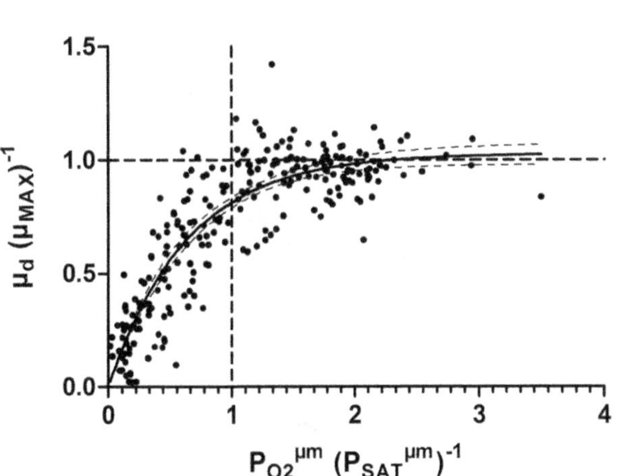

Figure 2. Responses of photosynthesis to light intensities; a. Oxygen production per biovolume ($P_{O2}{}^{\mu m}$) relative to oxygen production at saturation ($P_{SAT}{}^{\mu m}$) and achieved growth rate (μ_d) relative to maximal growth rate (μ_{MAX}) obtained for a gradient of growing light intensity (E) normalized to saturating light intensity of oxygen production ($E_K{}^{\mu m}$); b. relationship between obtained growth rate μ_d normalized to μ_{MAX} and oxygen production per biovolume ($P_{O2}{}^{\mu m}$) normalized to oxygen production at light saturation ($P_{SAT}{}^{\mu m}$). For both panels, the dashed line was set to 1 for all ratios and by definition corresponds to the point where the achieved value (μ_d, $P_{O2}{}^{\mu m}$ or E) equals the normalized coefficient value: μ_{MAX}, $P_{SAT}{}^{\mu m}$ or $E_K{}^{\mu m}$.

between 660 and 1376 µmol photons m^{-2} s^{-1}, while for $E_M{}^{\mu m}$ it varied between 287 and 979 µmol photons m^{-2} s^{-1} (Table 3).

Species differences in PE curve inevitably introduce noise when comparing light dependent variables such as pigments [9]. This problem was accounted for by dividing growth PFD (E) to the photosynthesis light saturation point per biovolume ($E_K{}^{\mu m}$) obtained for each species. This variable ($E:E_K{}^{\mu m}$) was utilised to compare light dependent variables between species and allowed to form 2 groups corresponding to light limitation ($E:E_K{}^{\mu m}<1$) or saturation ($E:E_K{}^{\mu m}>1$) [9]. We also extended this approach to other variables: $P_{O2}{}^{\mu m}$ using the rate of photosynthesis at saturation $P_{SAT}{}^{\mu m}$ and μ_d using μ_{MAX} (Fig. 2a and 2b). Comparing these variables showed that most species, regardless of their phylogeny, followed very similar trend with respect to saturation of photosynthesis and cell division rate (Fig. 2a). The relationships between μ_d: μ_{MAX} and $E:E_K{}^{\mu m}$ and between $P_{O2}{}^{\mu m}:P_{SAT}{}^{\mu m}$ and $E:E_K{}^{\mu m}$ showed for all species that cell division reached μ_{MAX} (μ_d: $\mu_{MAX} = 1$) and that photosynthesis reached saturation ($P_{O2}{}^{\mu m}:P_{SAT}{}^{\mu m} = 1$) at $E_K{}^{\mu m}$ (Fig. 2a). Similar results were obtained when comparing the relationship between $P_{O2}{}^{\mu m}:P_{SAT}{}^{\mu m}$ and μ_d: μ_{MAX} (Fig. 2b) and we observed that below saturation of photosynthesis, growth rate linearly increased, while it stabilized to μ_{MAX} above photosynthetic saturation (Fig. 2b).

Pigment content

For all species, Chl *a* content was higher in phytoplankton exposed to light limiting conditions (PFD < $E_K{}^{\mu m}$) compared to saturating conditions (PFD > $E_K{}^{\mu m}$), although it was not significant for *O. lacustris* (Fig. 3a). Our data also showed that bacillariophytes have less Chl *a* compared to most species. There was great variability in photoprotective carotenoid response since it tended to increase above light saturation for most cyanophytes but it decreased for most chlorophytes and bacillariophytes or remained unchanged in cryptophyte and one cyanophyte (Fig. 3b). Despite these different responses, the Car to Chl *a* ratio followed a similar trend for all species (except *O. lacustris*) and was significantly higher above light saturation (Fig. 3c). The sum of accessory pigments (PC, APC, PE, Chl *b*, Chl *c* and Chl *d*) relative to Chl *a*, reflecting the size of the light harvesting antennae, also varied in function of light intensity. This ratio was higher under light limiting conditions, while it was lower when phytoplankton grew under light saturating conditions (it was the opposite for cyanophytes) (Fig. 3d). In cyanophytes and cryptophyte this ratio was generally higher compared to the other species because of phycobiliproteins, important for light harvesting in these species [48]. Our results showed that phycobiliproteins content decreased above light saturation for *M. flos-aquae*, *Anabaena spiroïdes* and *Cryptomonas obovata* and that the ratio of PC to APC also decreased for both filamentous species, *A. flos-aquae* and *A. spiroïdes* (Fig. 3e and 3f). Finally, we observed that phycobiliproteins (PC and APC or PE) decreased on average by 30% from low to high light, while Chl *a* decreased by 40% for the same conditions.

Biooptical characteristics

Modifications of pigments content induced by photoacclimation processes also modified the biooptical properties of individual cells. For all conditions and studied species, the *in vivo* Chl *a*-specific light absorption coefficient in the red (a^*_φ (red)) varied between 0.002 and 0.025 m^2 mg chl a^{-1} with the lowest value(<0.005) obtained for *C. obovata* (Table 4). When averaged over the whole spectrum, the corresponding light absorption coefficient reflecting total light absorption by pigments (a^*_φ) or specific to PSII (a^*_{PSII}), varied between 0.001 and 0.032 m^2 mg Chl a^{-1} and 0.001 and 0.013 respectively (Table 4 and e.g. Fig. 4a). For most species, these coefficients were higher under photosynthetic light saturation conditions, except for *C. obovata* and *O. lacustris* (for a^*_φ, a^*_{PSII} and a^*_φ (red)) and *A. granulata* (for a^*_φ (red)) (Table 4). Increased light absorption efficiency in high light, as presented here, was counterbalanced by the lower Chl *a* content above light saturation (Fig. 3a). In fact, when normalizing light absorption coefficients to the Chl *a* content per biovolume, correcting for changes in Chl *a* quotas due to photoacclimation processes, our data showed a decrease of all coefficients (a^*_φ, a^*_{PSII} and a^*_φ (red)) above light saturation (e.g. $a^*_\varphi{}^{\mu m}$ in Fig. 4b; data not shown for other coefficients). The fraction of light absorption associated to PSII and LHCII relative to light absorption by the whole cell (fAQ_{PSII}) decreased by 8.9 to 26.1% for most species following high light acclimation (Fig. 4c). On the other hand, the cellular fraction of

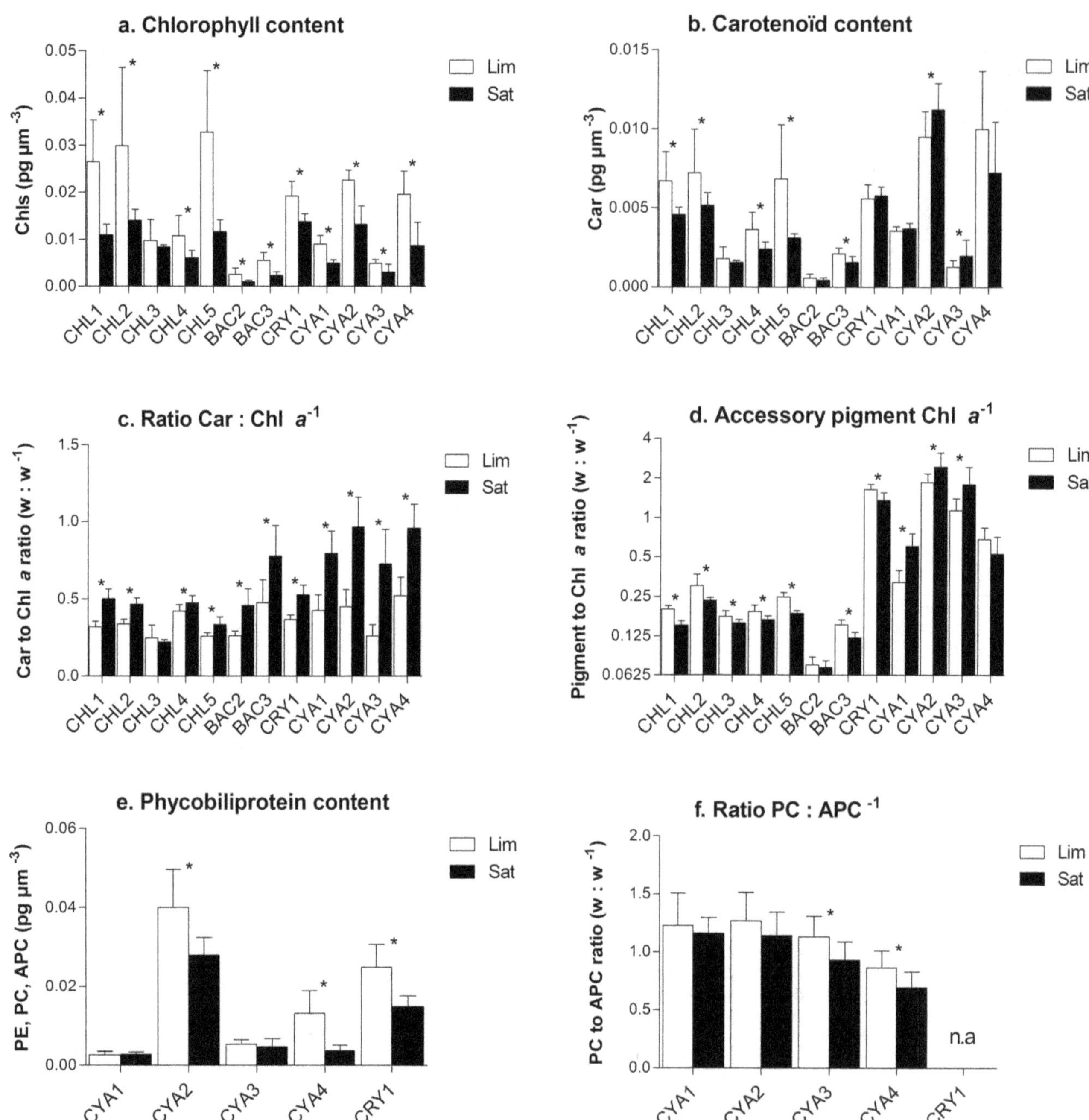

Figure 3. Comparison of the average pigment content normalized to biovolume (pg μm^{-3}) or the average pigment ratios obtained for each species grown under photosynthetic light limiting (Lim) or light saturating (Sat) conditions; a. total chlorophyll content, b. carotenoid content, c. Car to Chl *a* ratio, d. sum of accessory pigments (Chl *b*, *c*, d and phycobiliproteins), e. phycobiliprotein content for species having these pigments and f. phycocyanin (PC) to allophycocyanin (APC) ratio in cyanophytes. * Significant difference between treatment obtained for each species using t-test ($p<0.05$). See Table 1 for species list.

Chl *a* associated to PSII relative to that associated to PSI (F_{II}) was found to increase by 5.7 to 76.9% depending on species (Fig. 4d). Regardless of the light conditions, we also found that a significantly higher fraction of light absorption was directed toward PSII (fAQ_{PSII}) in chlorophytes (0.73±0.05) and bacillariophytes (0.72±0.05) compared to cryptophyte (0.63±0.08), while it was much lower (0.33±0.06) in cyanophytes (Fig. 4c). For all light conditions, cyanophytes were also characterized by very low F_{II} (0.14±0.07) compared to cryptophyte (0.35±0.01), chlorophytes (0.56±0.04) and bacillariophytes (0.61±0.10).

Photosynthetic electron transport and quantum requirement

In this section, we compared the effect of photoacclimation on photosynthesis through photosynthetic electron transport, light utilisation and dissipation and quantum requirement (QR). Decrease in the PSII operational quantum yield (Φ'_M) was observed with increasing light intensity for the 12 studied species (Fig. 5a). Our data showed that Φ'_M remained stable under light limiting conditions ($E:E_K{}^{\mu m}<1$) and decreased by 54 to 85% at

Table 4. Averaged data (% CV) obtained and compared between light limiting (Lim) and light saturating (Sat) intensity for photosynthesis of the 12 studied species.

	a^*_φ		a^*_{PSII}		a^*_φ (red)	
	Lim	**Sat**	**Lim**	**Sat**	**Lim**	**Sat**
A. falcatus (CHL1)	0.008 (19)	0.011 (8)*	0.005 (16)	0.007 (3)*	0.012 (19)	0.015 (4)*
P. morum (CHL2)	0.004 (17)	0.006 (14)*	0.003 (16)	0.004 (15)*	0.006 (13)	0.008 (14)*
O. lacustris (CHL3)	0.011 (24)	0.011 (4)	0.008 (27)	0.008 (7)	0.016 (20)	0.017 (4)
P. boryanum (CHL4)	0.009 (16)	0.012 (15)*	0.007 (16)	0.008 (10)*	0.012 (15)	0.014 (10)*
C. snowii (CHL5)	0.006 (17)	0.009 (12)*	0.004 (16)	0.006 (9)*	0.009 (17)	0.013 (9) *
A. granulata (BAC2)	0.014 (7)	0.016 (8)*	0.010 (6)	0.012 (9)*	0.021 (8)	0.022 (10)
F. crotonensis (BAC3)	0.007 (15)	0.012 (33)*	0.005 (12)	0.007 (20)*	0.009 (12)	0.012 (13)*
C. obovata (CRY1)	0.002 (18)	0.002 (10)	0.001 (18)	0.001 (17)	0.002 (18)	0.002 (14)
P. mucicola (CYA1)	0.010 (16)	0.016 (17)*	0.003 (13)	0.004 (8)*	0.012 (13)	0.014 (8)*
M. flos-aquae (CYA2)	0.014 (9)	0.025 (20)*	0.004 (6)	0.006 (17)*	0.015 (6)	0.017 (12)*
A. flos-aquae (CYA3)	0.014 (19)	0.023 (33)*	0.003 (15)	0.004 (26)*	0.014 (13)	0.017 (17)*
A. spiroïdes (CYA4)	0.010 (14)	0.013 (16)*	0.003 (10)	0.003 (13)*	0.011 (8)	0.013 (8)*

Pigment absorption was averaged over the whole light absorption spectrum (400 to 700 nm) for whole cell (a^*_φ) or specific to PSII (a^*_{PSII}), or was averaged in the red band (670 to 680 nm) for Chl *a* specific absorption (a^*_φ (red)).
*Significantly different by t-test ($p<0.05$), unequal variance assumed.

light intensities above $E_K{}^{\mu m}$ (Fig. 5a). The averaged Φ'_M was the highest for chlorophytes (0.19–0.75) followed by cryptophyte (0.28–0.66) and bacillariophytes (0.15 to 0.63), while it was the lowest in cyanophytes (0.06–0.51) (Fig. 5a). The decrease of Φ'_M under high light conditions also affected the number of photons required to evolve 1 O_2 molecule (oxygen quantum requirement: QR). As seen, QR increased (from 149 up to 279%) for all species when light intensity increased above saturation level (Fig. 5b). Under light limiting conditions, where Φ'_M was the highest, the QR was on average 13.2±1.4 mol e mol $O_2{}^{-1}$ for all algal species, while it was higher in cyanophytes with 19.4±2.0 mol e mol $O_2{}^{-1}$. Furthermore, we noticed that the decrease of Φ'_M and the concomitant increase in QR were accompanied by an increase in non-photochemical quenching (NPQ) and in the level of unquenched fluorescence (UQF_{rel}) (Fig. 5c and 5d). Group comparison showed that NPQ was higher in cyanobacteria (0.38±SE 0.02) and chlorophytes (0.33±SE 0.02) compared to cryptophyte (0.19±SE 0.02) and bacillariophytes (0.13±SE 0.01) and was surprisingly low in the latter group. We also demonstrated that NPQ tended to increase from limiting to saturating light conditions (average 220%) although that tendency was not significant for two of the cyanophytes studied in which NPQ was constant (Fig. 5c). Similarly, the unquenched fluorescence level (UQF_{rel}), which is proportional to the redox state of the photosynthetic electron transport chain [34], increased (average 255%) for all species between limiting and saturating light conditions (Fig. 5d). When averaged over all light conditions, the highest UQF_{rel} value was measured in bacillariophytes (0.21±SE 0.01) followed by cyanophytes (0.17±SE 0.01) and cryptophyte (0.15±SE 0.02), while for chlorophytes the value (0.12±SE 0.01) was significantly lower.

Discussion

Effect on growth

In this study, we showed that chlorophytes, bacillariophytes, cryptophyte and cyanophytes successfully acclimate and grow under a wide range of light conditions (from 14 to 1079 µmol photons m^{-2} s^{-1}) when given proper acclimation period. As expected, the achieved growth rate increased with light intensity and reached its maximal value between 187 and 605 µmol photons m^{-2} s^{-1} depending on the species. Above that point, growth inhibition was observed for most cyanophytes, corresponding to the response of low light adapted organisms [7,30], but also for *O. lacustris*, while growth of other species remained unaffected by high light up to 1079 µmol photons m^{-2} s^{-1}. Photoinhibition observed for *O. lacustris* was not surprising since we have shown low pigment plasticity and overall low photoacclimation driven responses (pigments, biooptic, photosystem ratio) for that species. As expected, this lack of response under high light conditions resulted in suboptimal growth [7,17]. Although not mandatory, our growth data suggested that being chlorophyte, flagellate and/or small organism are characteristics allowing higher than average growth rate (Table 2). Conversely, most of the species presenting low growth rates (*A. granulata*, *F. crotonensis* or *A. flos-aquae*) were colonial or filamentous and had large cell as seen by their high averaged biovolume of 484 to 3593 μm^3. It is well admitted that usually larger organisms have slower growth rates compared to smaller organisms due to their cell metabolism and higher package effect [21,49]. However, we noticed some exceptions to that trend as seen with *M. flos-aquae* and *P. morum*. For *M. flos-aquae*, we have measured low growth rate despite its small size(±27 μm^3), but this species also formed colonies. Another exception was observed for the colonial species *P. morum* for which high growth rate was measured despite its large size. For that species, individual cells are flagellated and we observed that their motion can actively position the colony in relation to available light and this may contribute to optimize its growth [27,50]. Knowing that capacity to acclimate to light and morphological characteristics are important to determine growth [24,31], our results tend to demonstrate that size, which affects the light absorption efficiency (package effect), and aptitude to movement (to optimize light harvesting) were relevant factors.

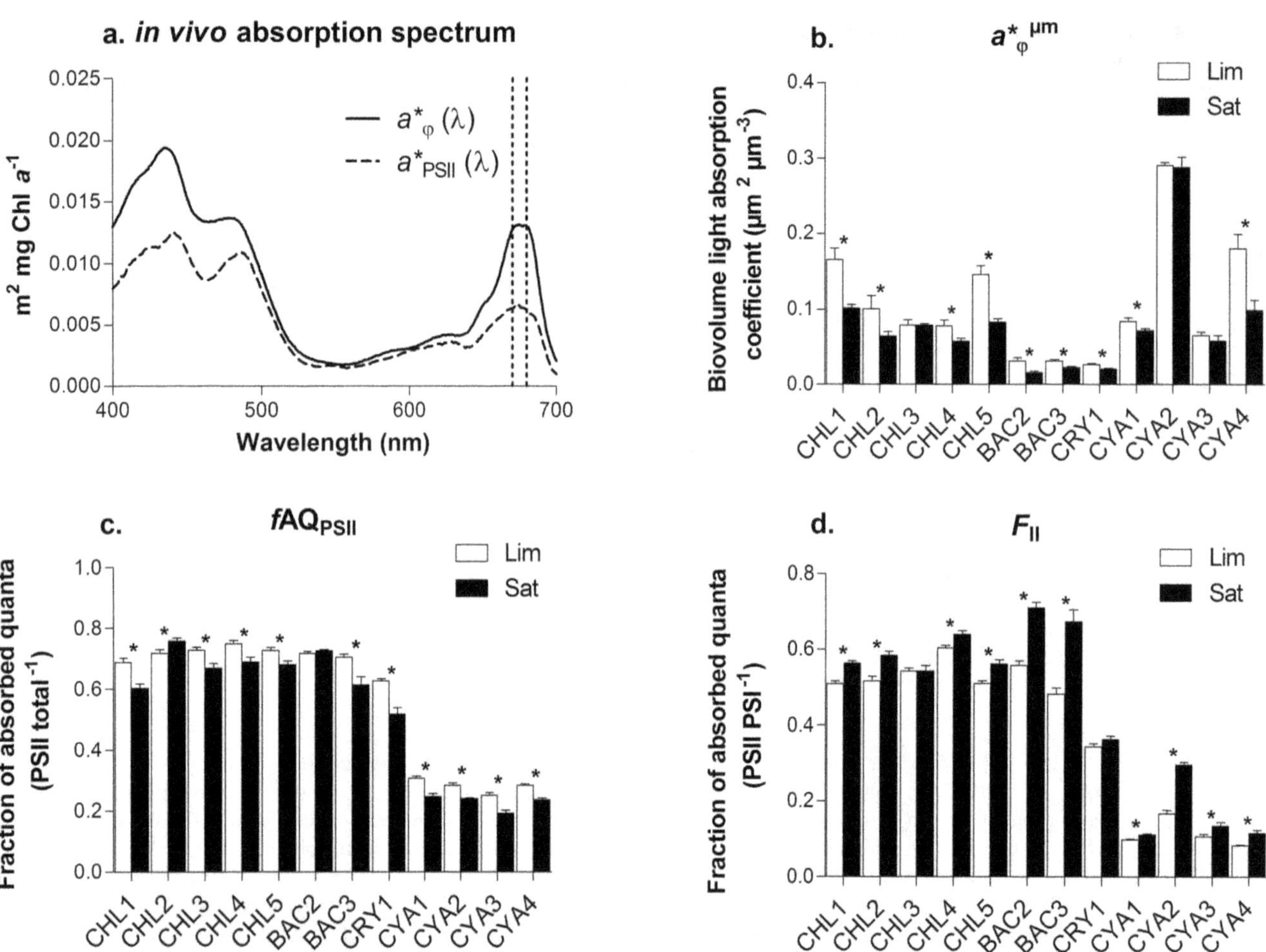

Figure 4. Responses of biooptical variables to light intensities; a. example of *in vivo* Chl *a* absorption spectrum (a^*_φ (λ) and a^*_{PSII} (λ)) obtained for *A. falcatus* acclimated to 76 µmol photons m^{-2} s^{-1}. Averaging a^*_φ (λ) or a^*_{PSII} (λ) over the whole spectrum (400 to 700 nm) yielded to a^*_φ and a^*_{PSII} respectively, while averaging the coefficient in the red band (670 to 680 nm) yielded to a^*_φ (red) and a^*_{PSII} (red) respectively. Other panels, comparison of averaged biooptical data obtained for each species grown under photosynthetic light limiting (Lim) or light saturating (Sat) conditions where b. is the averaged light absorption coefficient normalized to biovolume, c. the fraction of absorbed quanta to PSII ($fAQ_{PSII} = a^*_{PSII}/a^*_\varphi$) and d. the fraction of absorbed quanta associated to PSII relative to PSI ($F_{II} = a^*_{PSII}$ (red)/a^*_φ (red)). * Significant difference between treatment obtained for each species using t-test ($p<0.05$). See Table 1 for species list.

Pigment acclimation

Because light can be damaging for photosystems, by causing oxydative stress to individual cell, and is indispensable as a source of energy, phytoplankton capacity to acclimate to a limitation or excess in photon flux is critical [12]. In this study, we showed that most species presented similar response to light acclimation, but to varying degrees. We found a decrease of photosynthetic pigment content (chlorophylls and phycobiliproteins) in all species following high light acclimation confirming previous observations [2,8,9,13]. In most species, this decrease was accompanied by a decrease in carotenoid content and in the size of the light harvesting antennae as seen by lower accessory pigments to Chl *a* ratio (Fig. 3b and 3d). In cyanophytes, we observed a small reduction in the size of PBS under light saturation, but the proportion of PBS relative to Chl *a* increased suggesting an increase of LHC antenna size in that group (see below for more details). Lowering pigment content and antenna size is a typical response of high light acclimated cells [51]. These responses directly decrease the number of photons absorbed by the LHCs and decrease energy transfer to PSII and PSI RCs [51]. This adjustment results in a lower excitation pressure on the photosynthetic apparatus and is essential to minimize photoinhibition and cell damage induced by oxidative stress [52,53]. In low light environment, and as observed in this study, these pigment modifications also worked in the opposite direction. In fact, increased pigmentation and antenna size allowed to maximize light harvesting and thus, alleviated the energy deficit caused by surrounding light scarcity [7,9]. The light dependent variation in photosynthetic pigment content described here, was accompanied by modifications of the proportion of photoprotective pigments (Car) with respect to Chl *a* (Fig. 3c). This ratio was the highest under light saturation condition for all species and corresponded to previous finding showing that high Car to Chl *a* ratio increases protection against excess photon flux by allowing light energy dissipation through NPQ processes [13,54,55].

For all light conditions, cyanobacteria and cryptophyte had significantly more accessory pigments relative to Chl *a* and that was attributed to the presence of phycobiliproteins reflecting the dominance of these pigments for light harvesting in these species [2,47,56]. Surprisingly, this ratio was significantly higher under saturating light condition for three cyanobacteria, while it was lower in the other tested species of this study and others [9,57,58].

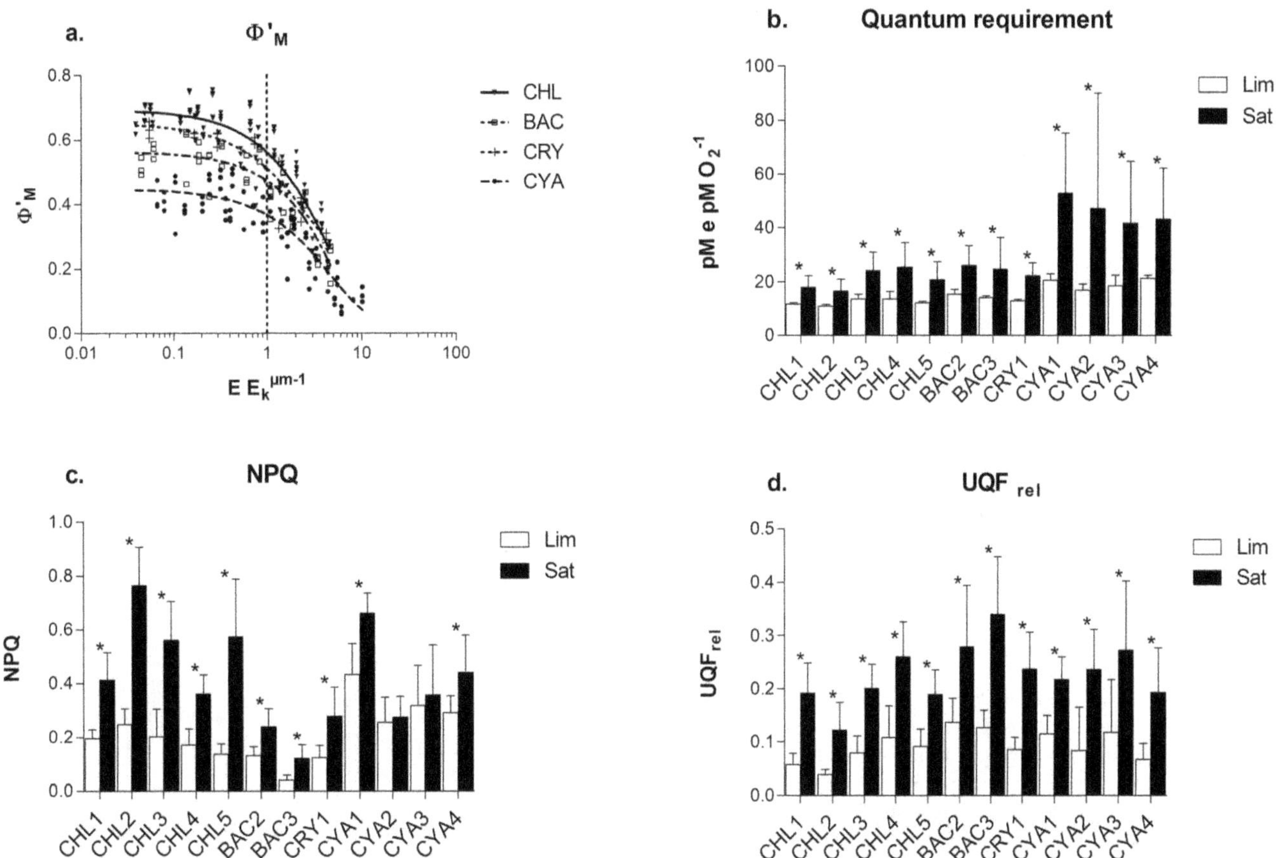

Figure 5. Responses of chlorophyll fluorescence parameters to light intensities; a. Group specific relationship between PSII operational quantum yield (Φ'_M) and growth light intensity normalized to photosynthetic light saturation point (E_K $^{\mu m}$). Other panels, comparison of averaged chlorophyll fluorescence paramaters obtained for each species grown under photosynthetic light limiting (Lim) or light saturating (Sat) conditions where b. is the quantum requirement, c. the non-photochemical quenching (NPQ) and d. the relative unquenched fluorescence parameter (UQF_{Rel}). * Significant difference between treatment obtained for each species using t-test ($p<0.05$). See Table 1 for species list.

Since this increase was accompanied by a decrease of Chl *a* and PBS individually, it indicates that when acclimated to high light, these species favour light harvesting through PBS relative to Chl *a*. Phycobilisomes are highly mobile pigment complexes that can unbind from the RC core when exposed to high irradiance and thus prevent energy funnelling under excess light condition [59,60]. Furthermore, previous studies have shown that in some cyanophytes, orange carotenoids interact with PBS when exposed to high light intensity in order to induce dissipation of excess energy [61,62,63]. Thus, the observed increase of PBS relative to Chl *a* may help to protect against high light as a complementary mechanism to energy dissipation through carotenoids. We can also hypothesis that favouring PBS over carotenoids is a strategy allowing higher light harvesting flexibility for organisms suddenly exposed to a lower light environment.

Biooptical acclimation

Analysis of the biooptical data showed that changes in pigment content and ratio reported here successfully modified light harvesting efficiency and energy allocation between PSI and PSII (Fig. 4 and Table 4). For most species, the Chl *a* specific light absorption coefficients (a^*_φ, a^*_φ (red) and a^*_{PSII}) significantly increased with light intensity. This increase was important (reaching up to 169%) for all species with minor exceptions (*O. lacustris*, *A. granulata* and *C. obovata*) and as was found previously, this was related to an increase in light absorption efficiency for high light acclimated cells [20]. This counterintuitive result may be attributed to an increased light absorption of Car (relative to Chl *a*) and assigned to Chl *a* in *a** calculation, but also to a lower pigment packaging (reduced self-shading) in high light acclimated cells [2,20,64,65,66]. According to our results, both phenomena occurred in our conditions since we observed an increase in the Car to Chl *a* ratio and the a^*_φ (red) values never reached 0.033 m^2 mg Chl a^{-1} (max value obtained was 0.025 m^2 mg Chl a^{-1}, see also Table 4) and this value is expected to be close to the absorption coefficient of Chl *a* embedded in thylakoid membrane without any package effect [64]. Nevertheless, the higher light absorption efficiency observed for high light cells was mitigated by a lower content in Chl *a* per biovolume (Fig. 3a). When taking that variable into account, we observed that the biovolume specific absorption cross section coefficient (a^*_φ $^{\mu m}$) decreased or remained stable as can be expected following acclimation to high light [2,20,48]. Very similar results and conclusions were drawn for PSII specific absorption coefficient (a^*_{PSII}) and Chl *a* absorption in the red (a^*_φ (red)) indicating that following high light acclimation, energy directly associated to Chl *a* and PSII tended to decrease (in most cases) or remained stable on a biovolume basis. These modifications observed under high light conditions minimized the excitation pressure on the photosynthetic apparatus despite the increased light availability [18].

Comparison in the partition of harvested energy showed that above light saturation of photosynthesis, a lower proportion of intercepted photon was directed toward PSII in almost all species (Fig. 4c). This can be explained by the increase of Car to Chl *a* ratio and associated increased proportion of energy dissipation through heat by carotenoids and/or PBS uncoupling in cyanophytes. Our data also showed that energy balance between PSII and PSI was modified under high light, since a higher fraction of the energy was associated to PSII compared to PSI (Fig. 4d). Thus, it indicates that photoacclimation process did not only decrease PSII and PSI excitation pressure under higher light intensity, but it also redirected light absorption toward PSII. This rebalance of energy between the photosystems is necessary to prevent any excess energy to one of the photosystems (minimize excitation pressure) and to optimize electron flow between photosystems [11,17,18,67].

Photoacclimation and photosynthesis

Our data clearly showed that the photosynthetic activity of PSII was also affected by photoacclimation processes. As seen, the PSII operational quantum yield (Φ'_M) changed in close relationship with light limitation to light saturation gradient (Fig. 5a). It remained high and stable under light limited conditions and it decreased when light intensity was above the saturation point. Interestingly, that tendency was similar for all studied species regardless of their taxonomic groups and despite different average Φ'_M or pigment composition. It also indicates that under light limitation, phytoplankton optimized light utilisation through high PSII quantum yield, while other biochemical or physiological factors became limiting in draining electrons under high light [9,10]. Concomitantly to these changes, we observed an increase of NPQ and UQF_{REL} above light saturation. Non-photochemical quenching and associated processes allowed the dissipation of excess energy and alleviated the excitation pressure on PSI and PSII [54]. The unquenched fluorescence reflects the redox state of the electron transport chain [34] and the high values obtained above light saturation indicates that the PSI and/or other electron sinks were less efficient to drain electrons under high light compared to low light conditions. This lower capacity to drain electrons from PSII may be induced by PSII:PSI energy imbalance or by a lack of available reductants (NADP+ and ADP) [10,18,52]. Finally, the variations observed for Φ'_M were also reflected in the quantum requirement (QR) which remained close to the theoretical value of 8 photons per O_2 molecule evolved with 13.4($\pm$1.4 mol e mol O_2^{-1}) for chlorophytes, bacillariophytes and cryptophyte under light limiting conditions. However, for the cyanophytes the average QR was higher (19.4$\pm$2.0 mol e mol O_2^{-1}) indicating that this group was less efficient to convert light energy into chemical energy. Under saturating conditions, QR increased for all species to more than 30(>50 in cyanophytes) indicating a lower photosynthetic efficiency compared to low light conditions. Despite the lower conversion efficiency under light saturation, phytoplankton cells were able to maintain high growth rates indicating that adjustments to their energy dissipation processes (high NPQ and UQF_{rel} and low Φ'M) under high light conditions were not disadvantageous. These differences between phytoplankton groups clearly indicate variations in photoacclimation processes, as was also observed for pigments and optical properties.

Primary production and growth uncoupling

The oxygen production estimates calculated from a combination of chlorophyll fluorescence and biooptical method [42] and normalized to Chl *a* (P_{O2}^{Chl}) varied between 0 and 2.2 mmol O_2 mg Chl a^{-1} hr^{-1} for all species (Fig. 1b). This was comparable to the range reported previously for different phytoplankton species [9,28,42,44,68]. Oxygen production was lower when normalized to biovolume ($P_{O2}^{\mu m}$) and this difference can be attributed to variation in the ratio of Chl *a* per biovolume specific to individual species following photoacclimation. When normalized to Chl *a*, oxygen production was informative of the photosynthetic apparatus efficiency where high values correspond to high photosynthetic efficiency. Oxygen production normalized to biovolume allows to relate photosynthetic efficiency to biomass, and therefore to the achieved growth rate. In our study, PE curves were reconstructed similarly to growth versus light curves since they were based on photosynthetic activity obtained at different growth light intensities. This method is slightly different to PE curves obtained by short term exposure to different light intensities of pre-acclimated phytoplankton [9]. Thus, our approach permits to directly estimate if there is a relationship between growth and photosynthesis when phytoplankton is acclimated to specific light conditions. Interestingly, when comparing the light intensity required to reach maximal photosynthesis for both variables (P_M^{Chl} and $P_M^{\mu m}$), we found that $E_M^{\mu m}$ (287–979 μmol photons m^{-2} s^{-1}) was always lower than E_M^{Chl} (660–1376 μmol photons m^{-2} s^{-1}) for all studied species (Table 2). This difference reflects a decoupling between photosynthesis and cellular investment in chlorophyll (see below). We may therefore advance that cellular investment in the photosynthetic components and Chl *a* relative to the other cellular constituents was sub optimal for the studied species and this suggests that phytoplankton cells, in our growth conditions, did not try to maximize their photosynthetic activity, otherwise $E_M^{\mu m}$ should tend toward E_M^{Chl}. A good example of that phenomenon was observed for *A. granulata* since this diatom has the highest P_M^{Chl} of all tested species (Fig. 1b) and thus high photosynthetic efficiency relative to cellular Chl *a* investment. However, this high photosynthetic efficiency was not reflected into a better growth rate. In fact, we found very low oxygen production on a biomass basis ($P_M^{\mu m}$) for this species, indicating that its strategy was not to invest in photosynthetic apparatus and Chl *a* (Fig. 1b and 1c). Consequently, this species presented one of the lowest μ_d and μ_{MAX} values of this study despite a potential of high photosynthetic efficiency. Our findings that growth rate approached its maximal value when oxygen production per biovolume reached saturation and the absence of change of μ_d above photosynthesic saturation and up to maximal photosynthesis was another indication of the decoupling between cell division and photosynthesis. This can be attributed to lower Chl *a* content in high light acclimated cells and can also be caused by an increase in the respiration processes relative to photosynthesis or by accumulation of compound that were not included in our growth rate estimates such as lipids.

Conclusions

The general response of phytoplankton to increased light intensity worked toward reducing the excitation pressure on the photosynthetic apparatus and also toward reducing their efficiency to utilize the absorbed energy. According to our results, these mechanisms induced a decoupling between photosynthesis and growth rate when light intensity was above photosynthetic saturation level indicating that photoacclimation processes do not necessarily optimize photosynthesis to maximize growth. Interestingly, all species of our study followed that tendency despite being of different functional groups (colonial, flagellated, different size) and of different phylogeny. Even if some species did reach higher growth rates in our conditions and thus, should dominate in natural environment with respect to light intensity, we cannot exclude that other environmental factors also influence the population dynamic making the outcome difficult to predict. Finally, the fact that morphologically distinct species isolated from the same community, but belonging to different phylogenic groups, were able to adjust to a wide range of light intensities (from 14 to 1079 μmol photons m^{-2} s^{-1}) demonstrates the great plasticity and adaptation ability of freshwater phytoplankton to their light environment and help to understand their ubiquity in natural environment.

Author Contributions

Conceived and designed the experiments: CPD PJ. Performed the experiments: CPD AM. Analyzed the data: CPD PJ. Contributed reagents/materials/analysis tools: PJ. Wrote the paper: CPD PJ.

References

1. Litchman E (2003) Competition and coexistence of phytoplankton under fluctuating light: experiments with two cyanobacteria. Aquat Microb Ecol 31: 241–48.
2. Dubinsky Z, Stambler N (2009) Photoacclimation processes in phytoplankton: mechanisms, consequences, and applications. Aquat Microb Ecol 56: 163–176.
3. Abeliovich A, Shilo M (1972) Photooxidative death in blue-green algae. J Bacteriol 111: 682–89.
4. Eloff JN, Steinit Y, Shilo M (1976) Photooxidation of cyanobacteria in natural conditions. Appl Environ Microbiol 31 (1): 119–26.
5. Gerber S, Häder DP (1995) Effects of enhanced solar irradiation on chlorophyll fluorescence and photosynthetic oxygen production of five species of phytoplankton. FEMS Microbiol Ecol 16: 33–42.
6. Schanz F, Senn P, Dubinsky Z (1997) Light absorption by phytoplankton and the vertical light attenuation: ecological and physiological significance. Oceanogr Mar Biol Annu Rev 35: 71–95.
7. Richardson K, Beardall J, Raven JA (1983) Adaptation of unicellular algae to irradiance: an analyses of strategies. New Phytol 93: 157–191.
8. Falkowski PG, La Roche J (1991) Acclimation to spectral irradiance in algae. J Phycol 27: 8–14.
9. MacIntyre HL, Kana TM, Anning T, Geider RJ (2002) Review: Photoacclimation of photosynthesis irradiance response curves and photosynthetic pigments in microalgae and cyanobacteria. J Phycol 38: 17–38.
10. Sukenik A, Bennett J, Falkowski PG (1987) Light saturated photosynthesis limitation by electron transport or carbon fixation? Biochim Biophys Acta 891: 205–215.
11. Fisher T, Schurtz-Swirski R, Gepstein S, Dubinsky Z (1989) Changes in the levels of ribulose-1,5-bisphosphate carboxylase/oxygenase (RUBISCO) in *Tetraedron minimum* (Chlorophyta) during light and shade adaptation. Plant Cell Physiol 30: 221–228.
12. Herzig R, Dubinsky Z (1992) Photoacclimation, photosynthesis, and growth in phytoplankton. Isr J Bot 41: 199–212.
13. Steiger S, Schaëfer L, Sandmann G (1999) High-light-dependent upregulation of carotenoids and their antioxidative properties in the cyanobacterium *Synechocystis* PCC6803. J Photochem Photobiol B, Biol 52: 14–18.
14. Grossman AR, Schaefer MR, Chiang GG, Collier JL (1993) The phycobilisome, a light harvesting complex responsive to environmental conditions. Microbiol Rev 57: 725–749.
15. Demmig-Adams B, Adams WW (1996) The role of xanthophyll cycle carotenoids in the protection of photosynthesis. Trends Plant Sci 1: 21–26.
16. Kana TM, Geider RJ, Critchley C (1997) Regulation of photosynthetic pigments in micro-algae by multiple environmental factors: A dynamic balance hypothesis. New Phytol 137 (4): 629–638.
17. Barber J, Anderson B (1992) Too much of good thing: light can be bad for photosynthesis. Trends Biochem Sci 17: 61–66.
18. Huner NPA, Öquist G, Sarhan F (1998) Energy balance and acclimation to light and cold. Trends Plant Sci 3 (6): 224–230.
19. Choudhury NK, Behera RK (2001) Photoinhibition of photosynthesis: role of carotenoids in photoprotection of chloroplast constituents. Photosynthetica 39: 481–488.
20. Johnsen G, Sakshaug E (2007) Bio-optical characteristics of PSII and PSI in 33 species (13 pigment groups) of marine phytoplankton, and the relevance for pulse-amplitude-modulated and fast-repetition-rate fluorometry. J Phycol 43: 1236–1251.
21. Raven JA (1998) The Twelfth Tansley Lecture, Small is Beautiful: The Picophytoplankton. Funct Ecol 12 (4): 503–513.
22. Beardall J, Allen D, Bragg J, Finkel ZV, Flynn KJ, et al. (2009) *Tansley review*: Allometry and stoichiometry of unicellular, colonial and multicellular phytoplankton. New Phytol 181: 295–309.
23. Reynolds CS (1998) What factors influence the species composition of phytoplankton in lakes of different trophic status? Hydrobiologia 369/370: 11–26.
24. Reynolds CS, Huszar V, Kruk C, Naselli-Flores L, Melos S (2002) Review: Towards a functional classification of the freshwater phytoplankton. J Plankton Res 24 (5): 417–428.
25. Agusti S, Phlips EJ (1992) Light absorption by cyanobacteria: Implications of colonial growth form. Limnol Oceanogr 32: 434–441.
26. Wilson AE, Kaul RB, Sarnelle O (2010) Growth rate consequences of coloniality in a harmful phytoplankter. PLoS One 5 (1): e8679.
27. Cullen JJ, MacIntyre JG (1998) Behavior, physiology and the niche of depth-regulating phytoplankton. In: Anderson DM, Cemballa AD, Hallegraeff GM, editors. Physiological ecology of harmful algal blooms: Springer-Verlag Heidelburg. pp. 559–580.
28. Dubinsky Z, Falkowski PG, Wyman K (1986) Light harvesting and utilization in phytoplankton. Plant Cell Physiol 27: 1335–1350.
29. Rolland A, Bird DF, Giani A (2005) Seasonal changes in composition of the cyanobacterial community and occurrence of hepatotoxic blooms in the eastern townships, Québec, Canada. J. Plankton Res 27 (7): 683–694.
30. Mur LR, Schreurs H (1995) Light as a selective factor in the distribution of phytoplankton species. Water Sci Technol 32 (4): 25–34.
31. Havens KE, Phlips EJ, Cichra MF, Li B-L (1998) Light availability as a possible regulator of cyanobacteria species composition in a shallow subtropical lake. Freshw Biol 39 (3): 547–556.
32. Liu S, Juneau P, Qiu B-S (2012) Effects of iron on the growth and minimal fluorescence yield of three marine *Synechococcus* strains (Cyanophyceae). Phycol Res 60 (1): 61–69.
33. Schreiber U, Schliwa U, Bilger W (1986) Continuous recording of photochemical and non-photochemical chlorophyll fluorescence quenching with a new type of modulation fluorometer. Photosynth Res 10: 51–62.
34. Juneau P, Green BR, Harrison PJ (2005) Simulated of Pulse-Amplitude-Modulated (PAM) fluorescence: limitations of some PAM-parameters in studying environmental stress effects. Photosynthetica 43 (1): 75–83.
35. Campbell D, Hurry V, Clarke AK, Gustafsson P, Öquist G (1998) Chlorophyll fluorescence analysis of cyanobacterial photosynthesis and acclimation. Microbiol Mol Biol Rev 62 (3): 667–83.
36. Ritchie RJ (2008) Universal chlorophyll equations for estimating chlorophylls a, b, c, and d and total chlorophylls in natural assemblages of photosynthetic organisms using acetone, methanol, or ethanol solvents. Photosynthetica 46 (1): 115–26.
37. Lichtenthaler HK, Wellburn AR (1985) Determination of total carotenoids and chlorophylls a and b of leaf in different solvents. Biol Soc Trans 11: 591–592.
38. Bennett A, Bogorad L (1973) Complementary chromatic adaptation in a filamentous blue-green alga. J Cell Biol 58: 419–35.
39. Jassby AD, Platt T (1976) Mathematical formulation of the relationship between photosynthesis and light for phytoplankton. Limnol Oceanogr 21: 540–547.
40. Zimmerman RC, Beeler SooHoo J, Kremer JN, D'Argenio DZ (1987) Evaluation of variance approximation techniques for non-linear photosynthesis-irradiance models. Marine Biol. 95: 209–215.
41. Kopf U, Heinze J (1984) 2,7-Bis(diethylamino)phenazoxonium chloride as a quantum counter for emission measurements between 240 and 700 nM. Anal Chem 56: 1931–1935.
42. Hancke TB, Hancke K, Johnsen G, Sakshaug E (2008) Rate of O_2 production devrived from pulse-amplitude-modulated fluorescence: testing three biooptical approaches against measured O_2-production rate. J Phycol 44: 803–813.
43. Gilbert M, Domin A, Becker A, Wilhelm C (2000) Estimation of primary productivity by chlorophyll a in vivo fluorescence in freshwater phytoplankton. Photosynthetica 38: 111–26.
44. Ritchie RJ (2008) Fitting light saturation curves measured using modulated fluorometry. Photosynth Res 96: 201–215.
45. Quinn P, Keough MJ (2003) Experimental design and data analysis for biologists. Cambridge press. 537 p. ISBN 0 521 00976 6.
46. Dunnett CW (1955) A multiple comparison procedure for comparing several treatments with a control. J Am Stat Assoc 50: 1096–1121.
47. Wetzel RG (2001) Limnology: Lake and River Ecosystems, 3rd ed.Springer-Verlag, New York, 1006 p.
48. Falkowski PG, Raven JA (2007) Aquatic photosynthesis. 2nd ed. Princeton University Press, Princeton, NJ. 484 p.
49. Raven JA, Kübler JE (2002) New light on the scaling of metabolic rate with the size of algae (Short survey). J Phycol 38: 11–16.
50. Fee EJ (1976) The vertical and seasonal distribution of chlorophyll in lakes of the experimental lakes areas, northwestern Ontario: implications for primary production estimates. Limnol Oceanogr 21: 767–783.

51. Behrenfeld MJ, Prasil O, Babin M, Bruyant F (2004) In search of a physiological basis for covariations in light-limited and light saturated photosynthesis. J Phycol 40 (1): 4–25.
52. Sonoike K, Hihara Y, Ikeuchi M (2001) Physiological significance of the regulation of photosystem stoichiometry upon high light acclimation of *Synechocystis* sp. PCC6803. Plant Cell Physiol 42 (4): 379–384.
53. Huner NPA, G Öquist, Melis A (2003) Photostasis in plants, green algae and cyanobacteria: The role of light harvesting antenna complexes. In: Green BR, Parson WW editors. Light-harvesting antennas in photosynthesis. Dordrecht, Kluwer Academic Publishers. pp. 402–421.
54. Müller P, Xiao-Ping L, Niyogi KK (2001) Non-photochemical quenching. A response to excess light energy. Plant Physiol 125: 1558–566.
55. Lavaud J, Rousseau B, Etienne AL (2004) General features of photoprotection by energy dissipation in planktonic diatoms (Bacillariophyceae). J Phycol 40: 130–137.
56. Gantt E, Conti SF (1966) Granules associated with the chloroplast lamellae of *Porphyridium cruentum*. J Cell Biol 29: 423–434.
57. Raps S, Wyman K, Siegelman HW, Falkowski PG (1983) Adaptation of the cyanobacterium *Microcystis aeruginosa* to light intensity. Plant Physiol 72: 829–832.
58. Kana TM, Glibert PM (1987) Effect of irradiances up to 2000 μE m-2 s-1 on marine *Synechococcus* WH7803. I. Growth, pigmentation, and cell composition. Deep-Sea Res 34: 479–495.
59. Subramaniam A, Carpenter EJ, Karentz D, Falkowski PG (1999) Bio-optical properties of the marine diazothrophic cyanobacteria *Trichodesmuim* spp. I. Absorption and photosynthetic action spectra. Limnol Oceanogr 44: 608–617.
60. Tamary E, Kiss V, Nevo R, Adam Z, Bernát G, et al. (2012) Structural and functional alterations of cyanobacterial phycobilisomes induced by high-light stress. Biochim Biophys Acta 1817: 319–327.
61. Wilson A, Ajlani G, Verbavatz JM, Vass I, Kerfeld CA, et al. (2006) A soluble carotenoid protein involved in phycobilisome-related energy dissipation in cyanobacteria. Plant Cell 18: 992–1007.
62. Karapetyan NV (2007) Non-photochemical quenching in cyanobacteria. Biochemistry (Mosc) 72 (10): 1127–1135.
63. Kirilovsky D, Kerfeld CA (2012) The orange carotenoid protein in photo-protection of photosystem II in cyanobacteria. Biochim Biophys Acta 1817 (1): 158–166.
64. Johnsen G, Prezelin BB, Jovine RVM (1997) Fluorescence excitation spectra and light utilization in two red tide dinoflagellates. Limnol Oceangr 42 (S. part 2) : 166–177.
65. Geider RJ, Platt T, Raven JA (1986) Size dependence of growth and photosynthesis in diatoms: a synthesis. Mar Ecol Prog Ser 30: 93–104.
66. Kirk JTO (1986) Optical properties of picoplankton suspensions. Can Bull Fish Aquat Sci 214: 501–520.
67. Suggett DJ, Le Floc'H E, Harris GN, Leonardos N, Geider RJ (2007) Different strategies of photoacclimation by 2 strains of *Emiliania huxleyi* (Haptophyta). J Phycol 43: 1209–1222.
68. Falkowski PG, Dubinsky Z, Wyman K (1985) Growth-irradiance relationships in phytoplankton. Limnol Oceanogr 30: 311–321.

8

Quantum Transport in Networks and Photosynthetic Complexes at the Steady State

Daniel Manzano[1,2]*

1 Instituto Carlos I de Fisica Teorica y Computacional, University of Granada, Granada, Spain, 2 Institute for Theoretical Physics, University of Innsbruck, Innsbruck, Austria

Abstract

Recently, several works have analysed the efficiency of photosynthetic complexes in a transient scenario and how that efficiency is affected by environmental noise. Here, following a quantum master equation approach, we study the energy and excitation transport in fully connected networks both in general and in the particular case of the Fenna–Matthew–Olson complex. The analysis is carried out for the steady state of the system where the excitation energy is constantly "flowing" through the system. Steady state transport scenarios are particularly relevant if the evolution of the quantum system is not conditioned on the arrival of individual excitations. By adding dephasing to the system, we analyse the possibility of noise-enhancement of the quantum transport.

Editor: Gerardo Adesso, University of Nottingham, United Kingdom

Funding: The research was funded by the Austrian Science Fund (FWF): F04011 and F04012 and by Spanish MEC-FEDER, project FIS2009-08451, together with the Campus de Excelencia Internacional and the Junta de Andalucia, project FQM-165 (Spain). The funders had no role in study design, data collection and analysis, decision to publish, or preparation of the manuscript.

Competing Interests: The author has declared that no competing interests exist.

* E-mail: daniel.manzano@uibk.ac.at

Introduction

In the last years, quantum transport in photosynthetic complexes has become an interesting field of study and debate. An important part of this research focusses on the excitation transfer from the antennae that harvest the sunlight to the reaction centre (RC) where the photosynthetic process takes place. More concretely, for the Fenna–Matthew–Olson (FMO) complex of green sulfur bacteria, empirical evidence suggests that such transport is coherent even at room temperature [1–3]. These experiments show that the transient behaviour takes place on time scales much shorter than the decoherence time due to the environment. Thus, most of the recent analysis has focussed on the single-excitation scenario in the transient regime obtained after pulsed photoexcitation [4–12].

Actually, there is a vivid debate about the validity of the single-excitation picture for modelling the photosynthetic process *in vivo*. Photosynthesis in nature is a continuous process of absorption of energy from a radiation field. As there is no specific measurement mechanism that determines when the quanta of energy are effectively absorbed, some authors have argued that the photosynthetic complex and the radiation field should evolve to a steady state where the energy is constantly flowing through the system [13,14]. Some of the conclusions of [14], regarding the importance of a steady state picture, are summarized in the following paragraph: 'The classical picture of the photon as a particle incident on the molecule, repeatedly initiating dynamics, also assumes a known photon arrival time. This too is incorrect and inconsistent with the quantum analysis insofar as no specific arrival time can be presumed unless the experiment itself is designed to measure such times'. Also, it has been shown that some conclusions regarding the presence of entanglement in this kind of system rely on the assumption that the system is excited by a single excitation Fock state. This state cannot be obtained just by weak illumination, and changing this assumption for a more realistic one changes dramatically the conclusions [15]. These arguments makes it reasonable to analyse the natural photosynthetic processes also in other regimes, such as a steady state scenario.

Moreover, quantum transport in a non-equilibrium steady state is an active field in condensed matter physics. For ordered systems composed of qubits or harmonic oscillators, it has been shown that it is possible to violate Fourier's law and thus achieve an infinite thermal conductivity in the absence of noise [16,17]. This ballistic transport turns into a diffusive one, with finite conductivity, if noise is added to the system as a dephasing channel, reducing therefore the energy transfer. That fact highlights the importance of the interaction with a dephasing environment for the energy transfer. The analysis of quantum transport can contribute to the design of artificial light-harvesting systems that are more efficient and robust [18].

Recently, quantum transport in photosynthetic complexes has been analysed through different models with different measures of the efficiency, principally in the single-excitation regime. In [10], the dynamics of the FMO complex was analysed by the use of a Markovian Redfield equation and by a generalized Bloch–Redfield equation [19]. The measure of efficiency that they use is the average time that a single excitation spends in the network before being absorbed by the sink. The results show that the Redfield approach correctly describes the dynamics of the system, but also that it fails to determine the optimal dephasing ratio that minimizes the trapping time. Moreover, this approach gives the unphysical results of a zero trapping time in the limit of strong dephasing $\gamma \to \infty$. An analogous model was considered in [9], with the difference that the efficiency was quantified by the population of the sink in the long time limit. Finally, Scholak *et al.* [11,12] have studied this problem in the absence of a sink, in such a way that the only incoherent dynamics was due to the presence of a dephasing environment. Here, the index for the quantification of the efficiency was the highest probability of finding the excitation in the outgoing qubit in a time interval $[0,\tau]$, with τ being related to the estimate of the duration of the excitation transfer in real systems. The authors conclude that the addition of noise can increase the efficiency, but mainly in configurations that initially performed poorly. Despite their differences, all these papers coincide in analysing only the transient behaviour, and not the steady state, and they use very different indexes for quantifying the efficiency of the system.

In this paper, we analyse the energy transfer in quantum networks and, specifically, in the FMO complex in a steady state. We show that the excitations move coherently through the system also in this regime. The addition of a dephasing environment reduces, but does not destroy, the coherent transport. We also analyse the change in efficiency due to such an environment. The model we consider here is based on a quantum network connected to a thermal bath, to model the absorption of energy from the radiation field, and to a sink, that delivers the energy quanta to the reaction centre. As a particular case, we analyse the FMO complex and similar fully connected networks. In this scenario, the system evolves to a non-equilibrium steady state, where all the observables remain constant. A similar framework has already been used to analyse entanglement in light-harvesting complexes in the transient regime [20].

As has been discussed before, several indexes of the efficiency are usually applied in order to calculate the efficiency of these kinds of system. Also, for the complete photosynthetic procces itself, there is an important difference between analysing it by the use of quantum efficiency, that is, the average number of absorbed photons that finally give rise to photosynthetic products, and the energy efficiency. The second one is considered a more appropriate measure for comparing the efficiency of photosynthetic complexes with artificial light harvesting systems and for analysing the global procedure [18]. Because of that, we will use the energy transfer per unit of time, that is, the power, as our principal index of the efficiency of the systems. This measure will be compared with the excitation transfer, which corresponds to the quantum efficiency. We will show that, in general, they behave in a very different way, especially under the effect of noise.

The present paper is organized as follows. In the next section, we introduce the details of the model and of the master equation which describes its dynamics. In Section III, we introduce two indexes for evaluating the efficiency, and we perform an analytical comparison between the two of them. Uniform and general networks are analysed in Section IV, while in Section V we focus our attention on the FMO complex and related Hamiltonians. Finally, in Section VI some conclusions are drawn.

Methods

Description of the Model and the Master Equation

The energy transfer in photosynthetic complexes, such as the FMO complex, can be described by exciton dynamics. Such systems can be modelled as fully connected networks of two-level systems (qubits), and several recent works have analysed photosynthetic processes by the use of this framework. Most of these works have used a single excitation framework, different from the one used here. Since the FMO complex is composed of seven chromophores, it should be modelled by a network of seven qubits. To describe the absorption of energy from the antennae and the decay to the reaction centre (RC), we use a Markovian quantum master equation in Lindblad form [21]. The validity of this master equation has been numerically verified in systems composed of harmonic oscillators [22], showing that it is accurate for small coupling even in the low and high temperature regimes. In the transient regime, similar master equations have been previously used to describe the dynamics of the FMO complex and to analyse the effects of noise on the quantum dynamics [9,10].

The quantum evolution of the network is determined by a Hamiltonian of the form

$$H=\sum_{i=1}^{7}\hbar\omega_i\sigma_i^+\sigma_i^- + \sum_{\substack{i,j=1\\ j\neq i}}^{7}\hbar g_{ij}\left(\sigma_i^+\sigma_j^- + \sigma_i^-\sigma_j^+\right), \qquad (1)$$

where $\sigma_i^{\pm}$ are the raising and lowering operators that act on qubit i, $\hbar\omega_i$ are the one-site energies, and g_{ij} represents the coupling between qubits i and j.

The interaction of the photosynthetic complex with the environment can be divided into four different processes, and each of them is modelled by a different Lindblad superoperator. According to the empirical results from [23], the absorption of energy from the antennae populates principally site 1, with a non-negligible population of site 6. For simplicity, this process is modelled in our paper by a thermal bath connected to site 1. The delivery of the excitation energy from the complex to the RC is mediated by site 3. As this is a irreversible process, we model it by a sink, a zero temperature thermal bath, which removes the excitations from this site in an incoherent way. Also, photosynthetic complexes are not isolated from the surrounding environment. They interact with other biological components, an interaction that *in vivo* happens at room temperature. That leads to two different effects on the system. First, the loss of coherence in the transport due to the dephasing induced in the system, and second, the absorption of excitations by the environment.

The injection of excitations by the thermal bath acting on the first qubit is modelled by the Linblad superoperator

$$\begin{aligned}\mathcal{L}_{\text{th}}\rho=&\Gamma_{\text{th}}(n+1)\left(\sigma_1^-\rho\sigma_1^+ - \frac{1}{2}\{\sigma_1^+\sigma_1^-,\rho\}\right)\\ &+\Gamma_{\text{th}}n\left(\sigma_1^+\rho\sigma_1^- - \frac{1}{2}\{\sigma_1^-\sigma_1^+,\rho\}\right),\end{aligned} \qquad (2)$$

where the parameter Γ_{th} represents the strength of the coupling between the quantum system and the environment. As there are no empirical estimates of this parameter, we take $\Gamma_{\text{th}}=1\ \text{cm}^{-1}$ through the paper. The parameter n is the mean number of excitations with frequency ω in the bath.

The delivery of energy from the system to the RC is modelled by a second Lindblad superoperator, which models a zero temperature thermal bath. This bath is usually referred to as a sink.

$$\mathcal{L}_{\text{sink}}\rho=\Gamma_{\text{sink}}\left(\sigma_3^-\rho\sigma_3^+ - \frac{1}{2}\{\sigma_3^+\sigma_3^-,\rho\}\right). \qquad (3)$$

This term describes the irreversible decay of the excitations to the reaction centre. When the excitation is absorbed by the sink, it triggers a charge separation event and can not go back to the complex. We assume that this process is faster than the system dynamics and, because of that, the sink does not saturate. So, it can be described by a Markovian approach. Again, the coupling strength Γ_{sink} has not been estimated from experimental data, and we choose $\Gamma_{sink} = 1\text{cm}^{-1}$.

The interaction between the complex and its surrounding environment has two different effects. First, it reduces the quantum coherence of the system. This is modelled in our master equation by the term

$$\mathcal{L}_{deph}\rho = \gamma \sum_{i=1}^{7} \left(\sigma_i^+ \sigma_i^- \rho \sigma_i^+ \sigma_i^- - \frac{1}{2}\{\sigma_i^+ \sigma_i^-, \rho\} \right). \quad (4)$$

This interaction does not change the mean number of excitations in the system but, as we will see in next section, it can affect the its mean energy, and so the energy flux. Also, this is the interaction that can improve the efficiency of the system by removing the destructive coherences that delay the transmission of the excitation to the third site. The parameter γ represents the strength of the interaction between the complex and the dephasing environment. This will be the free parameter we use in this paper in order to optimize the efficiency of the system.

Finally, the system is also susceptible to a radiative decay process that transfers the excitations from the complex to the environment. This process effectively reduces the mean number of excitations in the system together with the mean energy.

$$\mathcal{L}_{diss}\rho = \Gamma_{diss} \sum_{i=1}^{7} \left(\sigma_i^- \rho \sigma_i^+ - \frac{1}{2}\{\sigma_i^+ \sigma_i^-, \rho\} \right). \quad (5)$$

Again, the coupling parameter Γ_{diss} has not been inferred from experimental data, so we will check different values of it in order to analyse the noise-enhancement under different kinds of dissipative environments.

The complete time evolution of the density matrix of the system is described by the master equation

$$\dot{\rho} = -\frac{i}{\hbar}[H, \rho] + \mathcal{L}_{th}\rho + \mathcal{L}_{sink}\rho + \mathcal{L}_{deph}\rho + \mathcal{L}_{diss}\rho. \quad (6)$$

The steady state occurs in the long time limit, and satisfies the condition $\dot{\rho} = 0$, meaning that the density matrix is stationary.

Our analysis has been performed in two different regimes, depending on the rate of excitation injection. First, we study the low-temperature case, by choosing $n = 1$. This choice corresponds to a slow injection of excitations into the system, as should be the case for the FMO complex under weak illumination. In this case, the injection of energy is so slow that the probability of finding more than one excitation at the same time in the network is almost negligible (but not excluded). Second, we simulate a high temperature environment, $n = 100$, where there is a higher probability of finding more than one excitation inside the system at the same time. As we will see in the following sections, these different situations lead to different results.

Energy and Excitation Fluxes

In order to evaluate the efficiency of these systems, we consider two different indexes. First, we observe that the net energy transfer across the system is quantified by the time derivative of the expectation value of the Hamiltonian,

$$\dot{E} = \frac{d}{dt}\langle H \rangle = Tr(H\dot{\rho}). \quad (7)$$

By using the master equation (6) and the fact that, in the steady state, the mean energy of the system is conserved, we obtain an expression for the energy exchanged through the different environmental channels.

$$\begin{aligned} 0 &= Tr(H\mathcal{L}_{th}\rho) + Tr(H\mathcal{L}_{sink}\rho) + Tr(H\mathcal{L}_{deph}\rho) + Tr(H\mathcal{L}_{diss}\rho) \\ &=: J_{th} + J_{sink} + J_{deph} + J_{diss}, \end{aligned} \quad (8)$$

where J_{th} represents the energy flux from the thermal environment to the system, J_{sink} from the system to the sink, J_{deph} between the system and the dephasing environment, and J_{diss} is the energy loss due to the decay of the excitations.

As our main interest is to quantify the energy that flows from the system to the sink, we use $J \equiv J_{sink}$ as our first index to measure the efficiency. This expression has been applied in previous papers in order to analyse the efficiency of quantum refrigerators [25] and to study Fourier's law in quantum systems [16,17].

The second measure of the efficiency that we use is the excitation flux, defined as the number of excitations incoherently absorbed by the sink per unit time. For a time interval $[0, \tau]$, it is given by

$$p_{sink} = \Gamma_{sink} \int_0^{\tau} P dt, \quad (9)$$

where $P \equiv \langle \sigma_3^+ \sigma_3^- \rangle$ is the population of the third site, obviously time-independent in the steady state. This fact allows us to consider the population of the third site as a measure of the speed of the excitation transfer to the RC. These two measures, the energy transfer and the population of the third site, are not equivalent since they measure slightly different quantities. Indeed, the results about the efficiency of the systems are different depending on the measure that is used.

In order to relate these two quantities, it is is useful to decompose the Hamiltonian:

$$H = H_3 + \sum_{\substack{j=1 \\ j \neq 3}}^{7} H_{3j} + \tilde{H}, \quad (10)$$

where $H_3 = \hbar\omega_3 \sigma_3^+ \sigma_3^-$, $H_{3j} = \hbar g_{3j}\left(\sigma_3^+ \sigma_j^- + \sigma_3^- \sigma_j^+\right)$, and $\tilde{H}$ represents the part of the Hamiltonian that does not involve qubit three. Hence, J can be expressed by

$$J = Tr(H_3\mathcal{L}_{sink}) + \sum_{\substack{j=1 \\ j \neq 3}}^{7} Tr(H_{3j}\mathcal{L}_{sink}) + Tr(\tilde{H}\mathcal{L}_{sink}), \quad (11)$$

and the last term is null due to the fact that $Tr(\mathcal{L}_{sink}) = 0$. A straightforward evaluation of this expression allows of expressing the energy transfer to the sink as a function of the population of the third site and the coherences between this qubit and all the others.

$$J = -\Gamma_{\text{sink}}\hbar\left(-\omega_3 P - \sum_{\substack{i=1\\ i\neq 3}}^{7}\frac{g_{i3}}{2}\left(\langle\sigma_3^+\sigma_i^-\rangle + \langle\sigma_i^-\sigma_3^+\rangle\right)\right). \quad (12)$$

As $\langle\sigma_3^+\sigma_i^-\rangle = \langle\sigma_3^-\sigma_i^+\rangle^*$, the heat transfer will depend only on the real part of the next-neighbours coherences. It has been proved that in a linear chain, with equal one-site energies and couplings, these coherences are purely imaginary and the heat flux depends only on the population [16]. In the general case, these coherences will be nonzero and they can contribute in a positive or negative way to the energy flux. It is clear from Eq. (12) that there is a strong connection between the energy and the population fluxes. It is also clear that these measures are related but not equivalent.

In a similar way, the expression of the energy flux due to the dephasing environment can be calculated

$$J_{\text{deph}} = \sum_{\substack{i,j=1\\ j>i}}^{7} -\frac{\gamma\hbar g_{ij}}{2}\left(\langle\sigma_i^-\sigma_j^+\rangle + \langle\sigma_i^+\sigma_j^-\rangle\right). \quad (13)$$

Again, in the concrete case of a uniform chain, these next-neighbours coherences are purely imaginary and because of that this term vanishes. In a general fully connected network, these elements are in general complex and there is an interchange of energy between the network and the environment due to the dephasing channel. That effect happens because the environment projects the system onto a basis that is not composed of the Hamiltonian's eigenvectors. That means a reduction of the elements that are not eigenvectors of a single site basis. As the eigenvalues of the Hamiltonian are usually composed of these non-local terms, this interaction can effectively reduce or increase the energy inside the system. Recently, the energy cost of quantum projective measurements has been analysed and related to the work value of the acquired information [24]. Measurements can change the energy of the network and, in the steady state, that will lead to an energy flux. In a similar way, the presence of dissipation in the system reduces both the excitation and energy fluxes.

Results and Discussion

General Networks

First, we analysed the case of a general fully connected network where the excitations are injected by a thermal reservoir and delivered to a sink. For that, we have calculated the energy end excitation rates as functions of the dephasing ratio for a fully connected homogeneous network in both the low and high temperature regime. Both the one-site energies and the couplings between the qubits are equal and they have the value $\hbar\omega_i = \hbar g_{ij} = 1\text{cm}^{-1}\ \forall i,j$. The couplings with the thermal environment and the sink are $\Gamma_{\text{th}} = \Gamma_{\text{sink}} = 1\ \text{cm}^{-1}$. The analysis is made for different values of the dissipation rate Γ_{diss}, in order to analyse the effects of a dissipative environment on the transfer.

The results for a low temperature thermal bath ($n = 1$), and several dissipation rates ($\Gamma_{\text{diss}} = 0, 0.2, 0.5, 1\ \text{cm}^{-1}$), are shown in Fig. 1. The two different measures of efficiency, the energy and excitation transfer, exhibit very different behaviours when the dephasing parameter increases. The excitation transfer increases for small values of the dephasing ratio and decreases for higher ones. It is clear that the excitation transfer at the steady state can be improved by the addition of external noise to the system. This effect is due to the reduction of destructive interferences that inhibit the transport of the excitation to the qubit coupled to the sink, as is explained in [9]. Similar results have been obtained in the transient regime [9]. On the other hand, the energy transfer is always reduced if a dephasing environment acts on the system. That means that, even where the noise can enhance the number of particles arriving at the sink per unit time, it can at the same time reduce the energy transferred to it. This difference is due to the reduction of the coherences in Eq. (12). For high values of the dephasing rate γ, the transport is reduced, due to a quantum Zeno effect, that avoids the coherent transport. The optimal dephasing ratio that maximizes the excitation transfer is $\gamma^*_{\text{uniform}} = 1.25 \pm 0.1\ \text{cm}^{-1}$. For the energy transfer, the optimal rate is to have no dephasing at all.

For a high temperature bath ($n = 100$), we have similar results, as is displayed in Fig. 2. For this case, both fluxes are higher than in the low temperature regime, as is to be expected, but the effect of dissipation is different. If the system is under the effect of a highly dissipative environment, as $\Gamma_{\text{diss}} = 1\ \text{cm}^{-1}$, the efficiency of the excitation flux can not be improved by the addition of dephasing. The optimal dephasing ratio is independent of the temperature of the bath for no dissipation, and in the low temperature regime it is independent of the dissipation. That indicates that it could be a general property of the Hamiltonian. For a highly dissipative bath, the noise-enhancement of the excitation transfer progressively disappears and, because of that, there is no optimal dephasing rate.

For analysing a more general case, we have also generated 7000 random Hamiltonians, by Monte Carlo simulation, where the one-site energies and the couplings between the qubits are randomly selected from a uniform distribution, with $\hbar\omega_i \in [0,10]$ and $\hbar g_{ij} \in [-10,10]\ \text{cm}^{-1}$. Again, the heat and the excitation transfer exhibit very different behaviours. In Fig. 3, the heat and population fluxes are plotted for a small dephasing ratio, $\gamma = 1$, as a function of the fluxes for the same systems without dephasing, in the low temperature regime. For the heat transfer, the addition of noise to the system can either increase or decrease the efficiency of the system. This improvement is smaller when the system is highly efficient. That shows that the most efficient configurations are also the most robust against the addition of noise, and they are very difficult to improve. The excitation transfer to the sink is improved for all the analysed Hamiltonians, again this improvement is less when the efficiency of the system is greater. These results are compatible with the conclusions of [11], where it is shown that in the single excitation picture, systems with low efficiency are more suitable for improvement by a dephasing channel than are the highly efficient ones.

The results in the case of a high energy thermal bath are very similar, as is displayed in Fig. 4. There, the enhancement is smaller than for $n = 1$ and the population transfer can also be reduced, but only to a small degree. That implies that the energy transfer is more stable under the effects of noise for systems with high number of excitations than for ones in which this number is smaller. Again, the most efficient configurations are more robust against the effect of noise.

Both in the low and high temperature regimes, the effects of dissipation in these simulations are similar. The presence of dissipation reduces both the fluxes and the improvement possible with a dephasing noise. The results are very similar to the ones displayed in Figs. 3 and 4 and are not shown for simplicity.

The differences between the energy and heat transfer come from the fact that the energy transfer depends both on the population of the outgoing site and the coherences between it and the others qubits. Even if the dephasing channel increases the population of

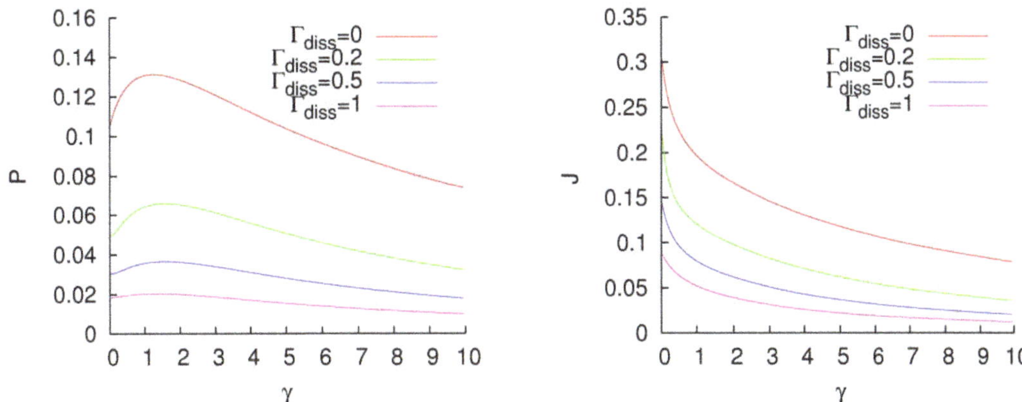

Figure 1. Excitation (left) and energy (right) fluxes for a homogeneous fully connected network, with $\hbar\omega_i = \hbar g_{ij} = 1\text{cm}^{-1}\ \forall i,j$, as functions of the dephasing ratio γ, for a low-temperature thermal bath, $n = 1$, and different dissipation rates Γ_{diss}. Units of cm^{-1}.

this qubit, increasing consequently the transfer of excitations to the sink, it also reduces the amount of coherence between qubits. These two effects compete to determine whether the energy transfer will be improved or depressed. The relation between the one-site energies and the couplings play an essential and non-trivial role in this effect.

Photosynthetic Complexes

To analyse light-harvesting biological systems, we study the FMO protein complex in green sulfur bacteria. This complex is assumed to have seven chromophores and, because of that, it can be modelled as a network of seven sites. We use the experimental Hamiltonian given in [23], tables 2 MEAD and 4 (trimer). In cm^{-1}, this Hamiltonian reads:

$$H_{\text{FMO}} = \begin{pmatrix} 12445 & -104.0 & 5.1 & -4.3 & 4.7 & -15.1 & -7.8 \\ -104.0 & 12450 & 32.6 & 7.1 & 5.4 & 8.3 & 0.8 \\ 5.1 & 32.6 & 12230 & -46.8 & 1.0 & -8.1 & 5.1 \\ -4.3 & 7.1 & -46.8 & 12355 & -70.7 & -14.7 & -61.5 \\ 4.7 & 5.4 & 1.0 & -70.7 & 12680 & 89.7 & -2.5 \\ -15.1 & 8.3 & -8.1 & -14.7 & 89.7 & 12560 & 32.7 \\ -7.8 & 0.8 & 5.1 & -61.5 & -2.5 & 32.7 & 12510 \end{pmatrix}. \tag{14}$$

As for this Hamiltonian the one-site energies are two orders of magnitude higher than the couplings, we can expect that they should play a more relevant role for the energy flux. In this concrete case, the energy and excitation transfer are very similar, in contrast to the general networks analysed before. As there are no empirical measurements of the coupling between the complex and the antennae or the RC, we choose $\Gamma_{\text{th}} = \Gamma_{\text{sink}} = 1\ \text{cm}^{-1}$.

In Fig. 5 the energy and excitation fluxes are plotted as a function of the dephasing ratio γ, for $n = 1$, and $\Gamma_{\text{diss}} = 0, 0.2, 0.5, 1\ \text{cm}^{-1}$. The addition of noise improves both the excitation and the energy fluxes in this system and both measures of efficiency have practically the same behaviour. Similar results arise for $n = 100$, and are omitted. The optimal value of the dephasing ratio is equal for both the less and the highly excited scenarios, and for the energy and excitation transfers, with an optimal value $\gamma^*_{\text{FMO}} = 60 \pm 1\ \text{cm}^{-1}$. It is also independent of the dissipation acting on the system. This result is of the same order, but quantitatively different, as the one obtained in [10] by using the mean trapping time as a measure of the efficiency and a global Redfield equation to describe the dynamics of the network. This optimal ratio is higher than in the case of the homogeneous network analysed before, due to the different order of magnitude of the ratio of the energies and the couplings. Again, the presence of dissipation in the system reduces the fluxes but it does not affect the qualitative behaviour. The similarity between both measures of

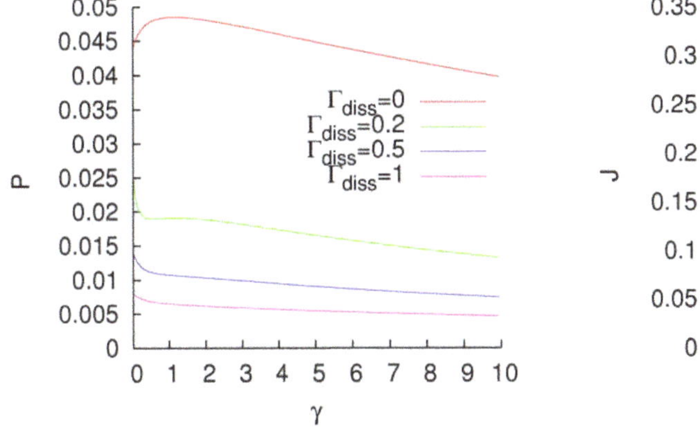

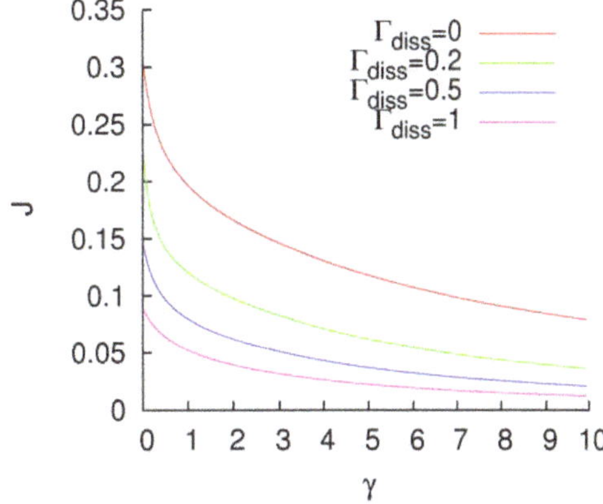

Figure 2. Excitation (left) and energy (right) fluxes for a homogeneous fully connected network, $\hbar\omega_i = \hbar g_{ij} = 1\text{cm}^{-1}\ \forall i,j$, with a thermal mean excitation number $n = 100$, and different dissipation rates Γ_{diss}. Units of cm^{-1}.

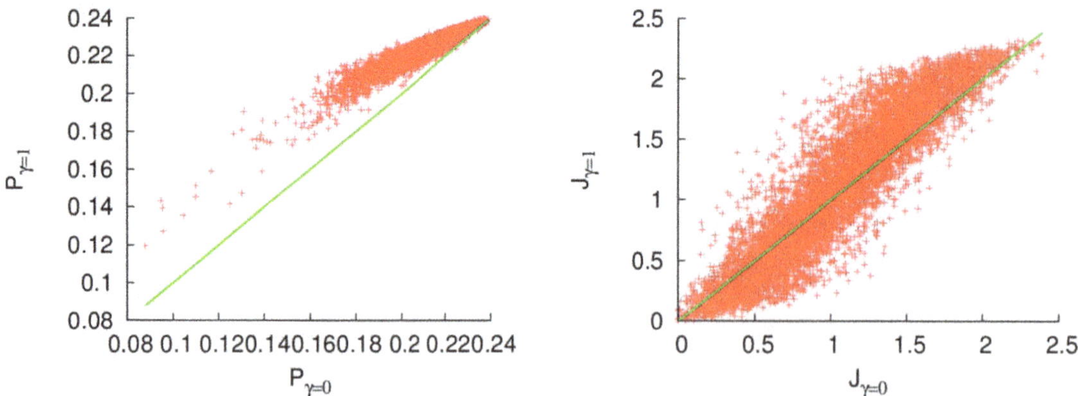

Figure 3. Excitation (left) and energy (right) fluxes for random Hamiltonians in the low-temperature regime ($n = 1$), for a dephasing ratio $\gamma = 1$ as a function of the fluxes without dephasing. The green line separates the configurations with enhancement and depression of the transfer. Units of cm^{-1}.

efficiency comes from the differences between the one site energies and the couplings, which make the population of the third site the dominant component of Eq. (12).

As the FMO Hamiltonian is inferred from experimental spectroscopy data, it is subject to experimental uncertainty. To check a more complete scenario, we performed a Monte Carlo simulation. For this simulation, 7000 random Hamiltonians were generated, where each parameter x corresponds to a Gaussian distribution with mean in the corresponding FMO parameter x_{FMO} and variance $var = 0.1\, x_{FMO}$. By this simulation we analyse random Hamiltonians with the same order of magnitude as that of the FMO.

For a low temperature bath, the results for the energy transfer are shown in Fig. 6. For small dephasing, most of the configurations are improved. This improvement is more important for configurations with low efficiency, and it is less for the most efficient ones. For higher values of γ, most of the configurations are degraded, principally only the ones with low efficiencies are enhanced. Again, this result is similar if we use the population of the sink as our index. This similarity can be understood by analysing Eq. (12). The energy flux depends on the population of the third site and on the coherences, modulated by the one site energy $\hbar\omega_3$ and the couplings $\hbar g_{3i}$, respectively. If the one site energy of the third qubit is much higher than the couplings, the population of this site becomes the dominant term and both fluxes exhibit a similar behaviour.

In this section, we proved that for the FMO complex, it is possible to improve both the energy and the excitation fluxes. That is due, principally, to the fact that the FMO Hamiltonian is a configuration with a very low efficiency. If the Hamiltonian is modified in order to obtain a similar but more efficient one, this improvement reduces drastically.

Conclusions

In this paper, we analysed the energy transfer in quantum networks and its behaviour when external noise is added to the system. Special emphasis has been placed on networks that model a real photosynthetic light-harvesting system, the FMO complex from green sulfur bacteria. This analysis has been performed in a steady state scenario. In this regime, the evolution of the system is not conditioned on the arrival of individual excitations, and the energy flows across the system continuously.

From our analysis we can conclude the following:

- Even in the non-equilibrium steady state, there are time independent coherences in the system. These coherences can contribute both in a positive or negative way to the energy transfer.

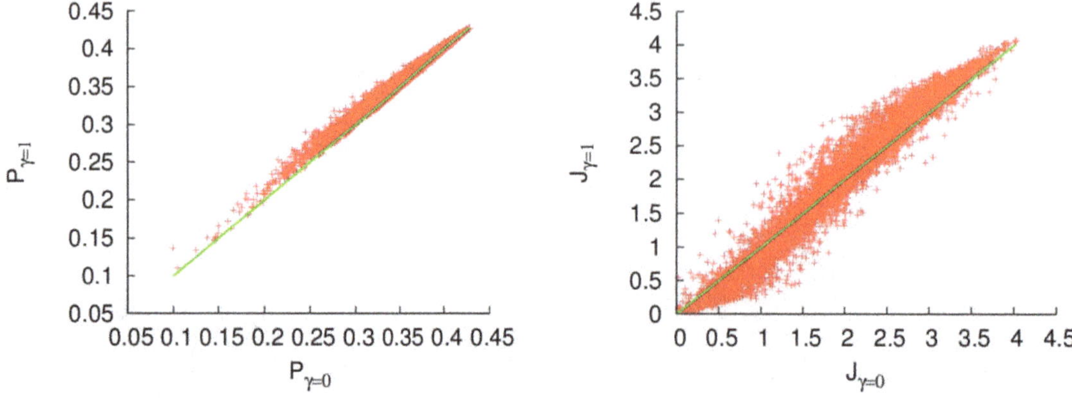

Figure 4. Excitation (left) and energy (right) fluxes for random uniformly distributed Hamiltonians in the low-temperature regime ($n = 100$), for a dephasing ratio $\gamma = 1$, as functions of the fluxes without dephasing. The green line separates the configurations with enhancement and depression of the transfer. Units of cm^{-1}.

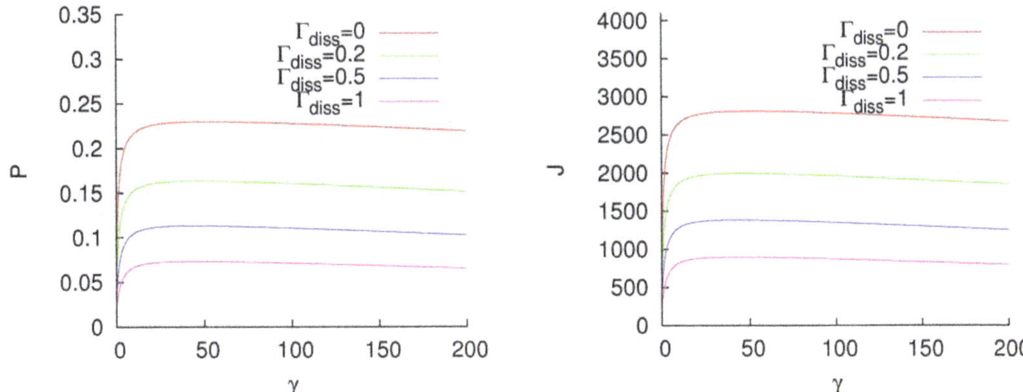

Figure 5. Excitation (left) and energy (right) fluxes from the network to the sink for the FMO Hamiltonian as functions of the dephasing ratio, $n=1$**.** Units of cm^{-1}.

- The power of the system behaves in a very different way from the population transfer. That is due to the fact that the energy depends on the population but also on the coherences of the system.
- The population transfer is more amenable to improvement than the power. That means that even in the case in which the dephasing channel reduces the destructive interference in the system, it also reduces the coherences inside it, and that reduces the energy transfer in most cases.
- Also, the systems with low efficiency are more amenable to enhancement by a dephasing channel than the highly efficient ones. That is due to the fact that the interferences an be both destructive and constructive, and dephasing reduces both of them. Because of that, it can rarely improve a well performing configuration.
- Finally, systems that are in a high illumination regime are more stable under fluctuations due to the environmental interaction.

From the biological point of view, and keeping in mind that this is only a simple model far from the real photosynthetic scenario, we conclude that the energy transfer of the FMO complex can be enhanced by the addition of dynamical noise. On the other hand, this result is based on a single Hamiltonian inferred from empirical data. Our analysis shows also that if we slightly move from this Hamiltonian there are configurations that are better performing than the FMO. Also, we conclude that their energy transfer is not much enhanced by a dephasing channel. So, in order to design a highly efficient light harvesting system, there are two possibilities: it can have low efficiency in a pure coherent dynamic and can be improved under the effect of decoherence, or it can be just more efficient for the coherent dynamics in the first place. The optimal choice depends on the environmental situation and on the practical constraints, but in most of the cases analysed in this paper, the second choice is the optimal one.

Acknowledgments

The author would like to thank M. Tiersch, G.G. Guerreschi, and H.J. Briegel for useful conversations.

Author Contributions

Conceived and designed the experiments: DM. Performed the experiments: DM. Analyzed the data: DM. Contributed reagents/materials/analysis tools: DM. Wrote the paper: DM.

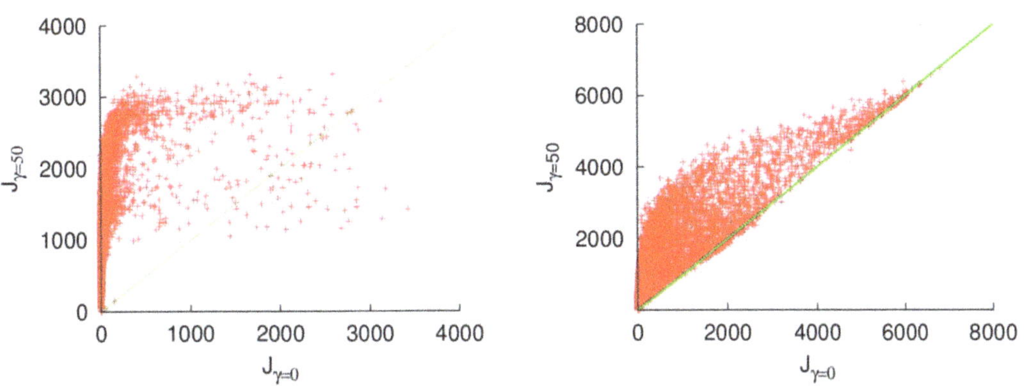

Figure 6. Energy flux for random Hamiltonians normally distributed around the FMO Hamiltonian, for $n=1$ **(left) and** $n=100$ **(right).** The y-axis represents the energy flux under a dephasing channel with $\gamma=50$ and the x-axis represents the energy flux without dephasing. The green line represents the space where both fluxes are equal. Units of cm^{-1}.

References

1. Engel GS, Calhoun TR, Read EL, Ahn TK, Mancal T, et al. (2007) Evidence for wavelike energy transfer through quantum coherence in photosyntetic systems. Nature 446: 782.
2. Collini E, Wong CY, Wilk KE, Curmi PMG, Brumer P, et al. (2010) Coherently wired light-harvesting in photosynthetic marine algae at ambient temperature. Nature 463: 644–646.
3. Panitchayangkoon G, Hayes D, Fransted KA, Caram JR, Harel E, et al. (2010) Long-lived quantum coherence in photosynthetic complexes at physiological temperature. Proc Natl Acad Sci 107: 12766.
4. Plenio S, Huelga M (2008) Dephasing-assisted transport: quantum networks and biomolecules. J New Phys 10: 113019.
5. Mohseni M, Rebentrost P, Lloyd S, Aspuru-Guzik A (2008) Enviroment-assisted quantum walks in photosynthetic energy transfer. Journal of Chemical Physics 129: 174106.
6. Olaya-Castro A, Lee CF, Olsen FF, Johnson NF (2008) Efficiency of energy transfer in a light-harvesting system under quantum coherence. PRB 78: 085115.
7. Caruso F, Chin AW, Datta A, Huelga SF, Plenio MB (2009) Highly efficient energy excitation transfer in light-harvesting complexes: The fundamental role of noise-assisted transport. J Chem Phys 131: 105106.
8. Rebentrost P, Mohseni M, Kassal I, Lloyd S, Aspuru-Guzik A (2009) Environment-assisted quantum transport. NJP 11: 033003.
9. Chin AW, Datta A, Caruso F, Huelga SF, Plenio MB (2010) Noise-assisted energy transfer in quantum networks and light harvesting complexes. J New Phys 12: 065002.
10. Wu J, Liu F, Shen Y, Cao J, Silbey RJ (2010) Efficient energy transfer in light-harvesting systems, I: Optimal temperature, reorganization energy and spatial-temporal correlations. NJP 12(105012).
11. Scholak T, de Melo F, Wellens T, Mintert F, Buchleitner A (2011) Efficient and coherent excitation transfer across disordered molecular networks. Phys Rev E 83(2): 021912.
12. Scholak T, Wellens T, Buchleitner A (2011) Optimal networks for excitonic energy transport. J Phys B: At Mol Opt Phys 44: 184012.
13. Mancal L, Valkunas T (2010) Exciton dynamics in photosynthetic complexes: excitation by coherent and incoherent light. NJP 12: 065044.
14. Brumer M, Shapiro P (2012) Molecular response in one photon absorption: Coherent pulsed laser vs. thermal incoherent source. Proc Natl Acad Sci 109: 19575.
15. Tiersch M, Popescu S, Briegel HJ (2012) A critical view on transport and entanglement in models of photosynthesis. Phil Trans R Soc A 370: 3771.
16. Manzano D, Tiersch M, Asadian A, Briegel HJ (2012) Quantum transport efficiency and Fourier's law. Phys Rev E 86: 061118.
17. Asadian A, Manzano D, Tiersch M, Briegel HJ (2013) Heat transport through lattices of quantum harmonic oscillators in arbitrary dimensions. Phys Rev E 87: 012109.
18. Blankenship RE, Tiede DM, Barber J, Brudvig GW, Fleming G, et al. (2011) Comparing photosynthetic and photovoltaic efficiencies and recognizing the potential for improvement. Science 332(6031): 805–809.
19. Cao JS (1997) A phase-space study of Bloch–Redfield theory. Journal of Chemical Physics 107: 3204.
20. Caruso F, Chin AW, Datta A, Huelga SF, Plenio MB (2010) Entanglement and entangling power of the dynamics in light-harvesting complexes. Phys Rev A 81(6): 062346.
21. Breuer F Petruccione HP (2002) The theory of open quantum systems. Oxford: Oxford University Press.
22. Rivas A, Plato AD, Huelga S, Plenio MB (2010) Markovian master equations: A critical study. J New Phys 12: 113032.
23. Adolphs T, Renger J (2006) How proteins trigger excitation energy transfer in the FMO complex of green sulfur bacteria. Biophys Journal 91(2778).
24. Jacobs K (2012) Quantum measurement and the first law of thermodynamics: The energy cost of measurement is the work value of the acquired information. Phys Rev E 86: 040106.
25. Linden N, Popescu S, Skrzypczyk P (2010) How small can thermal machines be? The smallest possible refrigerator. Phys Rev Lett 105(13): 130401.

9

Mutations of Photosystem II D1 Protein That Empower Efficient Phenotypes of *Chlamydomonas reinhardtii* under Extreme Environment in Space

Maria Teresa Giardi[1]🙤, Giuseppina Rea[1]🙤, Maya D. Lambreva[1], Amina Antonacci[1], Sandro Pastorelli[1], Ivo Bertalan[2], Udo Johanningmeier[2], Autar K. Mattoo[3]*

1 Institute of Crystallography, National Research Council of Italy, CNR, Rome, Italy, **2** Institute of Plant Physiology, Martin-Luther University Halle-Wittenberg, Halle (Saale), Germany, **3** The Henry A. Wallace Beltsville Agricultural Research Center, United States Department of Agriculture, Agricultural Research Service, Sustainable Agricultural Systems Laboratory, Beltsville, Maryland, United States of America

Abstract

Space missions have enabled testing how microorganisms, animals and plants respond to extra-terrestrial, complex and hazardous environment in space. Photosynthetic organisms are thought to be relatively more prone to microgravity, weak magnetic field and cosmic radiation because oxygenic photosynthesis is intimately associated with capture and conversion of light energy into chemical energy, a process that has adapted to relatively less complex and contained environment on Earth. To study the direct effect of the space environment on the fundamental process of photosynthesis, we sent into low Earth orbit space engineered and mutated strains of the unicellular green alga, *Chlamydomonas reinhardtii*, which has been widely used as a model of photosynthetic organisms. The algal mutants contained specific amino acid substitutions in the functionally important regions of the pivotal Photosystem II (PSII) reaction centre D1 protein near the Q_B binding pocket and in the environment surrounding Tyr-161 (Y_Z) electron acceptor of the oxygen-evolving complex. Using real-time measurements of PSII photochemistry, here we show that during the space flight while the control strain and two D1 mutants (A250L and V160A) were inefficient in carrying out PSII activity, two other D1 mutants, I163N and A251C, performed efficient photosynthesis, and actively re-grew upon return to Earth. Mimicking the neutron irradiation component of cosmic rays on Earth yielded similar results. Experiments with I163N and A251C D1 mutants performed on ground showed that they are better able to modulate PSII excitation pressure and have higher capacity to reoxidize the Q_A^- state of the primary electron acceptor. These results highlight the contribution of D1 conformation in relation to photosynthesis and oxygen production in space.

Editor: Rajagopal Subramanyam, University of Hyderabad, India

Funding: This work was supported by Italian Space Agency, German Aerospace Research Establishmen (DLR) and European Space Agency. Mention of trade names or commercial products in this publication is solely for the purpose of providing specific information and does not imply recommendation or endorsement by the U.S. Department of Agriculture. The funders had no role in study design, data collection and analysis, decision to publish, or preparation of the manuscript.

Competing Interests: The authors have declared that no competing interests exist.

* E-mail: autar.mattoo@ars.usda.gov

🙤 These authors contributed equally to this work.

Introduction

The primary photochemistry leading to photosynthetic oxygen evolution and electron flow in oxygenic organisms involves a supramolecular protein-pigment complex, photosystem II (PSII) [1–3]. PSII is dominated by the D1–D2 reaction center protein heterodimer, which is sensitive to environmental stresses including damage from light and UV irradiation [2,4]. Replacement of highly light-labile D1 protein is a primary event of the PSII repair cycle [5] to sustain crop productivity. It is also known that mutations or genetic substitution of some amino acids in D1 can lead to either an increase or a decrease in photosynthetic activity [6,7].

Oxygenic photosynthesis is a phenomenon prevalent on Earth [8] while little is understood about performance of PSII in space. In space and at high altitudes, the presence of microgravity and cosmic rays, the flow of high-energy particles and ionizing radiation, dominate the environment and are hazardous, especially when the protecting atmosphere is missing [9]. Previous studies on higher plants performance in space reported an overall decrease in photosynthetic activities, mainly due to the microgravity associated lack of convection forces, resulting in alteration in the gases, water and small molecule exchange [10–13]. Unicellular organisms sent on a space flight underwent alterations in growth rate, developmental cycle and morphological characteristics [14,15]. A few studies performed in low Earth orbit and on-ground facilities have focused on the impact of ionizing radiation on photosynthetic organisms [16,17], demonstrating the negative effect of UV or gamma-radiation or fast neutrons on the light-dependent reactions and photosynthetic apparatus [16,18–20]. Current state of knowledge presents space as an intrinsically complex environment, whose components interact to dramatically affect living matter while effects of individual components have been difficult to discern [21]. The importance of Earth magnetic field in maintaining photosynthetic efficiency, chlorophyll accu-

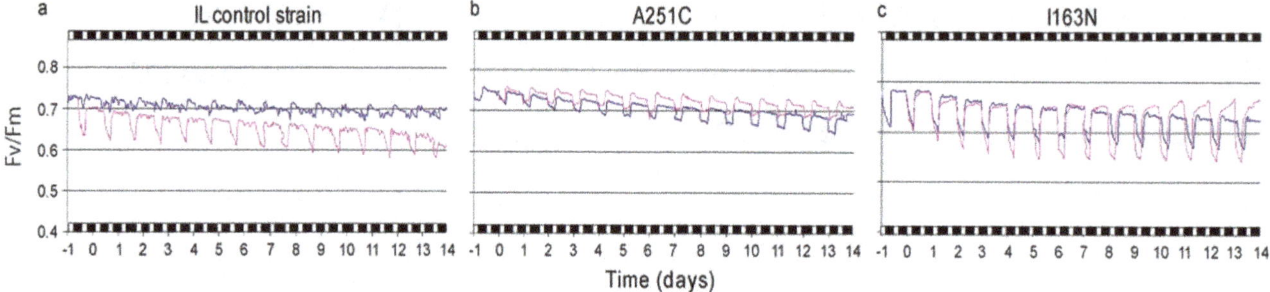

Figure 1. Real-time monitoring of fluorescence emission of *C. reinhardtii* strains during the Foton flight mission. The parental strain IL and its D1 mutants were sent into space inside the re-entry module of the Russian spacecraft Foton M2 capsule aboard Soyuz-U rockets launched from the Baikonur Cosmodrome in Kazakhstan. A multisensor fluorimeter specifically developed to accommodate the *Chlamydomonas* strains monitored fluorescence emission during the space flight and in ground control experiments as described in methods. The algal strains were grown in liquid TAP medium with 150 rpm agitation under a white light (50 µmol m^{-2} s^{-1}) at 25°C. Cells in mid-exponential growth phase (A_{750nm} = 0.4) were harvested by low speed centrifugation, re-suspended in 150 µL of TAP medium in order to reach A_{750nm} = 18 and layered on TAP medium containing 1.65% agar. The organisms were exposed to a 17 h darkness and 7 h 50 µmol $m^{-2}s^{-1}$ daily cycle in the biocells of the fluorometer. Time zero refers to the take-off. Days 0 to14 refer to the flight period. Day -1 refers to 1 day before the take off. Pink, in-flight analysis; blue, cells analyzed on ground under simulated temperature and light conditions of the space mission.

mulation levels, and protecting PSII functionality in photoinhibitory conditions has been alluded to [22–24].

To gain further insights into growth and PSII performance in response to the harsh environment of space, we sent engineered and mutated strains of the unicellular green alga *Chlamydomonas reinhardtii* into low Earth orbit. This unicellular green alga is widely used as a model in studies of oxygenic photosynthesis [25–27] and can adapt to environmental extremes on Earth [28]. Other factors that favoured this choice included the ease in making specific amino acid substitutions in the D1 protein [6] and its ancient origin [6,7,27]. The amino acid substitutions in D1 were made **in** the Q_B binding pocket (Ala 250 and Ala 251), and close to the redox-active Tyr 161 (Val 160 and Ile 163) [1]. The Ala 251 residue, located in the Q_B binding pocket, was previously recognized as a key residue for D1 [6]. Earlier reports on some mutations near these regions of D1 showed a potential in acclimation of *C. reinhardtii* to radiation pressure [27] or ambient temperatures [29] on Earth. Chlorophyll *a* fluorescence induction kinetics of the strains aboard the Foton M2 spacecraft were monitored in real time using the fluorescence sensor, Photo II device [30]. Here we show that two specific D1 mutants (A251C and I163N) of *C. reinhardtii* were capable of efficient photosynthesis in the harsh environment of space. The findings are discussed in relation to the role of D1 conformation in stabilizing (and enhancing) PSII function in photosynthesis and oxygen production in space.

Table 1. Fluorescence parameters and growth of parent strain (IL) and D1 mutants of *C. reinhardtii* after return from space.

Strains	F_0 (% change)[a]	$\Delta F_v/F_m$ (units modified)[b]	Growth[c] ($A_{750Flight}/A_{750Control}$)
IL	+24±1	−0.200±0.004	0.4±0.07
A250L	+50±6	−0.350±0.009	n.d.
A251C	−5±2	+0.004±0.002	3.0±0.04
V160A	+48±6	−0.580±0.010	n.d.
I163N	−10±2	+0.005±0.003	1.6±0.05

The fluorescence parameters, F_0 and F_v/F_m, were recorded by the multicell fluorescence sensor during the space-flight and after landing, representing photosynthetic activity of the indicated genotypes immobilized on TAP medium and placed in the biocells of the fluorometer. Following the space-flight and landing on earth, the algal cultures were transferred to liquid TAP medium and re-growth under continuous light. The growth was determined after re-growing the space-returned genotypes in fresh liquid media for 3 days and compared to the ground controls.

[a]Values are a ratio of F_0 measured one day after landing to values registered in space on the last day of the flight, using the equation: F_0 (% change) = $[(F_{0\ after\ landing} - F_{0\ in\ flight})/(F_{0\ in\ flight})]*100$.

[b]Values represent the difference between F_v/F_m on the day after landing and those registered in space on the last day of the flight, using the equation: $\Delta F_v/F_m = (F_v/F_{m\ after\ landing} - F_v/F_{m\ in\ flight})$.

[c]The growth is represented as the ratio of A_{750nm} values measured after 3 days of re-growth of the space-flown samples and the corresponding ground controls. Growth = $A_{750Flight}/A_{750Control}$.

Materials and Methods

Chlamydomonas reinhardtii culture conditions

C. reinhardtii stock cultures were maintained at 25°C on agar plates prepared with Tris-acetate-phosphate (TAP) medium [31] under continuous illumination (~50 µmol m^{-2} s^{-1}). Liquid cultures were similarly grown but on TAP liquid medium with agitation at 150 rpm. All experiments were carried out on cell cultures in the mid-exponential growth phase.

Production of *C. reinhardtii* D1 mutants

The intronless *C. reinhardtii* strain (IL) and its deletion mutant Dell have been previously described [27,32]. Dell was transformed to create mutations in the *psbA* gene encoding the PSII reaction center protein D1. Substitutions made were: Ala 250 with Leu (A250L) and Ala 251 with Cys (A251C) – all near the Q_B binding pocket; Val 160 with Ala (V160A) and Ileu 163 with Asn (I163N) near the protein environment surrounding the redox-active tyrosine 161 (Y_Z) (Figure S1). The pSH5 plasmid containing the complete intronless *psbA* gene and the 3′-flanking region was amplified in a total volume of 50 µl containing 100 ng plasmid DNA, 5 µl 10x Taq-polymerase buffer, 3 µl of 25 mM $MgCl_2$, 5 µl of 2 mM dNTPs, 20 pmoles of each primer and 1 unit Taq-polymerase using Trioblock (Biometra) with the appropriate primers (Table S1). The standard PCR protocol used was: 25 cycles of 94°C denaturation (1 min), 52°C annealing (1 min), 72°C extension (2 min), with a 5-min denaturation step at 94°C in the first cycle and a 10-min extension step at 72°C in the final cycle.

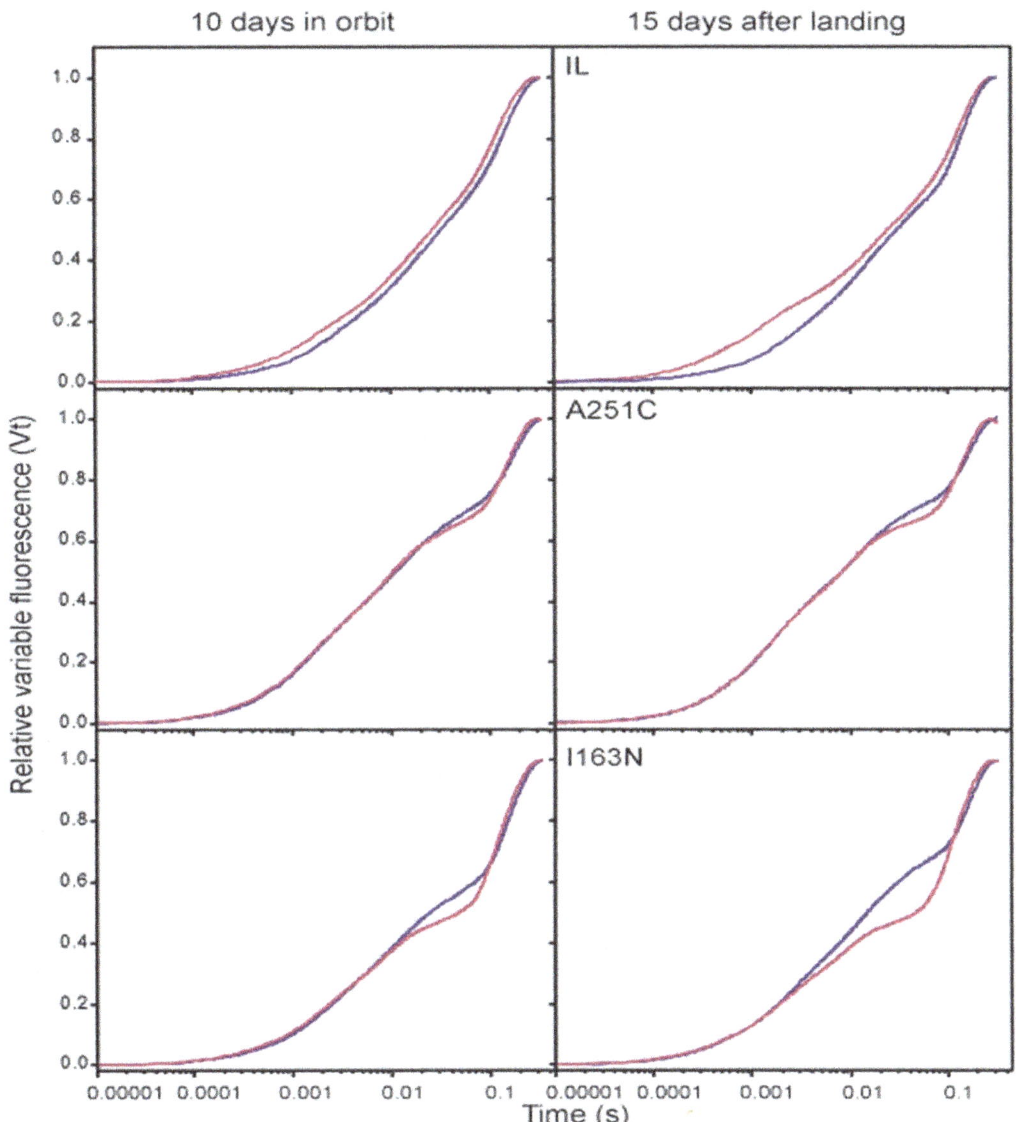

Figure 2. Flight-induced modifications of fluorescence transients in IL and D1-mutants (I163N and A251C) of *C. reinhardtii.* Curves in the left and right panels refer to measurements carried out in flight (10th day in orbit) and 15 days after return to Earth, respectively. Fluorescence transients were measured in samples during space flight and those on ground by Photo II device. The fluorescence curves recorded at the end of the light phase are normalized for both F_0 and F_m values. Curves from a representative experiment are reported – pink, in-flight analysis; blue, on-ground controls.

Amplified products were purified on agarose gels and directly used for algal biolistic transformation [33]. The point mutations were verified by gene sequencing (Seqlab, Göttingen, Germany). Only homoplasmic colonies of the *psbA* mutants, verified by standard PCR and agarose gel electrophoresis, were used.

The Fluorescence Sensor: In-Flight monitoring of Chlorophyll a fluorescence induction kinetics

An automatic bio-device, Photo II (Figure S2), designed by Biosensor.srl (www.biosensor.it), was used to measure and store chlorophyll fluorescence induction data (also known as fluorescence transient or Kautsky effect [34]) during the space flight as well as for the control experiments performed on the ground as previously described [30]. Different strains of *C. reinhardtii* were placed in 24 measuring cells in triplicate/quadruplicate and the fluorescence measurements were recorded hourly for 20 days. The instrument allowed simultaneous determination of the following chlorophyll fluorescence parameters: F_0, F_m, F_v, the ratio F_v/F_m (where $F_v = F_m - F_0$ is the variable fluorescence), the area below the fluorescence curve, and the time to reach F_m in each sample. F_0 was calculated by using an algorithm that determined the line of best fit for the initial data points recorded at the onset of illumination; the best-fit line was then extrapolated to time zero to determine F_0 [30]. In each measuring cell, two white light LEDs were programmed to switch on to provide light (50 µmol m^{-2} s^{-1}) for 7 h in a 24-h period, photoperiod necessary for the organisms to grow on Earth.

Space flight and sample preparation

For the space flight, the algal cultures within the multicell fluorescence sensor were transported under controlled temperature 23–25±1°C to Kazakhstan cosmodrome in Baikonour. The

Table 2. Dark respiration and light compensation point of the *C. reinhardtii* strains.

Strains	Dark respiration μmol O_2 mg Chl^{-1} h^{-1}				Light compensation point μmol photons m−2 s^{-1}			
	Ground control[3a]	Ground control[15a]	Flight[3b]	Flight[15b]	Ground control[3a]	Ground control[15a]	Flight[3b]	Flight[15b]
IL	10±0.2	10±1.0	6±1.8	10±0.2	19±1.1	19±3.0	12±0.1	20±1.0
A251C	12±1.4	12±1.6	9±0.7	19±1.2	15±0.8	25±2.9	15±0.3	36±0.6
I163N	11±1.3	11±0.7	12±1.5	20±1.3	19±2.8	24±2.6	19±0.5	38±0.7

Dark respiration and light compensation point were calculated as the oxygen consumption rate in the dark and the light intensity at which oxygen consumption equals its production, respectively.
[a]Ground control refers to cultures handled on ground under space-simulated conditions and transferred to liquid TAP medium on day 3 ([3a]) and day 15 ([15a]) after landing.
[b]The algal strains following landing of the capsule were treated as described in methods.
Flight[3b] and Flight[15b] refer to space-sent samples after landing and which were transferred to liquid TAP medium on day 3 and day 15 after landing, respectively.
Average values of one experiment in triplicate (n = 3) are shown ± SE.

box containing the algal strains was placed 24 hours before the launch in the UV-shielded internal part of the Biopan re-entry module of the Russian spacecraft Foton built by TsSKB-Samara (ESA Foton website). The spacecraft consisted of three modules – battery module, service module and re-entry module – of which only the latter was retrieved at landing. The Foton M2 capsule was launched in 2005, aboard a Soyuz-U rocket at a height of approximately 300 km. The mission lasted 15 days during periods of a quiet solar activity in the minimum phase of the 23rd solar cycle. The total cosmic dose present inside the aircraft was 3.5±0.25 mGy, mainly due to high-energy heavy ions (Z) particles (HZE) and the secondary radiation derived from neutrons. The temperature during the flight mission and prior to entry and analysis on ground, measured by the fluorescence sensor, varied between 15–21±2°C.

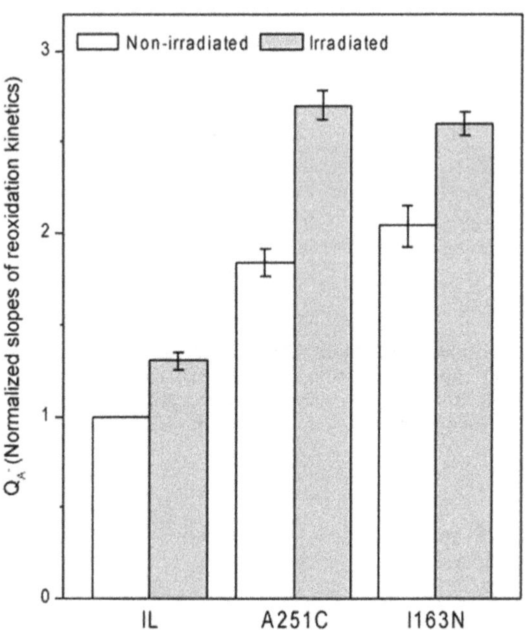

Figure 3. Relative Q_A reoxidation rates in response to on ground neutron irradiation in *C. reinhardtii* strains. To obtain the Q_A reoxidation kinetics slope of the first phase of the fluorescence decay curve for A251C and I163N strains, a logarithmic regression equation was used to find the best fit data sets. Then, the slope value of the fluorescence decay curve in the first 2 ms in each strain was divided by the slope value obtained for the IL control strain. The data are normalized to IL control without irradiation. Data of three independent experiments in duplicate (n = 6) ± SE are shown. The neutron dose rate given was 0.23 mSv h^{-1} for 24 h; energy varied in the range of 0–800 MeV. Measurements were performed under simulated 'space mission' conditions of temperature and light.

Flight and ground-control samples were prepared with small volumes of high-density cell cultures that were subsequently enclosed within the multicell container (multicells) of the fluorescence sensor (Figure S2). Before sample preparation, the multicells were sterilized and filled with TAP agar (1.65%) medium. Algal cells from liquid cultures at the mid-exponential growth phase were harvested by low speed centrifugation, re-suspended in 150 μl of TAP medium to A_{750nm} = 18 and layered on the TAP agar medium in the multicells. Under sterile conditions, after cells had adsorbed (approx. 30 min), the multi-cells were hermetically closed and mounted in the fluorescence sensor. The experimental device was sent into space as described **above.** Following the space flight and return to Earth, the fluorescence sensor was shipped to the CNR (Rome) laboratory on day 3 after landing. The first set of multicells was immediately opened under sterile conditions and the algal strains transferred to fresh liquid TAP medium. The second set of multicells was opened on day 15 after landing and handled like the first set. In both cases, the cell cultures were re-grown in liquid medium for three days to reach a mid-exponential growth phase and then analyzed.

Neutron irradiation and double modulation fluorescence experiments on the ground

The algal strains were irradiated with fast neutrons at a dose rate of 0.23 mSv h^{-1} for 24 h, similar to that in space, using a Super Proton Synchrotron at CERN (Conseil Européen pour la Recherche Nucléaire, Switzerland) as previously described [17]. Q_A^- reoxidation kinetics were determined using a dual-modulation kinetic fluorimeter (Photon Systems Instruments, Brno) [35]. Each algal suspension (20 mg Chl mL^{-1}) in TAP medium was dark-adapted for 10 min prior to each measurement. To analyze the Q_A reoxidation kinetics, a logarithmic regression was calculated to find the equations that best fitted the sets of data corresponding to the slope of the first phase of the fluorescence decay curve. The values were obtained by dividing the slope of the fluorescence decay curve in the first phase with the slope of the fluorescence decay relative to the IL control line. Three independent biological replicates were analyzed in triplicates.

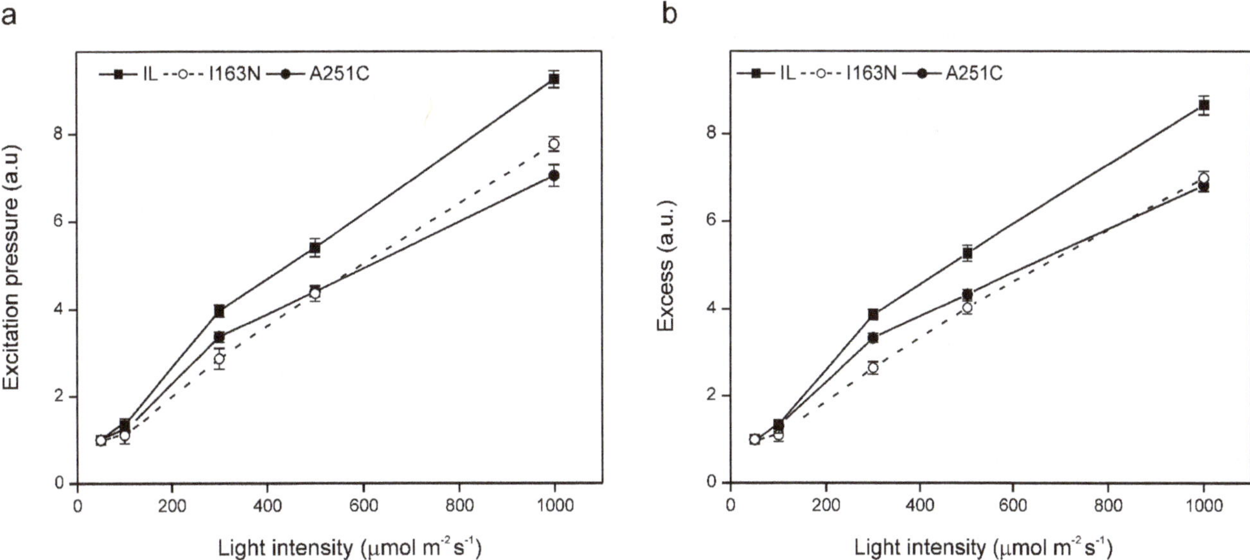

Figure 4. PSII sensitivity to photoinhibition of *C. reinhardtii* parent strain (IL) and D1 mutants (I163N and A251C). Parameters shown are: (**a**) Excitation pressure, $1\text{-qP} = 1\text{-}(F_m'\text{-}F_s)/(F_m'\text{-}F_o')$; (**b**) *Excess*, $(1\text{-qP})(F_v'/F_m')$, measured after 3 min exposure, at the indicated light intensities, as described [48]. Data for each strain are normalized to those obtained at the growth light intensity (50 μmol m^{-2} s^{-1}). Fluorescence measurements were performed on algal cultures at mid-log growth phase containing 98±1 μg chlorophyll ml^{-1} at 24°C in continuous stirring. Average values of three independent experiments in duplicate (n = 6) are shown ± SE.

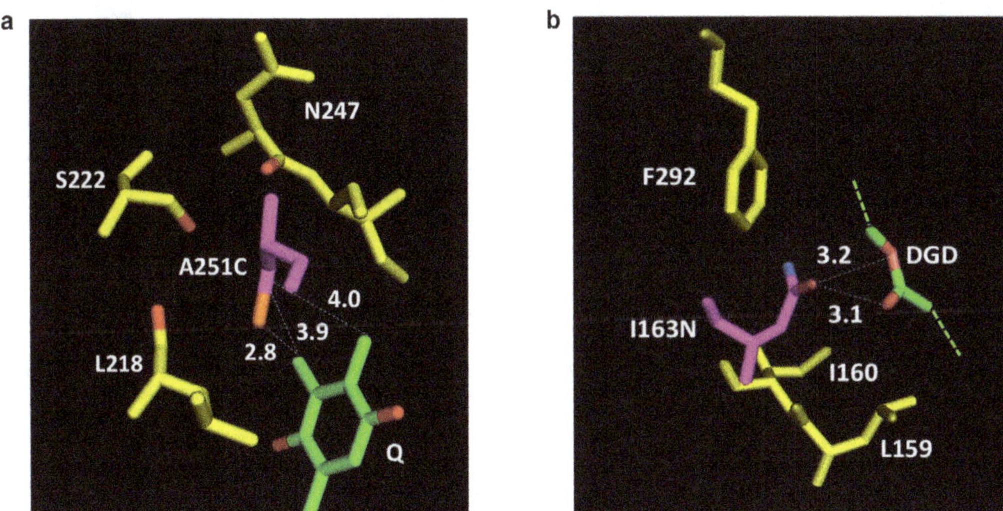

Figure 5. Environment of the D1 protein residue at positions 251 and 163 upon mutation to Cys and Asn, respectively*. A. Cys251 is colored in pink with the sulfur atom colored in orange. The quinone head of PQ is colored by atom type (carbon – green, oxygen – red). Residues having contact with hydrophobic Cβ atom of Ala251 are presented (yellow). Oxygen atoms of these residues that are discussed below are colored in red. Side chain placement of the mutated residue was performed with SCCOMP (19) with option to optimize positions of first-sphere residues. The Cβ atom of the residue at position 251 has two hydrophobic-hydrophobic contacts with carbon atoms of the quinone head (at 3.9 Å and 4.0 Å) in both wild type and the 3 mutants. Replacement of Ala with Cys does not cause steric problems. It provides additional stabilizing contacts (2.8 Å) to a carbon atom of the quinone head. Lengthening the marginally short distance to the quinone head can be achieved by small changes in dihedral angles. All contacts shown for this structure are stabilizing. Additional stabilization of the protein – quinone complex could result from a water mediated H-bond between the sulfur atom of Cys251 and the closest oxygen atom of the quinone head. Alternatively, stabilization of the protein structure itself could occur upon H-bond formation between the sulfur atom of Cys251 and OG atom of Ser222, assuming Cys takes a different rotamer. Either stabilizing effect could play an important role in the radiation resistance of the system. B. Asn163 is colored in pink with side chain oxygen and nitrogen atoms colored in red and blue, respectively. Residues having contact with the side chain atoms of Ile163 are presented (yellow). DGD (digalactosyl diacyl glycerol, partial representation) is colored by atom type (carbon – green; oxygen – red). Side chain placement of the mutated residue was performed with SCCOMP (19) with option to optimize positions of first-sphere residues. The side chain atoms of residue 163 have stabilizing contacts (mainly hydrophobic-hydrophobic) with Leu159 and Phe292. Replacement of Ile with Asn does not cause steric problems. It provides additional stabilizing contacts to two oxygen atoms of DGD (H-bond lengths of 3.2 and 3.1 Å). *After Sobolev V., Samish I., and Edelman M. (personal communication).

Table 3. List of amino acid residues in contact with mutated Cys251 and Asn163[#].

Residues in contact with mutated Cys251				**Residues in contact with mutated Asn163**			
		Dist. Å	**Surf Å²**			**Dist. Å**	**Surf Å²**
218	Leu	3.1	35.4	91A	Leu*	3.7	11.8
222	Ser	3.7	15.4	159A	Leu*	3.0	19.3
247	Asn	3.0	12.8	160A	Ile*	3.4	15.1
248	Ile	3.5	10.8	162A	Pro*	1.3	75.0
250	Ala	1.3	76.1	164A	Gly*	1.3	64.7
252	His	1.3	57.5	166A	Gly	3.2	9.0
254	Tyr	3.2	7.9	288A	Cys	3.6	25.1
255	Phe	2.9	30.1	292A	Phe*	3.3	34.7
713	PQ	2.8	42.8	657A	Dgd	3.0	44.1

Distances (Dist) and surfaces (Surf) are provided. See also Figure 5a, b.
[#]After Sobolev V., Samish I., and Edelman M. (personal communication).

Excitation pressure experiment on the ground

Fluorescence quenching parameters were measured on continuously stirred and uniformly illuminated liquid cultures at 24°C by combining Fluorescence Modulated System (FMS2, Hansatech Ins., Norfolk, UK) with OxyLab instrument (Hansatech Ins., Norfolk, UK). Algal cultures in the mid-log growth phase were concentrated by centrifugation and dark-adapted for 10 min prior to the measurement. Experimental protocol [36] and the nomenclature [37] for the fluorescence analysis were the same as described. Prior to irradiation, the minimum fluorescence (F_0) was measured by switching on the modulated light (0.004 µmol m^{-2} s^{-1}); the maximum (F_m) fluorescence was induced by a short saturating flash of 0.7 s of 9000 µmol m^{-2} s^{-1} of white actinic light. The levels of F_s, F_m' and F_0' (steady state, maximum and minimum levels of Chl fluorescence for light-adapted samples, respectively) were recorded after 3-min light exposure consecutively at indicated light intensities. F_m' was recorded during the saturating flash (0.7 s of 9000 µmol m^{-2} s^{-1}) and F_0' after the actinic light was temporarily switched off. Short-term far-red irradiation was applied to insure the oxidation of the PSII acceptor side for measuring F_0'.

Oxygen evolution measurements

Photosynthetic activity of the cells (20 mg mL^{-1} Chl) was measured using a Clark-type oxygen electrode (S1 electrode disk) connected to a Chlorolab 2 System and liquid-phase electrode chamber DW2/2 (Hansatech Ins., Norfolk, UK). To ensure that oxygen evolution was not limited by the carbon source available to the cells, 200 mL of a 50 mM sodium bicarbonate solution, pH 7.4, were added prior to the measurements [38]. The samples were illuminated under continuous stirring (70 *rpm*) at 24°C with increasing light intensities (from 0 to 350 µmol m^{-2} s^{-1}), provided by red LED Light Source LHII/2R with a maximum at 650 nm (Hansatech Ins., Norfolk, UK). The rate of oxygen evolution was recorded continuously for 2 min. The light compensation point was calculated as the light irradiance at which net gas exchange is zero when photosynthesis and respiration rates are balanced using regression analysis of the linear part of the curve (up to 100 µmol m^{-2} s^{-1}). At the end of the light exposure, the samples were dark adapted for 15 min and assayed for dark respiration.

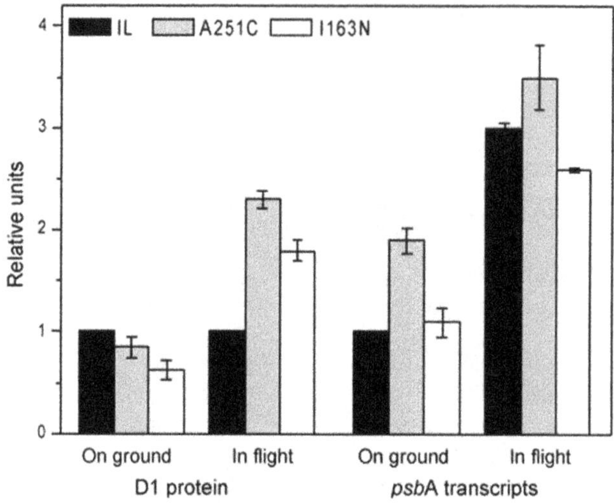

Figure 6. D1 content and relative abundance of *psbA* transcripts in *C. reinhardtii* strains after spaceflight. Upon return to Earth, the algal strains were transferred to liquid TAP medium for re-growth. Total protein and RNA were extracted three days later. D1 band strength was quantified by densitometric scanning of blots (n = 4). *psbA* transcripts were quantified by sqRT–PCR. Changes in D1 protein and *psbA* transcripts are expressed as fold-change in comparison to corresponding values of the IL ground control.

Protein extraction and immunoblotting

Total protein was extracted from the algal cells at the exponential growth phase (3–5×10^6 cells mL^{-1}, equivalent to 50 mg chlorophyll) by sonication on ice with a micro tip (Branson Soni®er 250, Branson Ultrasonics Corporation, Connecticut, USA) in 140 mL solution containing 30% sucrose, 5% SDS and 5% mercaptoethanol. Insoluble material was removed by centrifugation (15,000 g for 3 min) and solubilized proteins electrophoresed using discontinuous tricine/SDS-PAGE (15% acrylamide resolving gel and a 4% stacking gel) [39]. Two µg equivalent of Chl were loaded in each lane. For immunoblots, proteins separated by SDS/PAGE were electrotransferred onto nitrocellulose (BA-85, Schleicher & Schuell, Dassel, Germany) as described [40]. D1 protein was immunodecorated using the anti-*Psb*A (D1) C-terminal hen antibody (Agrisera AB, Sweden).

RNA extraction and sqRT-PCR

C. reinhardtii cultures (A_{750} =0.45, corresponding to: IL, 2.2×10^6 cells mL^{-1}; A251C, 2.5×10^6 cells mL^{-1} and I163N, 1.2×10^6 cells mL^{-1}) were harvested by centrifugation (10 min at 4500 *rpm*, 4°C). Total RNA was isolated [41], its quality verified by electrophoresis on a 1.2% formaldehyde-agarose gel, and quantified (NanoDropTM Spectrophotometer; Thermo Scientific). DNase treatment was given using RQ1 DNase 1U μL^{-1} (Promega). The enzyme was removed by phenol/chloroform extraction. *psb*A transcripts were quantified by sqRT–PCR. RNA from each sample was reverse transcribed and amplified using the SYBR Green PCR Master Mix and MuLV Reverse Transcriptase Reagents (Applied Biosystems), following the one-step RT-PCR protocol recommended by the manufacturer. Briefly, 100 ng of total RNA in 25 µL SYBR Green PCR Master Mix 1 with 0.25 U μL^{-1} MuLV Reverse Transcriptase, 100 nM forward and reverse primers (*psb*Afor 5′ ACACTTGGGCAGACATCA 3′; *psb*Arev 5′ GGAAGTTGTGAGCGTT 3′) were subjected to the following thermal profile: 42°C for 30 min, 95°C for 10 min, 40 cycles with a denaturation step at 95°C for 15 s and an annealing/extension step at 60°C for 1 min. PCRs were performed in the PE Biosystems GeneAmp 5700 Sequence Detection System using Frosted Subskirted optical tubes and Seal Film (Microbiotech). The amplification products were heat denatured over a 35°C temperature gradient at 0.03°C s^{-1} from 60 to 95°C. A negative control without the template was run alongside to assess the overall specificity. Relative abundance of each gene was determined by the $2\Delta^{DDCt}$ method [42]. RACK1 [43] was used as the endogenous control for calculating relative abundance. Each assay was duplicated. Primer design and their optimization in regard to primer dimerization, self-priming formation, and primer melting temperature (melting temperature of 59–60°C and product sizes between 90–150 bp) were done using Primer Quest (Integrated DNA Technologies, Coralville, IA. http://www.idtdna.com). Changes in D1 protein and *psb*A transcripts are expressed as fold-change in comparison to corresponding values of the IL ground control.

Data analysis

Triplicated/quadruplicated biological replicates of the different *C. reinhardtii* strains were placed in the 24 multicells of the Photo II device for the space flight as well as for the on-ground experiments. All analyses were performed in biological triplicates with at least 2 technical replicates. Statistical analyses were performed using analysis of variance (one way ANOVA), and significance of differences evaluated by p-level. Differences significant at $P\leq0.05$ are presented.

Results

Differential photosynthetic performance of Chlamydomonas IL parent and mutated D1 strains in low Earth orbit space Flight (LEO)

An intronless wild type *C. reinhardtii* (IL) [44,45] and its D1 mutants – two (A250L and A251C) near the Q_B binding pocket and two others (V160A and I163N) in the environment surrounding Tyr 161 (Y_Z) [1] (Figure S1) were sent into space aboard the re-entry module of the Foton M2 spacecraft enclosed into the Photo II biodevice (Figure S2). Photosynthetic performance of each strain in flight was monitored in real time and compiled as maximum quantum yield of photosystem II (PSII) photochemical efficiency (F_v/F_m) (Table 1 and Figure 1). In flight, PSII photochemical efficiency of the wild type IL declined daily, recording a total decrease of 0.1 F_v/F_m on the 14th (last) day of flight (Figure 1A). However, after landing (return to Earth) the F_v/F_m ratio of IL, A250L and V160A strains precipitously declined (Table 1). In contrast, two mutants A251C and I163N maintained high PSII activity in flight and showed a slight increase in F_v/F_m value toward the end of the flight (Figure 1B, C). The F_v/F_m ratio of A251C and I163N mutants continued to remain higher even after landing, being higher than even the ground controls (Table 1).

Differential modification of the PSII activity of the five strains in flight was also indicated by changes in the basal fluorescence level, F_0, whose increase signifies inactivation of PSII and decline an indication of increased non-photochemical quenching [44,45]. The F_0 value for the IL, A250L and V160A strains incrementally increased up to 24, 50 and 48%, respectively, while it decreased by 5–10% in the A251C and I163N mutants (Table 1). Thus, wild type IL and A250L and V160A mutants were negatively affected in space, and the A250L and V160A D1 mutants were incapable of re-growing and thus did not survive on return to Earth. In contrast, the growth of A251C and I163N D1 mutants was enhanced when re-grown on Earth upon return from space (Table 1). Because A250L and V160A mutants did not survive, the remaining analysis focused on the wild type, IL, and A251C and I163N mutants.

Space- and genotype-dependent modulation in the reduction of Q_A to Q_A^- and the redox state of the plastoquinone (PQ) pool, as shown above by changes in F_0 values, was further confirmed by analyzing the chlorophyll *a* fluorescence-rise kinetics, also called *OJIP* curve [45]. *OJIP* curve provides the minimal (F_0 or *O*) and maximal (F_m or *P*) fluorescence levels with two intermediate peaks at about 0.002 s (step *J*) and 0.030 s (step *I*). The *O-J* rise is diagnostic of the amount of reduced Q_A, while *J-I* and *I-P* phases denote closure of PSII reaction center and pool size of PQ and its fully reduced state [37]. The flight environment caused Q_A^- accumulation in IL but not in the A251C or I163N mutant, while reduced state of PQ pool of the mutants in space was actually lower than the ground control (Figure 2). These differences were more prominent after landing and, particularly, at the end of the recovery period (15th day after landing).

Light compensation point of A251C and I163N D1 mutants post LEO flight was higher than the parental strain

Oxygen production rates as a function of light and oxygen consumption levels in the dark of the three *Chlamydomonas* strains were measured post flight and compared to the ground controls (Table 2 and Figure S3). The rate of respiration of the IL strain was similar in the ground and post-flight samples but, as mirrored by photochemical activity, its post-flight values were significantly lower than the A251C and I163N mutants (Table 2). The light compensation point, which is a measure of light irradiance at which net gas exchange is zero when photosynthesis and respiration rates are balanced [46,10], was calculated from the light dependency curves, and found to be remarkably higher in post-flight samples of both mutants compared to the IL strain. Environmental conditions are known to influence the compensation point of photosynthetic organisms [46].

A251C and I163N D1 mutants are more efficient in reoxidizing Q_A than the parental strain in response to ionizing radiation exposure on the ground

The positive effect on Q_A reoxidation in space in the D1 mutants compared to the parent IL strain was tested on the ground under neutron radiation. Cosmic ionizing radiation is highly complex with energies between 1 MeV and 10^3 GeV, which can pass through the spacecraft shield, and are difficult to be mimicked on the ground [47]. Nonetheless, we exposed the three strains (IL and the two D1 mutants) to neutron irradiation at a dose of 0.23 mSv/h for 24 h, with energy of 800 MeV [27], at temperature and light conditions simulated to that in the space module. Higher order Q_A reoxidation kinetics was apparent in both the D1 mutants compared to the IL strain (Figure 3). Thus, mimicking only the neutron irradiation component of cosmic rays on Earth yielded similar results to those found in space-sent samples. One of the consequences of this interaction is the efficient and sustained ability to reoxidize Q_A^-, with a positive influence on the photosynthesis of the two D1 mutants.

Lesser susceptibilty of A251C and I163N mutants to photoinactivation than the parental strain

Acclimation of photosynthetic organisms, including the green algae, to photoinhibitory irradiances has been linked to their ability to modulate PSII excitation pressure (1-qP) [37,48]. We therefore tested the excitation pressure and "excess" parameters of I163N and A251C mutants against high irradiances in comparison to the IL strain using *in vivo* fluorescence quenching analyses [37]. The parameter excitation pressure (1-qP) is a measure of the amount of light energy absorbed by closed PSII reaction centers and is based on the reduction state of the primary quinone acceptor of PSII (Q_A) [37,48]. At 50 μmol m^{-2} s^{-1}, the 1-qP values of the A251C and I163N mutants were 0.087±0.004 and 0.071±0.006, respectively, which were slightly higher than the IL strain (0.066±0.002). As the irradiance was raised the 1-qP increased in all strains and this trend was less pronounced in the two D1 mutants. At irradiances >300 μmol m^{-2} s^{-1}, the mutants showed a smaller increase in the 1-qP levels and lower accumulation of PSII reaction centers with reduced Q_A than the parent strain (Figure 4). The 1-qP values at 300 μmol m^{-2} s^{-1} for IL, A251C and I163N were 4.0±0.05, 3.4±0.03 and 2.9±0.23 times higher, respectively, compared to those observed at 50 μmol m^{-2} s^{-1}. The parameter "excess" increased in I163N and A251C mutants as a function of light intensity but significantly slowly than in the IL strain (Figure 4). Since the parameter "excess" [E = (1-qP)(F_v'/F_m')] is a combined measure of the efficiency of the electron transport and non-photochemical dissipation and correlates linearly with the rate of PSII photoinactivation [48], the two D1 mutants seem less susceptible to photoinactivation than the IL strain.

Analysis of the likely environment of D1 Ala251 and Ile163 upon mutation predicts more stable D1 in mutants

The substitutions made to defined residues, Ile163 and Ala251, in the two functionally critical pockets of the D1 protein impacted the photochemical performance of PSII in space and growth stimulation upon return to Earth. We therefore analyzed the likely environment of Ala251 and Ile163 upon mutation. Analysis of the resolved PSII structure from *Thermosynechococcus vulcanus* [1] as per previous methodology [49; Sobolev V, Samish I, and Edelman M, personal communication] shows that the hydrophobic side chain of Ala251 in D1 is in contact with the quinone head of the PQ ligand and the hydrophobic part of Leu218 (Figure 5A; Table 3). Additional contacts are formed with the backbone oxygens of Leu218, Asn247 and Ile248, as well as the side-chain oxygen atom of Ser222. Addition of a sulfur atom to the Ala moiety upon mutation to Cys (A251C) would produce a side chain that is more hydrophilic and that can form weak H-bonds with the hydrophilic environment, resulting in a stabilization of the D1 protein structure. Similar analysis shows that the hydrophobic side chain of Ile163 has productive contacts with the hydrophobic side chain of Leu159 in the D1 protein and the aromatic ring of Phe292 from the D2 protein (Figure 5B; Table 3). It also has large contact with digalactosyl diacyl glycerol (a major chloroplast membrane lipid), including contact with two of the latter's oxygen atoms. Thus, replacement of Ile163 by the more hydrophilic Asn163 would stabilize the complex, similar to the effect of the viable Cys251 mutant.

The above analysis predicted that the D1 protein mutated at residues Ile163 and Cys251 in *Chlamydomonas* would be less prone to degradation. This was substantiated by immunoblot data (Figure 6), which showed higher accumulation of D1 in the mutants compared to the IL strain in space-sent samples after return to Earth. In this context, the higher accumulation of D1 in the mutants correlated with higher transcription of the corresponding *psbA* gene (Figure 6). On the contrary, there was no increase in D1 protein accumulation in the IL parental strain to match its increase of *psbA* mRNA. RNA accumulation levels increased in both the mutants as well as the IL strain in space-sent samples relative to ground controls (Figure 6). These results suggest that the I163N and A251C mutations render the D1 protein relatively stable to space conditions, enabling better PSII photochemistry and better viability of the mutants in space versus the IL strain.

Discussion

We demonstrate here that of the five strains including the control IL strain sent on a low Earth orbit space-flight, only two D1 *C. reinhardtii* mutants, I163N and A251C, efficiently carried out PSII photochemistry in space and were able to re-grow following return to Earth. Although the control IL strain survived the space flight but upon return to Earth it grew poorly. The acclimation of the I163N and A251C mutants was associated with modulation of the PSII excitation pressure (1-qP) and higher capacity to reoxidize QA−. In further evaluation of the reasons for better photochemistry of the I163N and A251C D1 mutants in space and their growth performance after return to Earth, a look at their physio-biochemical properties in comparison to the IL strain in the laboratory (on the ground) were revealing. Apart from the differences in the indicated amino acid substitutions, both D1 mutants accumulated ~50% less Chl *a* per cell (Figure S4) and exhibited slightly higher dark respiration (Table 2), the latter result being consistent with previous data on wheat grown in space [10]. The possibly smaller absorption cross section, reflected by 50% less Chl *a* per cell, would lead to lesser light energy capture and could be a factor in promoting higher tolerance of I163N and A251C mutants to photoinhibitory light.

The I163N and A251C substitutions in the two functionally critical pockets of the D1 protein, the redox-active Tyr 161 (Y_Z) and QB binding site (Figure S1), link the modifications to the photochemical performance of PSII and growth stimulation in space. The nature of these amino acid substitutions likely generates a more 'resistant' D1 protein conformation that is resilient to degradation in space. It has been shown that while there is some reduction in the rate of Q_A^- reoxidation and some destabilization of the $S_{2/3}$ states in A251C, its rate of growth and overall photosynthetic output on the ground is indistinguishable from those of wild type, possibly due to the reversibly reducible SH group of Cys exerting a positive effect on flow between the primary (Q_A) and secondary (Q_B) electron acceptors [6]. What remains then is the proposed increased stabilization of either the D1 protein structure or the quinine – protein complex in the Asn163 and Cys251 D1 mutants, which we speculate helps PSII to withstand the radiance rigors of space. This conclusion is supported by the non-viability of *C. reinhardtii* D1-protein mutant A250L sent up to space along with the I163N and A251C mutants. For A250L mutant, the immediate vicinity of position 251 in the protein is likely to be strongly destabilized by steric hindrance and/or hydrophobic – hydrophilic interactions (Figure 5; Table 3). As a result, it is expected that for these mutant proteins to fold and integrate into PSII, significant rearrangements in the structure of the molecule need to occur beyond the first-shell residues interacting with amino acids at positions 163 and 251. Significant structural rearrangements in the D1 protein can be expected to impinge on the photochemistry of electron transfer in the reaction center. In fact, in *Chlamydomonas* these two mutants are severely impaired in both acceptor side ($Q_A \rightarrow Q_B$) and donor side electron transfer, and show slower photoautotrophic growth under both low and high irradiance conditions [6].

Our data support an important role of D1 conformation in stabilizing (and enhancing) PSII function, and strengthen the concept of a dynamic role for D1 in mediating growth of algae and plants under normal and extreme environmental conditions [4,5]. Future studies on developing and characterizing specific mutants of D1, which are stable and more efficient in sustaining oxygenic photosynthesis, especially under extreme environmental conditions [27,29], are expected to have a great potential in increasing biomass. Their particular relevance falls in line with the growing demand for alternative sustainable energy, such as biofuels and food production.

Supporting Information

Figure S1 Structure of PSII D1 (blue) and D2 (yellow) proteins depicting mutated Val160, Ile163, Ala250 and Ala251 residues, the oxygen evolving complex (OEC, orange balls) and redox active cofactors: QA and QB (pink), redox-active Tyr161 (TyrZ, red), chlorophylls (green), and non-heme Fe (red ball) (based on Umena et al., 2011 [1]).

Figure S2 PHOTO II automatic biodevice.

Figure S3 Light dependency curves of oxygen evolution of the parent strain (IL) and D1 mutants (I163N and A251C) of *C. reinhardtii*. Following the space flight and landing on earth, the algal cultures were transferred to liquid TAP medium and re-grown under continuous light for 3 days. The measurements were performed on cultures containing 20 µg Chl mL-1 at 24°C, continuous stirring and in the presence of 10 mM sodium bicarbonate. The black and white symbols correspond to samples transferred to liquid TAP medium on day 3 and day 15 after landing, respectively. Average values of one experiment in triplicate (n = 3) are shown ± SE.

Figure S4 Chlorophyll (Chl) *a* content per cell in IL strain and D1 mutants (I163N and A251C) of *C. reinhardtii*. Average values of three biological replicates (n = 3) are shown ± SE.

Table S1 List of the DNA primers used in the 2-step PCR for site-directed mutagenesis experiments. The altered nucleotides at positions 163 (I163N) and 251 (A251C) are highlighted.

Acknowledgments

We thank R. Demets and P. Baglioni of ESA; R. Stalio, E. D'Aversa, J. Sabbagh and V. Cotronei of ASI; F. Turchet, Dania Esposito, Emanuela Pace, Agnese Serafini, Andrea Margonelli, Mario Damasso and Cecilia Ambrosi of CNR for their technical assistance and support; Fabio Polticelli for the D1–D2 figure; and Norman Huner (University of Western Ontario, Canada) for comments on the manuscript. We are grateful to Vladimir Sobolev, Ilan Samish and Marvin Edelman (Weizmann Institute of Science, Israel) for being gracious in sharing their unpublished analysis of the likely environment of Ala251 and Ile163 mutations in D1 as well as for their constructive edits on the manuscript. Support from the Italian Space Agency (to MTG), DLR (to UJ), and European Space Agency is acknowledged. Mention of trade names or commercial products in this publication is solely for the purpose of providing specific information and does not imply recommendation or endorsement by the U.S. Department of Agriculture.

Author Contributions

Conceived and designed the experiments: MTG UJ. Performed the experiments: MTG GR ML AA SP. Analyzed the data: MTG GR ML AKM. Contributed reagents/materials/analysis tools: IB. Wrote the paper: AKM. Coordinated the project, designed flight experiments, performed experiments and analyzed data: MTG. Performed the flight and ground-based experiments and analyzed data: ML GR. Contributed to flight and ground-based experiments: AA SP. Designed mutant strategy and supported the flight experiments: UJ. Produced mutants: IB. Contributed to manuscript preparation: MTG ML GR. Critically analyzed data, recommended excitation pressure analysis and wrote the paper: AKM.

References

1. Umena Y, Kawakami K, Shen J-R, Kamiya N (2011) Crystal structure of oxygen-evolving photosystem II at a resolution of 1.9Å. Nature 473: 55–60.
2. Mattoo AK, Marder JB, Edelman M (1989) Dynamics of the photosystem II reaction center. Cell 56: 241–246.
3. Rochaix Jean-David (2001) Assembly, function, and dynamics of the photosynthetic machinery in *Chlamydomonas reinhardtii*. Plant Physiol 127: 1394–1398.
4. Edelman M, Mattoo AK (2008) D1-protein dynamics in photosystem II: the lingering enigma. Photosynth Res 98: 609–620.
5. Yokthongwattana K, Melis A (2006) Photoinhibition and recovery in oxygenic photosynthesis: mechanism of a photosystem II damage and repair cycle. In: Demmig-Adams B, Adams WW III, Mattoo AK (eds). Photoprotection, Photoinhibition, Gene Regulation, and Environment. Springer: Dordecht, 175–191.
6. Lardans A, Förster B, Prásil O, Falkowski PG, Sobolev V, et al. (1998) Biophysical, biochemical, and physiological characterization of *Chlamydomonas reinhardtii* mutants with amino acid substitutions at the Ala251 residue in the D1 protein that result in varying levels of photosynthetic competence. J Biol Chem 273: 11082–11091.
7. Rea G, Polticelli F, Antonacci A, Scognamiglio V, Katiyar P, et al. (2009) Structure-based design of novel *Chlamydomonas reinhardtii* D1-D2 photosynthetic proteins for herbicide monitoring. Prot Sci 18: 2139–2151.
8. Nelson N, Ben-Shem A (2004) The complex architecture of oxygenic photosynthesis. Nature Rev Mol Cell Biol 5: 971–982.
9. Schulze-Makuch D, Irwin LN (2008) Life in the Universe: Expectations and Constraints. Academic Science Books, Springer: Berlin. 251 p.
10. Tripathy BC, Brown CS, Levine HG, Krikorian AD (1996) Growth and photosynthetic responses of wheat plants grown in space. Plant Physiol. 110: 801–806.
11. Monje O, Stutte G, Chapman D (2005) Microgravity does not alter plant stand gas exchange of wheat at moderate light levels and saturating CO_2 concentration. Planta 222: 336–345.

12. Stutte GW, Monje O, Goins GD, Tripathy BC (2005) Microgravity effects on thylakoid, single leaf, and whole canopy photosynthesis of dwarf wheat. Planta 223: 46–56.
13. Porterfield DM (2002) The biophysical limitations in physiological transport and exchange in plants grown in microgravity. J Plant Growth Regul 21: 177–190.
14. Wang GJ, Yan WY, Sun Y, Huang JH, Diao YL, et al. (2004) Study on breeding new line of wheat by space treatment. Heilongjiang Agri Sci 4: 1–4.
15. Lehto KM, Lehto HJ, Kanervo EA (2006) Suitability of different photosynthetic organisms for an extraterrestrial biological life support system. Res Microbiol 157: 69–76.
16. Esposito D, Faraloni C, Margonelli A, Pace E, Torzillo G, et al. (2006) The effect of ionising radiation on photosynthetic oxygenic microorganisms for survival in space flight revealed by automatic photosystem II-based biosensors. Microgravity Sci Technol 18: 215–218.
17. Rea G, Esposito D, Damasso M, Serafini A, Margonelli A, et al (2008) Ionizing radiation impacts photochemical quantum yield and oxygen evolution activity of Photosystem II in photosynthetic microorganisms. Int J Radiat Biol 84: 867–877.
18. Saakov VS (2003) Specific effects induced by gamma-radiation on the fine structure of the photosynthetic apparatus: evaluation of the pattern of changes in the high-order derivative spectra of a green leaf in vivo in the red spectral region. Dokl Biochem Biophys 388: 22–28.
19. Sarghein SH, Carapetian J, Khara J (2008) Effects of UV-radiation on photosynthetic pigments and UV absorbing compounds in *Capsicum longum* (L.). Int J Bot 4: 486–490.
20. Ivanova PI, Dobrikova AG, Taneva SG, Apostolova EL (2008) Sensitivity of the photosynthetic apparatus to UV-A radiation: role of light-harvesting complex II–photosystem II supercomplex organization. Rad Environ Biophys 47: 169–177.
21. Wolff SA, Coelho L, Zabrodina M, Brinckmann E, Kittang AI (2013) Plant mineral nutrition, gas exchange and photosynthesis in space: A review. Advances in Space Research 51: 465–475.
22. Hakala-Yatkin M, Sarvikas P, Paturi P, Mäntysaari M, Mattila H, et al. (2011) Magnetic field protects plants against high light by slowing down production of singlet oxygen. Physiol Plant. 142: 26–34.
23. Belyavskaya NA (2004) Biological effects due to weak magnetic field on plants. Adv Space Res 34: 1566–1574.
24. Jovanic BR, Jevtovic R (2002) Effect of a permanent magnetic field on the optical and physiological properties of green plant leaves. Int J Environ Stud 59: 599–606.
25. Pazour GJ, Agrin N, Walker BL, Witman GB (2006) Identification of predicted human outer dynein arm genes: candidates for primary ciliary dyskinesia genes. J Med Genet 43: 62–73.
26. Merchant SS, Prochnik SE, Vallon O, Harris EH, Karpowicz SJ, et al. (2007) The Chlamydomonas genome reveals the evolution of key animal and plant functions. Science 318: 245–250.
27. Rea G, Lambreva M, Polticelli F, Bertalan I, Antonacci A, et al. (2011) Directed evolution and in silico analysis of reaction centre proteins reveal molecular signatures of photosynthesis adaptation to radiation pressure. PLoS ONE 6: e16216.
28. Grossman AR, Croft M, Gladyshev VN, Merchant SS, Posewitz MC, et al. (2007) Novel metabolism in Chlamydomonas through the lens of genomics. Curr Opin Plant Biol 10: 190–198.
29. Shlyk-Kerner O, Samish I, Kaftan D, Holland N, Sai PS, et al. (2006) Protein flexibility acclimatizes photosynthetic energy conversion to the ambient temperature. Nature 442: 827–830.
30. Cano JB, Giannini D, Pezzotti G, Rea G, Giardi MT (2011) Space impact and technological transfer of a biosensor facility to earth application for environmental monitoring. Recent Patents on Space Tech 1: 18–25.
31. Harris EH (1989) The Chlamydomonas Sourcebook. A Comprehensive Guide to Biology and Laboratory Use. Academic Press: San Diego. 780 p.
32. Johanningmeier U, Heiß S (1993) Construction of a *Chlamydomonas reinhardtii* mutant lacking introns in the psbA gene. Plant Mol Biol 22: 91–99.
33. Dauvillee D, Hilbig L, Preiss S, Johanningmeier U (2004) Minimal extent of sequence homology required for homologous recombination at psbA locus in *Chlamydomonas reinhardtii* chloroplasts using PCR-generated DNA fragments. Photosynth Res 79: 219–224.
34. Govindjee, Sesta?k Z, Peters WR (2002) The early history of 'Photosynthetica', 'Photosynthesis Research', and their publishers. Photosynthetica 40: 1–11.
35. Nedbal L, Trtílek M, Kaftan D (1999) Flash fluorescence induction: a novel method to study regulation of Photosystem II. J Photochem Photobiol B: Biology 48: 154–157.
36. Schreiber U, Schliwa U, Bilger W (1986) Continuous recording of photochemical and non-photochemical chlorophyll fluorescence quenching with a new type of modulation fluorometer. Photosynth Res 10: 51–62.
37. Baker NR (2008) Chlorophyll fluorescence: a probe of photosynthesis in vivo. Annu Rev Plant Biol 59: 89–113.
38. Melis A, Neidhardt J, Benemann JR (1999) *Dunaliella salina* (Chlorophyta) with small chlorophyll antenna sizes exhibit higher photosynthetic productivities and photon use efficiencies than normally pigmented cells. J Appl Phycol 10: 515–525.
39. Schagger H, Borchart U, Aquila H, Link TA, von Jagow G (1985) Isolation and amino acid sequence of the smallest subunit of beef heart bc1 complex. FEBS Lett 190: 89–94.
40. Towbin H, Staehelin T, Gordon J (1979) Electrophoretic transfer of proteins from polyacrylamide gels to nitrocellulose sheets: procedure and some applications. Proc Natl Acad Sci USA 76: 4350–4354.
41. Johanningmeier U, Howell SH (1984) Regulation of light harvesting chlorophyll-binding protein mRNA accumulation in *Chlamydomonas reinhardtii*: possible involvement of chlorophyll synthesis precursors. J Biol Chem 259: 13541–13549.
42. Livak KJ, Schmittgen TD (2001) Analysis of relative gene expression data using real-time quantitative PCR and the $2-\Delta\Delta CT$ method. Methods 25: 402–408.
43. Schloss A (1990) Chlamydomonas gene encodes a G protein beta subunit-like polypeptide. Mol Gen Genet 221: 443–452.
44. Strasser RJ, Srivastava A, Govindjee (1995) Polyphasic chlorophyll a fluorescence transient in plants and cyanobacteria. Photochem Photobiol 61: 32–42.
45. Tóth S, Schansker G, Strasser R (2007) A non-invasive assay of the plastoquinone pool redox state based on the OJIP-transient. Photosynth Res 93: 193–203.
46. Tenhunen JD, Lange OL, Gebel J, Beyschlag W, Weber JA (1984) Changes in photosynthetic capacity, carboxylation efficiency, and CO_2 compensation point associated with midday stomatal closure and midday depression of net CO_2 exchange of leaves of *Quercus suber*. Planta 162: 193–203.
47. Miroshnichenko LI (2003) Radiation Hazard in Space. Kluwer Academic Publishers: Dordrecht, 238 p.
48. Kornyeyev D, Logan BA, Holaday AS (2010) Excitation pressure as a measure of the sensitivity of photosystem II to photoinactivation. Funct Plant Biol 37: 943–951.
49. Eyal E, Najmanovich R, McConkey BJ, Edelman M, Sobolev V (2004) Importance of solvent accessibility and contact surfaces in modeling side chain conformations in proteins. J Comput Chem 25: 712–724.

10

Does Chloroplast Size Influence Photosynthetic Nitrogen Use Efficiency?

Yong Li[1,2], Binbin Ren[1], Lei Ding[1], Qirong Shen[1], Shaobing Peng[2], Shiwei Guo[1]*

1 College of Resources and Environmental Sciences, Nanjing Agricultural University, Nanjing, Jiangsu, China, **2** National Key Laboratory of Crop Genetic Improvement, MOA Key Laboratory of Crop Ecophysiology and Farming System in the Middle Reaches of the Yangtze River, College of Plant Science and Technology, Huazhong Agricultural University, Wuhan, Hubei, China

Abstract

High nitrogen (N) supply frequently results in a decreased photosynthetic N-use efficiency (PNUE), which indicates a less efficient use of accumulated Ribulose-1,5-bisphosphate carboxylase/oxygenase (Rubisco). Chloroplasts are the location of Rubisco and the endpoint of CO_2 diffusion, and they play a vital important role in photosynthesis. However, the effects of chloroplast development on photosynthesis are poorly explored. In the present study, rice seedlings (*Oryza sativa* L., *cv.* 'Shanyou 63', and 'Yangdao 6') were grown hydroponically with three different N levels, morphological characteristics, photosynthetic variables and chloroplast size were measured. In Shanyou 63, a negative relationship between chloroplast size and PNUE was observed across three different N levels. Here, plants with larger chloroplasts had a decreased ratio of mesophyll conductance (g_m) to Rubisco content (g_m/Rubisco) and a lower Rubisco specific activity. In Yangdao 6, there was no change in chloroplast size and no decline in PNUE or g_m/Rubisco ratio under high N supply. It is suggested that large chloroplasts under high N supply is correlated with the decreased Rubisco specific activity and PNUE.

Editor: Ive De Smet, University of Nottingham, United Kingdom

Funding: This work was supported by the National Basic Research Program of China (2013CB127403) and the National Natural Science Foundation of China (NSFC-31272236). The funders had no role in study design, data collection and analysis, decision to publish, or preparation of the manuscript.

Competing Interests: The authors have declared that no competing interests exist.

* E-mail: sguo@njau.edu.cn

Introduction

The high grain yields of most crops are dependent upon the supply of nitrogen (N) from fertilizers. The increasing cost and high energy requirement of such fertilizer, together with the adverse environmental effects of N pollution have stimulated much research activity that aiming towards enhancing the efficiency of its use. An important variable is the intrinsic N-use efficiency (NUE) in plants. A key component of NUE is the photosynthetic N-use efficiency (PNUE), defined as net photosynthetic rate (A) per unit leaf N content. Approximately 75% of N is allocated to chloroplasts [1,2] and about 27% of this is in Ribulose–1,5–bisphosphate carboxylase/oxygenase (Rubisco) [3,4], which carries out the primary fixation of CO_2 in the Benson-Calvin cycle. Thus, Rubisco plays a pivotal role in PNUE as a major repository of N and an enzyme that limits photosynthetic rate under various conditions.

Due to the low concentration of CO_2 in the atmosphere and the low affinity for CO_2, the catalytic effectiveness of Rubisco is poor under ambient conditions [4–7]. Rubisco may operate significantly below its potential catalytic capacity in C_3 plants, suggesting that under high N supply or in high N content leaves, there is excess Rubisco protein serving only as a N storage and not contributing to photosynthesis [8–12], especially under limiting light. The lower relative Rubisco activity in high N content leaves may thus contribute to a decreased PNUE in such leaves.

In full sunlight, photosynthesis in C_3 plants is mainly limited by Rubisco activity [11,13,14]. Rubisco activity is related to CO_2 concentration in chloroplasts [15], and therefore it has been suggested that the decreased Rubisco activity in high N content leaves is due to an insufficient supply of CO_2 [16]. In the diffusion pathway from atmosphere to chloroplasts, CO_2 diffuses across a boundary layer above the leaf surface, and then through the stomata into the substomatal cavity. In the substomatal cavity, CO_2 dissolves in the water-filled pores of the cell wall and then diffuses through the cell wall, the plasma membrane, the cytosol, and the chloroplast envelope to enter the chloroplast. The rate of CO_2 diffusion from the intercellular spaces to the carboxylation sites in chloroplasts is referred to as the mesophyll conductance, g_m. It has been demonstrated that g_m markedly limits chloroplast CO_2 concentration relative to intercellular CO_2 concentration (C_i) [17–20].

It is thought that chloroplast size would probably affect g_m [21]. The conductance in the liquid phase in mesophyll cells is the dominant component of g_m [17,22], especially the conductance through the inner chloroplast envelope membrane, which constitutes about one half of total internal resistance [23]. Thus, g_m depends upon the conductance per unit of chloroplast surface area and the surface area of chloroplasts facing the intercellular air spaces [22]. Larger chloroplasts are usually correlated with higher N content [24] and would potentially increase g_m [25]. A larger chloroplast would also store more leaf N and Rubisco. However, it is not clear whether the extent of the increase in g_m is sufficient to provide enough CO_2 for activating the increased amount of Rubisco, and thus whether an imbalance between the increases in g_m and in Rubisco content contributes to the decrease in PNUE observed in high N leaves.

Few studies have specifically investigated the relationship between chloroplast ultrastructure and PNUE that aiming at testing whether larger chloroplasts are related to lowered Rubisco activity and PNUE. We have studied the responses of two rice varieties that respond differently to N supply and provide evidence that links changes in chloroplast size with a deficiency in g_m that can explain reduced PNUE and Rubisco activity. Hence, we propose a novel explanation for decreased PNUE under high N supply, and suggest an approach to plant breeding to increase N productivity.

Results

Growth response to N supply

The response of both Shanyou 63 and Yangdao 6 to the N supply was as predicted (Table 1). Increases in plant biomass were observed at high N in both cases. There was a decrease in root mass ratio (RMR, = root biomass /whole plant biomass), and an increase in leaf mass ratio (LMR, = leaf biomass /whole plant biomass) in both varieties with increasing N supply. Leaf sheath and culm mass ratio (SCMR, = leaf sheath and culm biomass/ whole plant biomass) was unresponsive to N supply, except for a decrease under high N supply in Yangdao 6. SLW was also unresponsive to N supply, indicating no alterations in leaf thickness.

Photosynthetic variables

In both rice cultivars, A, N, NO_3^- and relative Rubisco content were higher under high N supply compared with low N supply (Table 2). However, the responses of these varieties differed markedly when other variables were measured. Most importantly, PNUE (calculated as A/N) decreased with increasing N supply in Shanyou 63, but did not change significantly in Yangdao 6. The same trends (decrease in Shanyou 63 and no change in Yangdao 6) were observed in A/Rubisco. This phenomenon can also be observed from the relationships between A and leaf N content, and relative Rubisco content (Fig. 1). A/N and A/Rubisco were much lower in high N or Rubisco content leaves in Shanyou 63, with no significant decrease in Yangdao 6 (Fig. 1). Similarly, with increasing N supply, both the initial and maximum Rubisco activities were lower in Shanyou 63, but there were no significant differences in Yangdao 6.

In Shanyou 63, stomatal conductance (g_s) was independent of N supply, while in Yangdao 6 it increased at high N supply (Table 2). The values of g_m were higher in high N in both varieties. The ratio g_m/Rubisco declined markedly with increasing N supply in Shanyou 63 but remained constant in Yangdao 6. A/C_i response curves showed that photosynthesis was more responsive to N supply in Yangdao 6 than in Shanyou 63 (Fig. 2).

Chloroplast development

Chloroplast size also increased with increasing N supply but only in Shanyou 63, with no significant difference observed in Yangdao 6 (Table 2; Fig. 3). Chloroplast length (L_{chl}) was less sensitive than chloroplast thickness (D_{chl}) to N supply: L_{chl} increased by 32% compared to a 65% increase in D_{chl}. Single chloroplast volume (V_{chl}) and single chloroplast surface area (S_{chl}) increased with increasing N supply in Shanyou 63, but S_{chl}/V_{chl} decreased in high N supply. There were no significant differences in V_{chl}, S_{chl} and S_{chl}/V_{chl} among N supply levels in Yangdao 6 (Table 2). Compared with high N supply, chloroplasts in low N supply showed an accumulation of starch granules in both cultivars (Fig. 3).

Discussion

The effect of nitrogen supply on chloroplast development

As much as 75% of leaf N is invested to chloroplasts to synthesis photosynthetic apparatus, including thylakoid membranes and photosynthetic enzymes. Thereby, chloroplast development, such as chloroplast division, chloroplast grana and stroma lamellae stacking, is highly dependent on nitrogen supply. Sufficient N will significantly enlarge chloroplast size, increase chloroplast number,

Table 1. Growth variables of rice seedlings grown at different N supplies.

Variables	Shanyou 63			Yangdao 6		
	Low-N	Int-N	High-N	Low-N	Int-N	High-N
Plant dry mass (g plant^{-1})	2.92±0.73^b	4.27±0.60^a	4.89±0.49^a	1.80±0.18^c	2.21±0.45^b	3.21±.082^a
RMR (g plant^{-1})	0.27±0.02^a	0.19±0.01^b	0.12±0.02^c	0.27±0.02^a	0.22±0.03^b	0.15±0.01^c
SCMR (g plant^{-1})	0.48±0.02^a	0.51±0.02^a	0.48±0.04^a	0.47±0.01^a	0.47±0.04^a	0.40±0.03^b
LMR (g plant^{-1})	0.26±0.01^c	0.30±0.02^b	0.41±0.02^a	0.26±0.01^c	0.31±0.02^b	0.45±0.02^a
leaf area (cm^2 plant^{-1})	256±53^c	444±80^b	668±44^a	162±13^c	235±42^b	494±116^a
SLW (g m^{-2})	28.98±2.73^a	29.23±1.15^a	29.87±2.02^a	29.19±1.46^a	29.19±2.06^a	29.29±1.97^a

Rice plants (*cv.* Shanyou 63 and Yangdao 6) were supplied with N at three different levels (low: 20 mg L^{-1} N, intermediate: 40 mg L^{-1} N, and high: 100 mg L^{-1} N). Data are means ± SD of 5 individual plants. Variables were determined 40 days after the start of treatment.

Notes: Significant differences ($P<5\%$) between N supplies or varieties were indicated by different lowercase letters or different uppercase letters, respectively. RMR, SCMR and LMR represent root mass ratio, leaf sheath and culm mass ratio and leaf mass ratio, respectively. They were calculated as the ratio of separate dry mass to whole plant dry mass. SLW represents specific leaf weight, and was calculated as the ratio of leaf fresh weight to leaf area.

Table 2. Effects of N supply level on leaf photosynthesis in rice seedlings.

Variables	Shanyou 63			Yangdao 6		
	Low-N	Int.-N	High-N	Low-N	Int.-N	High-N
A (μmol CO_2 m^{-2} s^{-1})	16.32±0.09[b]	18.76±2.88[ab]	22.68±0.86[a]	10.33±2.32[c]	15.56±2.80[b]	23.36±2.89[a]
g_s (mol CO_2 m^{-2} s^{-1})	0.16±0.03[a]	0.19±0.04[a]	0.18±0.03[a]	0.10±0.04[b]	0.13±0.02[b]	0.21±0.03[a]
$g_{m\ (Harley)}$ (mol CO_2 m^{-2} s^{-1})	0.18±0.05[b]	0.21±0.07[ab]	0.25±0.03[a]	0.11±0.03[c]	0.18±0.02[b]	0.26±0.05[a]
$g_{m\ (Ethier)}$ (mol CO_2 m^{-2} s^{-1})	0.21±0.05[b]	0.24±0.02[ab]	0.28±0.04[a]	0.13±0.01[c]	0.19±0.03[b]	0.28±0.08[a]
Γ^* (μmol CO_2 mol^{-1})	38.29±0.24[b]	40.80±1.05[ab]	42.26±1.89[a]	38.53±0.68[b]	39.91±1.34[ab]	42.58±1.23[a]
R_d (μmol CO_2 m^{-2} s^{-1})	1.00±0.05[a]	0.67±0.026[b]	0.61±0.06[b]	1.37±0.06[a]	1.23±0.17[a]	0.83±0.09[b]
N (mmol m^{-2})	87.14±0.93[c]	106.43±4.14[b]	143.57±0.50[a]	61.43±4.29[c]	92.14±2.86[b]	139.29±8.57[a]
NO_3^- (mmol m^{-2})	0.20±0.02[b]	0.27±0.07[b]	0.60±0.06[a]	0.20±0.06[b]	0.30±0.07[b]	0.42±0.04[a]
Relative Rubisco content (μmol m^{-2})	35.89±2.32[c]	46.79±2.32[b]	61.79±4.29[a]	26.61±2.86[c]	40.89±2.68[b]	60.71±4.11[a]
Chloroplast length (μm)	4.20±0.59[b]	4.69±0.62[b]	5.55±0.82[a]	4.51±0.61[a]	4.72±0.63[a]	4.87±0.67[a]
Chloroplast thickness (μm)	1.84±0.33[c]	2.34±0.34[b]	3.03±0.49[a]	2.40±0.44[a]	2.30±0.45[a]	2.33±0.46[a]
Initial Rubisco activity (mol mol^{-1} Rubisco s^{-1})	1.05±0.12[a]	0.87±0.04[b]	0.67±0.10[c]	0.69±0.11[a]	0.77±0.21[a]	0.83±0.18[a]
Max Rubisco activity (mol mol^{-1} Rubisco s^{-1})	1.41±0.13[a]	1.29±0.13[ab]	1.07±0.23[b]	1.31±0.36[a]	1.38±0.25[a]	1.17±0.06[a]
A/Rubisco (mol CO_2 mol^{-1} Rubisco s^{-1})	0.45±0.03[a]	0.40±0.04[ab]	0.36±0.05[b]	0.36±0.06[a]	0.35±0.05[a]	0.38±0.04[a]
PNUE (A/N) (μmol CO_2 mol^{-1} N s^{-1})	190±9[a]	178±23[ab]	164±5[b]	154±51[a]	171±27[a]	168±24[a]
Chloroplast volume (V_{chl}, $μm^3$)	7.44	13.44	26.67	13.59	13.07	13.84
Chloroplast surface area (S_{chl}, $μm^2$)	18.43	27.33	43.16	27.54	26.82	27.87
g_m/Rubisco $_{(Harley)}$ (mol CO_2 $mmol^{-1}$ Rubisco s^{-1})	5.02	4.49	4.05	4.13	4.4	4.28
g_m/Rubisco $_{(Ethier)}$ (mol CO_2 $mmol^{-1}$ Rubisco s^{-1})	5.85	5.13	4.53	4.89	4.65	4.61

Rice plants (*cv.* Shanyou 63 and Yangdao 6) were supplied with N at three different levels (low: 20 mg L^{-1} N, intermediate: 40 mg L^{-1} N, and high: 100 mg L^{-1} N). Data are means ± SD of more than 20 individual chloroplasts for their length and thickness, and 5 individual plants for other variables.
Notes: Significant differences (P<5%) between N supplies or varieties were indicated by different lowercase letters or different uppercase letters, respectively. A, g_s, g_m, N, Rubisco and PNUE represent leaf photosynthetic rate, stomatal conductance to CO_2, mesophyll conductance to CO_2, leaf nitrogen content, leaf Rubisco content, photosynthetic N-use efficiency, respectively.

and enhance grana aggregation [26–29]. In the present study, chloroplasts were enlarged under high N supply in Shanyou 63, with chloroplast thickness more responsive than chloroplast length (Table 2 and Fig. 3). In contrast, chloroplast size, both chloroplast length and thickness, was insensitive to N supply in Yangdao 6 (Table 2 and Fig. 3).

Under full sunlight, photosynthetic assimilates should be translocated quickly out of chloroplasts to sites with high carbon sink activity to avoid starch granules formation, which will in turn inhibit leaf photosynthesis [28,30]. It is reported that there are more and larger starch granules under high N supply for their high leaf photosynthetic capacity [27]; however, there are also numerous studies showed less and smaller starch granules under high N supply [26,28,30]. In the present study, starch granules were much larger under low N than under high N supply (Fig. 3). The reason is probably that high N supply can stimulate the translocation of assimilates from chloroplasts to sites with high carbon sink activity [26].

The relationship between Rubisco content and total chloroplast volume

Although the level of Rubisco is sometimes excessive for photosynthesis [12,19,31], the increase in leaf N content at high N supply is generally accompanied by a higher Rubisco content. In the present study, this was observed in both the rice varieties, Shanyou 63 and Yangdao 6. A higher amount of Rubisco

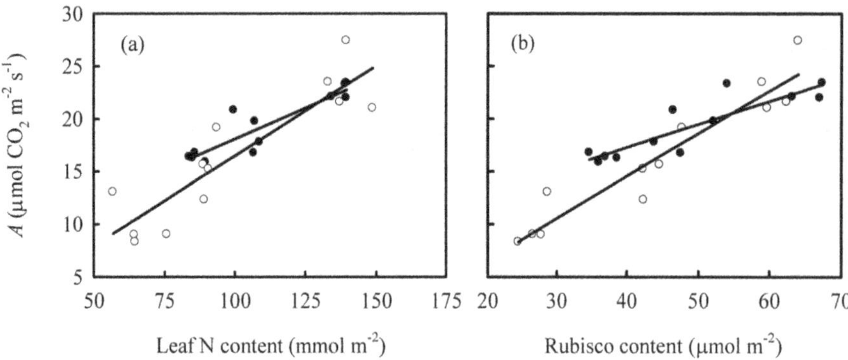

Figure 1. The relationships between leaf photosynthetic rate (*A*) and (a) leaf N content, and (b) Rubisco content and in Shanyou 63 (closed cycles) and Yangdao 6 (open cycles). The lines represent the following regression equations: (a) $y=0.12x+6.33$ $R^2=0.82$ $P<0.01$ for Shanyou 63; $y=0.17x-0.60$ $R^2=0.79$ $P<0.01$ for Yangdao 6; (b) $y=0.22x+8.59 R^2=0.78$ $P<0.01$ for Shanyou 63; $y=0.40x-1.53$ $R^2=0.91$ $P<0.01$.

potentially be associated with either a larger total chloroplast volume (V_{T-chl}) or a higher Rubisco concentration in chloroplasts (Rub_{chl}). Analysis of data from a number of different plant species reveals a linear relationship between Rubisco content and V_{T-chl} expressed on a leaf area basis (Fig. 4). This points to a constant value of Rub_{chl}, at approximately 44 mg cm^{-3} (the slope of the regression equation in Fig. 4). Thus, the higher Rubisco content in high N leaves should be associated with a larger V_{T-chl}, this hypothesis is also speculated by Evans et al. [32]. The increase in V_{T-chl} could arise from an increase in either V_{chl} or chloroplast number per leaf area (n_{chl}).

In Yangdao 6, there was no difference in chloroplast size (Table 2 and Fig. 3), despite the higher Rubisco content in leaves with higher leaf N. In this case, it is suggested that the number of chloroplasts should increase. In contrast, in Shanyou 63, the thickness and the length of chloroplasts were larger in the high N leaves compared to those with low N supply, and the higher Rubisco content at high N supply in Shanyou 63 was hence associated with a larger chloroplast volume. Wider and longer chloroplasts not only have a larger S_{chl}, but also a larger V_{chl} and a decreased S_{chl}/V_{chl} compared to smaller ones.

The effect of chloroplast size on mesophyll conductance, g_m

The total surface area of chloroplasts facing the intercellular space (S_c) is the key determinant of g_m, rather than the total chloroplast surface area *per se*. S_c is given by $S_c = \alpha \times S_{chl} \times n_{chl}$, where α is the ratio of chloroplast surface area facing the intercellular space to total chloroplast surface area. S_c is therefore also higher when there are larger chloroplasts, the higher S_c with larger chloroplasts would potentially increase g_m. The increase in S_c is again not proportional to the increase in total chloroplast volume, V_{T-chl} ($=V_{chl} \times n_{chl}$) and therefore the ratio S_c/V_{T-chl} (which can be given as $(\alpha \times S_{chl} \times n_{chl})/(V_{chl} \times n_{chl})$, and further as $\alpha \times S_{chl}/V_{chl}$) also decreases in large chloroplasts (Table 2). In fact, α is also likely to decrease when chloroplasts are thicker, so amplifying the decrease in this ratio. Hence, it can be concluded that the increase in chloroplast size on Shanyou 63 would be associated with an increase in g_m, while it would not be proportional to the increase in total chloroplast volume and Rubisco content (Table 2). In contrast, in Yangdao 6, where chloroplast size did not change, the increase in g_m at high leaf N would be attributable to an increase in chloroplast number, with no change in the S_c/V_{T-chl} ($=\alpha \times S_{chl}/V_{chl}$) and g_m/Rubisco ratio (Table 2).

Recent studies demonstrated that photorespiration can efficiently affect the precision of g_m estimation [33,34]. But it is now still difficult for the methods of both simultaneous measurement of gas exchange and chlorophyll fluorescence, and A/C_i reponse curve-fitting method to exclude photorespiration's effect. It should be clear that g_m in the present study was estimated without eliminating photorespiration effects. In the present study, two

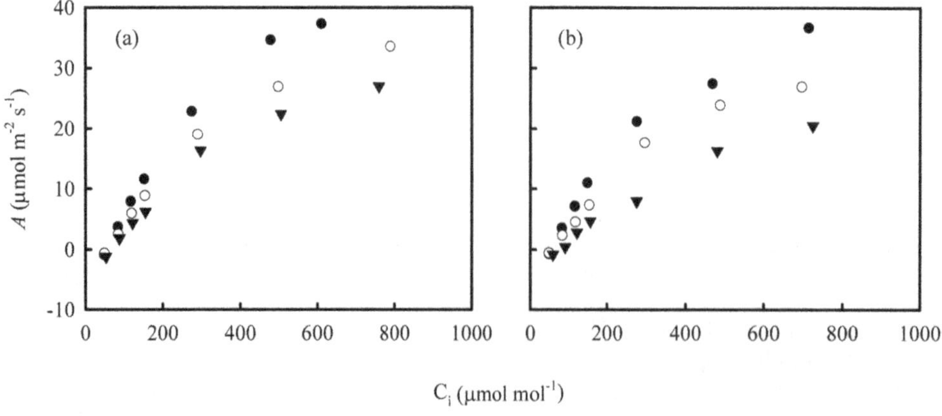

Figure 2. *A*/C_i response curves of newly expanded leaves in Shanyou 63 (a) and Yangdao 6 (b). The symbols of solid cycles, open cycles and solid triangles represent high, intermediate and low N supply, respectively.

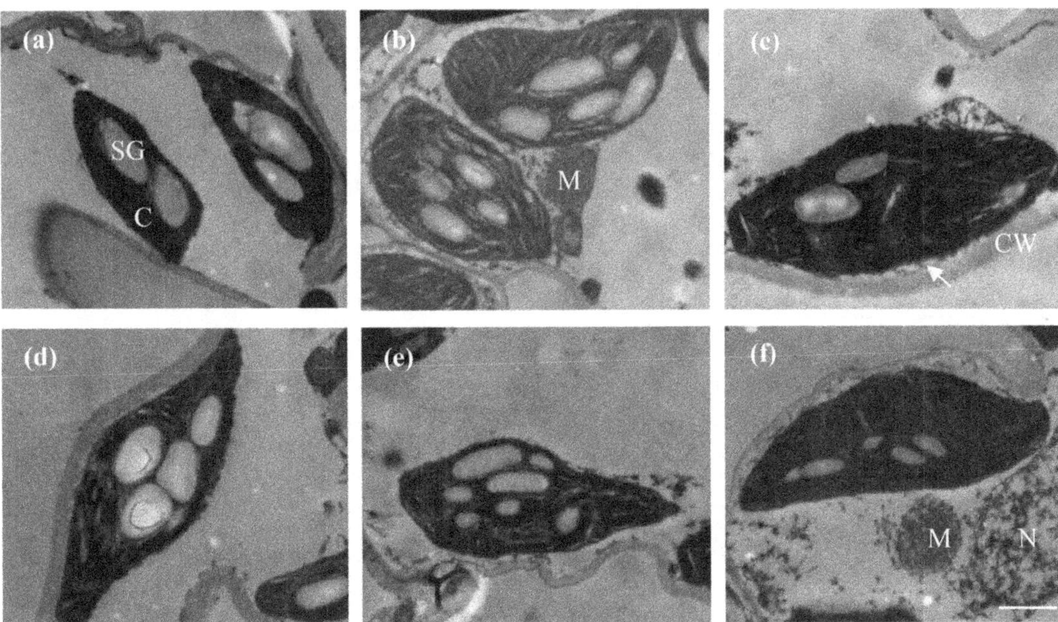

Figure 3. Electron micrographs of chloroplasts in newly expanded leaves of Shanyou 63 (a, low-N supply; b, intermediate.-N supply; c, high-N supply) and Yangdao 6 (d, low-N supply; e, intermediate.-N supply; f, high-N supply). Bar = 1 μm. C, chloroplasts; M, mitochondrion; N, nucleus; SG, starch granules; CW, cell wall; arrows point to plasma membrane.

independent methods were conducted to improve the accuracy for g_m estimation. It should be proposed that, further efforts should be done to improve accuracy for g_m estimation in the method of simultaneous measurement of gas exchange and chlorophyll fluorescence to continue its convenience in g_m estimation.

It is illustrated that more than 95% of mesophyll cell periphery in rice plants is covered by chloroplasts in well-grown and N sufficient leaves [35]. So, when mesophyll surface is mainly covered by chloroplasts under high N supply, g_m and photosynthesis would probably be insensitive to chloroplast development if C_{liq} is similar. This phenomenon was observed in the present study, where photosynthesis and g_m were similar between the two cultivars under high N supply while chloroplast size was substantially larger in Shanyou 63 (Table 2). When less mesophyll surface is covered by chloroplasts under low N supply, small chloroplasts would be benefit for g_m and photosynthesis because they can more efficiently cover mesophyll cell surface and enhance g_m (Table 2).

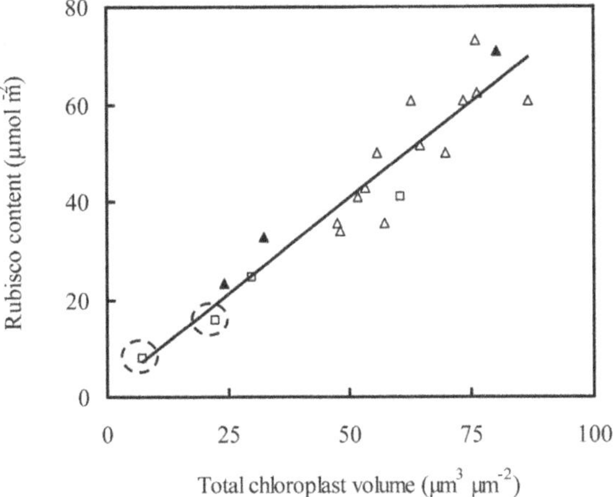

Figure 4. The relationship between Rubisco content and total chloroplast volume per leaf area. The line represents the regression equation: $y = 0.79x + 1.36$, $R^2 = 0.88$, $P < 0.01$. Data sources: *Nicotiana tabacum* □ [21]; *Chenopodium album* ▲ [48]; *Aucuba japonica* Thunb. △ [24]. The two data points in dotted cycles are from transgenic tobacco with a reduced Rubisco content.

Would chloroplast size affect PNUE at high N?

Rubisco activity can be measured under both *in vivo* and *in vitro* conditions. Because of different synthesis conditions, *in vitro* Rubisco activity can not always reveal its *in vivo* activity especially under drought stress [36]. Nevertheless, there are numerous studies showed the positive relationship between them [31,37,38]. The correlation was also checked in the present study (Fig. 5), and the positive relationship revealed that the slowed-down Rubisco turnover rate with increasing N supply in Shanyou 63 was probably the reason for the decreased *A*/Rubisco and PNUE.

In large chloroplasts, the ratio of S_c to Rubisco content ($= Rub_{chl} \times V_{T\text{-}chl}$) is lower, and hence g_m per unit Rubisco content also lower, compared to the values in smaller chloroplasts (Fig. 6). Because of the correlation between g_m and total conductance (g_t), responses of g_t/Rubisco to chloroplast size were similar with those of g_m/Rubisco (Data not shown). The lower g_m/Rubisco and g_t/Rubisco ratio would result in an insufficient supply of CO_2 in chloroplasts and a consequent decrease in Rubisco activity (Fig. 7). A decrease in g_m/Rubisco and a reduction in Rubisco activity at high leaf N were observed in Shanyou 63 but not in Yangdao 6 (Table 2). As a consequence, in the former variety, there was a decreased *A*/Rubisco and PNUE, whereas these variables were unchanged in the latter variety. Therefore, the lower g_m/Rubisco ratio induced by chloroplast enlargement under high N supply in Shanyou 63 at least partially explains the lowered Rubisco efficiency and decline in PNUE (Fig. 8). However, in Yangdao 6, the constant g_m/Rubisco would ensure a sufficient supply of CO_2, and thus Rubisco efficiency and PNUE

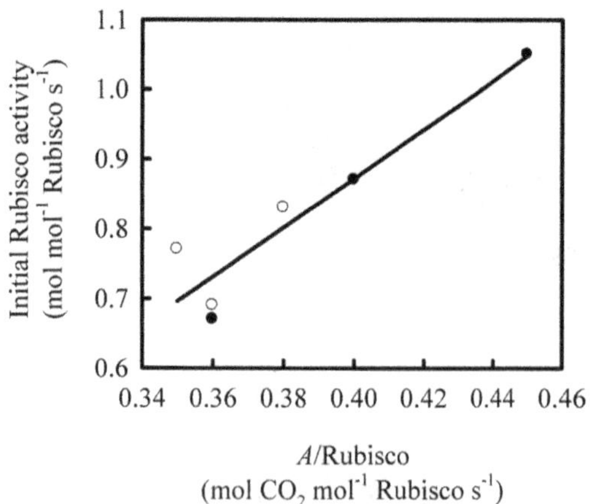

Figure 5. The relationship between initial Rubisco activity and the ratio of leaf photosynthetic rate (*A*) to Rubisco content on Shanyou 63 (solid cycles) and Yangdao 6 (open cycles). The line represents linear regression: $y = 3.51x - 0.53$ $R^2 = 0.88$ $P < 0.01$.

would not decrease under high N supply. It is suggested that whether PNUE decreases under high N would, at least partially, depends on the mechanism by which the Rubisco content increases i.e. whether through increasing V_{chl} or by increasing n_{chl}. Thus, we hypothesize that a high N-dependent increase in chloroplast size would cause a decrease in g_m per unit Rubisco that results in a fall in PNUE, an effect that does not occur if the response to high N is an increase in chloroplast number.

This hypothesis was tested for its general applicability in rice plants grown under different conditions (Fig. 9). The results showed that PNUE and *A*/Rubisco were negatively related to chloroplast size. The relationships were again stronger with chloroplast thickness than chloroplast length. With similar Rubisco content, leaves with smaller chloroplasts have a higher CO_2 assimilation rate. Thus, the lack of balance between g_m and Rubisco content as chloroplasts increase in size is a tenable explanation of observed decreases in Rubisco turnover rate and the lowered PNUE.

Materials and Methods

Plant material and growth conditions

After germination on moist filter paper, rice seeds (*Oryza sativa* L., ssp. indica hybrid, *cv.* 'Shanyou 63', and ssp. indica inbred, *cv.* 'Yangdao 6') were transferred to 2.0 mM $CaSO_4$ for germination at 25±5°C. After 3 days, the rice seedlings were transferred to 6-L rectangular containers (30×20×10 cm) and ¼-strength nutrient solution (for composition, see below). Three days later, the seedlings were transferred to a ½-strength nutrient solution, and after 5 days, the seedlings were supplied with full-strength nutrient solution for 1 week. Seedlings were supplied with nutrient solution containing three different N levels: low N (20 mg L^{-1}), intermediate N (40 mg L^{-1}), and high N (100 mg L^{-1}). The N sources were equimolar amounts of $(NH_4)_2SO_4$ and $Ca(NO_3)_2$. In addition, the macronutrients in the solution were as follows (mg L^{-1}): 10 P as KH_2PO_4, 40 K as K_2SO_4 and KH_2PO_4, and 40 Mg as $MgSO_4$. The micronutrients were (mg L^{-1}): 2.0 Fe as Fe-EDTA, 0.5 Mn as $MnCl_2{\cdot}4H_2O$, 0.05 Mo as $(NH_4)_6Mo_7O_{24}{\cdot}4H_2O$, 0.2 B as H_3BO_3, 0.01 Zn as

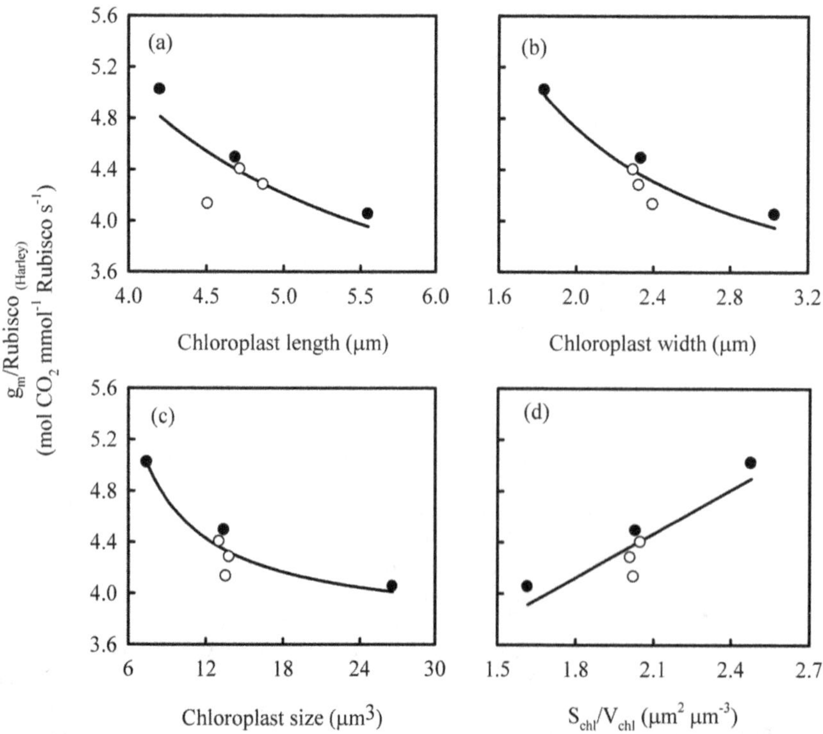

Figure 6. The relationships between chloroplast size and the ratio of mesophyll conductance (g_m) to Rubisco content on Shanyou 63 (solid cycles) and Yangdao 6 (open cycles). Chloroplast surface area (S_{chl}) and volume (V_{chl}) were calculated from the Cesaro formula. The lines represent the following regressions: (a) $y = 2.54x/(x-1.98)$ $R^2 = 0.63$ $P > 0.05$; (b) $y = 2.99x/(x-0.73)$ $R^2 = 0.88$ $P < 0.01$; (c) $y = 3.72x/(x-1.91)$ $R^2 = 0.89$ $P < 0.01$; (d) $y = 1.15x + 2.06$ $R^2 = 0.82$ $P < 0.05$.

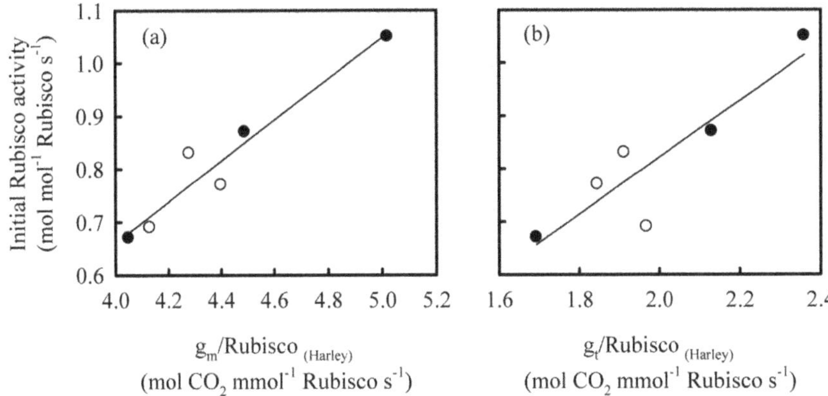

Figure 7. The relationships between initial Rubisco activity and (a) the ratio of mesophyll conductance (g_m) to Rubisco content, and (b) the ratio of total conductance (g_t) to Rubisco content on Shanyou 63 (solid cycles) and Yangdao 6 (open cycles). The lines represent the following regressions: (a) $y=0.39x-0.89$ $R^2=0.93$ $P<0.01$; (b) $y=0.54x-0.25$ $R^2=0.80$ $P<0.05$.

$ZnSO_4\cdot 7H_2O$, 0.01 Cu as $CuSO_4\cdot 5H_2O$, and 0.0028 Si as $Na_2SiO_3\cdot 9H_2O$. To compensate for the lower Ca, additional $CaCl_2$ was added to the low N and intermediate N treatments. A nitrification inhibitor (dicyandiamide) was added to each nutrient solution to prevent the oxidation of ammonium. Nutrient solutions were changed every 2 days, and the pH was adjusted to 5.50±0.05 every day with HCl or NaOH. All treatments were planted in 5 individual containers and were placed in a completely randomised design.

Plants were grown in a greenhouse at 25/18°C day/night temperature. Light was supplied by SON–T AGRO 400W bulbs, with the light intensity maintained at a minimum of 1000 μmol photons m^{-2} s^{-1} (PAR) at the leaf level and a 14 h photoperiod.

Gas exchange and fluorescence measurements

Forty days after the start of treatment, photosynthesis and chlorophyll fluorescence were simultaneously measured on light–adapted leaves using a Li–Cor 6400 infrared gas analyzer. Leaf temperature during the measurement was maintained at 30.5±1.1°C, with a photosynthetic photon flux density (PPFD) of 1500 μmol photons m^{-2} s^{-1}. The CO_2 concentration in the cuvette was adjusted to the ambient CO_2 concentration (424.3±1.9 μmol mol^{-1}), and the relative humidity was maintained at 50%. After equilibration to a steady state, the fluorescence was measured (F_s) and a 0.8 s saturating pulse of light (approx. 8000 μmol m^{-2} s^{-1}) was applied to measure the maximum fluorescence (F_m'). Gas exchange variables were also recorded simultaneously. The efficiency of photosystem II (Φ_{PSII}) was calculated as $\Phi_{PSII} = 1-F_s/F_m$'.

Total electron transport rate (J_T) was calculated as $J_T = \Phi_{PSII} \times PPFD \times \alpha_{leaf} \times \beta$, where α_{leaf} and β were leaf absorption and the proportion of quanta absorbed by photosystem II, respectively. The product $\alpha_{leaf} \times \beta$ was determined from the slope of relationship between Φ_{PSII} and the quantum efficiency of CO_2 uptake (Φ_{CO2}), obtained by varying light intensity under non–photorespiratory conditions at less than 2% O_2 [39]. The variable J_T method [17] was used to calculate g_m using the equation $g_m = A/\{C_i-\Gamma^*[J_T+8(A+R_d)]/[J_T-4(A+R_d)]\}$, where A is the rate of leaf photosynthetic CO_2 uptake per unit leaf area, Γ^* is CO_2 compensation point and R_d is the rate of dark respiration. Γ^* and R_d were measured following Laisk's method [40], as described by Brooks and Farquhar [41] with minor modifications. The process was described in detail in our previous study [42]. g_t was calculated as $g_t = g_s \times g_m/(g_s+g_m)$ [43].

After the above gas exchange measurement, A/C_i response curves were conducted on the same leaves. Leaf temperature, PPFD, and relative humidity during measurements were maintained as mentioned above. Prior to measurements, leaves were

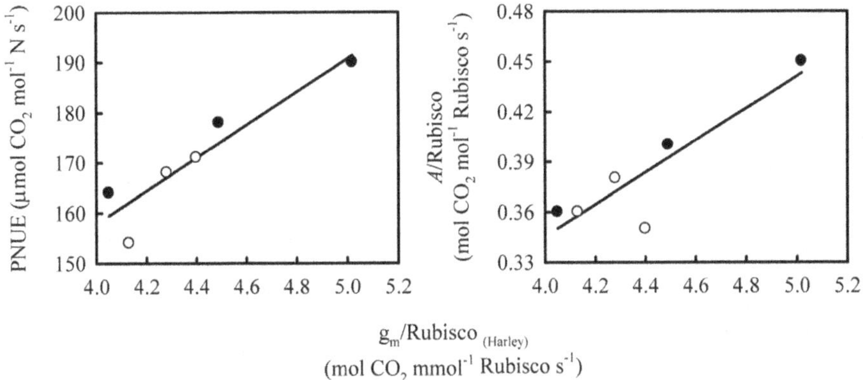

Figure 8. The relationships between the ratio of mesophyll conductance (g_m) to Rubisco content and (a) photosynthetic N-use efficiency (PNUE) and (b) the ratio of leaf photosynthetic rate (*A*) and Rubisco content on Shanyou 63 (solid cycles) and Yangdao 6 (open cycles). The lines represent the following regressions: (a) $y=32.89x+26.28$ $R^2=0.86$ $P<0.01$; (b) $y=0.096x-0.038$ $R^2=0.80$ $P<0.05$.

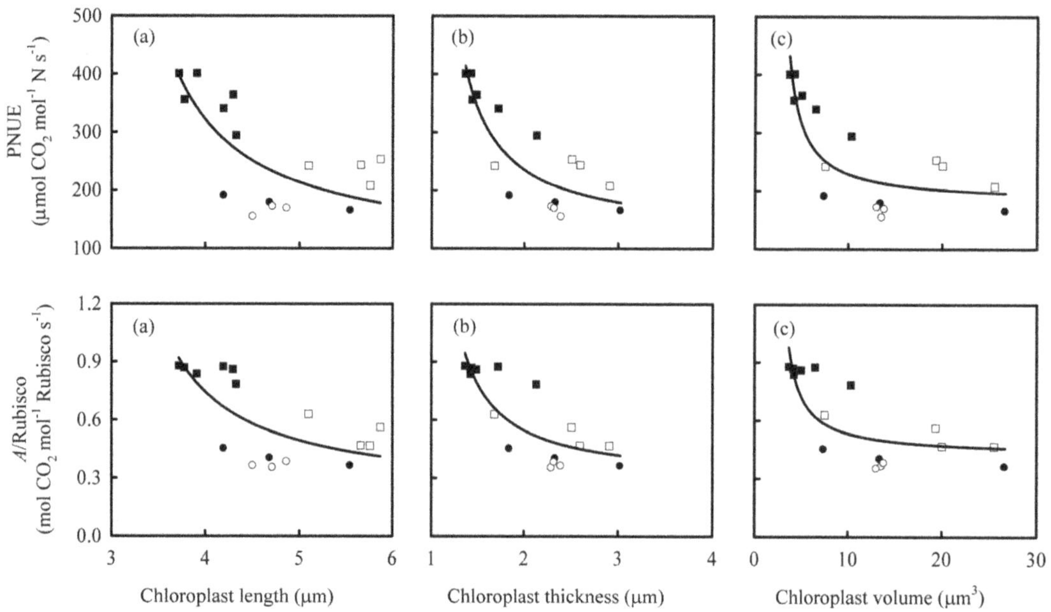

Figure 9. The relationships between chloroplast size and photosynthetic N-use efficiency (PNUE), and the ratio of leaf photosynthetic rate (*A*) to Rubisco. The lines represent the following regression equations: (a) $y=91.04x/(x-2.87)$, $R^2=0.53$, $P<0.01$; (b) $y=121.76x/(x-0.96)$, $R^2=0.75$, $P<0.01$; (c) $y=180.56x/(x-2.12)$, $R^2=0.72$, $P<0.01$; (d) $y=0.21x/(x-2.86)$, $R^2=0.53$, $P<0.01$; (e) $y=0.29x/(x-0.95)$, $R^2=0.69$, $P<0.01$; (f) $y=0.42x/(x-2.06)$, $R^2=0.63$, $P<0.01$. Data sources: data of solid squares were collected from Wuyujing 3 (*Oryza sativa* L. ssp. japonica) with different N supplies; data of open squares were from Shanyou 63 with different N forms and water supply [42]; data of solid and open cycles were from Shanyou 63 and Yangdao 6 with different N supplies.

placed in the cuvette at a PPFD of 1500 μmol photons $m^{-2}\ s^{-1}$; CO_2 concentration in the cuvette was maintained at 400 μmol $CO_2 mol^{-1}$ with a CO_2 mixer. Ten minutes later, CO_2 concentration in the cuvette was controlled across a series of 1000, 800, 600, 400, 200, 150, 100, and 50 μmol $CO_2\ mol^{-1}$. After equilibration to a steady state, data were recorded automatically. g_m was then calculated based on gas exchange measurements themselves according to the method in Ethier and Livingston [44], which is modified from that in Farquhar et al. [13].

Relative Rubisco content and activity measurements

The Rubisco content of newly expanded leaves was determined according to the method of Makino et al. [45,46]. Briefly, samples of newly expanded leaves were immersed in liquid N, and then stored at −70°C. For analysis, 0.5 g were ground in a solution containing 50 mM Tris–HCl (pH 8.0), 5 mM β–mercaptoethanol, and 12.5% glycerol (v/v), and then centrifuged at 1500 g for 15 min at 4°C. The supernatants were mixed with a solution containing 2% (w/v) SDS, 4% (v/v) β–mercaptoethanol and 10% (v/v) glycerol, boiled in a water bath for 5 min before SDS–PAGE using a 4% (w/v) stacking gel, and a 12.5% (w/v) separating gel. After electrophoresis, the gels were stained with 0.25% Commassie Blue for 12 h, and destained. Gel slices containing the large subunits and small subunits of Rubisco were transferred to a 10-mL cuvette containing 2 ml of formamide and incubated at 50°C in a water bath for 8 h. The absorbance of the wash solution was measured at 595 nm. Protein concentrations were determined using bovine serum albumin as a standard.

Rubisco activity was measured according to Jin et al. [7] with minor modification. Briefly, about 0.1 g of newly expanded leaves were ground with 2 ml of a solution containing 50 mM Tris–HCl (pH 7.5), 10 mM β–mercaptoethanol, 12.5% (v/v) glycerol, 1 mM EDTA–Na_2, 10 mM $MgCl_2$, and 1% (m/v) PVP–40. After centrifugation at 15,000 *g* for 1 min at 4°C, the activity in the supernatants were assayed. The initial Rubisco specific activity was measured at 30°C by adding 100 μL of the supernatant to 900 μL of assay solution containing 56 mM HEPES–NaOH (pH 7.5), 1 mM EDTA–Na_2, 20 mM $MgCl_2$, 3 mM DTT, 11 mM $NaHCO_3$, 6 mM ATP, 6 mM creatine phosphate, 0.2 mM NADH, 11 units of phosphocreatine kinase, 11 units of glyceraldehyde–3–phosphate dehydrogenase, 11 units of phosphoglycerate kinase, 11 mM Tris–HCl (pH 7.5), and 0.7 mM RuBP. The absorbance at 340 nm was recorded at 3 s intervals for 30 s. To measure total Rubisco specific activity, 100 μL of the supernatant were added to 200 μL of activation medium containing 1 mM EDTA–Na_2, 50 mM $MgCl_2$, 15 mM $NaHCO_3$, and 50 mM Tris–HCl (pH 7.5), and incubated at 30°C for 10 min. Following the addition 700 μL of assay solution, the activity was determined by recording the absorbance at 340 nm at 3 s intervals for 30 s.

Chloroplast ultrastructure

Leaf pieces of approximately 1–2 mm^2 were cut from the middle of newly expanded leaves using two razor blades, then they were fixed in 2.5% glutaraldehyde in 0.1 M phosphate buffer, pH 7.4, and post–fixed with 2% osmium tetroxide. Specimens were dehydrated in a graded acetone series and embedded in Epon 812. Leaf sections, 70 nm thick, were cut with a Power Tome–XL ultramicrotome, stained with 2% uranyl acetate, and examined under an H–7650 transmission. Chloroplast length and thickness were calculated from at least 20–30 chloroplasts. Chloroplasts were assumed to be ellipsoids of revolution, which is a shape that is generated by rotating an ellipse around one of its axes. According to the Cesaro formula [47], S_{chl} was calculated as $S_{chl}=4\times\pi\times(a\times b^2)^{2/3}$, where $a=L_{chl}/2$, $b=D_{chl}/2$. V_{chl} was calculated from the Cesaro formula, $V_{chl}=(4/3)\times\pi\times a\times b^2$.

Measurement of biomass, leaf nitrate and N content

Nitrate from newly expanded leaves was extracted in boiling water for 30 min and reacted with salicylic acid, and the color was determined at a wavelength of 410 nm. 45 days after treatments were started when all measurements had been completed, plants were harvested and separated into root, leaf sheath and culm, and leaf fractions. Leaf area and leaf fresh weight were determined and specific leaf weight (SLW) was calculated as the ratio of leaf fresh weight to leaf area. All samples were oven-dried at 105°C first for 30 min, and then at 70°C to constant weight. The dried leaves were digested with H_2SO_4–H_2O_2 at 260–270°C, and the total leaf N concentration was determined using a digital colorimeter (AutoAnalyzer 3; Bran+Luebbe).

Statistics

To test the differences between varieties and N supplies, data were analyzed using two-way analysis of variance (ANOVA) and the least significant difference (LSD) test with the SAS 9.0 statistical software package. Different lowercase letters were used to indicate significant differences ($P<5\%$) among N supplies, and different uppercase letters were used between varieties.

Acknowledgments

We wish to thank Professor Peter Horton FRS (University of Sheffield, UK) for assistance in preparation of the manuscript.

Author Contributions

Conceived and designed the experiments: YL SWG. Performed the experiments: YL BBR LD. Analyzed the data: YL. Contributed reagents/materials/analysis tools: YL SWG. Wrote the paper: YL SWG QRS SBP.

References

1. Evans JR, Terashima I (1987) Effects of nitrogen nutrition on electron transport components and photosynthesis in spinach. Aust J Plant Physiol 14: 59–68.
2. Poorter H, Evans JR (1998) Photosynthetic nitrogen-use efficiency of species that differ inherently in specific leaf area. Oecologia 116: 26–37.
3. Evans JR (1989) Photosynthesis and nitrogen relationships in leaves of C_3 plants. Oecologia 78: 9–19.
4. Makino A, Shimada T, Takumi S, Kaneko K, Matsuoka M, et al. (1997) Does decrease in Ribulose-1,5-bisphosphate carboxylase by antisense RbcS lead to a higher N-use efficiency of photosynthesis under conditions of saturating CO_2 and light in rice plants? Plant Physiol 114: 483–491.
5. Jensen RG (2000) Activation of Rubisco regulates photosynthesis at high temperature and CO_2. Proc Natl Acad Sci USA 97: 12937–12938.
6. Spreitzer RJ, Salvucci ME (2002) Rubisco: structure, regulatory interactions, and possibilities for a better enzyme. Ann Rev Plant Biol 53: 449–479.
7. Jin SH, Hong J, Li XQ, Jiang DA (2006) Antisense inhibition of Rubisco activase increases Rubisco content and alters the proportion of Rubisco activase in stroma and thylakoids in chloroplasts of rice leaves. Ann Bot 97: 739–744.
8. Stitt M, Schulze D (1994) Does Rubisco control the rat eof photosynthesis and plant growth? An exercise in molecular ecophysiology. Plant Cell Environ 17: 465–487.
9. Cheng LL, Fuchigami LH (2000) Rubisco activation state decreases with increasing nitrogen content in apple leaves. J Exp Bot 51: 1687–1694.
10. Murchie EH, Hubbart S, Chen Y, Peng S, Horton P (2002) Acclimation of rice photosynthesis to irradiance under field conditions. Plant Physiol 139: 1999–2010.
11. Manter DK, Kerrigan J (2004) A/C_i curve analysis across a range of woody plant species: influence of regression analysis parameters and mesophyll conductance. J Exp Bot 55: 2581–2588.
12. Warren CR, Dreyer E, Adams MA (2003) Photosynthesis-Rubisco relationships in foliage of Pinus sylvestris in response to nitrogen supply and the proposed role of Rubisco and amino acids as nitrogen stores. Trees 17: 359–366.
13. Farquhar GD, von Caemmerer S, Berry JA (1980) A biochemical model of photosynthetic CO_2 assimilation in leaves of C_3 species. Planta 149: 78–90.
14. Sage RF (1990) A model describing the regulation of ribulose-1,5-bisphosphate carboxylase, electron transport, and triose phosphate use in response to light intensity and CO_2 in C_3 plants. Plant Physiol 94: 1728–1734.
15. Flexas J, Ribas-Carbó M, Bota J, Galmés J, Henkle M, et al. (2006) Decreased Rubisco activity during water stress is not induced by decreased relative water content but related to conditions of low stomatal conductance and chloroplast CO_2 concentration. New Phytol 172: 73–82.
16. Warren CR (2004) The photosynthetic limitation posed by internal conductance to CO_2 movement is increased by nutrient supply. J Exp Bot 55: 2313–2321.
17. Harley PC, Loreto F, Marco GD, Sharkey TD (1992) Theoretical considerations when estimating the mesophyll conductance to CO_2 flux by analysis of the response of photosynthesis to CO_2. Plant Physiol 98: 1429–1436.
18. Loreto F, Harley PC, Marco GD, Sharkey TD (1992) Estimation of mesophyll conductance to CO_2 flux by three different methods. Plant Physiol 98: 1437–1443.
19. Eichelmann H, Laisk A (1999) Ribulose-1,5-bisphosphate carboxylase/oxygenase content, assimilatory charge, and mesophyll conductance in leaves. Plant Physiol 119: 179–189.
20. Bernacchi CJ, Portis AR, Nakano H, von Caemmerer S, Long SP (2002) Temperature response of mesophyll conductance. Implications for the determination of Rubisco enzyme kinetics and for limitations to photosynthesis in vivo. Plant Physiol 130: 1992–1998.
21. Evans JR, von Caemmerer S, Setchell BA, Hudson GS (1994) The relationship between CO_2 transfer conductance and leaf anatomy in transgenic tobacco with a reduced content of Rubisco. Aust J Plant Physiol 21: 475–495.
22. Hanba YT, Shibasaka M, Hayashi Y, Hayakawa T, Kasamo K, et al. (2004) Overexpression of the barley aquaporin HvPIP2;1 increases internal CO_2 conductance and CO_2 assimilation in the leaves of transgenic rice plants. Plant Cell Physiol 45: 521–529.
23. Uehlein N, Otto B, Hanson DT, Fischer M, McDowll N, et al. (2008) Function of nicotiana tabacum aquaporins as chloroplast gas pores challenges the concept of membrane CO_2 permeability. Plant Cell 20: 648–657.
24. Muller O, Oguchi R, Hirose T, Werger MJA, Hikosaka K (2009) The leaf anatomy of a broad-leaved evergreen allows an increase in leaf nitrogen content in winter. Physiol. Plantarum 136: 299–309.
25. Li Y, Gao YX, Xu XM, Shen QR, Guo SW (2009) Light-saturated photosynthetic rate in high-nitrogen rice (Oryza sativa L.) leaves is related to chloroplastic CO_2 concentration. J Exp Bot 60: 2351–2360.
26. Ariovich D, Cresswell CF (1983) The effect of nitrogen and phosphorus on starch accumulation and net photosynthesis in two variants of Panicum maximum Jacq. Plant Cell Environ 6: 657–664.
27. Bondada BR, Syvertsen JP (2003) Leaf chlorophyll, net gas exchange and chloroplast ultrastructure in citrus leaves of different nitrogen status. Tree Physiol 23: 553–559.
28. Bondada BR, Syvertsen JP (2005) Concurrent changes in net CO_2 assimilation and chloroplast ultrastructure in nitrogen deficient citrus leaves. Environ Exp Bot 54: 41–48.
29. Antal T, Mattila H, Hakala-Yatkin M, Tyystjärvi T, Tyystjärvi E (2010) Acclimation of photosynthesis to nitrogen deficiency in Phaseolus vulgaris. Planta 232:887–898.
30. Doncheva S, Vassileva V, Ignatov G, Pandev S (2001) Influence of nitrogen deficiency on photosynthesis and chloroplast ultrastructure of pepper plants. Agric Food Sci in Finland 10: 59–64.
31. Quick WP, Schurr U, Fichtner K, Schulze ED, Rodermel SR, et al. (1991) The impact of decreased Rubisco on photosynthesis, growth, allocation and storage in tobacco plants which have been transformed with antisense rbcS. Plant J 1: 51–58.
32. Evans JR, Kaldenhoff R, Genty B, Terashima I (2009) Resistances along the CO_2 diffusion pathway inside leaves. J Exp Bot 60: 2235–2248.
33. Tholen D, Zhu XG (2011) The mechanistic basis of internal conductance: A theoretical analysis of mesophyll cell photosynthesis and CO_2 diffusion. Plant Physiol 156: 90–105.
34. Tholen D, Ethier G, Genty B, Pepin S, Zhu XG (2012) Variable mesophyll conductance revisited. Theoretical background and experimental implication. Plant Cell Environ 35: 2087–2103.
35. Sage TL, Sage RF (2009) The functional anatomy of rice leaves: Implications for refixation of photorespiratory CO2 and efforts to engineer C4 photosynthesis into rice. Plant and Cell Physiology 50: 756–772.
36. Bota J, Medrano H, Flexas J (2004) Is photosynthesis limited by decreased Rubisco activity and RuBP content under progressive water stress? New Phytol 162: 671–681.
37. Jiang CZ, Rodermel SR, Shibles RM (1993) Photosynthesis, Rubisco activity and amount, and their regulation by transcription in senescing soybean leaves. Plant Physiol 101: 105–112.
38. Law RD, Crafts-Brandner SJ (1999) Inhibition and acclimation of photosynthesis to heat stress is closely correlated with activation of ribulose-1,5-bisphosphate carboxylase/oxygenase. Plant Physiol 120: 173–181.
39. Valentini R, Epron D, Angelis PD, Matteducci G, Dreyer E (1995) In-situ estimation of net CO_2 assimilation, photosynthetic electron flow and photorespiration in turkey oak (Q. cerris L) leaves. Diurnal cycles under different levels of water-supply. Plant Cell Environ 18, 631–640.
40. Laisk AK (1977) Kinetics of photosynthesis and photorespiration in C_3-plants. (In Russian) Nauka, Moscow.

41. Brooks A, Farquhar GD (1985) Effect of temperature on the CO_2/O_2 specificity of ribulose–1,5–bisphosphate carboxylase/oxygenase and the rate of respiration in the light. Planta 165: 397–406.
42. Li Y, Ren BB, Yang XX, Xu GH, Shen QR, et al. (2012) Chloroplast downsizing under nitrate nutrition restrained mesophyll conductance and photosynthesis in rice (Oryza sativa L.) under drought conditions. Plant Cell Physiol 53: 892−900.
43. Grassi G, Magnani F (2005) Stomatal, mesophyll conductance and biochemical limitations to photosynthesis as affected by drought and leaf ontogeny in ash and oak trees. Plant Cell Environ 28: 834−849.
44. Ethier GJ, Livingston NJ (2004) On the need to incorporate sensitivity to CO_2 transfer conductance into the Farquhar-von Caemmerer-Berry leaf photosynthesis model. Plant Cell Environ 27: 137−153.
45. Makino A, Mae T, Chira K (1985) Photosynthesis and rubulose–1,5–bisphosphate carboxylase/oxygenase in rice leaves from emergence through senescence. Planta 166: 414–420.
46. Makino A, Mae T, Chira K (1986) Colorimetric measurement of protein stained with coomassie brilliant blue ron sodium dodecyl sulfate–polyacrylamide gel electrophoresis by eluting with formamide. Agric Biol Chem 50: 1911–1912.
47. Ivanova LA, P'yankov VI (2002) Structural adaptation of the leaf mesophyll to shading. Russian J Plant Physiol 49: 419–431.
48. Oguchi R, Hikosaka K, Hirose T (2003) Does the photosynthetic light–acclimation need change in leaf anatomy? Plant Cell Environ 26: 505–512.

11

Trait Values, Not Trait Plasticity, Best Explain Invasive Species' Performance in a Changing Environment

Virginia Matzek*

Department of Environmental Studies and Sciences, Santa Clara University, Santa Clara, California, United States of America

Abstract

The question of why some introduced species become invasive and others do not is the central puzzle of invasion biology. Two of the principal explanations for this phenomenon concern functional traits: invasive species may have higher values of competitively advantageous traits than non-invasive species, or they may have greater phenotypic plasticity in traits that permits them to survive the colonization period and spread to a broad range of environments. Although there is a large body of evidence for superiority in particular traits among invasive plants, when compared to phylogenetically related non-invasive plants, it is less clear if invasive plants are more phenotypically plastic, and whether this plasticity confers a fitness advantage. In this study, I used a model group of 10 closely related *Pinus* species whose invader or non-invader status has been reliably characterized to test the relative contribution of high trait values and high trait plasticity to relative growth rate, a performance measure standing in as a proxy for fitness. When grown at higher nitrogen supply, invaders had a plastic RGR response, increasing their RGR to a much greater extent than non-invaders. However, invasive species did not exhibit significantly more phenotypic plasticity than non-invasive species for any of 17 functional traits, and trait plasticity indices were generally weakly correlated with RGR. Conversely, invasive species had higher values than non-invaders for 13 of the 17 traits, including higher leaf area ratio, photosynthetic capacity, photosynthetic nutrient-use efficiency, and nutrient uptake rates, and these traits were also strongly correlated with performance. I conclude that, in responding to higher N supply, superior trait values coupled with a moderate degree of trait variation explain invasive species' superior performance better than plasticity per se.

Editor: Mark van Kleunen, University of Konstanz, Germany

Funding: The author was supported by a NASA Earth Systems Science Fellowship while doing this experiment. The funders had no role in study design, data collection and analysis, decision to publish, or preparation of the manuscript.

Competing Interests: The author has declared that no competing interests exist.

* E-mail: vmatzek@scu.edu

Introduction

Invasive species have long proved puzzling to the ecologist: Why do some species become invasive outside their native range, and others do not? One line of reasoning common to many investigations of invasive plants is that invaders have particular traits that make them superior competitors in the invaded habitat. Invasive species may possess novel traits that are poorly represented in the native flora, such as N fixation [1], or they may exhibit more extreme values of competitively advantageous traits than do the local species [2–4]. Multispecies studies comparing phylogenetically related invaders to non-invaders have begun to yield insights into which traits are typically associated with invasion success, which may boost efforts to screen plants for invasiveness before introduction [5–8].

A separate line of reasoning with regard to plant traits is that invaders have higher phenotypic plasticity, which has long been theorized to promote invasion by permitting introduced species to colonize a broader range of environments, or escape extinction in the early period of invasion when the number of available genotypes is small [9–11]. Empirical studies comparing plasticity in invasive and non-invasive plants are now so numerous that they have been subjected to meta-analysis twice–but the meta-analyses came to different conclusions [12,13], with one concluding that invaders showed higher plasticity and the other finding no evidence for such a trend. Likewise, plasticity may be adaptive if it increases fitness (or permits smaller declines in fitness in response to harsher conditions) [14,15], but whether higher plasticity has resulted in higher fitness or invasion success in invasive species has not been made entirely clear by literature reviews or by models [13,16,17].

Studies of phenotypic plasticity have special significance for understanding the response of known invaders to global change. The possible shrinkage or expansion of invaders' range as a consequence of nitrogen deposition, climate warming, increased atmospheric CO_2, or other aspects of environmental change is an issue of serious consequence to land and resource managers, and these responses may be partly mediated by plasticity [18–20]. It would therefore be useful to have more multispecies, phylogenetically controlled comparisons to evaluate the relative contribution of competitively advantageous traits, and plasticity in those traits, as mechanisms of invasion success. As several authors have pointed out [8,13,14,21], even low plasticity may be adaptive in a species that has high values of traits that confer a competitive advantage. Unfortunately, comparisons between invaders and their native or non-invasive counterparts have been performed according to a wide variety of (sometimes problematic) experimental designs [7], including comparing invaders and native species of radically different phylogenetic history, or comparing invaders to indigenous species that have never been introduced outside their native range and therefore have unknown potential for invasiveness [22].

In this study, I investigate the trait values and trait plasticity of a group of 10 species of known invasiveness, all in the genus *Pinus*, grown at different levels of nitrogen fertilization. *Pinus* has been suggested as an ideal model for invasion studies because pine species of tremendous ecological variety have been widely introduced and invasiveness (or lack thereof) is well documented for many members of the genus [23]. In lieu of a measure of reproductive fitness, which was impractical to consider in these long-lived trees, I used the relative growth rate, a performance measure which encompasses many aspects of plant function and has important effects on competitive ability and recruitment [24]. Proxies for fitness relying on biomass or size are common in plasticity studies because of the practical difficulty of measuring fitness [25]. To generalize within this group of closely related congeners, I examined species-level plasticity–i.e., plasticity expressed across an environmental gradient by individuals from the same population [15], as opposed to genotype-level plasticity, the expression of different phenotypes in different environments by a single genotype [9]. I sought to answer the following questions:

1) How do invasive pines compare to non-invasive pines in functional traits at different levels of N supply?
2) Are invasive species more plastic than non-invasive species in responding to increased N? Which traits are most plastic as N supply changes?
3) Which trait values correlate with the fitness proxy, RGR? For which traits does phenotypic plasticity itself appear to be adaptive?

Methods

Experimental design

Study species comprised five invasive pines (*Pinus banksiana, P. halepensis, P. muricata, P. pinaster,* and *P. radiata*) and five non-invasive pines (*P. cembra, P. flexilis, P. lambertiana, P. sabiniana,* and *P. torreyana*). "Invasive" pines have a record of invasiveness on at least two continents, while "non-invasive" pines have no reports of invasiveness after planting on at least three continents [2,26]. Most of the species are in the subgenus *Pinus*, but two of the non-invaders, *P. cembra* and *P. flexilis*, are in the subgenus *Strobus* and are therefore more distantly related to the others [27].

Pine seeds obtained from commercial suppliers were germinated in a sand-vermiculite mixture after a species-specific cold stratification period [28]. Seedlings were transplanted into nutrient treatments when their second set of true leaves emerged, to minimize ontogenetic differences between species. To avoid effects of environmental gradients within the greenhouse, the experiment was blocked, with each of the 9 blocks containing both nutrient treatments and 1 randomly selected individual of each species per treatment. Nine seedlings of each species (except *P. cembra*; n = 7) were grown in 35L pots with nitrogen supply of 50 mg N pot^{-1} wk^{-1} (high N treatment) or 1 mg N pot^{-1} wk^{-1} (low N treatment). Phosphorus was supplied at 10 mg P pot^{-1} wk^{-1} and all other nutrients were supplied in abundance as a half-strength N- and P-free Hoagland's solution. Seedlings grew for 12 weeks after transplant into the treatments. Plants were watered freely and monitored with a soil probe to ensure that moisture was not limiting to plant growth. Average midday PAR at plant height was approximately 1350 μmol/m^2/s. Greenhouse temperatures were ~25°C (day) and ~15°C (night), with daylength set at 12 h.

Twenty additional seedlings of each species (except *P. cembra*; n = 13) were randomly selected for destructive harvest at transplant size and used to estimate initial seedling weight for calculation of the relative growth rate (RGR; total plant dry biomass per unit initial seedling dry weight per day) and the specific absorption rate (SAR; net gain of nutrient per unit root mass per day), integrated over the harvest interval of 12 weeks [29].

Measurement of physiological and morphological traits

Seventeen traits related to biomass allocation, resource capture, leaf construction costs, and nutrient use and uptake efficiency were selected for analysis, because variation in these traits may confer fitness advantages, and because they provide useful points of comparison to other plasticity studies. The photosynthetic rate of each individual was measured immediately prior to final harvest by gas exchange, on a detached shoot tip enclosed in the conifer chamber of a portable infrared gas analyzer (LiCor LI-6400, Lincoln, NE). Measurements were made on several different days outdoors in full sun, when temperatures were moderate (20–30°C), relative humidities were in the range of 70%–75%, and light intensities ranged from 1400–1600 μmol/m^2/s. CO_2 input was fixed at 400 ppm and airflow through the chamber was 500 μmol/s. Measurements were made as soon as the CO_2 concentration in the chamber stabilized, typically <2 minutes. Self-shading was minimized by orienting the chamber so that the shoot tip was maximally illuminated. The maximal photosynthetic rate was calculated on an area basis using a leaf mass/area conversion (methods below). Photosynthetic nitrogen-use efficiency (PNUE) was calculated as the ratio of A_{area} and N_{area} (μmol CO_2 g N^{-1} s^{-1}), using the photosynthetic rate of individuals in each species and the value for N or P concentration from the species tissue composite (below). Instantaneous water-use efficiency (WUEi) was calculated as the ratio of photosynthesis to transpiration (μmol CO_2 mmol H_2O^{-1} s^{-1}).

Chemical analyses were performed on the two youngest fully expanded whorls of needles at each shoot tip, which were removed, weighed, and flash frozen in liquid nitrogen at the time of harvest. The frozen leaf tissue was bulked by species, ground in liquid nitrogen with a mortar and pestle, and either stored at −80°C for use in protein and chlorophyll measurements, or oven-dried at 55°C for assays of N and P concentration by Kjeldahl digest. The remaining root, stem, and leaf tissue were separately weighed and oven-dried at 55°C for measurement of biomass allocation traits (LMR = leaf mass ratio, SMR = stem mass ratio, RMR = root mass ratio, LAR = leaf area ratio) and leaf dry mass fraction (DMF, ratio of dry mass to fresh mass). Specific leaf area (SLA) was calculated from the projected leaf area of fresh needles from a subset of 3–4 harvested individuals that were scanned on a flatbed scanner before being dried and weighed. Protein, chlorophyll, and nutrient content results are expressed on an area basis.

Soluble protein content was determined by a Lowry assay compatible with detergents and reducing agents, using an extract of frozen leaf tissue heated to 55°C in a buffer of 5% sucrose, 5% sodium dodecyl sulfate, and 5% β-mercaptoethanol [30]. For chlorophyll measurements, absorbance of an extract of frozen leaf tissue in 100% acetone was measured at 662 and 645 nm to determine chl *a* and *b* concentrations [31].

Data analysis

To increase the generalizability of the results, and because the small amount of biomass available from individuals in the low-nutrient treatment required bulking tissue from different individuals into a species composite, species is the replicate for this study, not individuals. Mean trait values for all 17 measured traits, plus RGR, were calculated for each species in each nitrogen treatment. To assess the effects of nitrogen availability and invasiveness on plant traits, I performed a mixed-model, nested ANOVA with "nitrogen level" and "invasive status" as fixed effects and "species" nested in "invasive status" as a random effect, for all traits except biomass allocation traits. Biomass allocation patterns can be influenced by plant size, so leaf-, stem-, and root-mass ratios (LMR, SMR, and RMR, respectively) were analyzed as mixed-model nested ANCOVAs with final harvest biomass as a covariate. Because species is the replicate in this experiment, variation among individuals of a species and among greenhouse blocks does not figure into the ANOVA and ANCOVA analyses. Using species as a replicate comes at a sacrifice of some statistical power, but has the advantage of using the same level of data resolution for the trait value analysis as the plasticity index analysis (below).

To quantify trait plasticity, I calculated a plasticity index (PI_v) [32] as the difference between the maximum and minimum values of the trait mean, divided by the maximum value, for each species. There are many different ways to calculate relative plasticity [33]; this one approximates a reaction norm and has the advantage of being insensitive to differences in variance between samples in the two environments [21,32]. Student's t-test for unequal variances was used to distinguish invader and non-invader groups for each trait.

Though useful for calculating averages and correlations, a disadvantage of an absolute-value measure like PI_v is that it obscures the direction of the response. Invasive species and non-invasive species may differ in whether they increase or decrease trait values in response to increased N. Therefore, a second plasticity index, the relative trait range (RTR) [34], was calculated to see whether systematic differences existed between invaders and non-invaders in the sign of change for any trait. To determine the RTR, I subtracted the mean trait value in the low N treatment from the mean trait value in the high N treatment and divided it by the maximum of these two values. Positive RTR values mean that the trait value increased in response to higher N supply. RTR values were not used in statistical calculations, but are identical to PI_v values except for sign.

The relative growth rate (RGR) under high resource conditions was used as a measure of performance and a proxy for fitness. For each trait, I evaluated whether species' mean trait values across nutrient treatments, or their mean plasticity indices, were significantly correlated with mean RGR across treatments, by calculating Pearson product-moment correlation coefficients for the association.

Because analysis of 17 separate traits involves performing multiple comparisons on the same dataset, it is necessary to adjust the probability of Type I error for the large number of statistical tests. Rather than using the sequential Bonferroni correction, which has the drawback of greatly inflating the probability of Type II error, I instead report all exact p-values and control the false discovery rate using the procedure of Benjamini and Hochberg [35], which is suitable when tested variables lack independence from each other [36], as is true here.

Results

Comparison of mean trait values for invasive and non-invasive pines at different N levels

The N treatments had a strong effect on plant performance, as represented by relative growth rate (RGR). Relative growth rate

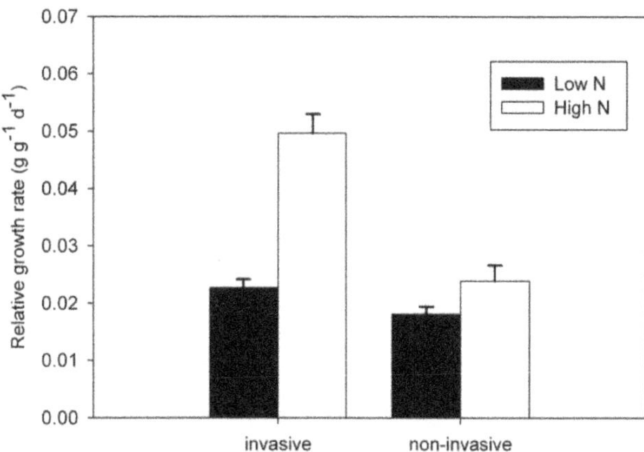

Figure 1. Performance (RGR) of invasive and non-invasive species across nutrient treatments. Values are means ± standard error of invasive and non-invasive species groups in low-N and high-N treatment.

was increased both by high nitrogen and invasive status, and there was a significant interaction whereby invasive species benefited more from the high nitrogen levels than did non-invaders in increasing their growth rate (Figure 1).

Descriptions and units for the 17 functional traits measured in the experiment are shown in Table 1. All of the biomass allocation traits differed significantly between the invasive group and the non-invasive group (Table 2 and 3). Invaders had higher leaf mass ratio (LMR) and leaf area ratio (LAR), but lower root mass ratio (RMR) and stem mass ratio (SMR), than non-invaders. Nitrogen level also affected allocation to biomass; increased nitrogen caused pine seedlings to increase leaf mass and leaf area ratios, and decrease root mass ratios.

Of leaf-level traits, invasives had higher specific leaf area (SLA) and lower dry-mass fraction (DMF) than did non-invasives. Also, photosynthetic capacity (A_{area}) and stomatal conductance (gs) were higher, but leaf nitrogen (Narea) was lower, in invaders than non-invaders. High N supply increased leaf nitrogen content, chlorophyll content (chl_{area}), photosynthesis, and stomatal conductance, and decreased leaf DMF.

All the whole-plant traits associated with nutrient use and uptake showed greater efficiency in the invader group. However, the specific root absorption rates for nitrogen and phosphorus ($SAR_{nitrogen}$ and $SAR_{phosphorus}$) also were affected by N supply, with a significant interaction between invasive status and nitrogen level for $SAR_{nitrogen}$ whereby invaders increased their N uptake rate to a greater degree than non-invaders when N supply increased. Photosynthetic nitrogen-use efficiency (PNUE) was increased only by invasive origin, not by nitrogen supply. Instantaneous water-use efficiency (WUEi) was higher in non-invaders than invaders, and was unaffected by N level.

Reaction norms for traits of invaders and non-invaders responding to the increase in N supply are summarized in Figure 2. To see mean values for every trait in each species, consult the Supplemental Information (Table S1).

Comparison of mean plasticity values for invasive and non-invasive pines

No significant difference in the plasticity index, PI_v, between the groups of invasive species and non-invasive species was apparent for any trait, after correction for multiple comparisons (Table 4). Relative trait range index (RTR) values, which are identical to PI_v

Table 1. Descriptions of traits and performance measure (fitness proxy).

Abbreviation	Description	Units
RGR	Relative growth rate	g plant g^{-1} init wt d^{-1}
LMR	Leaf mass ratio	g leaf g^{-1} plant
RMR	Root mass ratio	g root g^{-1} plant
SMR	Stem mass ratio	g stem g^{-1} plant
LAR	Leaf area ratio	cm^2 leaf g^{-1} plant
SLA	Specific leaf area	cm^2 mg^{-1} leaf
DMF	Dry mass/fresh mass ratio	
P_{area}	Phosphorus content per unit area	g p m^{-2} leaf
N_{area}	Nitrogen content per unit area	g N m^{-2} leaf
chl_{area}	Chlorophyll (a+b) content per unit area	μg chl cm^{-2} leaf
$protein_{area}$	Protein content per unit area	mg protein cm^{-2} leaf
A_{area}	Photosynthetic rate per unit area	μg mol CO_2 m^{-2} leaf s^{-1}
g_s	Stomatal conductance	mol CO_2 m^{-2} leaf s^{-1}
PNUE	Photosynthetic nitrogen-use efficiency	mmol CO_2 g N^{-1} s^{-1}
WUEi	Instantaneous water-use efficiency	mmol CO_2 $mmol^{-1}$ H_2O s^{-1}
$SAR_{nitrogen}$	Specific absorption rate of N	mg N gain g^{-1} root d^{-1}
$SAR_{phosphorus}$	Specific absorption rate of P	mg P gain g^{-1} root d^{-1}

values except that they indicate (by their sign) whether trait values increased or decreased in response to higher N, are reported for all 10 species and 17 traits in the Supplemental Information (Table S2).

Generally, plasticity indices decreased in the order whole-plant > leaf-level > biomass allocation. The most plastic traits in response to nitrogen supply were those associated with nutrient uptake ($SAR_{nitrogen}$ and $SAR_{phosphorus}$), leaf nitrogen (N_{area} and

Table 2. Trait values for invasive and non-invasive species at two nitrogen levels.

Trait	Low nitrogen		High nitrogen	
	invasive	*non-invasive*	*invasive*	*non-invasive*
LMR	505±013(a,b)	452±030(a)	583±014(c)	510±025(b)
RMR	348±022(a)	375±027(a)	256±013(c)	308±023(b)
SMR	172±013(b)	193±024(a,b)	188±012(a,b)	199±023(a)
LAR	21.020±1.890(b)	14.290±1.185(c)	23.913±1.124(a)	16.275±1.379(b,c)
SLA	42.519±3.494(a)	32.319±2.238(b)	42.112±2.089(a)	32.669±2.359(b)
DMF	255±016(b)	303±016(a)	201±007(c)	262±022(a,b)
P_{area}	1.359±248(a)	1.196±132(a)	1.213±129(a)	1.312±140(a)
N_{area}	3.327±305(c)	5.051±1.059(b)	6.681±841(a,b)	8.182±1.373(a)
chl_{area}	19.57±2.33(a)	33.11±11.95(a)	41.07±6.67(a)	43.97±4.43(a)
Chl a/b	3.342±090(a)	3.472±141(a)	3.704±189(a)	3.441±106(a)
$protein_{area}$	1.109±266(a)	1.406±393(a)	1.713±233(a)	1.361±179(a)
A_{area}	16.795±1.740(a)	13.599±0.768(a)	34.761±4.442(b)	20.219±4.139(a,b)
g_s	229±034(b)	191±038(b)	449±077(a)	262±068(a,b)
PNUE	5.090±368(a)	3.142±647(a)	5.557±1.086(a)	2.723±569(a)
WUEi	1.769±214(b)	2.899±518(a)	2.386±205(a,b)	2.732±350(a)
$SAR_{nitrogen}$	807±252(b)	136±029(b)	5.993±697(a)	2.056±326(b)
$SAR_{phosphorus}$	840±190(a)	551±193(a)	2.856±266(b)	1.179±460(a)

Trait abbreviations as in Table 1. Values are means ± standard errors. Different lower-case letters within rows denote significant differences ($\alpha = .05$) from mixed-model, nested ANOVAs using "nitrogen level" and "invasive status" as fixed effects and "species" nested in "invasive status" as a random effect. Biomass allocation traits (top 4 rows) were analyzed as mixed-model nested ANCOVAs with final harvest biomass as a covariate.

Table 3. ANOVA and ANCOVA statistics for trait values.

Trait	Nutrient		Status		Nutrient × Status	
	F	*p*	*F*	*p*	*F*	*p*
LMR	19.2101	**0.0032§**	69.4507	**<.0001§**	0.9853	0.354
RMR	29.9966	**0.0009§**	19.1447	**0.0333§**	2.8382	0.1359
SMR	4.2337	0.0786	13.6685	**0.0077§**	1.9603	0.2042
LAR	14.5617	**0.0066§**	115.4058	**<.0001§**	3.493	0.1038
SLA	0.0002	0.9878	29.0339	**0.0007§**	0.0433	0.8404
DMF	26.8405	**0.0008§**	35.6139	**0.0003§**	0.5429	0.4823
P_{area}	0.0075	0.9332	0.0337	0.859	0.5677	0.4728
N_{area}	48.8286	**0.0001§**	12.0779	**0.0084§**	0.0577	0.8163
chl_{area}	8.5546	**0.0191§**	2.2099	0.1754	0.9245	0.3645
Chl a/b	1.2486	0.2963	0.2029	0.6644	1.7647	0.2207
$protein_{area}$	1.4839	0.2579	0.0141	0.9083	1.9935	0.1957
A_{area}	12.606	**0.0075§**	6.5621	**0.0336§**	2.6849	0.1399
g_s	10.6068	**0.0116§**	6.3525	**0.0358§**	2.7816	0.1339
PNUE	0.0014	0.9709	14.4137	**0.0053§**	0.4958	0.5013
WUEi	1.7596	0.2213	18.9281	**0.0024§**	5.3531	**0.0494**
$SAR_{nitrogen}$	67.8128	**<.0001§**	28.497	**0.0007§**	14.3216	**0.0054§**
$SAR_{phosphorus}$	19.423	**0.0023§**	10.7436	**0.0112§**	5.3555	**0.0494**

Trait abbreviations as in Table 1. Test statistics are for mixed-model, nested ANOVAs using "nitrogen level" and "invasive status" as fixed effects and "species" nested in "invasive status" as a random effect, except biomass allocation traits (top 4 rows), which were analyzed as mixed-model nested ANCOVAs with final harvest biomass as a covariate. Boldface denotes p-values less than.05; § denotes p-values significant at $\alpha = .05$ when corrected for 17 comparisons by the Benjamini-Hochberg procedure.

chl_{area}), and photosynthesis (A_{area} and g_s). The least plastic traits were chlorophyll a/b ratio, stem mass ratio, and SLA (Table 4).

When trait plasticity indices were averaged together in groups of traits (biomass allocation: LMR, SMR, RMR, and LAR; leaf-level: SLA, DMF, P_{area}, N_{area}, chl_{area}, $protein_{area}$, A_{area}, and g_s; whole-plant: PNUE, WUEi, $SAR_{nitrogen}$ and $SAR_{phosphorus}$), greater plasticity in the invasive species was apparent for all the groupings, but still did not rise to the level of statistical significance after correction for three simultaneous comparisons (Table 4, Figure 3). Whole-plant trait plasticity was the most different between invasives and non-invasives, followed by leaf-level trait plasticity and then by plasticity in biomass allocation traits.

Correlation with performance (RGR)

Trait values were much more frequently and strongly correlated with performance than plasticity indices (Table 5). Only the plasticity index for SMR was significantly correlated with RGR; however, closer examination of the data revealed that SMR plasticity for one species, *P. cembra,* was an extreme outlier (defined as <1.5 times the interquartile range for the distribution). Deletion of the outlying data point resulted in a non-significant correlation, though still a strong one ($r = .76557$, $p = .0448$). Among trait values in the biomass allocation group, LMR and LAR were significantly correlated with RGR, as were SLA and dry-mass fraction among the leaf-level traits. The strongest performance correlates were at the whole-plant level, where variation in PNUE and $SAR_{nitrogen}$ and $SAR_{phosphorus}$ each explained >80% of the variation in RGR.

When trait plasticity indices were combined according to the groupings above, leaf-level trait plasticity was significantly associated with RGR, but biomass allocation and whole-plant trait plasticity were not.

Discussion

Trait values

Thirteen of the seventeen traits measured in this study differed between invaders and non-invaders, and ran the gamut of plant function from biomass allocation and leaf morphology to nitrogen uptake and photosynthetic efficiency. In a recent meta-analysis of trait comparisons related to invasiveness, invaders were found to be significantly different from non-invaders in an equally wide range of functional categories: shoot allocation, leaf-area allocation, physiology, size, growth rate, and fitness [37].

Some of my results supported the idea that invaders tend to occupy the "quick-return" end of the so-called leaf economics spectrum [38], where high carbon fixation rates and nutrient contents are associated with shorter leaf lifespans and thinner, less dense leaves. Several studies have attributed higher SLA, LAR, leaf nutrients, and/or photosynthetic capacity to invasives, including a large-dataset study of local and global leaf traits [39] and some studies of invasive-native congener pairs [40–44]. Leaves with higher SLA and lower DMF present more surface area for gas exchange relative to their investment in biomass and construction costs, so this trait syndrome may confer a strong competitive advantage to invaders.

My results also indicated that invasive species are "leafier," investing more heavily in leaf tissue at the expense of stems and roots. In some phylogenetically controlled studies in low-resource environments, invaders have been shown to invest more than non-invaders in root mass [45,46], but in this case, invaders allocated less biomass to root mass. Investing in additional biomass to more thoroughly mine the soil might be adaptive in a low-nutrient environment, but my results suggest invaders compensate with other efficiencies. For instance, invaders were more photosynthet-

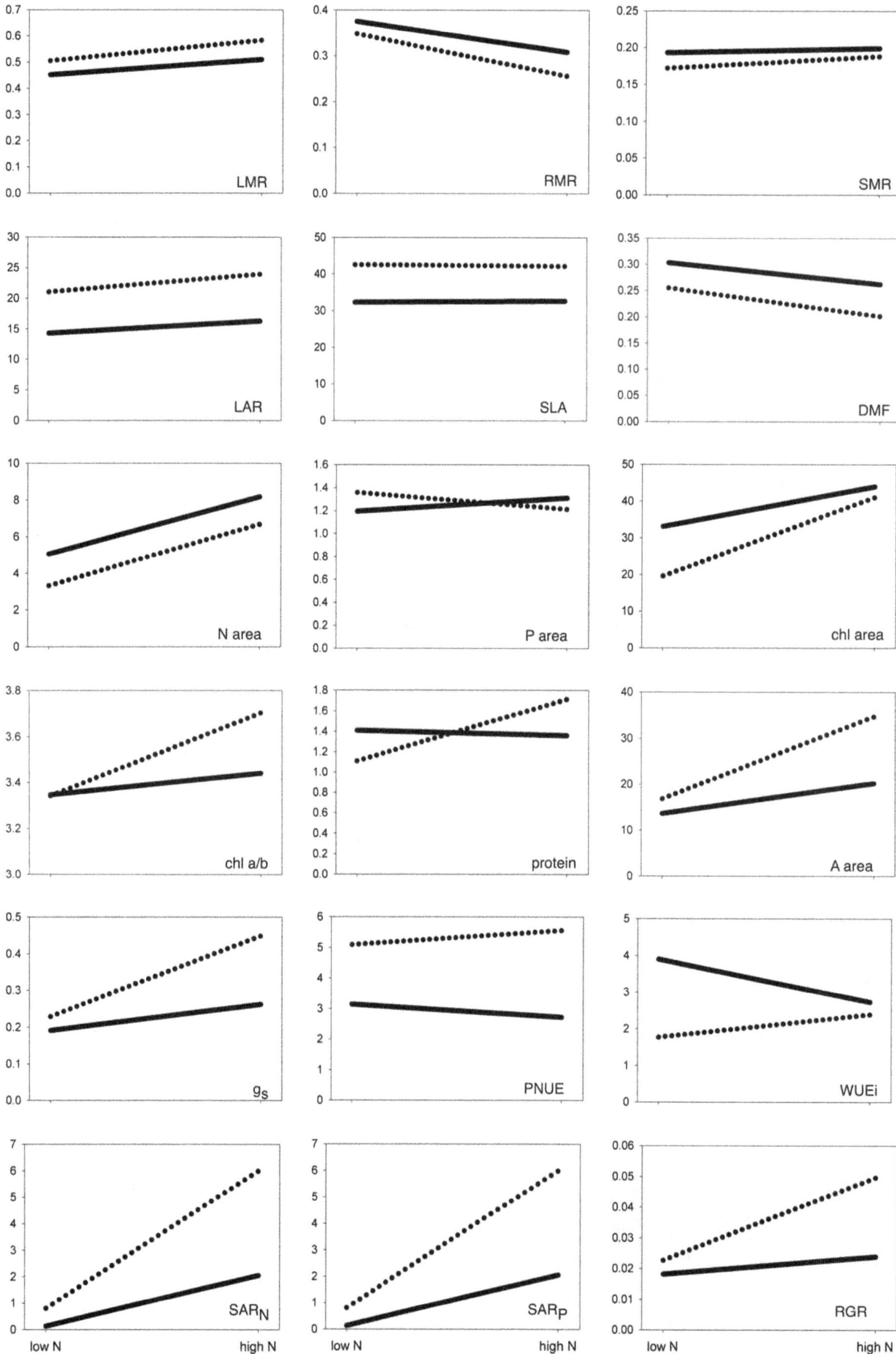

LMR
RMR
SMR
LAR
SLA
DMF
N area
P area
chl area
chl a/b
protein
A area
g_s
PNUE
WUEi
SAR_N
SAR_P
RGR
low N
high N

Figure 2. Trait reaction norms for invasive and non-invasive species across nutrient treatments. Dotted line = invasive species; solid line = non-invasive species. Trait abbreviations are as in Table 1. For each trait, the line links the mean in the low-N treatment to the mean in the high-N treatment, so steeper slopes indicate greater relative responses to the change in nutrient supply.

ically nitrogen-use efficient–able to photosynthesize on less leaf N–as well as more efficient at taking up nutrients across the root surface (SAR_N and SAR_P). Photosynthetic nutrient-use efficiency frequently correlates positively with SLA and negatively with DMF, because denser leaves with thicker cell walls have lower internal conductance and may require more N allocation to structural proteins rather than the photosynthetic apparatus [47,48]. Several investigators have associated higher PNUE with invaders in phylogenetically constrained pairings of invaders and non-invaders [41,49,50]. Water-use efficiency, a trait that has been shown to trade off with PNUE [51], was higher in non-invaders. A few empirical studies have recorded either higher or lower WUEi in invaders when congeneric pairs are compared [43,49,50], suggesting that its relationship with invasiveness may depend on the environment invaded.

Plasticity

Contrary to expectation, no significant plasticity differences between invaders and non-invaders were found for individual traits, nor for groupings of traits. However, all of the groupings, and all but two of the individual traits, showed (nonsignificant) trends of higher plasticity in invaders. The plasticity indices calculated here are based on species mean values, so this result may be partly owing to a lack of statistical power in a study with only 10 species to compare. However, the trait value analysis (above) was also performed on species mean values, but resulted in nearly all the traits being highly significantly different between invaders and non-invaders. Therefore plasticity differences related to invasion status are much weaker than trait value differences.

Other studies have used meta-analysis to harness greater statistical power in answering this question, although most of the studies they synthesize compared invaders to native species, not necessarily non-invasive ones. The results have been mixed. One, a meta-analysis of invasive-native pairs [13], concluded that invaders showed greater plasticity overall; it also identified six individual traits for which invaders were significantly more plastic. Of these traits, three have analogs in the present study (PNUE, WUEi, and root:shoot ratio) but were not found to be more plastic in invasive pines. Four other traits represented in both studies (N content, P content, photosynthesis, and SLA) exhibited no plasticity relationships to invasiveness in either study. A second

Table 4. Plasticity indices (PI_v) for invasive and non-invasive species.

Traits	Invasive	Invasive response	Non-invasive	Non-invasive response	t	p
LMR	133±016	Increase	116±021	increase	0.6696	0.5231
RMR	260±032	Decrease	182±010	decrease	2.2934	0.0732
SMR	116±012	Mixed	041±015	mixed	3.86	**0.0053**
LAR	142±051	mixed	121±024	increase	0.3789	0.7186
SLA	0.134±093	mixed	037±037	mixed	1.46255	0.1437
DMF	205±03	decrease	135±10	decrease	2.161	0.0807
P_{area}	281±058	mixed	276±055	mixed	1.26	0.2469
N_{area}	489±037	increase	380±071	increase	0.0573	0.9557
chl_{area}	483±084	increase	411±100	mixed	1.36	0.2224
Chl a/b	091±043	increase	102±031	mixed	0.5544	0.5949
$protein_{area}$	395±106	mixed	323±134	mixed	0.2093	0.8399
A_{area}	462±117	increase	324±072	mixed	0.4242	0.6832
g_s	491±092	mixed	271±051	mixed	1.008	0.3488
PNUE	286+.072	mixed	209±071	mixed	0.7607	0.4687
WUEi	267±083	mixed	158±058	mixed	1.216	0.274
$SAR_{nitrogen}$	856±049	increase	933±010	increase	1.57	0.1867
$SAR_{phosphorus}$	733±078	increase	559±052	mixed	1.86	0.1149
all allocation	163±018		115±009		2.3623	0.0585
all leaf-level	337±026		251±023		2.4928	**0.0378**
all whole-plant	535±021		457±018		2.807	**0.0237**

Trait abbreviations as in Table 1. Test statistics (t) and p-values (p) are from Student's t-test for unequal variances; p-values in boldface are <.05. The bottom three rows represent mean values for combined indices with several traits' plasticity indices averaged together by species (see also Figure 3). No plasticity indices for individual traits were significantly different between invasive and non-invasive groups after Benjamini-Hochberg correction for 17 trait comparisons, nor were grouped indices significantly different after correction for three comparisons. Means, standard errors, and test statistics were calculated using the PI_v, which represents the absolute value of the change in trait value between N treatments, but a second index, the RTR, was used to determine the directionality of the response, which is indicated to the right of each PI_v column for individual traits. "Increase" means that all species in the group increased the trait value in response to increased N; "decrease" means that all species in the group decreased the trait value; and "mixed" means that at least one increase and one decrease were observed in the species group. Table S2 in the Supplementary Information shows RTR values for all traits and all species.

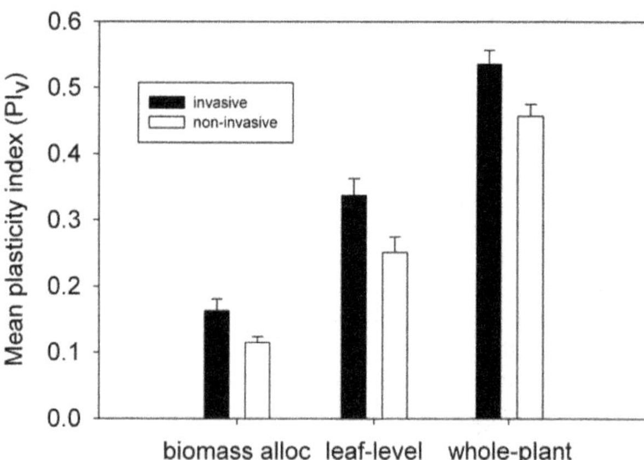

Figure 3. Differences in mean plasticity between invasive and non-invasive species by trait grouping. Values are means ± standard error for invasive and non-invasive species, where the plasticity indices of several traits in a grouping have been averaged together for each of the 5 invasive and 5 non-invasive species. For the biomass allocation grouping, each species' value is its mean plasticity index for the traits LMR, SMR, RMR and LAR; for the leaf-level grouping, each species' value is its mean plasticity index for the traits SLA, DMF, P_{area}, N_{area}, chl_{area}, $protein_{area}$, A_{area}, and g_s; and for the whole-plant grouping, each species' value is its mean plasticity index for the traits PNUE, WUEi, SAR_N, and SAR_P. After correction for multiple comparisons, no differences were statistically significant.

recent meta-analysis [12] covered some of the same papers, but restricted the trait differences studied to a narrower range of environmental conditions (e.g., SLA plasticity was only examined in studies where light was a variable). It found that there was no general trend of higher plasticity in invaders, and individual trait plasticities were not reported. The authors reached the conclusion that invaders' success must be due more to constitutive factors–i.e., trait values–than to trait plasticity, or else that higher plasticity is only characteristic of invaders in the early phase of invasion, and is gradually eroded by genetic assimilation [12]. A third meta-analysis [52] found that more widely distributed alien species had higher plasticity than less-widespread aliens in the response of biomass to increases of light, water, and nutrients, but that this plasticity trend did not extend to the individual traits of SLA or root:shoot ratio. In sum, the evidence for greater plasticity in particular functional traits is weak even when large numbers of species comparisons are considered.

Relationship between performance and trait values or trait plasticity

As a group, invasive pines in this study clearly outperformed their non-invasive congeners in response to an increase in N supply, growing at a slightly higher rate when N was low, but nearly twice as fast when N was high. This significant phenotype-by-environment interaction is evidence that invasives conform to a "Master-of-some" strategy, i.e., a superior ability to increase fitness in response to a favorable environment [15]. Several previous studies of pines have found that invasive pines have higher RGR than non-invasive ones [46,50], but this result indicates that invaders may be most advantaged in high-resource environments.

Godoy and colleagues [21] suggested two mechanisms, not mutually exclusive, for explaining how functional traits might underlie invasives' superior fitness gains (or smaller fitness losses) in a changing environment: 1) higher trait plasticity in invaders that

Table 5. Correlation of trait values and plasticity indices with performance.

Trait	Trait value		Plasticity index	
	r	*P*	*r*	*p*
LMR	0.75556	**0.0115**§	−0.03584	0.9217
RMR	−0.54992	0.0996	0.53532	0.1108
SMR	−0.23861	0.5067	0.83433	**0.0027**§
LAR	0.90241	**0.0004**§	−0.08539	0.8146
SLA	0.80569	**0.0049**§	0.64762	**0.0429**
DMF	−0.80043	**0.0054**§	0.47629	0.164
N_{area}	−0.63011	0.0509	0.54719	0.1016
P_{area}	−0.28759	0.4204	−0.11245	0.7571
chl a/b	−0.02989	0.9347	−0.23981	0.5046
chl_{area}	−0.57344	0.0831	0.40295	0.2483
$protein_{area}$	−0.10995	0.7624	0.10118	0.7809
A_{area}	0.63557	**0.0483**	0.57877	0.0796
g_s	0.654	**0.0402**	0.75942	**0.0108**
PNUE	0.93944	**.0001**§	0.25531	0.4765
WUEi	−0.49952	0.1416	0.26716	0.4555
$SAR_{nitrogen}$	0.87841	**0.0008**§	−0.46466	0.176
$SAR_{phosphorus}$	0.75985	**0.0108**§	0.59434	0.07
all allocation			0.42409	0.2219
all leaf-level			0.84134	**0.0023**§
all whole-plant			0.57544	0.08176

Trait abbreviations as in Table 1. Pearson product-moment correlations (r) and p-values (p) for the relationship between mean trait values or mean plasticity index and mean RGR. Mean values are the average of the two nutrient treatments. The bottom three rows represent correlations between mean RGR and a combined plasticity index drawn from several traits (see text). Boldface denotes p-values less than.05; § denotes p-values significant at $\alpha = .05$ when corrected for 17 multiple comparisons by the Benjamini-Hochberg procedure for individual traits or indices, and for 3 multiple comparisons for grouped plasticity indices. After elimination of an outlier for SMR plasticity (see text), the correlation coefficient $r = .76557$ and $p = .0448$ (non-significant).

results in greater fitness than non-invaders; and 2) similar levels of plasticity among both groups, coupled with constantly superior values for fitness-related traits in invaders. Evidence to support the first mechanism can come from regressions of trait plasticity against fitness [13,20]. For pines in this study, only one trait showed a significant correlation between its plasticity and the performance measure RGR, indicating that, in general, higher plasticity in functional traits is not the best explanation for the performance response observed. Moreover, the biological significance of the sole performance-plasticity correlation (for SMR) is called into question, first by the existence of an influential outlier whose removal decreased its statistical significance, and second by the fact that SMR trait values themselves were not significantly correlated with RGR. This reveals one of the drawbacks of using a plasticity index in the correlation–namely, plasticity is represented by any change, whether an increase of decrease. For SMR, both invasives and non-invasives sometimes increased and sometimes decreased the stem mass ratio when N supply (and growth rates) increased. This makes SMR a poor indicator of growth rate and a poor candidate for an important functional trait, but because the size of the change in either direction is generally larger for invaders and smaller for non-invaders, the plasticity index itself is correlated with performance. In short, there is little evidence from regressions of individual trait plasticities against RGR to support the contention that invaders have higher adaptive plasticity.

Nonetheless, if invaders grow faster at higher N, there must be some underlying trait plasticity that explains why. One clue comes from regrouping the plasticity indices and correlating a combined set of traits with RGR, to see what general categories of plant function may have the greatest adaptive plasticity. There was a significant plasticity-performance correlation for the grouping of traits related to leaf structure and metabolism, whose combined plasticity explained ~84% of the variance in RGR, but not for the other groupings. Another way of answering the question is to look for traits where both the trait value itself and its plasticity index maximize their relationship with RGR. Traits for which the plasticity index and the trait value itself each explain at least half the variation in RGR in this study (i.e., both r >50) are the biomass allocation trait RMR, the whole-plant trait SAR_P, and the leaf-level traits SLA, N_{area}, A_{area}, and g_s. These two lines of evidence suggest that, on the whole, traits related to leaf chemistry, morphology, and photosynthetic ability are the best candidates for adaptively plastic traits in these species–with the caveat that this study only considered differences in N supply, and that other gradients might have produced a different suite of traits. It is also possible that the differential RGR response of invaders is due to plasticity in traits not measured here.

Overall, though, the relationship between functional trait plasticity and performance in this study is not strong. Some traits instead conform to the second possible mechanism for higher fitness–consistently superior values in invaders of traits that either lack plasticity, or have lower plasticity in invaders than non-invaders [21]. The traits LAR, LMR, SLA, and PNUE all fit this description. Contrary to the idea of objectively superior plasticity conferring invasion success, Godoy and colleagues posit the existence of a "general purpose phenotype," characterized by high mean values of fitness-related traits coupled with sufficient plasticity to compete in a wide variety of environments. This phenomenon may explain the results of other multispecies invasive-noninvasive comparisons that found invaders' higher plasticity in biomass production unaccompanied by higher plasticity in presumably related functional traits [8,53] as well as the mixed results from the various meta analyses [12,13,52].

It is important to note, too, that fast growth, the fitness proxy here, is not necessarily the best strategy in every environment [13]; in low-resource systems, building structurally sound or heavily defended tissues may be more important in the long run than high RGR [49]. Also, the use of commercially sourced, rather than wild-sourced, seeds may have introduced a bias toward faster growth that is not representative of introduced pine populations. Other limitations of the study include ignoring reproductive traits like seed mass or fecundity that have proved to be powerful predictors of invasion success in pines and other woody species [54], and failing to account for phylogenetic distances among pairs of species, as well as potential differences among invaders in niche breadth and the width of distribution in their invaded ranges. However, this work provides a rare picture of the comparative functional significance of plant traits and plasticity that is lacking in many invasive-noninvasive comparisons. Future work on this topic could expand the range of traits and environments studied, especially with regard to mimicking probable future conditions under climate change, and comparing traits and fitness proxies between invaders and co-occurring natives in the invaded range.

Acknowledgments

The advice of Peter Vitousek, David Ackerly, and Chris Field were helpful in designing the study, and the logistical support of Tom deHoog was critical to executing it. The author thanks Mark van Kleunen and two anonymous reviewers for comments that greatly improved the final manuscript.

Author Contributions

Conceived and designed the experiments: VM. Performed the experiments: VM. Analyzed the data: VM. Wrote the paper: VM.

References

1. Vitousek PM, Walker LR (1989) Biological invasion by Myrica faya in Hawaii-plant demography, nitrogen-fixation, ecosystem effects. Ecological Monographs 59: 247–265.
2. Grotkopp E, Rejmanek M, Rost TL (2002) Toward a causal explanation of plant invasiveness: Seedling growth and life-history strategies of 29 pine (Pinus) species. American Naturalist 159: 396–419.
3. Leishman MR, Thomson VP, Cooke J (2010) Native and exotic invasive plants have fundamentally similar carbon capture strategies. Journal of Ecology 98: 28–42.
4. Mason RAB, Cooke J, Moles AT, Leishman MR (2008) Reproductive output of invasive versus native plants. Global Ecology and Biogeography 17: 633–640.
5. Pyšek P, Richardson DM (2007) Traits associated with invasiveness in alien plants: where do we stand? In: Nentwig W, editor. Biological Invasions: Springer Berlin Heidelberg. 97–125.
6. Rejmánek M (2000) Invasive plants: approaches and predictions. Austral Ecology 25: 497–506.
7. van Kleunen M, Dawson W, Schlaepfer D, Jeschke JM, Fischer M (2010) Are invaders different? A conceptual framework of comparative approaches for assessing determinants of invasiveness. Ecology Letters 13: 947–958.
8. van Kleunen M, Schlaepfer DR, Glaettli M, Fischer M (2011) Preadapted for invasiveness: do species traits or their plastic response to shading differ between invasive and non-invasive plant species in their native range? Journal of Biogeography 38: 1294–1304.
9. Sultan SE (2001) Phenotypic plasticity for fitness components in Polygonum species of contrasting ecological breadth. Ecology 82: 328–343.
10. Callaway RM, Pennings SC, Richards CL (2003) Phenotypic plasticity and interactions among plants. Ecology 84: 1115–1128.
11. Bossdorf O, Lipowsky A, Prati D (2008) Selection of preadapted populations allowed Senecio inaequidens to invade Central Europe. Diversity and Distributions 14: 676–685.
12. Palacio-López K, Gianoli E (2011) Invasive plants do not display greater phenotypic plasticity than their native or non-invasive counterparts: a meta-analysis. Oikos 120: 1393–1401.
13. Davidson AM, Jennions M, Nicotra AB (2011) Do invasive species show higher phenotypic plasticity than native species and, if so, is it adaptive? A meta-analysis. Ecology Letters 14: 419–431.
14. Van Kleunen M, Fischer M (2005) Constraints on the evolution of adaptive phenotypic plasticity in plants. New Phytologist 166: 49–60.
15. Richards CL, Bossdorf O, Muth NZ, Gurevitch J, Pigliucci M (2006) Jack of all trades, master of some? On the role of phenotypic plasticity in plant invasions. Ecology Letters 9: 981–993.
16. Daehler CC (2003) Performance comparisons of co-occurring native and alien invasive plants: implications for conservation and restoration. Annual Review of Ecology, Evolution, and Systematics 34: 183–211.
17. Peacor SD, Allesina S, Riolo RL, Pascual M (2006) Phenotypic plasticity opposes species invasions by altering fitness surface. Plos Biology 4: 2112–2120.
18. Chown SL, Slabber S, McGeoch MA, Janion C, Leinaas HP (2007) Phenotypic plasticity mediates climate change responses among invasive and indigenous arthropods. Proceedings of the Royal Society B-Biological Sciences 274: 2531–2537.
19. Dukes JS, Pontius J, Orwig D, Garnas JR, Rodgers VL, et al. (2009) Responses of insect pests, pathogens, and invasive plant species to climate change in the forests of northeastern North America: What can we predict? Canadian Journal of Forest Research 39: 231–248.

20. Nicotra AB, Atkin OK, Bonser SP, Davidson AM, Finnegan EJ, et al. (2010) Plant phenotypic plasticity in a changing climate. Trends in Plant Science 15: 684–692.
21. Godoy O, Valladares F, Castro-Díez P (2011) Multispecies comparison reveals that invasive and native plants differ in their traits but not in their plasticity. Functional Ecology 25: 1248–1259.
22. Muth NZ, Pigliucci M (2006) Traits of invasives reconsidered: phenotypic comparisons of introduced invasive and introduced noninvasive plant species within two closely related clades. American Journal of Botany 93: 188–196.
23. Richardson DM (2006) Pinus: a model group for unlocking the secrets of alien plant invasions? Preslia 78: 375–388.
24. Poorter H, Garnier E, editors (1999) Ecological significance of inherent variation in relative growth rate and its components. New York: Marcel Dekker, Inc.
25. Hunt J, Hodgson J, editors (2010) What is fitness, and how do we measure it? Oxford: Oxford University Press. 46–70 p.
26. Rejmanek M, Richardson DM (1996) What attributes make some plant species more invasive? Ecology 77: 1655–1661.
27. Gernandt DS, López GG, García SO, Aaron L (2005) Phylogeny and classification of Pinus. Taxon 54: 29–42.
28. Young JA, Young CG (1992) Seeds of woody plants in North America. Portland, OR: Dioscorides Press. 407 p.
29. Causton D, Venus J (1981) The biometry of plant growth. London: Edward Arnold.
30. Ekramoddoullah AKM (1993) Analysis of needle proteins and N-terminal amino-acid-sequences of 2 photosystem-II proteins of Western white-pine (Pinus-monticola D Don). Tree Physiology 12: 101–106.
31. Lichtenthaler HK, Wellburn A (1983) Determination of total carotenoids and chlorophyll a and b of leaf extracts in different solvents. Biochemical Society Transactions 11: 591–592.
32. Valladares F, Wright SJ, Lasso E, Kitajima K, Pearcy RW (2000) Plastic phenotypic response to light of 16 congeneric shrubs from a Panamanian rainforest. Ecology 81: 1925–1936.
33. Valladares F, Sanchez-Gomez D, Zavala MA (2006) Quantitative estimation of phenotypic plasticity: bridging the gap between the evolutionary concept and its ecological applications. Journal of Ecology 94: 1103–1116.
34. Richardson AD, Ashton PMS, Berlyn GP, Mcgroddy ME, Cameron IR (2001) Within-crown foliar plasticity of western hemlock, Tsuga heterophylla, in relation to stand age. Annals of Botany 88: 1007–1015.
35. Benjamini Y, Hochberg Y (1995) Controlling the false discovery rate - a practical and powerful approach to multiple testing. Journal of the Royal Statistical Society Series B-Methodological 57: 289–300.
36. Benjamini Y, Yekutieli D (2001) The control of the false discovery rate in multiple testing under dependency. Annals of Statistics 29: 1165–1188.
37. Van Kleunen M, Weber E, Fischer M (2010) A meta-analysis of trait differences between invasive and non-invasive plant species. Ecology Letters 13: 235–245.
38. Wright IJ, Reich PB, Westoby M, Ackerly DD, Baruch Z, et al. (2004) The worldwide leaf economics spectrum. Nature 428: 821–827.
39. Leishman MR, Haslehurst T, Ares A, Baruch Z (2007) Leaf trait relationships of native and invasive plants: community- and global-scale comparisons. New Phytologist 176: 635–643.
40. Deng X, Ye WH, Feng HL, Yang QH, Cao HL, et al. (2004) Gas exchange characteristics of the invasive species Mikania micrantha and its indigenous congener M. cordata (Asteraceae) in South China. Botanical Bulletin of Academia Sinica 45: 213–220.
41. Feng YL, Fu GL (2008) Nitrogen allocation, partitioning and use efficiency in three invasive plant species in comparison with their native congeners. Biological Invasions 10: 891–902.
42. Feng YL, Fu GL, Zheng YL (2008) Specific leaf area relates to the differences in leaf construction cost, photosynthesis, nitrogen allocation, and use efficiencies between invasive and noninvasive alien congeners. Planta 228: 383–390.
43. McDowell SCL (2002) Photosynthetic characteristics of invasive and noninvasive species of Rubus (Rosaceae). American Journal of Botany 89: 1431–1438.
44. Wilson SB, Wilson PC (2004) Growth and development of the native Ruellia caroliniensis and invasive Ruellia tweediana. Hortscience 39: 1015–1019.
45. Funk JL (2008) Differences in plasticity between invasive and native plants from a low resource environment. Journal of Ecology 96: 1162–1173.
46. Grotkopp E, Rejmanek M (2007) High seedling relative growth rate and specific leaf area are traits of invasive species: Phylogenetically independent contrasts of woody angiospernis. American Journal of Botany 94: 526–532.
47. Hikosaka K (2004) Interspecific difference in the photosynthesis-nitrogen relationship: patterns, physiological causes, and ecological importance. Journal of Plant Research 117: 481–494.
48. Takashima T, Hikosaka K, Hirose T (2004) Photosynthesis or persistence: nitrogen allocation in leaves of evergreen and deciduous Quercus species. Plant Cell and Environment 27: 1047–1054.
49. Funk JL, Vitousek PM (2007) Resource-use efficiency and plant invasion in low-resource systems. Nature 446: 1079–1081.
50. Matzek V (2011) Superior performance and nutrient-use efficiency of invasive plants over non-invasive congeners in a resource-limited environment. Biological Invasions 13: 3005–3014.
51. Field C, Merino J, Mooney HA (1983) Compromises between water-use efficiency and nitrogen-use efficiency in five species of California evergreens. Oecologia 60: 384–389.
52. Dawson W, Rohr RP, van Kleunen M, Fischer M (2012) Alien plant species with a wider global distribution are better able to capitalize on increased resource availability. New Phytologist 194: 859–867.
53. Schlaepfer DR, Glattli M, Fischer M, van Kleunen M (2010) A multi-species experiment in their native range indicates pre-adaptation of invasive alien plant species. New Phytologist 185: 1087–1099.
54. Rejmanek M, Richardson DM, Higgins SI, Pitcairn MJ, Grotkopp E, editors (2005) Ecology of invasive plants: state of the art. Washington, D.C.: Island Press. 104–161 p.

12

Economic Analysis of Greenhouse Lighting: Light Emitting Diodes vs. High Intensity Discharge Fixtures

Jacob A. Nelson, Bruce Bugbee*

Crop Physiology Laboratory, Department of Plant Soils and Climate, Utah State University, Logan, Utah, United States of America

Abstract

Lighting technologies for plant growth are improving rapidly, providing numerous options for supplemental lighting in greenhouses. Here we report the photosynthetic (400–700 nm) photon efficiency and photon distribution pattern of two double-ended HPS fixtures, five mogul-base HPS fixtures, ten LED fixtures, three ceramic metal halide fixtures, and two fluorescent fixtures. The two most efficient LED and the two most efficient double-ended HPS fixtures had nearly identical efficiencies at 1.66 to 1.70 micromoles per joule. These four fixtures represent a dramatic improvement over the 1.02 micromoles per joule efficiency of the mogul-base HPS fixtures that are in common use. The best ceramic metal halide and fluorescent fixtures had efficiencies of 1.46 and 0.95 micromoles per joule, respectively. We also calculated the initial capital cost of fixtures *per photon delivered* and determined that LED fixtures cost five to ten times more than HPS fixtures. The five-year electric plus fixture cost per mole of photons is thus 2.3 times higher for LED fixtures, due to high capital costs. Compared to electric costs, our analysis indicates that the long-term maintenance costs are small for both technologies. If widely spaced benches are a necessary part of a production system, the unique ability of LED fixtures to efficiently focus photons on specific areas can be used to improve the photon capture by plant canopies. Our analysis demonstrates, however, that the cost per photon delivered is higher in these systems, regardless of fixture category. The lowest lighting system costs are realized when an efficient fixture is coupled with effective canopy photon capture.

Editor: Douglas Andrew Campbell, Mount Allison University, CANADA

Funding: This work was supported by the Utah Agricultural Experiment Station, Utah State University. Approved as journal paper number 8661. http://uaes.usu.edu/ JAN BB. The funders had no role in study design, data collection and analysis, decision to publish, or preparation of the manuscript.

Competing Interests: The authors have declared that no competing interests exist.

* E-mail: bruce.bugbee@usu.edu

Introduction

Rapid advances in lighting technology and fixture efficiency provide an expanding number of options for supplemental lighting in greenhouses, including numerous LED fixtures (light emitting diode, see [1,2] for a history of LED lighting in horticulture). Significant improvements have been made in all three high intensity discharge (HID, which includes high pressure sodium, HPS, and ceramic metal halide, CMH) fixture components: the lamp (often referred to as the bulb), the luminaire (often referred to as the reflector) and the ballast. High pressure sodium fixtures with electronic ballasts and double-ended lamps are now 1.7 times more efficient than older mogul-base HPS fixtures.

Lighting technologies vary widely in how radiation is distributed (Figure 1). There is no ideal pattern of radiation distribution for every application. In large greenhouses with small aisles and uniformly spaced plants, the broad, even output pattern typically emitted from HPS fixtures provides uniform (little variation over a large area) light distribution and increased capture of photosynthetic photons. In smaller greenhouses with spaced benches, the more focused pattern typically found in LED fixtures can maximize radiation transfer to plant leaves. As the area (height of width) covered by plants increases, the need for more focused radiation decreases (Figure 2).

In greenhouse applications, selection among lighting options should primarily be made based on the cost to deliver photons to the plant canopy surface. This analysis includes two parameters: 1) the fundamental fixture efficiency, measured as micromoles of photosynthetic photons per joule of energy input, and 2) the canopy photosynthetic (400–700 nm) photon flux (PPF) capture efficiency, which is the fraction of photons transferred to the plant leaves.

Electrical efficiency for plant growth is best measured as µmoles per Joule

The electrical efficiency of lamps is often expressed using units for human light perception (efficacy; lumens or foot-candles out per watt in) or energy efficiency (radiant watts out per electrical watt in). Photosynthesis and plant growth, however, is determined by moles of photons. It is thus important to compare lighting efficiency based on photon efficiency, with units of micromoles of photosynthetic photons per joule of energy input. This is especially important with LEDs where the most electrically efficient colors are in the deep red and blue wavelengths. A dramatic example of this is the comparison of red, blue, and cool white LEDs (Table 1). The lower radiant energy content of red photons allows more photons to be delivered per unit of input energy (radiant energy is inversely proportional to wavelength, Planck's Equation). Conversely, blue LEDs can have a 53% higher energy efficiency (49% vs. 32%) but only a 9% higher photon efficiency (1.87 vs. 1.72).

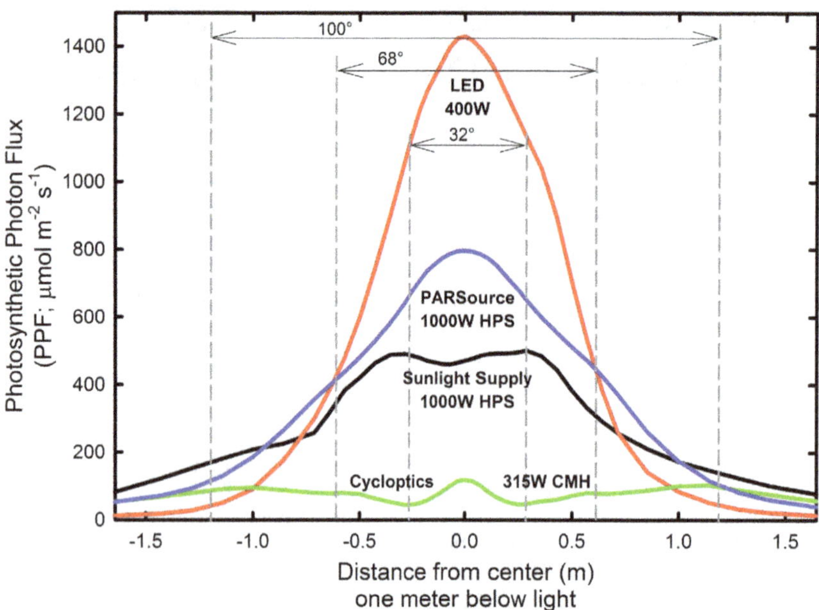

Figure 1. The photon distribution of four fixtures with similar photon efficiency. Each line represents a cross section of the photon intensity below the fixture. The LED fixture (Lighting Sciences Group) uses optics to achieve a narrow distribution, with the majority of the photons falling in a concentrated pattern directly below the fixture. Conversely, the Cycloptics ceramic metal halide fixture is designed for even light distribution, and therefore casts uniform radiation over a large surface area. Since the area increases exponentially as the distance from the center increases, an equal photon flux farther from the center represents a larger quantity of total photons.

Effect of light quality

There is considerable misunderstanding over the effect of light quality on plant growth. Many manufacturers claim significantly increased plant growth due to light quality (spectral distribution or the ratio of the colors). A widely used estimate of the effect of light quality on photosynthesis comes from the Yield Photon Flux (YPF) curve, which indicates that orange and red photons between 600 to 630 nm can result in 20 to 30% more photosynthesis than blue or cyan photons between 400 and 540 nm (Figure 3)[3,4]. When light quality is analyzed based on the YPF curve, HPS lamps are equal to or better than the best LED fixtures because they have a high photon output near 600 nm and a low output of blue, cyan, and green light [5].

The YPF curve, however, was developed from short-term measurements made on single leaves in low light. Over the past 30 years, numerous longer-term studies with whole plants in higher light indicate that light *quality* has a much smaller effect on plant growth rate than light *quantity* [6,7]. Light quality, especially the fraction of blue light, has been shown to alter cell expansion rate, leaf expansion rate[8], plant height and plant shape in several species [9–11], but it has only a small direct effect on photosynthesis. The effects of light quality on fresh or dry mass in whole plants typically occur under low or no sunlight conditions, and are caused by changes in leaf expansion and radiation capture during early growth [6].

Unique aspects of LED fixtures

The most electrically efficient colors of LEDs, based on moles of photosynthetic photons per joule, are blue, red, and cool white, respectively (Figure 4), so LED fixtures generally come in combinations of these colors. LEDs of other colors can be used to dose specific wavelengths of light to control aspects of plant growth [12], due to their monochromatic nature (see [13] for a review of unique LED applications). Ultraviolet (UV) radiation is typically absent in LED fixtures because UV LEDs significantly reduce fixture efficiency. Sunlight has 9% UV (percent of PPF), and standard electric lights have 0.3 to 8% UV radiation (percent of PPF)[5]. A lack of UV causes disorders in some plant species (e.g. Intumescence; [14]) and this is a concern with LED fixtures when used without sunlight. LED fixtures for supplemental photosynthetic lighting also have minimal far-red radiation (710 to 740 nm), which decreases the time to flowering in several photoperiodic species [15]. Green light (530 to 580 nm) is low or absent in most LED fixtures and these wavelengths better penetrate through the canopy and are more effectively transmitted to lower plant leaves [16]. The lack of UV, green, and far-red wavelengths, however, should be minimal when LEDs are used in greenhouses, because most of the radiation comes from broad spectrum sunlight.

Our objective is to help growers and researchers select the most cost effective fixture options for supplemental lighting in greenhouses. To achieve this goal we measured two fundamental components of each fixture: 1) the efficiency of conversion of electricity to photosynthetic photons that are delivered to a horizontal surface below the lamp, and 2) the distribution pattern of these photons below the fixture.

Materials and Methods

Fixture efficiency

Measurements of fixture efficiency (lamp, luminaire, and ballast) were made by integrating sphere and flat-plane integration techniques. The integrating sphere measurements were made by a certified testing laboratory (TÜV SÜD America) that specializes in the measurement of the efficiency of lighting fixtures using the IES LM79-08 measurement standard [17]. Radiometric output

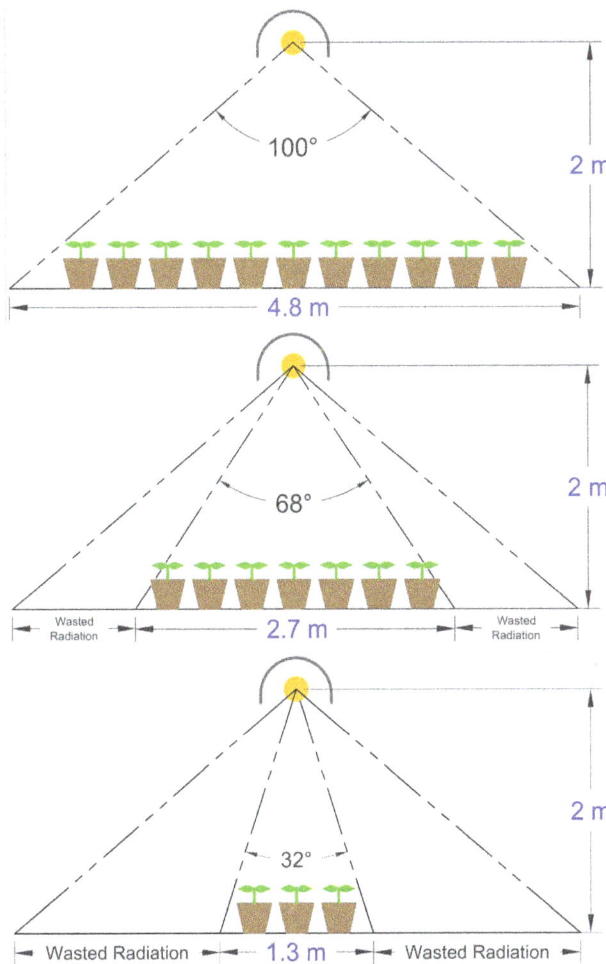

Figure 2. Canopy photon capture efficiency. As the plant growth area under the fixture gets smaller, wasted radiation often increases. This figure illustrates the concept of canopy photon capture efficiency. Two meters was chosen as a typical mounting height, but this can be scaled as a unit-less ratio. Multiple overlapping fixtures are typically used to minimize PPF variation over a large area.

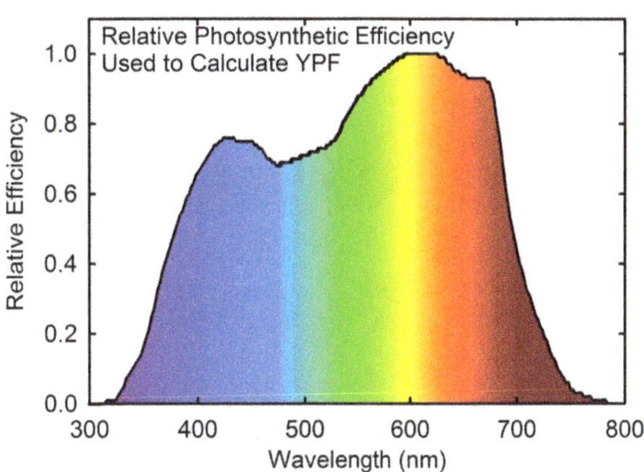

Figure 3. Yield photon flux curve. Effect of wavelength on relative photosynthesis per incident photon for a single leaf in low light (less than 150 μmol m^{-2} s^{-1}) [4].

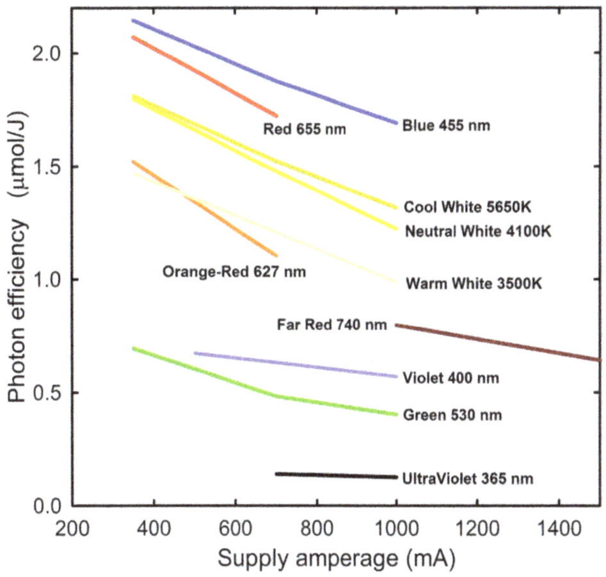

Figure 4. Effect of drive amperage and color on photon efficiency of LEDs. Data for Philips Lumileds LEDs (May 2014), courtesy of Mike Bourget, Orbitec.

Table 1. Efficiency of individual LEDs at a drive current of 700 mA.

LED Color	Peak wavelength or color temperature	Photon efficiency[z] (μmol/J)	Electrical efficiency[y] (%)	Luminous efficiency[x] (lm/W)
Cool white	5650 Kelvin	1.52	33	111
Red	655 nm	1.72	32	47
Blue	455 nm	1.87	49	17

[z]-Photon efficiency is the most appropriate measure for photosynthesis.
[y]-The relationship between electrical efficiency and photon efficiency is dependent on wavelength (Plank's equation $E = hc/\lambda$).
[x]-Luminous efficiency is shown to demonstrate how inappropriate it is as an indicator of lighting efficiency for plants.

Table 2. Efficiency of fixtures using integrating sphere measurements compared with flat-plane integration.

	TÜV SÜD America integrating sphere			USU[z] flat plane integration			flat plane/ integrating sphere[y]
Fixture	**Elec. input (W or J/s)**	**Photon output (μmol/s)**	**Photon efficiency (μmol/J)[x]**	**Elec. input (W or J/s)**	**Photon output (μmol/s)**	**Photon efficiency (μmol/J)[x]**	**(μmol/J)/(μmol/J)**
Gavita Pro 1000DE	1033	1751	1.70	1041	1814	1.74	2.7%
ePapillion 1000W	1041	1767	1.70	1037	1937	1.87	9.1%
LSG violet	384	653	1.70	391	628	1.61	−6.0%
SPYDR 600	326	541	1.66	332	575	1.73	4.4%
LSG red/white	390	634	1.63	397	601	1.51	−7.5%
Illumitex NeoSol	279	390	1.40	281	386	1.38	−1.8%
ParSource GLXII	1026	1334	1.30	1008	1433	1.42	8.6%
Lumigrow Pro 325	304	390	1.29	304	355	1.17	−10.1%
California Light Works SOLARSTORM	337	350	1.04	343	331	0.96	−7.7%
Black Dog BD360U	339	339	1.00	346	323	0.93	−7.2%
Apache AT120WR	169	163	0.96	167	150	0.90	−7.2%
iGrow 400W	394	374	0.95	397	354	0.89	−6.5%
Lumigrow es330	318	284	0.90	317	270	0.85	−5.1%
Hydrogrow Sol 9	423	378	0.89	430	396	0.92	2.9%

[z]-Utah State University

[y]-The flat-plane integration may have made an inadequate number of measurements to fully characterize the output of some of the lamps. The electric consumption (watts) by the fixture was nearly identical among test sites and did not likely have a significant effect of efficiency.

[x]- Photon Output per Electrical Input (μmol per second divided by joules per second).

was converted to photon output at each nanometer interval using Plank's Equation and then integrated from 400 to 700 nanometers. The radiation measurements were calibrated to NIST reference standards. These measurements of fixture efficiency are considered repeatable to within 1%.

Flat plane integration

Measurements were made in a dark room with flat black walls using a quantum sensor (LI-COR model LI-190, Lincoln, NE, USA), that was calibrated for each fixture with an NIST-traceable calibrated spectroradiometer (model PS-200, Apogee Instruments, Logan, UT, USA). This calibration is necessary to correct for small spectral errors (±3%) in the quantum sensor that occur because of imperfect matching of the ideal quantum response [18]. Measurements were made in three radial, straight lines below a level fixture and spatially integrated over a flat plane below the fixture to determine total photon output. Measurements were made 2.5 cm apart near the center, increasing to 10 cm near the perimeter as PPF variation decreased (121 measurements total). Fixtures were mounted 0.7 meters above the surface and measurements were made up to a 1.5 meter radius from the center and extrapolated to infinity using an exponential decay function. Fixture height is optional, depending on the size of the room and measurement area as long as measurement resolution captures the spatial variation in fixture output. The flat-plane integration measurements were used to quantify the pattern of photon distribution from the fixture. Total fixture output from these measurements was similar to measurements made using an integrating sphere (Table 2). When redundant measurements were available, the integrating sphere measurements were used to quantify fixture efficiency. Power draw and electrical characteristics were measured using a multimeter and a current clamp (Fluke model 289, Everett, WA, USA).

Cost of electricity

In the United States, commercial electric rates vary widely by region, ranging from $0.07 in Idaho to $0.17 in New York, with residential rates averaging $0.02 higher, and industrial rates $0.02 lower. Electric rates in Europe, and many other countries, can be more than double the rates in the United States. As electricity becomes more expensive, improved lighting becomes more valuable. See U.S. Energy Information Administration for a summary of current electric rates by state and region (accessed April 2014). We used a discounted cash flow model assuming a 5% per year cost of capital on future electrical costs.

Results

The photon efficiency (micromoles per joule) and cost per mole of photons for four categories of lighting technologies (HPS, LED, ceramic metal halide, and fluorescent), in 22 fixtures, are shown in Table 3. One fixture of each model was tested. This table also shows the five-year electric plus fixture costs per mole of photons. Most fixtures (lamp, luminaire and ballast) are now more efficient than the common 1000-W magnetic-ballast, mogul-base HPS fixtures (i.e. Sunlight Supply, 1.02 μmol per joule). If photons coming out of the fixture at all downward angles are considered (180°), the capital cost of the most efficient 400-W LED fixtures we tested is five to seven times more per photon than the 1000-W, double-ended, electronic ballast HPS fixtures (Gavita, ePapillion, Table 3). The high capital cost of LEDs makes the five year cost per mole of photons more than twice that of HPS fixtures (Table 3 and Figure 5A).

Table 3 assumes that all of the photons emitted from the fixture are absorbed by plant leaves. In Table 4, the area under the fixture in which the photons are considered captured by plants is progressively reduced, and the cost per mole of photons increases as more photons are lost around the perimeter. When only highly focused radiation is considered useful (34°), some LED fixtures have a lower cost per photon than the best HPS fixtures (Table 4, Figure 1, Figure 5B and Figure 6), but because photons are lost around the perimeter at this narrow angle, the cost per photon absorbed by plants is much greater. The lowest cost per photon is realized when a large canopy can be arranged to capture the photons.

Discussion

Importance of photon capture

As reviewed in the introduction, lighting system efficiency is the combined effect of efficient fixtures and efficient canopy photon capture efficiency. Precision luminaires, lenses (e.g. model vivid white, Lighting Sciences Group inc.), or adjustable angle LEDs (e.g. model SPYDR 600, BML inc.) can be used to apply highly focused lighting specifically to the plant growth areas. This is valuable in small greenhouses with widely spaced benches. Canopy photon capture efficiency can be maximized, to above 90%, for large greenhouses with narrow aisles regardless of fixture type. The use of LED intracanopy lighting can increase capture rates to near 100%, and may have other beneficial effects such as increased light sharing with intracanopy leaves [19,20]. The concentration of heat from HID fixtures makes intracanopy lighting infeasible with high wattage HPS fixtures. Just as precision irrigation can improve water efficiency, precision lighting can improve electrical efficiency.

Effect of fixture shadow

All fixtures block radiation from the sun, and the shadow is proportional to the size of the fixture. For the same photon output, 400-W HPS, ceramic metal halide, fluorescent, and LED fixtures block significantly more sunlight than 1000-W HPS fixtures. We did not include the effect of the shadow in this analysis, but this effect significantly favors the more energy dense, higher wattage HPS fixtures. In the long-term, LEDs can take advantage of innovative design options like mounting along greenhouse support structures, which could provide light without extra shading. Longer, narrower LED fixtures may be preferable to rectangular fixtures because the duration of the shadow is shorter. Fluorescent fixtures, including induction fluorescent, have large shadows relative to their photon output (and have low photon efficiencies) and are therefore generally not economical for greenhouse lighting.

Installation, annual maintenance costs, and life expectancy

Installation costs include wiring for fixtures and physically hanging the fixture. In our experience, the cost of installation is similar for both fixture types, although installation costs can be reduced by fewer, higher wattage fixtures. The annual maintenance costs are small relative to the cost of the electricity, and these costs are better established for HPS fixtures than for LED fixtures. Maintenance costs are largely determined by the life expectancy of the fixture.

Double-ended HPS lamps (1000-W) have a life expectancy of 10,000 hours to 90% survival (based on manufacturer literature), or 3.3 years when used an average of 8 hours per day or 3,000 hours per year (traditional mogul-base lamps have industry

Table 3. Photon efficiency and cost per mole of photons, assuming all photons (180°) are captured by plants.

Lamp type and Ballast	Fixture producer[z]	Electrical input (J/s or watts)	Photon output[y] (μmol/s)	Photon efficiency[x] (μmol/J)	Cost of one fixture[w] ($)	Fixtures needed per millimol/s[v]	Fixture cost per mol/s $/(mol/s)	Electric cost per μmol photons[u] $/(μmol/s)yr	Five year electric cost per μmol photons[t] $/(μmol/s)yr
High Pressure Sodium									
400 W magnetic	Sunlight Supply	443	416	0.94	$200	2.40	$0.48	$0.35	$0.40
1000 W magnetic	Sunlight Supply	1067	1090	1.02	$275	0.92	$0.25	$0.32	$0.33
1000 W magnetic	PARsource GLXI	1004	1161	1.16	$350	0.86	$0.30	$0.29	$0.31
1000 W electronic	PARsource GLXI	1024	1333	1.30	$380	0.75	$0.29	$0.25	$0.28
1000 W electronic	PARsource GLXII	1026	1334	1.30	$310	0.75	$0.23	$0.25	$0.27
1000 W electronic	Gavita	1033	1751	1.70	$500	0.57	$0.29	$0.19	$0.23
1000 W electronic	ePapillon	1041	1767	1.70	$600	0.57	$0.34	$0.19	$0.24
LED									
red/blue	LSG	384	653	1.70	$1,200	1.53	$1.84	$0.19	$0.54
red/white	BML	326	541	1.66	$1,000	1.85	$1.85	$0.20	$0.54
red/white	LSG	390	634	1.63	$1,200	1.58	$1.89	$0.20	$0.55
red/white	Illumitex	279	390	1.40	$1,400	2.56	$3.59	$0.24	$0.92
red/white/blue	Lumigrow (Pro 325)	304	390	1.29	$1,000	2.56	$2.56	$0.26	$0.73
red/white	California Lightworks	337	350	1.04	$1,000	2.85	$2.85	$0.32	$0.85
multiple	Black Dog	339	339	1.00	$950	2.95	$2.80	$0.33	$0.85
red/white	Apache	169	163	0.96	$860	6.14	$5.28	$0.34	$1.35
red/blue	Lumigrow (ES330)	318	284	0.90	$1,200	3.52	$4.22	$0.37	$1.16
red/white	Hydrogrow	423	378	0.89	$1,300	2.64	$3.44	$0.37	$1.01
Ceramic Metal Halide									
315 W 3100 K	Cycloptics	337	491	1.46	$640	2.04	$1.30	$0.23	$0.46
315 W 4200 K	Cycloptics	340	468	1.38	$640	2.14	$1.37	$0.24	$0.48
2@315 W 3100 K	Boulderlamp	651	817	1.25	$1,000	1.22	$1.22	$0.26	$0.47
Fluorescent									
400 W induction	iGrow	394	374	0.95	$1,200	2.68	$3.21	$0.35	$0.94
60 W	T8	58	48	0.84	$40	20.77	$0.83	$0.40	$0.51

[z]- See Table S1 for a list of fixture manufacturers and model numbers.
[y]- Integrated total photon output of fixture.
[x]- Photon Output per Electrical Input (μmol per second divided by joules per second).
[w]-Cost of fixtures as of April 2014.
[v]-The number of fixtures to get a total photon output of one millimol (1000 μmol) of photons per second.
[u]-Assumes 3000 hours per year operation and $0.11/kWh.
[t]-Cost of fixture (multiplied by fixtures needed) plus cost of electricity over 5 years. We used a discounted cash flow model assuming a 5% per year cost of capital. Installation and maintenance costs were assumed to be similar for all lamp types and were not included in this calculation.

Table 4. Cost per mole photons for four capture assumptions.

		Assuming all radiation (180°) is captured		Assuming radiation within a 1 to 2.38 height to width ratio (100°) is captured		Assuming radiation within a 1 to 1.35 height to width ratio (68°) is captured	
Lamp type and Ballast	**Fixture producer[z]**	**Fixtures needed per mmol/s[y]**	**Five year electric cost per µmol photons[x] \$/(µmol/s)yr**	**Fixtures needed per mmol/s[y]**	**Five year electric cost per µmol photons[x] \$/(µmol/s)yr**	**Fixtures needed per mmol/s[y]**	**Five year electric cost per µmol photons[x] \$/(µmol/s)yr**
High Pressure Sodium							
400 W magnetic	Sunlight Supply	2.40	$0.40	3.99	$0.66	8.51	$1.42
1000 W magnetic	Sunlight Supply	0.92	$0.33	1.71	$0.61	3.60	$1.30
1000 W magnetic	PARsource GLXI	0.86	$0.31	1.31	$0.47	2.82	$1.01
1000 W electronic	PARsource GLXI	0.75	$0.28	1.14	$0.42	2.49	$0.92
1000 W electronic	PARsource GLXII	0.75	$0.27	1.33	$0.47	2.81	$1.00
1000 W electronic	Gavita	0.57	$0.23	0.96	$0.38	2.12	$0.84
1000 W electronic	ePapillon	0.57	$0.24	1.46	$0.61	3.47	$1.45
LED							
red/blue	LSG	1.53	$0.54	1.62	$0.57	2.03	$0.71
red/white	BML	1.85	$0.54	2.13	$0.62	3.17	$0.93
red/white	LSG	1.58	$0.55	1.67	$0.59	2.09	$0.73
red/white	Illumitex	2.56	$0.92	2.66	$0.96	3.82	$1.37
red/white/blue	Lumigrow (Pro 325)	2.56	$0.73	3.05	$0.87	4.95	$1.42
red/white	California Lightworks	2.85	$0.85	3.09	$0.92	4.92	$1.46
multiple	Black Dog	2.95	$0.85	4.43	$1.27	8.64	$2.48
red/white	Apache	6.14	$1.35	6.58	$1.45	8.21	$1.81
red/blue	Lumigrow (ES330)	2.64	$1.01	2.82	$1.07	4.33	$1.65
red/white	Hydrogrow	3.52	$1.16	5.05	$1.67	10.70	$3.54
Ceramic Metal Halide							
315 W 3100 K	Cycloptics	2.04	$0.46	5.43	$1.22	19.55	$4.38
315 W 4200 K	Cycloptics	2.14	$0.48	5.72	$1.29	20.71	$4.66
2@315 W 3100 K	Boulderlamp	1.22	$0.47	1.56	$0.60	2.90	$1.12
Fluorescent							
400 W induction	iGrow	2.68	$0.94	4.69	$1.65	10.17	$3.58
60 W	T8	20.77	$0.51	38.03	$0.93	83.81	$2.05

[z]- See Table S1 for a list of fixture manufacturers and model numbers.
[y]-The number of fixtures to get 1 millimol (1000 µmol) of photons per second.
[x]-Cost of fixture (multiplied by fixtures needed) plus cost of electricity over 5 years. We used a discounted cash flow model assuming a 5% per year cost of capital. Installation and maintenance costs were assumed to be similar for all lamp types and were not included in this calculation.

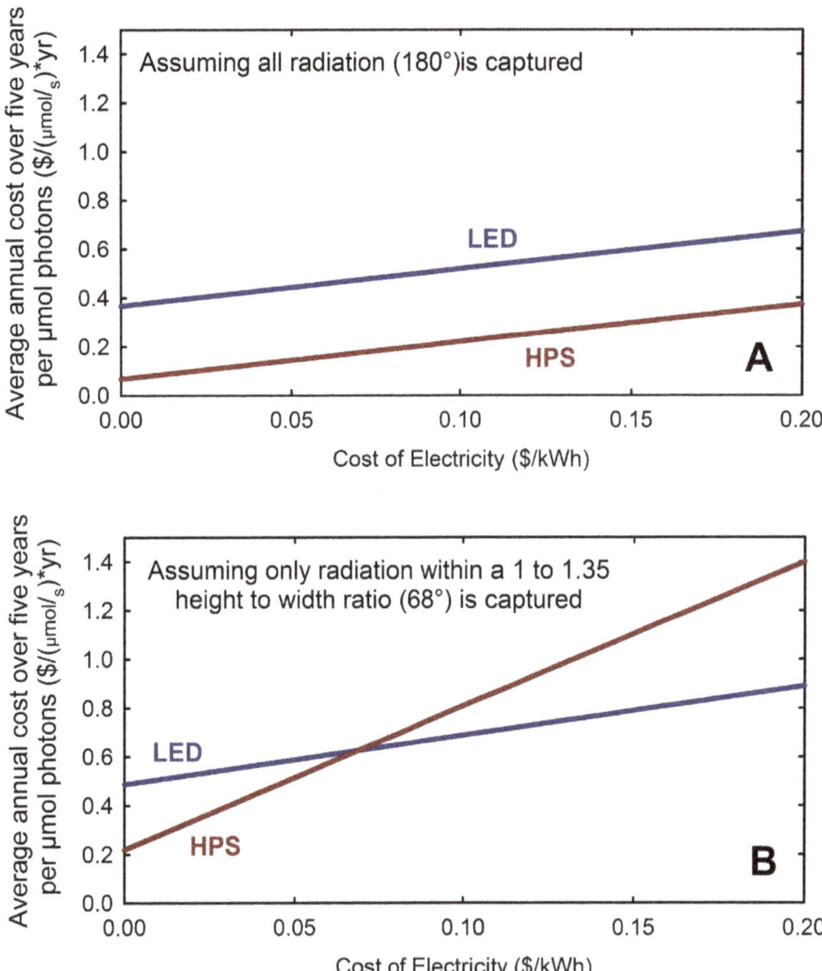

Figure 5. Effect of electricity price on average annual cost over five years for two capture scenarios. (A) When all radiation is assumed captured, the most efficient HPS fixture (Gavita) has a lower average annual five-year cost per photon than the most efficient LED fixture (Red/Blue fixture, Lighting Sciences Group). (B) When only a narrow region below the fixture (68°) is considered to be captured (e.g. on benches), the LEDs can have a lower cost per photon then HPS fixtures, but the cost per photon increases for both fixtures.

reported life expectancies of 10,000 to 17,000 hours, to 90% survival, and cost approximately $40). The cost of a 1000-W, double-ended replacement lamp is about $140, which averages to $28 per year if we assume a lamp will be replaced once in the first five years. This lamp replacement cost can increase to $30 to $35 per year when the labor to replace the bulb is included, but this is a small amount compared to the approximately $600 per year annual electric cost to operate the fixture. Adding the cost of lamp replacement increases the five-year cost of operation by approximately 5%.

When operated at favorable temperatures, individual LEDs generally have a predicted lifetime (to 70% of the initial light output) of up to 50,000 hours, about 16.7 years when used an average 8 hours per day or 3000 hours per year. The economic life for LED fixtures for plant lighting has not been established, but it depends on the value of the product being produced. The high capital cost of replacement means that LED fixtures would be operated longer, in spite of diminished photon output. Replacement of individual LEDs is more expensive than replacing an HID lamp. The life expectancy of LEDs is reduced if they are driven by higher amperage to achieve a higher output, or exposed to high temperatures. Fixtures may be warmed by radiation from sunlight. The cooler the LED temperature, the longer they last. Power supplies, fans, and other components in LED fixtures can fail well before the LEDs themselves. Fan failure would increase LED temperature and may not be immediately noticed by the user. These components are replaceable, but the labor costs to change fixture components increases operating costs.

For these reasons we have not included a differential operating cost between LED and HPS fixtures. We assumed that maintenance costs will be minimal during the first five years for all types of fixtures. Electronic ballasts for 1000-W HPS lamps are still a relatively new technology, and fixtures vary in quality. We have experienced premature failures of LED power supplies, LED circuit boards, HPS lamps, and electronic HPS ballasts in our greenhouse operations. LED fixtures with improved power supplies and optimized operating amperages are available from reputable manufacturers. Improvements in these new technologies are occurring rapidly.

Importance of PPF uniformity

PPF uniformity is critical in many greenhouse applications, especially in floriculture. It is easier to achieve uniformity with fixtures that have broad distribution of photons. Economically, the

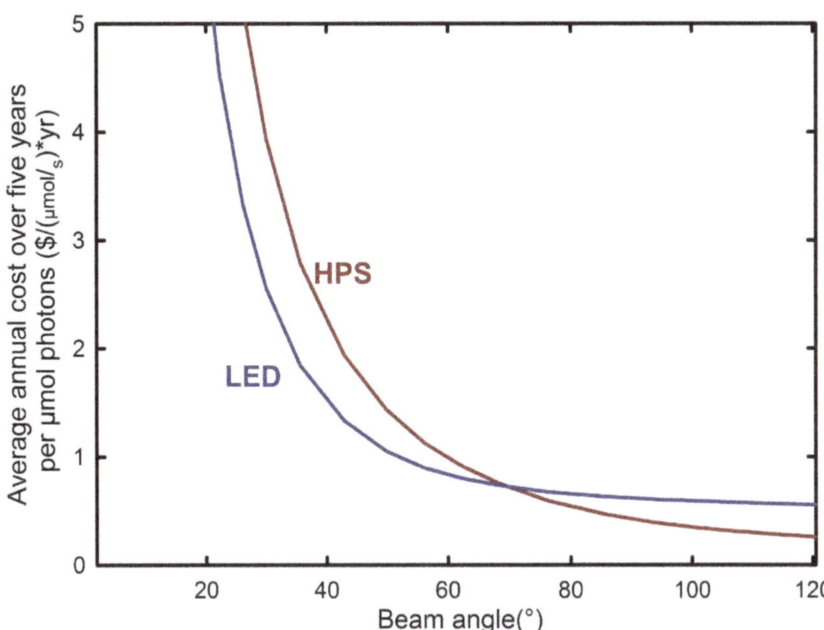

Figure 6. Effect of canopy capture efficiency on average annual cost over five years. The cost per mole of photons for LEDs (Red/Blue LED from Lighting Sciences Group) becomes more favorable than the best HPS fixtures (Gavita) when the lighting area is less than 68° from center, assuming $0.11 per kWh cost of electricity and 3000 hours per year use (approximate cumulative operation time at latitudes from 40 to 50 degrees).

value of uniform plants may outweigh the cost of wasted photons. Uniformity has been well characterized and modeled with HID lights [21,22], but these techniques have not yet been rigorously applied to LED fixtures. Ciolkosz et al. [23] showed that uniform light on the perimeter of a greenhouse requires higher fixture densities in the outer rows, and consequentially may increase the amount of radiation lost beyond the edge of the growing area, decreasing canopy photon capture. HPS fixtures with narrower focus luminaires tend to have lower photon efficiencies.

Effect of fixture efficiency on heating and cooling costs

Improved electrical efficiency reduces the cooling load in a greenhouse, which increases the value of efficient fixtures when cooling is required. The best HPS and LED fixtures have nearly identical efficiency, so cooling costs are similar for both fixture categories. The ability to cycle LED fixtures, which prematurely ages other fixture types, could be used to stabilize the heating and cooling load in a greenhouse during partly cloudy days, which could improve temperature control and increase the lifetime of cooling system equipment.

Additional thermal radiation is useful in warming the plant canopy during the heating season, but is detrimental if the canopy is too warm. When sunlight supplies adequate PPF, supplemental lighting is usually turned off.

Conclusions

The most efficient HPS and LED fixtures have equal efficiencies, but the initial capital cost per photon delivered from LED fixtures is five to ten times higher than HPS fixtures. The high capital cost means that the five-year cost of LED fixtures is more than double that of HPS fixtures. If widely spaced benches are a necessary part of a production system, LED fixtures can provide precision delivery of photons and our data indicate that they can be a more cost effective option for supplemental greenhouse lighting.

Manufacturers are working to improve all types of lighting technologies and the cost per photon will likely continue to decrease as new technologies, reduced prices, and improved reliability become available.

Supporting Information

Table S1 Fixture manufacturer and model numbers. A table containing the mixture manufactuere and model numbers of all fixtures referenced in this study.

Acknowledgments

Disclaimer: Mention of products or vendors does not imply endorsement by Utah State University to the exclusion of other products or vendors that also may be suitable.

We thank Peter Nelson and Alec Hay for their dedicated technical work, A.J. Both and Erik Runkle for conscientious technical review, and Paul Jakus and John F. Burr for analysis of the economics.

Author Contributions

Conceived and designed the experiments: JAN BB. Performed the experiments: JAN. Analyzed the data: JAN. Contributed reagents/materials/analysis tools: JAN BB. Contributed to the writing of the manuscript: JAN BB.

References

1. Bourget CM (2008) An introduction to light-emitting diodes. HortScience 43: 1944–1946.
2. Morrow RC (2008) LED lighting in horticulture. HortScience 43: 1947–1950.
3. Inada K (1976) Action spectra for photosynthesis in higher plants. Plant Cell Physiol 17: 355–365.

4. McCree KJ (1972) The action spectrum, absorptance and quantum yield of photosynthesis in crop plants. Agric Meteorol 9: 191–216.
5. Nelson JA, Bugbee B (2013) Spectral characteristics of lamp types for plant biology West Lafayette, IN. Available: http://cpl.usu.edu/files/publications/poster/pub__6740181.pdf. Accessed 2014 MAr 27.
6. Cope KR, Snowden MC, Bugbee B (2014) Photobiological Interactions of Blue Light and Photosynthetic Photon Flux: Effects of Monochromatic and Broad-Spectrum Light Sources. Photochem Photobio. doi:10.1111/php.12233.
7. Johkan M, Shoji K, Goto F, Hahida S, Yoshihara T (2012) Effect of green light wavelength and intensity on photomorphogenesis and photosynthesis in Lactuca sativa. Environ Exp Bot 75: 128–133. doi:10.1016/j.envexpbot.2011.08.010.
8. Dougher TA, Bugbee B (2004) Long-term blue light effects on the histology of lettuce and soybean leaves and stems. J Am Soc Hortic Sci 129: 467–472.
9. Cope KR, Bugbee B (2013) Spectral effects of three types of white light-emitting diodes on plant growth and development: Absolute versus relative amounts of blue light. HortScience 48: 504–509.
10. Dougher TA, Bugbee B (2001) Differences in the Response of Wheat, Soybean and Lettuce to Reduced Blue Radiation¶. Photochem Photobiol 73: 199–207.
11. Yorio NC, Goins GD, Kagie HR, Wheeler RM, Sager JC (2001) Improving spinach, radish, and lettuce growth under red light-emitting diodes (LEDs) with blue light supplementation. HortScience 36: 380–383.
12. Yang Z-C, Kubota C, Chia P-L, Kacira M (2012) Effect of end-of-day far-red light from a movable LED fixture on squash rootstock hypocotyl elongation. Sci Hortic 136: 81–86. doi:10.1016/j.scienta.2011.12.023.
13. Massa GD, Kim H-H, Wheeler RM, Mitchell CA (2008) Plant productivity in response to LED lighting. HortScience 43: 1951–1956.
14. Morrow RC, Tibbitts TW (1988) Evidence for involvement of phytochrome in tumor development on plants. Plant Physiol 88: 1110–1114.
15. Craig DS, Runkle ES (2013) A Moderate to High Red to Far-red Light Ratio from Light-emitting Diodes Controls Flowering of Short-day Plants. J Am Soc Hortic Sci 138: 167–172.
16. Kim H-H, Goins GD, Wheeler RM, Sager JC (2004) Green-light supplementation for enhanced lettuce growth under red-and blue-light-emitting diodes. HortScience 39: 1617–1622.
17. IESNA Testing Procedures Committee (2008) IES Approved Method for the Electrical and Photometric Measurements of Solid-state Lighting Products. Illuminating Engineering Society.
18. Blonquist M, Bugbee B (2013) Analysis of spectral and cosine errors in quantum sensors West Lafayette, IN. Available: http://www.apogeeinstruments.com/content/Quantum-Sensor-Poster-Park-City-April-2009.pdf. Accessed 2014 Mar 27.
19. Frantz JM, Joly RJ, Mitchell CA (2000) Intracanopy lighting influences radiation capture, productivity, and leaf senescence in cowpea canopies. J Am Soc Hortic Sci 125: 694–701.
20. Gómez C, Morrow RC, Bourget CM, Massa GD, Mitchell CA (2013) Comparison of intracanopy light-emitting diode towers and overhead high-pressure sodium lamps for supplemental lighting of greenhouse-grown tomatoes. HortTechnology 23: 93–98.
21. Both AJ, Ciolkosz DE, Albright LD (2000) Evaluation of light uniformity underneath supplemental lighting systems. Acta Hort ISHS: 183–190.
22. Ferentinos KP, Albright LD (2005) Optimal design of plant lighting system by genetic algorithms. Eng Appl Artif Intell 18: 473–484. doi:10.1016/j.engappai.2004.11.005.
23. Ciolkosz DE, Both AJ, Albright LD (2001) Selection and placement of greenhouse luminaires for uniformity. Appl Eng Agric 17: 875–882.

An Invasive Clonal Plant Benefits from Clonal Integration More than a Co-Occurring Native Plant in Nutrient-Patchy and Competitive Environments

Wenhua You, Shufeng Fan, Dan Yu*, Dong Xie, Chunhua Liu*

The National Field Station of Lake Ecosystem of Liangzi Lake, College of Life Science, Wuhan University, Wuhan, P.R. China

Abstract

Many notorious invasive plants are clonal, however, little is known about the different roles of clonal integration effects between invasive and native plants. Here, we hypothesize that clonal integration affect growth, photosynthetic performance, biomass allocation and thus competitive ability of invasive and native clonal plants, and invasive clonal plants benefit from clonal integration more than co-occurring native plants in heterogeneous habitats. To test these hypotheses, two stoloniferous clonal plants, *Alternanthera philoxeroides* (invasive), *Jussiaea repens* (native) were studied in China. The apical parts of both species were grown either with or without neighboring vegetation and the basal parts without competitors were in nutrient- rich or -poor habitats, with stolon connections were either severed or kept intact. Competition significantly reduced growth and photosynthetic performance of the apical ramets in both species, but not the biomass of neighboring vegetation. Without competition, clonal integration greatly improved the growth and photosynthetic performance of both species, especially when the basal parts were in nutrient-rich habitats. When grown with neighboring vegetation, growth of *J. repens* and photosynthetic performance of both species were significantly enhanced by clonal integration with the basal parts in both nutrient-rich and -poor habitats, while growth and relative neighbor effect (RNE) of *A. philoxeroides* were greatly improved by clonal integration only when the basal parts were in nutrient-rich habitats. Moreover, clonal integration increased *A. philoxeroides*'s biomass allocation to roots without competition, but decreased it with competition, especially when the basal ramets were in nutrient-rich sections. Effects of clonal integration on biomass allocation of *J. repens* was similar to that of *A. philoxeroides* but with less significance. These results supported our hypothesis that invasive clonal plants *A. philoxeroides* benefits from clonal integration more than co-occurring native *J. repens*, suggesting that the invasiveness of *A. philoxeroides* may be closely related to clonal integration in heterogeneous environments.

Editor: Fei-Hai Yu, Beijing Forestry University, China

Funding: This research was supported by the National Natural Science Foundation of China (30930011 and 31170339). The funders had no role in study design, data collection and analysis, decision to publish, or preparation of the manuscript.

Competing Interests: The authors have declared that no competing interests exist.

* E-mail: yudan01@public.wh.hb.cn (DY); liuchh@163.com (CL)

Introduction

Clonal integration, through which connected ramets of clonal plants can share water, carbohydrates,nutrients and other substances such as pollutants, diseases, etc. [1–3], may improve plants' exploitation of ubiquitous heterogeneous resources, help plants invade new environments and facilitate plants' spatial occupation of new habitats at a local scale [4]. Previous studies have shown that clonal integration may facilitate the colonization and growth of the ramets in heterogeneous habitats with stressful conditions [5,6], help genets to survive and to recover after severe environmental change [7,8] and allow for occupation of new space [9–11]. These positive effects of clonal integration may increase the performance of clonal plants over non-clonal plants or other clonal plants with little integration [12]. Therefore, increases in performance of clonal plants by clonal integration may affect the growth and reproduction of their co-existence species, and thus influence community structure and ecosystem function [13,14].

Plant invasions pose a great threat to biodiversity and global ecosystem stability [15,16]. Many of the most notorious alien invasive plants have the capacity for vigorous clonal propagation [17,18]. Some studies have suggested that the invasiveness of alien clonal plants may be closely correlated to clonal integration [4,19,20]. However, to our best of knowledge, few studies have focused on how clonal integration affects invasion of alien invasive clonal plants to native plant communities, but see [9,21–24]. Therefore, a better understanding of different clonal integration effects between alien invasive and native clonal plants when competing with each other is both scientific and practical interests.

Previous studies about the effects of clonal integration on performance of clonal plants when competing with neighbors were with inconsistent results [9]. Clonal integration had no significant effects on competitive ability of several terrestrial or amphibious plants [9,14,21,25], but it did increase growth of several salt marsh plants for below-ground resources, the competitive ability of *Solidago canadensis* against interspecific neighbors and the invasion of smooth brome clones to northern fescue prairies [26–28]. However, none of these studies had investigated how clonal integration affected the growth of the neighboring vegetation (competitors), but see [9]. In the study of Pennings and Callaway

(2000), the results showed that physiological integration played different roles in six salt marsh clonal plants with recipient ramets in different microhabitats or competing with co-existing neighbors in different status (clipped or kept intact). Moreover, several studies have addressed the effects of clonal integration on intra- or interspecific competition of recipient ramets [9,29,30]. Unfortunately, all these studies ignored the status of other connected ramets (donor ramets) and the environments which in, as clonal plants often experience small-scale spatial heterogeneity duo to large clonal systems [31]. For instance, clonal integration may increase the competitive ability of ramets when other connected ones were in resource-rich patches, because more subsidy could be supplied by donor ramets in non-limiting resource environments, which may facilitate the invasion of the clonal plants to neighboring communities. But to our knowledge, no studies have tested this. Furthermore, in a field experiment, Peltzer (2002) found that clonal integration did not alter the effects of competition from neighboring vegetation for *Populus tremuloides*, however, competition greatly improved the survivorship of *Populus* ramets after 2 years. Therefore, the importance of clonal integration in competition urgently needed for further research [21].

In heterogeneous habitats consisting of a mixture of rich and poor resource patches, via clonal integration, clonal plants can alter biomass allocation and divert more biomass to shoots or roots for acquisition of more abundant resource, and exploration of more favourable space, a phenomenon called 'division of labour' [2,32,33]. This pattern of biomass allocation is different from that used by non-clonal plants, or clonal plants grown in homogeneous conditions [2,32]. In particular, the relationship between plant photosynthetic efficiency and clonal integration has not been widely studied [9,31]. Photosynthetic efficiency can be estimated by measuring chlorophyll fluorescence [34]. A sensitive indicator of plant photosynthetic performance derived from the parameters of chlorophyll fluorescence is the maximum quantum yield of photosystem II (F_v/F_m), which usually significantly decreases when plants are faced with environmental stress [31,35]. Environmental stress on ramets may be alleviated by clonal integration, which may markedly lower the negative effects of stress on F_v/F_m [9,11]. Moreover, photosynthetic activity, measured in terms of the effective quantum yield of PSII (Yield), is closely related to plant performance. Biomass allocation and photosynthetic efficiency can both contribute to the performance of clonal plants when exposed to competitive stress, however, our understanding of their responses to clonal integration for invasive plants remains limited [9,11].

Therefore, to test different effects of clonal integration on one exotic invasive and one native clonal plants, we conducted a greenhouse experiment to investigate responses of growth, biomass allocation, photosynthetic performance and relative neighbor effects (RNE, used to indicate the plants' competitive ability) of two stoloniferous clonal plants, *Alternanthera philoxeroides* (invasive), *Jussiaea repens* (native) to clonal integration when competing with competitors (neighboring vegetation) in China. We used a factorial design with resource availability, stolon severing and competition with neighboring vegetation as factors. Specifically, we hypothesize 1) that clonal integration will improve growth and competitive ability of the two clonal plants, especially when donor ramets are in nutrient-rich habitats, 2) that clonal integration will modify biomass allocation of the two plants grown with competitors. Based on the theory of labour division [32], we predict that, through clonal integration, plants will allocate more biomass to leaves if the belowground competition is more severe and will allocate more biomass to roots if aboveground competition is more severe, 3) that clonal integration will enhance the photosynthetic performance and buffer the decrease in F_v/F_m of the two plants in competitive environments, especially when donor ramets are in nutrient-rich habitats, 4) that *A. philoxeroides* will benefit from clonal integration more than *J. repens*, in terms of competitive ability, photosynthetic performance and capacity of labor division, and 5) that clonal integration of these two clonal species will suppress the growth of neighboring vegetation due to competition with apical ramets.

Materials and Methods

Plant material

A. philoxeroides is a serious economic and environmental clonal weed which originates from Parana River region of South America and now invades may countries in the world [36,37]. In China, *A. philoxeroides* has invaded varied ecosystems and caused great economic and environmental problems, and it is listed as one of the 16 worst alien invasive weeds [38]. *J. repens* is a rooted emergent stoloniferous clonal plants and a fast-proliferating species in wetlands, naturally distributed in central and south China [39]. In natural environments, these two species often co-exist in diverse habitats that from wet to aquatic in south China.

In early May 2010, source material of *A. philoxeroides* and *J. repens* was collected from at least five locations at least 15 m apart in each of two wetlands in Liangzi Lake in the Hubei province of China (N 30°05′–30°18′, E 114°21′–114°39′). Given that genetic diversity of wetland clonal plants is relatively low [40], especially for *A. philoxeroides* in China [41], different populations of each species were assumed to belong to the same genet. Then plants from different locations were mixed and propagated in the greenhouse. After two weeks of adaptive culture, about 600 tip cuttings of each plant were collected and planted vertically into 12 plots (30 cm diameter ×15 cm height) with soil (TN 2.94 mg/g, TP 0.13 mg/g) from the lake side of Liangzi Lake. Ten days later, a homogeneous subset of 480 vigorously growing plants of each species were selected for this experiment.

Ethics Statement

Plant material used in this experiment was collected from natural plant populations at the National Field Station of Freshwater Ecosystem of Liangzi Lake (N 30°05′–30°18′, E 114°21′–114°39′). Both of the plant species were common and naturally distributed in this area. No specific permissions were required for these locations. This study did not involve endangered or protected species.

Experimental design

The growth experiment was conducted in a glasshouse under natural sunlight (about 14/10 day/night cycle) and ambient temperature at the National Field Station of the Lake Ecosystem of Liangzi Lake, Wuhan University. The experiment was conducted with a factorial design involving competition (without or with vegetation for control or competition treatment) and integration treatments (stolon connections were severed, intact or intact and with basal ramets in nutrient-rich patches, for severed, intact or nutrient treatment) (Fig. 1). The tested plants used in this experiment were 24 similar-sized clonal fragments (tip cuttings, 14.33±0.15 cm in length, 0.62±0.034 g in dry mass for *A. philoxeroides*; 15.02±0.21 cm in length, 0.78±0.041 g in dry mass for *L.repens*; means ± SE), each consisting of a stolon with five ramets for each species. No differences between treatments were detected in initial size of this plants ($P>0.05$ for both species, One-way ANOVA). Each clonal fragment was divided into two parts,

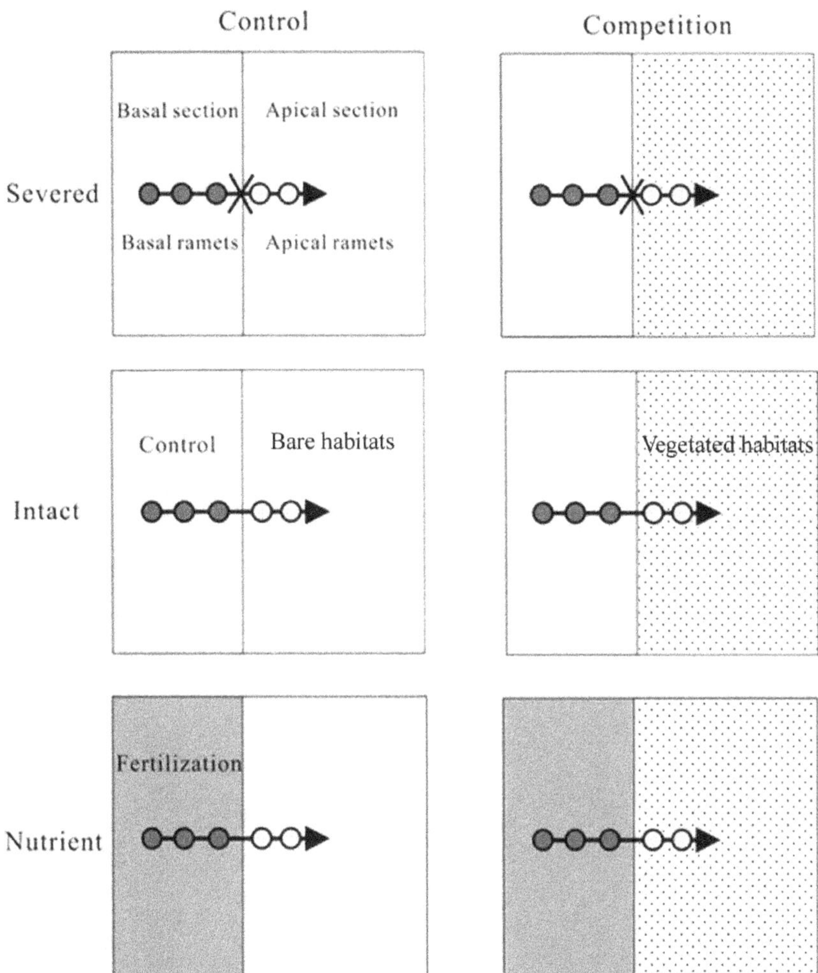

Figure 1. Schematic representation of the experimental design. Clonal fragments of the invasive plant *A. philoxeroides* or native plant *J. repens*, each consisting of three basal ramets (dark grey circles) and two apical ramets (light gray circles) with a stolon apex (horizontal arrow), were grown either with (competition) or without (control) competitive vegetation (*J. repens* or *A. philoxeroides*, spot-shadow) and with stolon connections between basal and apical ramets were either intact or severed (fork). Three integration treatments were used as follows: severed (stolon connections severed by the scissors), intact (stolon connections kept intact) and nutrient (stolon connections kept intact and with basal ramets in fertilized habitats).

one termed as 'basal part' consisting of three relatively old ramets (close to the mother ramets) and the other as 'apical part' consisting two relatively young ramets (distal to the mother ramets) and a stolon apex.

There were 28 plastic containers (50×50×25 cm; length×width×height), each having two separated sections in this experiment for each species (see Fig. 1). The basal section was 20 cm long and the apical section was 30 cm long. Resources (nutrients and water) and roots in the two sections did not interfere with each other. All the containers in both sections were filled with a mixture of sand and lake mud at a volume ratio of 3:1. To create highly fertilized soil patches, 12 containers were filled with the same mixture and 5 g slow-release fertilizer (Osmocote®, N–P–K: 16–8–12, 6 month) in the basal section. On June 10th 2011, the apical sections of 16 containers were planted vertically with cultured plant fragments of each species (monoculture) in the glasshouse to mimic natural plant populations (vegetated habitats), with a density of 200 plants m^{-2} (30 plants in each apical section) for each species. The remaining 12 containers were kept with apical sections bare.

On July 5th 2011, 24 clonal fragments of each species were horizontally positioned in 24 containers (12 with and 12 without competitive vegetation in apical sections), the remaining 4 containers with competitive vegetation were used as a control for plant population growth without competition. For each clonal fragment, three ramets of basal part were placed within basal section of a container and the other two ramets and apex of the apical part were within the apical section of the same container. The stolon of the apical ramets was anchored to the soil surface to facilitating rooting. Six days later, when the clonal fragments were successfully rooted, the stolon connections between the apical and basal parts were severed in 8 containers, while the other 16 ones were kept intact (see Fig. 1). The experiment was ended on September 10th 2011. The experimental units were randomly repositioned every two weeks to avoid the effects of possible environmental heterogeneity (such as light), and watered every other day to maintain the soil in the containers at wet condition. The mean light intensity in the greenhouse was 800–1200 μmol m^{-2} s^{-1}, and the mean air temperature was 20–28°C during the experimental period.

Measurements

One week before harvesting the plants, the minimum (F_0) and the maximum (F_m) fluorescence yield were measured for a fully developed, healthy leaf on the second-youngest of the ramets in each apical plant after a dark adaptation (shaded by leaf folders) of at least 20 minutes sufficient for photosystem II (PSII) reaction centers to open by a portable chlorophyll fluorometer (DIVING-PAM, Walz, Effeltrich, Germany) with the saturation pulse method [34]. The maximum quantum yield of PSII (F_v/F_m) was calculated as $(F_m - F_0)/F_m$. The effective quantum yield of PS II (Yield) was calculated as $(F_m'-F_t)/F_m'$, where F_m' is defined as the maximal fluorescence yield reached in a pulse of saturating light after a actinic light pulse of 120 μmol m^{-2} s^{-1} for 10 seconds, and F_t is the fluorescence yield of the leaf at that photosynthetic photon flux density [31,35].

At harvest, the number of ramets and leaves were counted, and the total stolon length, total leaf area (Li-3100 Area Meter, Li-Cor, USA) were measured for the apical parts of all treatments. The ramets in the apical part of the two clonal species were then harvested and separated into leaves, stolons and roots, and their biomass was determined after drying at 70 °C for 72 h. Neighboring vegetation (entire plants including roots) in the apical sections of the container for each species were also harvested and their dry mass was also determined in the same way.

The relative neighbor effect (RNE) was calculated to measure the competitive intensity [42]. The RNE of plant was calculated as $(C - A)/\max(C, A)$, where is A the mean biomass of plant across replicates without competition, C is biomass of plant with competition, and max (C, A) is the larger value between A and C. Usually. The values of RNE range from -1 to 0, and the greater the values are, the smaller the neighbor's effects is [9]. So, a significantly larger RNE with than without stolon connection treatments indicates clonal integration facilitates plant's competitive ability.

Data analysis

All data were log transformed to meet assumptions of normality and homoscedasticity before analysis. One-way ANOVA was used to test whether total biomass of vegetation (competitors) in the apical section for each species differed among the four treatments (no competition; competition with severed stolon connection; competition with intact stolon connection; competition with intact stolon connection and basal parts in nutrient-rich sections). Two-way ANOVA was used to assess the effects of integration treatments (severed, intact and nutrient) and competition on photosynthetic performance (F_v/F_m and Yield) of the two species in the apical section. Two-way multivariate analysis of variance (MANOVA) was employed to investigate the global effects of integration treatments and competition on growth measures (total biomass, ramet number, stolon length, leaf number and leaf area) and biomass allocation pattern (biomass allocation to leaves, stolons and roots) of both species in the apical parts, and corresponding univariate analyses were also conducted. If a significant treatment effect was detected, post-hoc pair-wise comparisons of means were made to examine differences between treatments using Studentized Tukey's HSD for multiple comparisons. The differences of RNE values among the integration treatments were tested by one-way ANOVA followed by Duncan tests. Statistical significance was assigned at a $P<0.05$. All data analyses were performed using SPSS 18.0 (SPSS, Chicago, IL, USA).

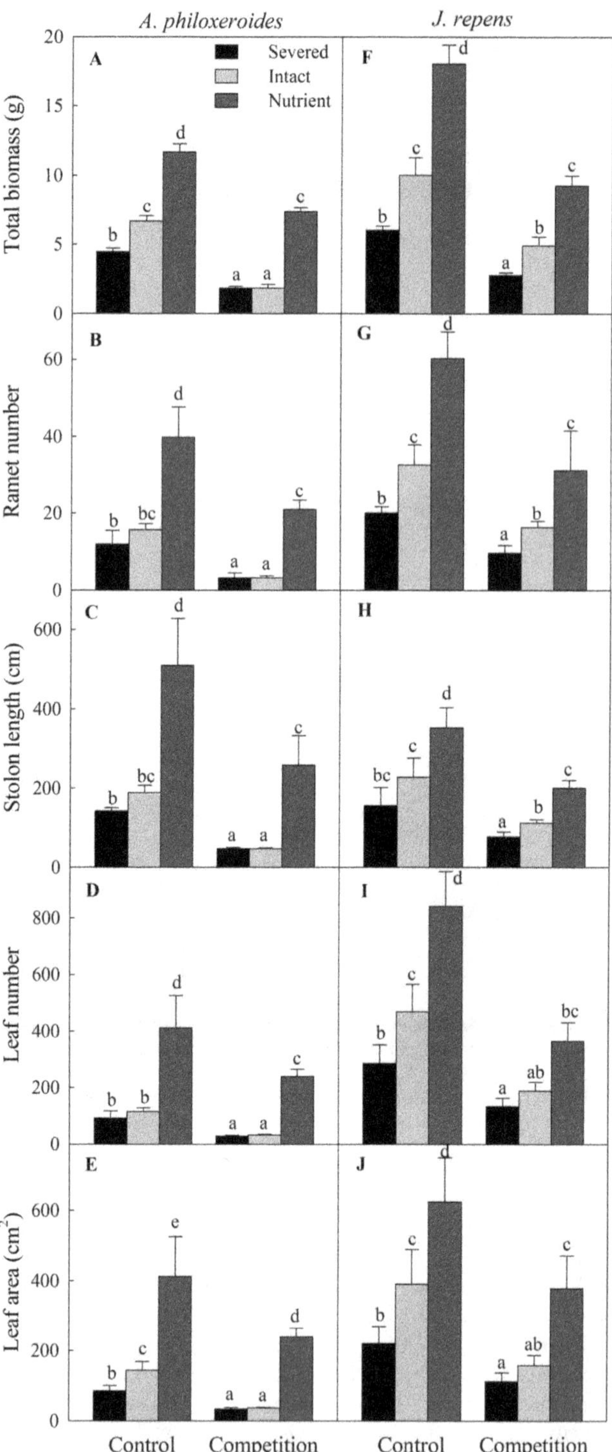

Figure 2. Effects of integration treatments and competition on growth measures of the two clonal plants. Total biomass, ramet number, stolon length, leaf number and total leaf area of the invasive plant *A. philoxeroides* (left: A, B, C, D, E) or native plant *J. repens* (right: F, G, H, I, J) in the apical sections, grown either with or without competitive vegetation (*J. repens* or *A. philoxeroides*) in three integration treatments. Data indicate the means ± SE. Bars sharing the same letter are not significantly different at $P=0.05$.

Table 1. Summary of MANOVA and univariate ANOVA for effects of integration treatments and competition on growth measures of the two clonal plants in the apical sections.

Multivariate test statistics

	A. philoxeroides				*J. repens*			
Source	**Wilk's Lambda**	*F*	**d.f.**	*P*	**Wilk's Lambda**	*F*	**d.f.**	*P*
Integration (I)	0.008	29.37	10,28	**<0.001**	0.018	17.93	10,28	**<0.001**
Competition (C)	0.015	180.23	5,14	**<0.001**	0.037	73.37	5,14	**<0.001**
I×C	0.079	7.14	10,28	**<0.001**	0.403	1.61	10,28	0.16

Univariate test statistics

	A. philoxeroides					*J. repens*				
Source	**Total biomass**	**Stolon length**	**Ramet number**	**Leaf number**	**Total leaf area**	**Total biomass**	**Stolon length**	**Ramet number**	**Leaf number**	**Total leaf area**
Integration (I)	611.68***	101.10***	202.22***	250.56***	275.19***	288.89***	71.32***	59.50***	60.74***	47.53***
Competition(C)	88.55***	146.37***	260.16***	181.63***	208.87***	404.84***	95.78***	134.55***	113.31***	52.91***
I×C	67.03***	8.79**	9.54**	10.15**	18.45***	3.22	0.40	2.45	0.32	1.35
d.f.	5,18	5,18	5,18	5,18	5,18	5,18	5,18	5,18	5,18	5,18

Significant *P*-values are presented in bold.
Values give *F*; symbols give *P*: * $P<0.05$; ** $P<0.01$; *** $P<0.001$.

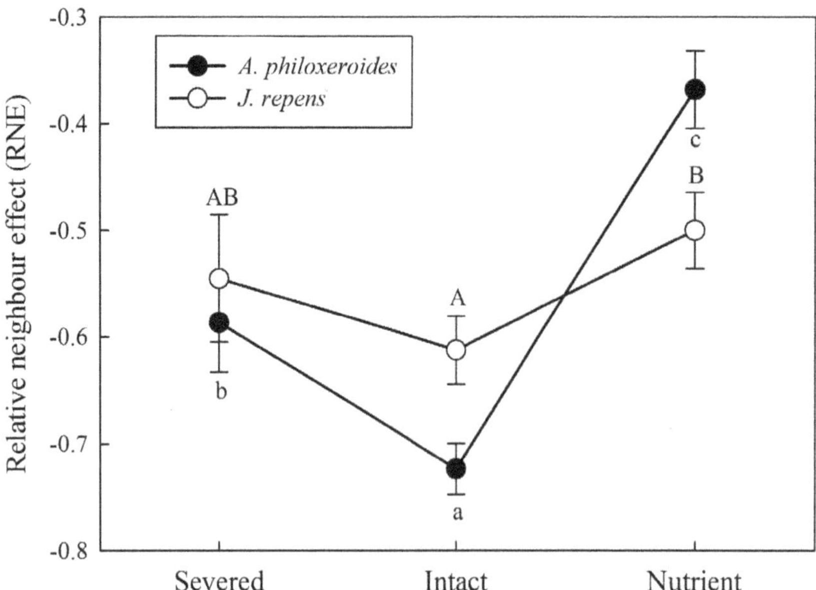

Figure 3. Effects of integration treatments on the relative neighbour effect (RNE) of the two clonal plants. The relative neighbour effect (RNE) of the invasive plant *A. philoxeroides* and native plant *J. repens* in the apical sections in three integration treatments. Data indicate the means ± SE. Bars sharing the same letter are not significantly different at $P = 0.05$.

Results

Growth and the relative neighbor effect (RNE)

Integration treatments and competition had significant effects on growth of both clonal species in the apical sections, and their interaction was also significant for *A. philoxeroides* but not for *J. repens* (Table 1). Competition greatly reduced the growth measures (including total biomass, number of ramets and leaves, stolon length and total leaf area) of the two clonal species (Table 1, Fig 2). Without competition, clonal integration greatly improved the growth of both of these two species in the apical sections, especially when the basal parts of the clonal fragments were in nutrient-rich patches (Fig 2A, B, C, D, E). However, with competition, clonal integration had no significant effect on the growth of *A. philoxeroides* but greatly enhanced that when its basal parts were in nutrient-rich sections (Fig 2A, B, C, D, E). For *J. repens*, the responses of the growth to integration treatments with competition were similar to that without competition (Fig 2F, G, H, I, J).

Integration treatments (severed, intact and nutrient) significantly affected the relative neighbor effect (RNE) of the two clonal species in the apical parts ($F_{2,11} = 96.14$, $P < 0.001$ for *A. philoxeroides*; $F_{2,11} = 6.24$, $P = 0.02$ for *J. repens*). Clonal integration significantly decreased the RNE of *A. philoxeroides* but greatly increased that in nutrient treatment (Fig. 3). The RNE of *J. repens* had a decreasing trend with the stolon connection intact and an increasing trend in nutrient treatment but not significantly (Fig. 3).

Biomass allocation pattern

Integration treatments, competition and their interaction significantly affected the biomass allocation of both species in the apical sections (Table 2). Clonal integration significantly increased biomass allocation of *A. philoxeroides* to the roots and decreased that to the leaves without competition, whereas it decreased biomass allocation to the roots and increased that to the leaves with competition (Table 2, Fig. 4A, C). However, when the basal parts of *A. philoxeroides* were in nutrient-rich sections, clonal integration increased biomass allocation to the leaves and decreased that to the roots whether when the apical parts of plants were with competition or not (Table 2, Fig. 4A, C). Biomass allocation to the stolons of both species were not affected by clonal integration but was significantly larger when the apical ramets were grown with rather than without competition (Table 2, Fig. 4B, E). Effects of integration treatments and competition on apical parts of *J. repens* were similar to that of *A. philoxeroides*, although the trend was less obvious (Table 2, Fig. 4D, F).

Photosynthetic performance

Integration treatments and competition significantly affected the photosynthetic performance (F_v/F_m and Yield) of both of the two clonal plants, and their interaction was also significant in F_v/F_m but not in Yield (Table 3). Competition greatly reduced the value of F_v/F_m of the two species in the apical sections, especially when the stolon connections were severed (Fig. 5A, C). Clonal integration markedly increased the value of F_v/F_m of the two species, especially when their basal parts were in nutrient-rich sections (Fig. 5A, C). The Yields of the two plants were both significantly reduced by competition (Fig. 5B, D), but greatly enhanced by clonal integration, especially when their basal parts were in nutrient-rich sections (Fig. 5B, D).

Growth of neighboring vegetation

Total neighboring vegetation biomass of the two species both had no significant differences among all the treatments ($F_{3,15} = 0.87$, $P = 0.39$ for *A. philoxeroides*; $F_{3,15} = 0.48$, $P = 0.70$ for *J. repens*). Total biomass in the apical sections for each species in four treatments (no competition; competition with severed stolon connection; competition with intact stolon connection; competition with intact stolon connection and basal parts in rich-patches) were 74.72±2.13 g, 76.33±3.21 g, 74.54±1.94 g and 73.52±3.17 g for vegetation of *A. philoxeroides*; and 89.38±3.67 g, 91.87±0.70 g, 93.59±2.97 g and 89.81±2.99 g (means ± SE) for that of *J. repens* respectively.

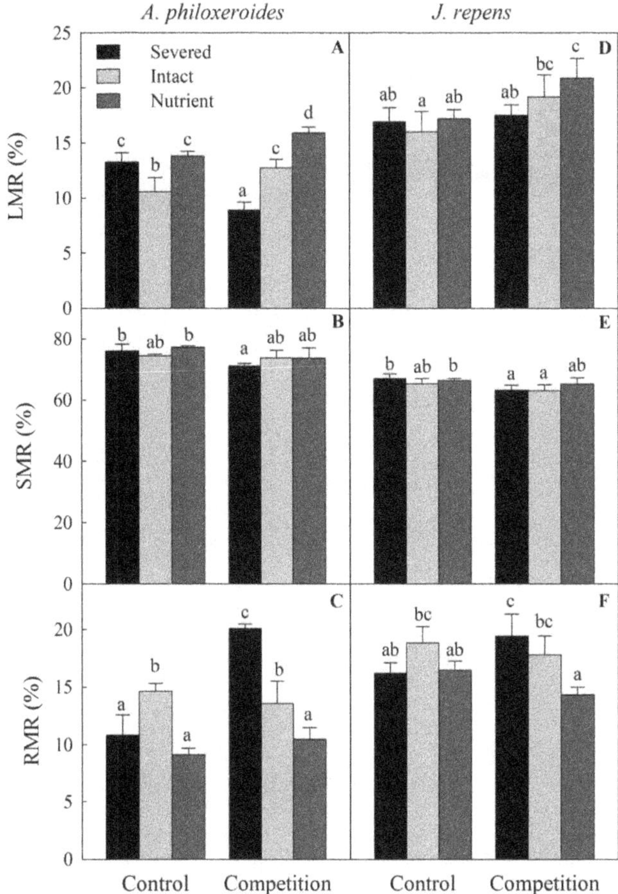

Figure 4. Effects of integration treatments and competition on biomass allocation of the two clonal plants. Biomass allocation (LMR, leaf mass ratio; SMR, stolon mass ratio; RMR, root mass ratio) of the invasive plant *A. philoxeroides* (left: A, B, C) or native plant *J. repens* (right: E, F, G) in the apical sections, grown either with or without competitive vegetation (*J. repens* or *A. philoxeroides*) in three integration treatments. Data indicate the means ± SE.

Discussion

Clonal integration improved the growth and photosynthetic performance, modified biomass allocation of both the introduced, invasive species *A. philoxeroides* and the co-occurring native species *J. repens* in nutrient-patchy and competitive environments. These results suggest that clonal integration is very important for both species when faced with competition in heterogeneous habitats [9]. However, some differences were observed in these two stoloniferous clonal plants.

Effects of clonal integration on growth and competitive ability

Without competition, clonal integration significantly improved the growth of both clonal species in the apical sections, especially when their basal parts were in nutrient-rich sections. This result occurred most likely because the well-established ramets in the basal sections supported the growth of the interconnected young apical sections and facilitated the production of new tissue due to acropetal (from basal ramets to apical ramets) translocation of carbohydrates via clonal integration [9]. The results agree with those obtained in previous studies on several other clonal plants including terrestrial plants [43,44], amphibious plants [9] and submerged aquatic plants [10], which showed that clonal integration facilitates establishment of newly produced ramets, improves growth of adult ramets and helps genets to occupy open space. This phenomenon was more pronounced when basal ramets were in nutrient-rich patches, probably because more subsidy was provided by the basal parts via clonal integration due to source-sink relationship [31,45,46]. These observations indicate that clonal integration is critical in allowing these two clonal species to explore new open space and rapid expansion, especially in heterogeneous habitats, which is a well-known mechanism to allow stoloniferous and rhizomatous plants to forage for resources over large areas [45,46].

When competing with neighboring vegetation, growth measures of both plants were markedly suppressed by competition, suggesting that strong interspecific competition in this experiment occurred in the apical parts of both plants. Interestingly, stolon connection had no effect on growth of *A. philoxeroides* with competition, while it greatly increased growth of that in the nutrient treatment. Moreover, clonal integration decreased the RNE of *A. philoxeroides* in intact treatment and greatly increased that in nutrient treatment, suggesting that clonal integration improved the competitive ability of *A. philoxeroides* only when the basal parts were in nutrient-rich conditions. The reason might be that ramets of *A. philoxeroides* were sophisticated (selective) and relatively independent. Therefore, few ramets were placed in the habitat with severe competition (less resources available in apical sections) and more ramets were placed in relatively more favourable conditions (more resources available in bare basal sections) [47,48]. Actually, more branches and biomass were observed in the basal sections (data not shown). However, when the basal parts of the plant were in nutrient-rich sections, strong intra-competition existed because of dense plant population due to vigorous growth in nutrient-rich habitats at the end of experiment. Nutrient-rich habitats may have equal or even less suitability compared to poor habitats because overcrowding reduces suitability [31]. To avoid self-shading, clonal integration enhanced the competitive ability and facilitated the invasion of the apical ramets of *A. philoxeroides* into neighboring vegetation [49]. For *J. repens*, with competition of *A. philoxeroides* vegetation, clonal integration promoted the growth of apical parts of plant in both stolon connection treatments (intact and nutrient treatments). In addition, clonal integration had no significant effects on the RNE of the apical ramets whether in intact treatment or in nutrient treatment. These results suggest that native *J. repens* may be more dependent on physiological integration and may share resources to a higher degree [50]. Thus, when faced with severed competition, growth and spread of the invasive *A. philoxeroides* would in general benefit more from clonal integration than native co-occurring *J. repens*, because: 1) in relatively poor habitats, clonal integration may preferentially allow the ramets of *A. philoxeroides* to escape from competitive stress and explore other open space to rapid expansion [9]; 2) in resource-rich patchy habitats, clonal integration may enhance the its competitive ability and facilitate the invasion of *A. philoxeroides* to neighboring vegetation [22].

Effects of clonal integration on biomass allocation pattern

Biomass allocations of both plants were significantly influenced by clonal integration which is consistent with previous findings for many other clonal plants [9,32,51]. Without competition, clonal integration increased biomass allocation of *A. philoxeroides* to roots at the expense of that to leaves, however, when its basal parts were in nutrient-rich sections, clonal integration reversed this trend of

Table 2. Summary of MANOVA and univariate ANOVA for effects of integration treatments and competition on biomass allocation of the two clonal plants in the apical sections.

Multivariate test statistics

	A. philoxeroides				*J. repens*			
Source	**Wilk's Lambda**	*F*	**d.f.**	*P*	**Wilk's Lambda**	*F*	**d.f.**	*P*
Integration (I)	0.074	14.30	6,32	**<0.001**	0.414	2.95	6,32	**0.021**
Competition (C)	0.302	12.35	3,16	**<0.001**	0.472	5.96	3,16	**0.006**
I×C	0.080	13.50	6,32	**<0.001**	0.457	1.61	6,32	**0.039**

Univariate test statistics

	A. philoxeroides			*J. repens*		
Source	**LMR**	**SMR**	**RMR**	**LMR**	**SMR**	**RMR**
Integration (I)	50.54***	1.98	47.00***	3.17	1.87	11.23**
Competition(C)	0.10	14.17**	41.07***	15.89**	12.57**	0.05
I×C	43.33***	2.45	39.13***	2.38	1.21	9.26**
d.f.	5,18	5,18	5,18	5,18	5,18	5,18

Significant *P*-values are presented in bold.
LMR: leaf mass ratio, SMR: stolon mass ratio, RMR: root mass ratio.
Values give *F*; symbols give *P*: * $P<0.05$; ** $P<0.01$; *** $P<0.001$.

biomass allocation. The results occurred most likely because soil resources were relatively more limiting for expansion of the ramets in the apical sections without competition [9]. For the connected apical ramets, the required carbohydrates could be transported efficiently from the basal ramets, so that relatively more biomass could be allocated to roots in order to improve the growth of the whole ramet system in the apical parts [9]. Additionally, when the donor ramets were in nutrient-rich habitats, due to source-sink

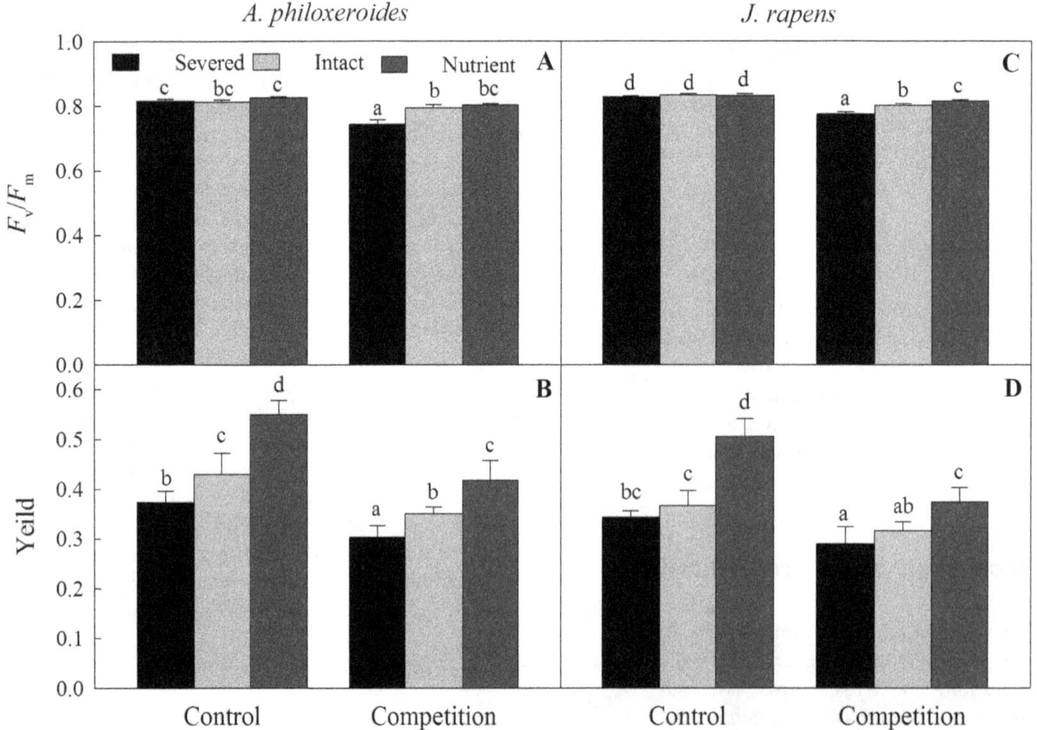

Figure 5. Effects of integration treatments and competition on photosynthetic performance of the two clonal plants. The maximum quantum yield of photosystem II (F_v/F_m) and the effective quantum yield of PSII (Yield) of the invasive plant *A. philoxeroides* (left: A, B) or native plant *J. repens* (right: C, D) in the apical sections, grown either with or without competitive vegetation (*J. repens* or *A. philoxeroides*) in three integration treatments. Data indicate the means ± SE.

Table 3. Two-way ANOVA for effects of integration treatments and competition on the maximum quantum yield of photosystem II (Fv/Fm) and the effective quantum yield of PSII (Yield) of the two clonal plants in the apical sections.

	A. philoxeroides		*J. repens*	
Source	F_v/F_m	**Yield**	F_v/F_m	**Yield**
Integration (I)	$F_{2,42}=62.74^{***}$	$F_{2,42}=55.34^{***}$	$F_{2,42}=85.56^{***}$	$F_{2,42}=75.09^{***}$
Competition (C)	$F_{1,42}=142.21^{***}$	$F_{1,42}=78.63^{***}$	$F_{1,42}=262.33^{***}$	$F_{1,42}=64.72^{***}$
I×C	$F_{2,42}=50.24^{***}$	$F_{2,42}=2.21$	$F_{2,42}=50.70^{***}$	$F_{2,42}=3.04$

Values give *F*; symbols give *P*: * P<0.05; ** P<0.01; *** P<0.001.

relationship [31,52], nutrient is not limiting resources and ramets in apical parts can allocate more fraction of biomass to aboveground (leaves and stolons) to rapid spread and occupation of new habitat.

With competition, however, clonal integration greatly increased biomass allocation to leaves and decreased that to roots, especially when basal parts of plants were in nutrient-rich sections. This might be because under severe competition, allocating more biomass to leaves can help apical ramets to harvest relatively more abundant light (by placing above the canopy of the dense vegetation), whereas poor soil resources (due to dense roots of competitive vegetation) could be compensated by basal ramets through clonal integration [9,53], especially when basal ramets were placed in resourceful conditions. These observations suggest that biomass allocation of invasive plant *A. philoxeroides* in present study agrees with the theory of labor division theory in clonal plants [32]. That is, young ramets in the apical sections explore locally most abundant resource (light) and receive mineral nutrients and water from older ramets in the basal sections via xylem, while carbohydrates can be imported or produced locally and even exported [46]. In this situation, environmentally induced labor division occurred in the apical ramets, as two essential resources (light and nutrients and/or water) negatively correlated due to competition by neighboring vegetation [54]. Biomass allocation pattern of *J. repens* was similar to that of *A. philoxeroides* but with less significance, suggesting that different integration strategies may occur in these two clonal plants when faced with competition. For instance, differences in extent of integration or degree of physiological integration in these two species may exist in heterogeneous environments [45,55]. Indeed, invasive clonal plants with stoloniferous perennial growth are considered to be conferred with the ability of rapidly covering areas via changing biomass allocation through clonal integration [9,11]. However, this needs further in-depth investigation.

Effects of clonal integration on photosynthetic performance

In favourable conditions, the value of F_v/F_m for most plant species ranges from 0.8 to 0.84 [9,11]. Without competition, F_v/F_m values of ramets of both plants in all clonal treatments were within the normal range of healthy plants and exhibited no significant differences among all the treatments, while growing with competitive vegetation greatly decreased F_v/F_m of both plants to the degree outside the normal range, suggesting that severe competition imposed stress on them. However, the decrease of plants' F_v/F_m values was markedly alleviated by stolon connections, especially when the donor ramets were in nutrient-rich sections, allowing the ramets to maintain F_v/F_m values within the normal range. Therefore, the results suggest that clonal integration significantly buffered plants against competitive stress and significantly increased plant photosynthetic performance. Previous studies [9,31] also found that clonal integration significantly alleviated the decrease in F_v/F_m of ramets grown in soils with heavy metals or with severe competition by neighboring plants. Moreover, photosynthetic capacity, measured in terms of the effective quantum yield of PS II (Yield), was significantly improved by clonal integration and nutrient addition in both plants. The responses of Yield values were closely correlated with the growth of apical ramets in response to clonal integration under competition, suggesting that clonal integration improved growth of both plants in patchy habitats when competing with competitors mainly by increasing photosynthetic efficiencies [31]. The benefit of clonal integration in terms of physiological traits (photochemical activity determined by F_v/F_m and Yield) supported our hypothesis and reinforced the capacity of division labor in these two plants, as an increase in F_v/F_m (and Yield) and a reduction of biomass allocation to the roots were observed in integration treatments, which could be interpreted as a specialization for aboveground resources. These results suggest that division of labor in stoloniferous clonal plants can happen both at morphological and physiological level [54]. However, no differences were observed in these two plants, indicating that differences in growth and division of biomass between the two species may due to different resource-sharing strategies mediated by clonal integration [45,55].

Effects of clonal integration on neighboring vegetation

Interestingly, total biomass of neighboring vegetation of both species was not affected by the presence of apical ramets, suggesting that competition treatments in present study did not suppress growth of plant populations. This is most likely because competition between apical ramets and competitive vegetation was asymmetrical because of low density of apical ramets in this experiment and their biomass was too small to influence the plant community [10]. This observation was supported by the fact that biomass of apical ramets in both plants with competition was sharply decreased to less than 30% as compared with that without competition. Therefore, even though clonal integration greatly improved the growth of plants in the apical sections under competition, their relatively small biomass contributed little to affect the plant community due to relatively short experimental duration (12 weeks). It can be expected that roles of clonal integration may be more important with longer experimental duration.

Conclusions

Overall, when competing with neighboring vegetation, clonal integration greatly improved growth and photosynthetic performance of both species when the connected basal ramets were in nutrient-rich habitats, suggesting that clonal integration is important for both species in nutrient-patchy and competitive environments. However, ramets of the invasive *A. philoxeroides* were more sophisticated and independent than the co-occurring native *J. repens* when faced with competition. Moreover, under competitive environments, changes in biomass allocation of *A. philoxeroides* through clonal integration was more significant than that of *J. repens*, although biomass allocation of both species well conformed to the theory of labor division, suggesting that different integration strategies may occur in these two clonal plants. These observations supported our hypothesis that invasive *A. philoxeroides* may benefit from clonal integration more than co-occurring native *J. repens*, indicating that invasiveness of *A. philoxeroides* may be closely related to clonal integration in heterogeneous environments. However, future comparative research is needed on additional species pairs in order to assemble conclusive evidence on the importance of integration for invasive and native species.

Acknowledgments

We thank Dr. KY Xiao for critical comments on an early version of the manuscript, Dr. LF Yu for help with the data analysis, and CM Han, DY Ma, J Chen, and YQ Han for assistance with plant harvest.

Author Contributions

Conceived and designed the experiments: WHY DY SFF CHL. Performed the experiments: WHY. Analyzed the data: WHY DX. Contributed reagents/materials/analysis tools: WHY DY. Wrote the paper: WHY DY CHL.

References

1. Alpert P, Mooney HA (1986) Resource sharing among ramets in the clonal herb, *Fragaria chiloensis*. Oecologia 70: 227–233.
2. Stuefer JF, Kroon HD, During HJ (1996) Exploitation of environmental heterogeneity by spatial division of labor in a clonal plant. Funct Ecol 10: 328–334.
3. Alpert P, Holzapfel C, Slonimski C (2003) Differences in performance between genotypes of *Fragaria chiloensis* with different degrees of resource sharing. J Ecol 91: 27–35.
4. Maurer DA, Zedler JB (2002) Differential invasion of a wetland grass explained by tests of nutrients and light availability on establishment and clonal growth. Oecologia 131: 279–288.
5. Chidumayo EN (2006) Fitness implications of clonal integration and leaf dynamics in a stoloniferous herb, *Nelsonia canescens* (Lam.) Spreng (Nelsoniaceae). Evol Ecol 20: 59–73.
6. Roiloa SR, Retuerto R (2007) Responses of the clonal *Fragaria vesca* to microtopographic heterogeneity under different water and light conditions. Environ Exp Bot 61: 1–9.
7. Yu FH, Wang N, He WM, Chu Y, Dong M (2008) Adaptation of rhizome connections in drylands: increasing tolerance of clones to wind erosion. Ann Bot 102: 571–577.
8. Moola FM, Vasseur L (2009) The importance of clonal growth to the recovery of *Gaultheria procumbens* L. (Ericaceae) after forest disturbance. Plant Ecol 201: 319–337.
9. Wang N, Yu FH, Li PX, He WH, Liu FH, et al. (2008) Clonal integration affects growth, photosynthetic efficiency and biomass allocation, but not the competitive ability, of the alien invasive *Alternanthera philoxeroides* under severe stress. Ann Bot 101: 671–678.
10. Xiao KY, Yu D, Wang LG, Han YQ (2011) Physiological integration helps a clonal macrophyte spread into competitive environments and coexist with other species. Aquat Bot 95: 249–253.
11. You WH, Yu D, Liu CH, Xie D, Xiong W (2013) Clonal integration facilitates invasiveness of the alien aquatic plant *Myriophyllum aquaticum* L. under heterogeneous water availability. Hydrobiologia 718: 27–39.
12. Herben T (2004) Physiological integration affects growth form and competitive ability in clonal plants. Evol Ecol 18: 493–520.
13. Wilsey B (2002) Clonal plants in a spatially heterogeneous environment: effects of integration on Serengeti grassland response to defoliation and urine-hits from grazing mammals. Plant Ecol 159: 15–22.
14. Březina S, Koubek T, Munzbergova Z, Herben T (2006) Ecological benefits of integration of *Calamagrostis epigejos* ramets under field conditions. Flora 201: 461–467.
15. Mack RN, Simberloff D, Lonsdale WM, Evans H, Clout M, et al. (2000) Biotic invasions: causes, epidemiology, global consequences, and control. Ecol Appl 10: 689–710.
16. Yurkonis KA, Meiners SJ, Wachholder BE (2005) Invasion impacts diversity through altered community dynamics. J Ecol 93: 1053–1061.
17. Kolar CS, Lodge DM (2001) Progress in invasion biology: predicting invaders. Trends Ecol Evol 16: 199–204.
18. Liu J, Dong M, Miao S, Li Z, Song M, et al. (2006) Invasive alien plants in China: role of clonality and geographical origin. Biol Invasions 8: 1461–1470.
19. Song YB, Yu FH, Keser LH, Dawson W, Fischer M, et al. (2013) United we stand, divided we fall: a meta-analysis of experiments on clonal integration and its relationship to invasiveness. Oecologia 171: 317–327.
20. Roiloa SR, Susana RE, Helena F (2013) Effect of physiological integration in self/non-self genotype recognition on the clonal invader *Carpobrotus edulis*. J Plant Ecol doi: 10.1093/jpe/rtt045.
21. Peltzer DA (2002) Does clonal integration improve competitive ability? A test using aspen (*Populus tremuloides* [Salicaceae]) invasion into prairie. Am J Bot 89: 494–499.
22. Yu F, Wang N, Alpert P, He W, Dong M (2009) Physiological integration in an introduced, invasive plant increases its spread into experimental communities and modifies their structure. Am J Bot 96: 1983–1989.
23. Wang N (2010) Clonal integration increased the competitive ability of invasive plant *Alternanthera philoxeroides* to *Plantago virginica*. Ecol Environ Sci 19: 2302–2306.
24. Roiloa SR, Rodríguez-Echeverria S, de la Peña E, Freitas H (2010) Physiological integration increases the survival and growth of the clonal invader *Carpobrotus edulis*. Biological Invasions 12: 1815–1823.
25. Price EAC, Hutchings MJ (1996) The effects of competition on growth and form in *Glechoma hederacea*. Oikos 75: 279–290.
26. Hartnett DC, Bazzaz FA (1985) The integration of neighbourhood effects by clonal genets of *Solidago canadensis*. J Ecol 73: 415–428.
27. Pennings SC, Callaway RM (2000) The advantages of clonal integration under different ecological conditions: a community-wide test. Ecology 81: 709–716.
28. Otfinowski R, Kenkel NC (2008) Clonal integration facilitates the proliferation of smooth brome clones invading northern fescue prairies. Plant Ecol 199: 235–242.
29. Schmid B, Bazzaz F (1987) Clonal integration and population structure in perennials: effects of severing rhizome connections. Ecology 68: 2016–2022.
30. de Kroon H, Hara T, Kwant R (1992) Size hierarchies of shoots and clones in clonal herb monocultures: do clonal and non-clonal plants compete differently? Oikos 63: 410–419.
31. Roiloa SR, Retuerto R (2006) Small-scale heterogeneity in soil quality influences photosynthetic efficiency and habitat selection in a clonal plant. Ann Bot 98: 1043–1052.
32. Hutchings MJ, Wijesinghe DK (1997) Patchy habitats, division of labour and growth dividends in clonal plants. Trends Ecol Evol 12: 390–394.
33. Ikegami M, Whigham DF, Werger MJA (2008) Optimal biomass allocation in heterogeneous environments in a clonal plant-spatial division of labor. Ecol Model 213: 156–164.
34. Schreiber U, Bilger W, Hormann H, Neubauer C (1998) Chlorophyll fluorescence as a diagnostic tool: basics and some aspects of practical relevance. In: Raghavendra AS (editor), Photosynthesis: a comprehensive treatise. Cambridge University Press, Cambridge, pp 320–336.
35. Björkman O, Demmig B (1987) Photon yield of O_2 evolution and chlorophyll fluorescence characteristics at 77 K among vascular plants of diverse origins. Planta 170: 489–504.
36. Julien MH, Skarratt B, Maywald GF (1995) Potential geographical distribution of alligator weed and its biological control by *Agasicles hygrophila*. J Aquat Plant Manage 33: 55–60.
37. Gunasekera L, Bonila J (2001) Alligator weed: tasty vegetable in Australian backyards? J Aquat Plant Manage 39: 17–20.
38. Ma R, Wang R (2005) Invasive mechanism and biological control of alligator weed, *Alternanthera philoxeroides* (Amaranthaceae), in China. Chin J Appl Environ Biol 11: 246–250.
39. Li M, Zhang LJ, Tao L, Li W (2008) Ecophysiological responses of *Jussiaea repens* to cadmium exposure. Aquat Bot 88: 347–352.
40. Sosnová M, van Diggelen R, Macek P, Klimesova J (2011) Distribution of clonal growth traits among wetland habitats. Aquat Bot 95: 88–93.
41. Wang B, Li W, Wang J (2005) Genetic diversity of *Alternanthera philoxeroides* in China. Aquat Bot 81: 277–283.
42. Kikvidze Z, Khetsuriani L, Kikodze D, Callaway RM (2006) Seasonal shifts in competition and facilitation in subalpine plant communities of the central Caucasus. J Veg Sci 17: 77–82.
43. Hartnett DC, Bazzaz FA (1983) Physiological integration among intraclonal ramets in *Solidago canadensis*. Ecology 64: 779–788.
44. Yu F, Chen Y, Dong M (2002) Clonal integration enhances survival and performance of *Potentilla anserina*, suffering from partial sand burial on Ordos plateau, China. Evol Ecol 15: 303–318.

45. Jónsdóttir IS, Watson MA (1997) Extensive physiological integration: An adaptive trait in resource-poor environments? In - de Kroon H.; van Groenendael J (editors). The ecology and evolution of clonal plants, p. 109–136.
46. D'Hertefeldt T, Jónsdóttir IS (1999) Extensive physiological integration in intact clonal systems of *Carex arenaria*. J Ecol 87: 258–264.
47. Hutchings MJ, de Kroon H (1994) Foraging in plants: the role of morphological plasticity in resources acquisition. Adv Ecol Res 25: 159–238.
48. Day KJ, John EA, Hutchings MJ (2003) The effects of spatially heterogeneous nutrient supply on yield, intensity of competition and root placement patterns in *Briza media* and *Festuca ovina*. Funct Ecol 17: 454–463.
49. Birch CPD, Hutchings MJ (1994) Exploitation of patchily distributed soil resources by the clonal herb *glechoma-hederacea*. J Ecol 82 (3): 653–664.
50. Watson MA (1986) Integrated physiological units in plants. Trends Ecol Evol 1: 119–123.
51. Roiloa SR, Alpert P, Tharayil N, Hancock G, Bhowmik PC (2007) Greater capacity for division of labour in clones of *Fragaria chiloensis* from patchier habitats. J Ecol 95: 397–405.
52. de Kroon H, Kreulen R, van Rheenen JWA, van Dijk A (1998). The interaction between water and nitrogen translocation in a rhizomatous sedge (*Carex flacca*). Oecologia 116: 38–49.
53. Nilsson J, D'Hertefeldt T (2008) Origin matters for level of resource sharing in the clonal herb *Aegopodium podagraria*. Evol Ecol 22: 437–448.
54. Roiloa SR, Rodríguez-Echeverría S, Freitas H, Retuerto R (2013) Developmentally-programmed division of labour in the clonal invader *Carpobrotus edulis*. Biol Invasions 15: 1859–1905.
55. D'Hertefeldt T, Falkengren-Grerup U (2002) Extensive physiological integration in *Carex arenaria* and *Carex disticha* in relation to potassium and water availability. New Phytol 156: 469–477.

14

Seagrass Proliferation Precedes Mortality during Hypo-Salinity Events: A Stress-Induced Morphometric Response

Catherine J. Collier[1]*, Cecilia Villacorta-Rath[1], Kor-jent van Dijk[1,2], Miwa Takahashi[1], Michelle Waycott[2]

1 School of Marine and Tropical Biology, James Cook University, Townsville, Australia, **2** School of Earth and Environmental Science, Australian Centre for Evolutionary Biology and Biodiversity, University of Adelaide, Adelaide, Australia

Abstract

Halophytes, such as seagrasses, predominantly form habitats in coastal and estuarine areas. These habitats can be seasonally exposed to hypo-salinity events during watershed runoff exposing them to dramatic salinity shifts and osmotic shock. The manifestation of this osmotic shock on seagrass morphology and phenology was tested in three Indo-Pacific seagrass species, *Halophila ovalis*, *Halodule uninervis* and *Zostera muelleri*, to hypo-salinity ranging from 3 to 36 PSU at 3 PSU increments for 10 weeks. All three species had broad salinity tolerance but demonstrated a moderate hypo-salinity stress response – analogous to a stress induced morphometric response (SIMR). Shoot proliferation occurred at salinities <30 PSU, with the largest increases, up to 400% increase in shoot density, occurring at the sub-lethal salinities <15 PSU, with the specific salinity associated with peak shoot density being variable among species. Resources were not diverted away from leaf growth or shoot development to support the new shoot production. However, at sub-lethal salinities where shoots proliferated, flowering was severely reduced for *H. ovalis*, the only species to flower during this experiment, demonstrating a diversion of resources away from sexual reproduction to support the investment in new shoots. This SIMR response preceded mortality, which occurred at 3 PSU for *H. ovalis* and 6 PSU for *H. uninervis*, while complete mortality was not reached for *Z. muelleri*. This is the first study to identify a SIMR in seagrasses, being detectable due to the fine resolution of salinity treatments tested. The detection of SIMR demonstrates the need for caution in interpreting in-situ changes in shoot density as shoot proliferation could be interpreted as a healthy or positive plant response to environmental conditions, when in fact it could signal pre-mortality stress.

Editor: John F. Valentine, Dauphin Island Sea Lab, United States of America

Funding: This work was funded by the Australian Government's National Environmental Research Program (NEEP) Tropical Ecosystem Hub. The funders had no role in study design, data collection and analysis, decision to publish, or preparation of the manuscript.

Competing Interests: The authors have declared that no competing interests exist.

* E-mail: Catherine.collier@jcu.edu.au

Introduction

Seagrasses are a group of angiosperms (flowering plants), within the monocotyledon order Alismatales [1,2]. Seagrasses evolved along four separate lineages but are considered a single functional group because of similar adaptive traits, principally their tolerance to seawater salinities [1]. Their preferred salinity ranges from 20 practical salinity units (PSU) through to 42 PSU, except for *Ruppia* spp which frequently inhabit fresh water (0 PSU) [3].

Seagrasses predominantly occur in estuaries and coasts where salinity can be affected by watershed run-off leading to hypo-saline conditions [4], or it can become hyper-saline in shallow embayments with high rates of evaporation [5,6] and at sites of desalinisation discharge [7]. In tropical and monsoonal climates, wet season depressions in salinity can reach 0 PSU during extreme runoff events [8]. Run-off can be associated with widespread declines in seagrass abundance, with significant consequences for the broader ecosystem [9,10]. A number of studies have described the effects of hypo-salinity on northern hemisphere seagrass species in Europe and the USA [4,11–13]; however, sensitivity to hypo-salinity is not known for most Indo-Pacific seagrass species. Furthermore, previous seagrass studies, with some exceptions [13], have lacked the treatment and temporal resolution to determine hypo-salinity thresholds whereby extreme mortality occurs. Without these thresholds it is difficult to determine what role hypo-salinity stress has during mortality associated with watershed run-off.

Salinity affects water uptake, plant water potential and cellular ion concentrations, and when plants become salinity-stressed there are damaging consequences for cellular integrity, biochemical processes and ultimately, plant fitness [14,15]. Seagrasses are halophytes, that is, they maintain high intracellular osmotic potentials in saline environments, by ion sequestration and the generation of osmotically-active solutes [3,15]. These osmolytes enable seagrasses to exclude Na+ and Cl- ions even at very high concentrations [3,15]. Exceedance of optimum salinity and disruption of cellular processes affects photosynthetic efficiency and leads to reduced growth rates and morphological changes and eventual mortality [11,15–19].

The duration of exposure affects the level of impact on plant fitness and seagrasses may recover following brief levels of exposure to salinity stress but may fail to recover after prolonged stress [15,19]. Furthermore, the rate of salinity change affects plant

Figure 1. ***Halophila ovalis*** **(A),** ***Halodule uninervis*** **(B),** ***Zostera muelleri*** **in the experimental units after 3 weeks exposure to 9 PSU (C) and a** ***Zostera muelleri*** **meadow in Gladstone Harbour, Australia where experimental plants were collected (D).**

health, with incremental salinity change increasing plant survivorship [11,12,19]. This is an important consideration when testing plant survivorship as hypo-salinity changes are rarely sudden – even though experimental approaches frequently assume so – but rather they occur gradually as flood waters mix with saline waters [5,12].

We tested response to hypo-salinity of three seagrass species that inhabit estuarine and coastal environments where marine salinity is typical, but seasonal hypo-salinity events are common [8]. We mimicked the gradual reduction in salinity that would be expected as flood waters emerge from watersheds and flood into estuaries and coasts. This detailed approach revealed not just broad salinity tolerance but also a stress-induced morphogenic response (SIMR) [20–22] in which shoot proliferation occurred – a stress response not previously reported for seagrass.

Materials and Methods

All plants were collected under permit MTB41, issued by the School of Marine and Tropical Biology, at James Cook University, in accordance with low impact research guidelines in the Great Barrier Reef Marine Park.

Experimental conditions

Hypo-salinity exposure experiments were conducted on three species of seagrass, which are ubiquitous throughout the Indo-Pacific, except *Zostera muelleri* Irmisch ex Ascherson, which is widespread in Australia and New Zealand only (Fig 1). *Halodule uninervis* (Forsskål) Ascherson is a tropical species that occurs throughout the Indo-West Pacific in coastal and reef habitats, while *Halophila ovalis* R. Brown is one of the most broadly distributed seagrass species occurring throughout the Indo-West Pacific, including temperate regions, and can be found in estuarine, reef and deepwater habitats [23]. Their habitats are periodically exposed to flood plumes of reduced salinity [8]. Both *Z. muelleri* and *H. uninervis* are species with linear leaf blades (blady),

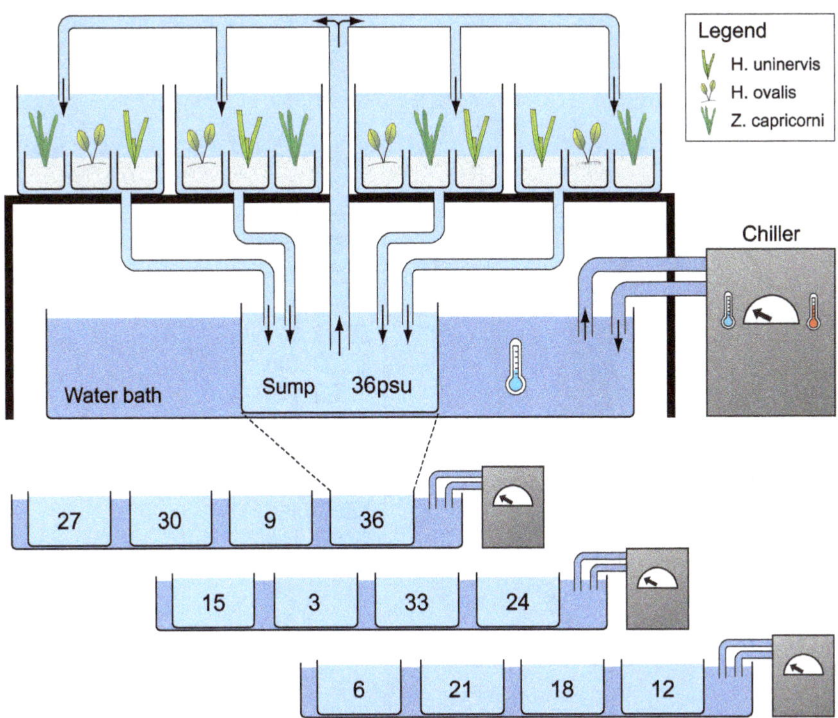

Figure 2. Experimental set-up showing three chilled water baths each with four randomly allocated sumps immersed within them. Each sump contained one of the 12 salinity treatments. Water was piped from the sump to the four replicate tanks and back again on closed-circulation.

Table 1. Results of single factor repeated measures analysis of variance (RM-ANOVA) for change in shoot density at salinity treatments of 3 to 36 PSU in three seagrass species: *Halophila ovalis, Halodule uninervis* and *Zostera muelleri*.

	H. ovalis			*H. uninervis*			*Z. muelleri*		
	df	F	p	df	F	p	df	F	p
Within-subjects effects									
Time	2.580	9.432	**<0.001**	2.364	4.338	n.s.	3.131	78.380	**<0.001**
Time x salinity	28.384	21.949	**<0.001**	26.004	6.556	**<0.001**	34.445	11.848	**<0.001**
Between-subjects effects									
Salinity	11	44.170	**<0.001**	11	7.366	**<0.001**	11	19.531	**<0.001**
Transformations	4^{th}Rt (x+101)			SqRt (x+101)			SqRt (x+101)		
Significance level (p)	0.01			0.01			0.01		

Transformations performed to meet assumptions of ANOVA and significance level used for interpretation of results are also indicated for each species.

whereas *H. ovalis* has pairs of ovate leaves arising from the rhizome on petioles (Fig 1).

Zostera muelleri plants were collected from Pelican Banks, Gladstone (23°45.895′S, 151°18.244′E) during low tide three months before the experiments started. The plants were collected using a 10 cm corer, with sediment and rhizome and roots collected intact. The cores were placed in plastic-lined pots, the plastic bag sealed over the top of the seagrass with 2–3 cm of water during transport to the experimental facility. *Halodule uninervis* and *H. ovalis* plants were collected from Cockle Bay, Magnetic Island (19°10.612S, 146°49.737E) using the same technique two months prior to the experiments. The plants were kept in 1000L aquaria at the Aquaculture facility in James Cook University on a closed circulation system in seawater piped from Bowling Green Bay seawater intake under a 30% light-reducing roof.

The experiment consisted of 12 salinity treatments, starting from 3 PSU and increasing by 3 PSU to 36 PSU (approximate marine seawater). Salinity treatments were obtained by diluting the seawater with de-chlorinated freshwater. Every salinity treatment consisted of four replicate tanks (65L KiTab clear plastic containers) with one pot of each species per tank (i.e. n = 4). All treatments started at 36 PSU and salinity was reduced by 25% each day over four days to the target treatment salinity to mimic the more gradual decline in salinities that occur during run-off events and to minimize potential impacts from shock osmotic changes. Throughout the experiment, salinity was measured every

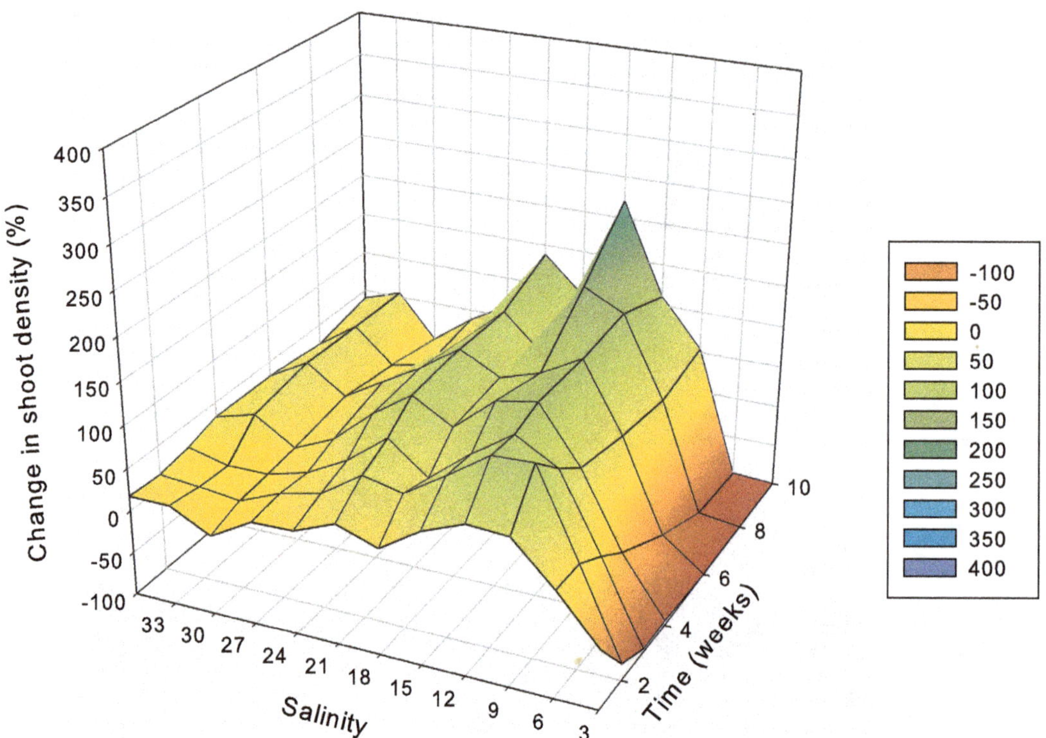

Figure 3. Change in *Halophila ovalis* shoot density relative to pre-treatment (week 0) (y-axis) as indicated by colour shading from 100% loss (red) through to 400% increase (blue), at salinities 3 to 36 PSU (x-axis) after 1 through to 10 weeks of exposure to treatment salinity (z-axis). n = 4.

Table 2. Summary of Tukeys Post-hoc comparisons for each week for change in shoot density.

Week	*H. ovalis*	*H. uninervis*	*Z. muelleri*
1	3<9,12,15,21	n.s.	n.s.
	12>30		
2	3<all others	n.s.	6>15,21-36
3	3<all others	n.s.	3>21-30, 36
	6<9,12,15,21		6>18-36
	12>30,33		9>15-36
			12>18-36
			15>36
4	3<all others	3<9-18	3>15-36
	6<12-24		6>12-36
	12>30		9>15-36
			12>33-36
6	3<6<all others	3<9-36	3<6, 3>21-36
	12>30	6<12,18	6>3, 12-36
			9>15-36
			12>36
8	3=6<all others	3<9-36	3<6, 3>21-36
		6<9-18,30-33	6>3, 12-36
			9>15-36
			12>21-30,36
10	3=6<all others	3<9-36	3<6, 3>21-36
	15>27-36	6<9-15	6>3, 12-36
			9>15-36
			12>21-30,36

Differences among treatments are indicated for each species at each measuring time

1 to 3 days using a digital salinity/conductivity/temperature meter (YSI, model 63) and salinity was adjusted when necessary to maintain salinity within 0.5 PSU of target salinity. Plant responses to these salinities were monitored for 10 weeks. Previous salinity studies indicate that seagrass changes settle down by this time [5,12], and furthermore, this experimental duration is approximately equal to or more likely exceeds the length of individual hypo-salinity events in the region.

The experiments were conducted outdoors during summer/autumn months (February to April) when high ambient temperatures occur, thus chilling units were installed to moderate temperature fluctuations within the treatment tanks throughout the experiment. There were three chilled freshwater baths (1000L tanks) that were cooled using external water chillers. Each of the 12 salinity treatments had one 60L sump (60L plastic bin) that was placed randomly in one of the 3 chilling baths, each bath containing 4 sumps (Fig 2). The chilled baths with sumps were held underneath tables that held the experimental tanks. From each sump, water with corresponding salinity was pumped into four replicate tanks resulting in a total of 48 tanks (4 replicate tanks ×12 sumps/salinity treatments =48 tanks in total). Each tank contained one pot of each of the three species (48 tanks ×3 species/pots = 144 pots). Temperature was recorded every 30 mins using iBCod 22L model of iBTag in six randomly selected tanks for the duration of the experiment. Water temperature was 26°C on average and ranged from 22°C to 34°C reaching these temperature extremes for short periods (1–2 h) on some days. Nitrogen (N) as NH_4Cl and phosphate (P) as KH_2PO_4 were added to the water column at very low concentrations to increase concentrations within each system by 0.05 μMol of P and 1.0 μMol of N every 2 weeks. Nutrient concentration was measured after six weeks and was found to be 0.8 μMol (±0.2) NH_3, 0.4 μMol (±0.1) NOx, and 0.2 μMol (±0.8) PO_4. Average light intensity under the 30% light-reducing roof was 17 mol photons m^{-2} d^{-1} of Photosynthetically Active Radiation (PAR), measured with an Odyssey 2Pi quantum sensor (Dataflow, Odyssey photosynthetic recording system) recording every 30 mins throughout the experimental period. The tanks were periodically cleaned by syphoning out sediment and organic matter accumulating at the bottom of tanks and plants were inspected every week for signs of grazing by amphipods. Amphipods were removed to prevent an outbreak, which could lead to overgrazing of the plants. Although signs of grazing were observed at times, this cleaning regime was sufficient to avoid outbreaks.

Plant growth and survival

The number of shoots in each pot for *Z. muelleri* and *H. uninervis*, or the number of leaf pairs for *H. ovalis* were counted prior to the experiment and then weekly during the first four weeks of the experiment and fortnightly from the sixth week up to and including the tenth week. Change in shoot density (ΔSht) was calculated as a percentage change in each week relative to pre-treatment for each individual replicate:

$$\Delta Sht = \left[\frac{(Sht\,t_x - Sht\,t_0)}{Sht\,t_0}\right] \times 100 \qquad (1)$$

where ΔSht is change in shoot density, Sht t_x is shoot density at time x (weeks 1 through to 10) and Sht t_0 is shoot density at week zero (pre-treatment).

Leaf morphometrics (width and height) of *Z. muelleri*, *H. uninervis* and *H. ovalis* and number of leaves per shoot for the two blady species were measured after 10 weeks at treatment salinity. These data were used to calculate foliar surface area (SA) as follows:

$$SA = shoot\ density \times (leaves\ per\ shoot) - 0.5 \times leaf\ length \times leaf\ width \qquad (2)$$

for blady species (*H. uninervis and Z. muelleri*); and,

$$SA = leaf\ density \times \pi \times \left(\frac{leaf\ length}{2}\right) \times \left(\frac{leaf\ width}{2}\right) \qquad (3)$$

for the ovate species *H. ovalis* where SA is the foliar leaf area (cm^2), shoot density are leaves per experimental pot, leaves per shoot are the mean number of leaves (usually 1 to 4) per seagrass shoot and leaf length (cm) and leaf width (cm) of the youngest fully mature leaf. A half leaf was subtracted from the total number of leaves per shoot in calculating LA of blady species to account for one leaf on each shoot being in development and therefore not full sized [24].

Halophila ovalis was the only species to flower throughout the experimental period. Flowering had commenced prior to the initiation of the experiment and continued throughout. New leaf pairs are produced in *H. ovalis* every 3 or 4 days at experimental water temperatures of approximately 27–27°C [25] and *H. ovalis* typically had 4 to 5 leaf pairs per branch. Flowering is initiated in

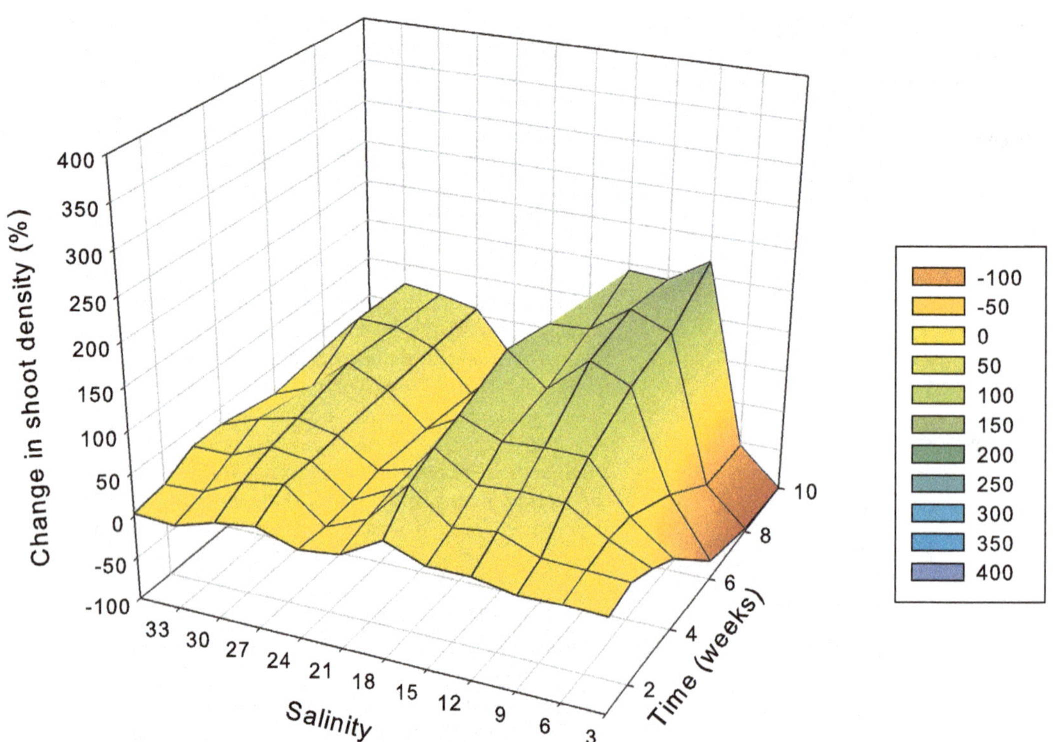

Figure 4. Change in *Halodule uninervis* shoot density relative to pre-treatment (week 0) (y-axis) as indicated by colour shading from 100% loss (red) through to 400% increase (blue), at salinities 3 to 36 PSU (x-axis) after 1 through to 10 weeks of exposure to treatment salinity (z-axis). n = 4.

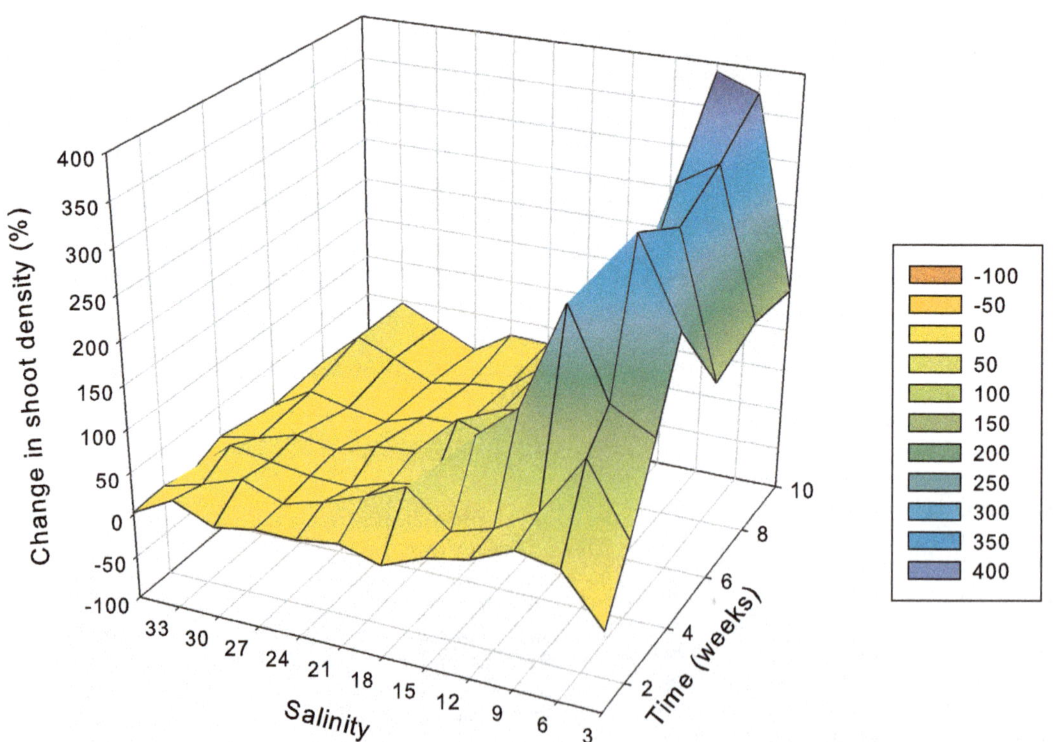

Figure 5. Change in *Zostera muelleri* shoot density relative to pre-treatment (week 0) (y-axis) as indicated by colour shading from 100% loss (red) through to 400% increase (blue), at salinities 3 to 36 PSU (x-axis) after 1 through to 10 weeks of exposure to treatment salinity (z-axis). n = 4.

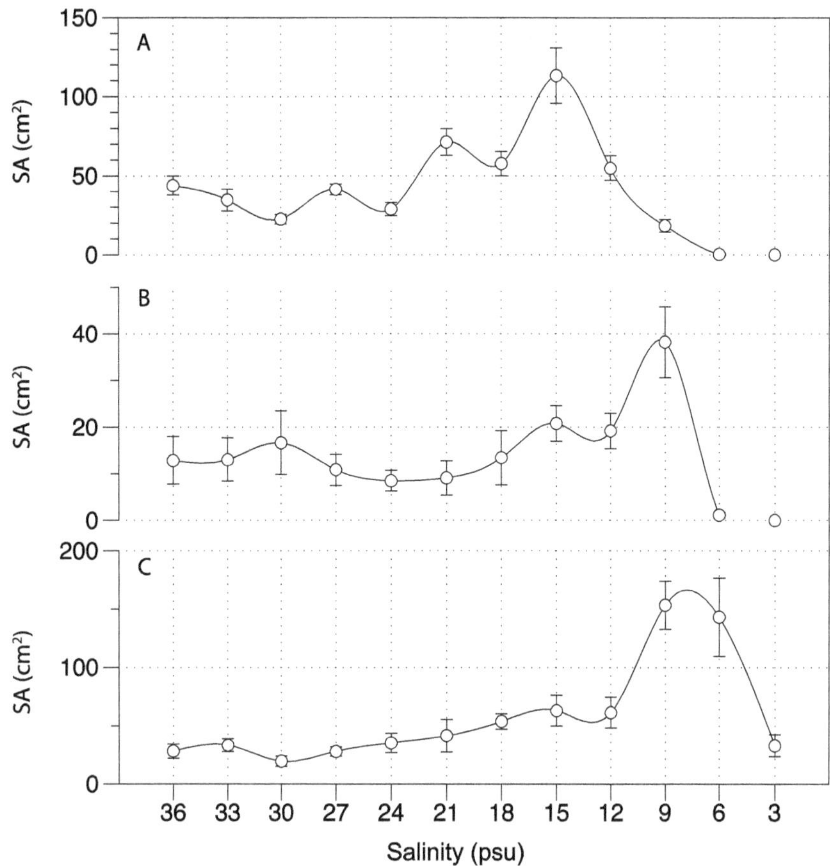

Figure 6. Foliar surface area (SA, cm^2) calculated from shoot density, leaves per shoot and leaf length and width of *H. ovalis* (A), *H. uninervis* (B) and *Z. muelleri* (C) after 10 weeks at treatment salinity. n = 4 ± SE

young leaf pairs, with more advanced reproductive structures away from the growing apex. Assuming a leaf pair production rate of 4 days, we conservatively assumed that all reproductive structures present after 4 weeks (28 d) were initiated under treatment conditions. We counted all reproductive structures (male and female flowers, as well as fruits) in each pot at weeks 4, 6, 8 and 10. We calculated reproductive potential – the highest number of reproductive structures occurring under treatment conditions as follows:

$$\textit{Reproductive potential} = max[R_4, R_6, R_8, R_{10}] \quad (4)$$

where R_4 is mean structures in week 4 of treatment salinity through to R_{10}, which is mean structures in week 10. We also present the total number of reproductive structures against shoot density for each replicate.

Leaf growth rate was measured in week 10 on the two blady species (*Z. muelleri* and *H. uninervis*) using the leaf hole punch method [26]. Holes were punched using a hypodermic needle in the top of the sheath of each shoot, and after 5–7 days we measured the distance between the mark in the sheath and the mark on the leaves. We aimed to measure up to 10 shoots per replicate pot, though the actual number measured in each pot was variable depending on shoot density and visibility of marks.

Statistical analyses

Shoot density data was analysed using a one-way repeated measures analysis of variance (RM ANOVA) with salinity as a fixed factor between-subjects effect and time (weeks) as the within-subjects effect. Data were first checked for homogeneity of variances using Levene's test, and transformed if failing this assumption of ANOVA. Transformation was not successful at improving variances at all times, typically one or two measuring times failed these tests ($p<0.05$) in which case the ANOVA was still performed on transformed data as the ANOVA is relatively robust to violations of assumptions in large experiments such as this; however, the significance level was set to 0.01 to minimize the risk of a Type II error [27]. Data were also checked for sphericity (correlations among time) and the degrees of freedom was adjusted using the Greenhouse-Geiser epsilon adjustment where necessary. Where a significant interaction between time and salinity was observed, post-hoc analyses to explore differences among treatments were performed for each measuring time separately. For single time data, single factor ANOVA's were performed with salinity as a fixed factor. Data were tested and treated as described above, and post-hoc analyses were conducted using S-N-K comparisons. All statistical analyses were performed using SPSS v20.0. Key statistical results are described in text with detailed statistical results in Tables.

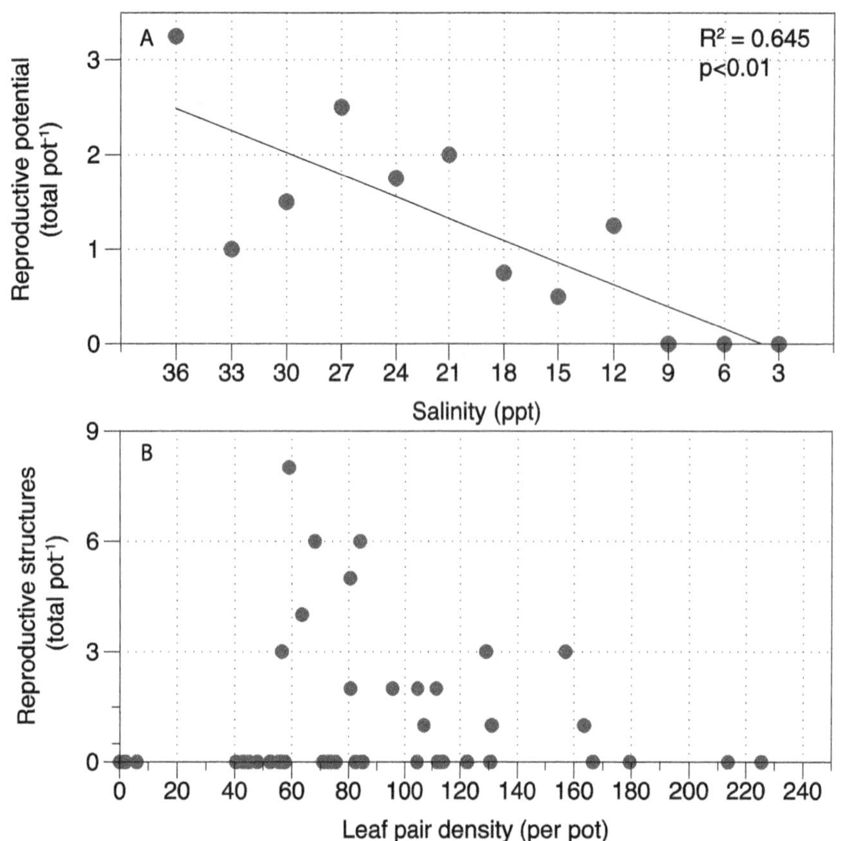

Figure 7. Sexual reproduction in *Halophila ovalis* under salinity treatment conditions showing (A) reproductive potential which is the highest mean (total number of flowers and fruits) recorded for each treatment in weeks 6–10; and, (B) reproductive output (total number of flowers and fruits) correlated with shoot density at 10 weeks. n = 4 ± SE.

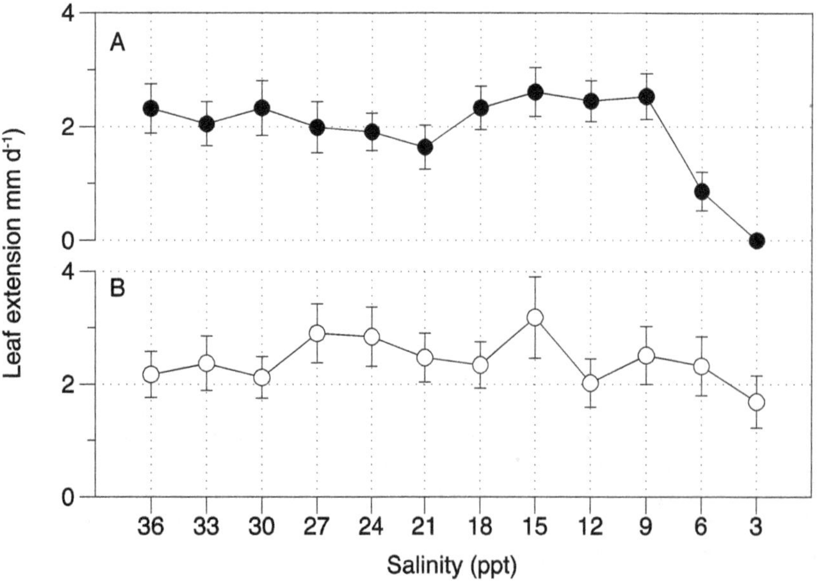

Figure 8. Leaf extension rate (mm d^{-1}) for *H. uninervis* (A) and *Z. muelleri* (B) after 10 weeks exposure to hypo-salinity. n = 4 ± SE.

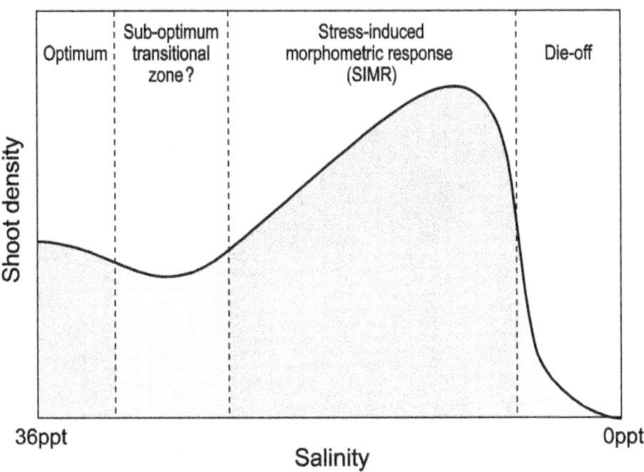

Figure 9. Conceptual summary of the seagrass responses to hypo-salinity. High (marine, 36 PSU) salinities are "optimum", as shoot density steadily increased throughout the experimental period at this salinity while sexual reproduction (for *H. ovalis*) was at its "peak". At slightly depressed salinities (30–33 PSU) there appeared to be a "sub-optimal" transition zone as shoot density showed minimal increase and, furthermore, sexual reproduction (for *H. ovalis*) was low. With further hypo-salinity (<30 PSU), a stress-induced morphometric response was associated with a re-prioritisation of resources that saw massively increased shoot density (and leaf area) and reduced sexual reproduction. At extreme hypo-salinity (3–6 PSU) plant mortality occurred. The cut-off for each response phase moved to higher salinities with increased duration of exposure.

Results

Shoot density

Initial shoot densities were on average 55 (± SE 8) leaf pairs for *H. ovalis*, 8 (± SE 1) shoots for *H. uninervis* and 30 (± SE 5) shoots for *Z. muelleri*. Changes in shoot density in response to hypo-salinity generally followed the same trends among species with a salinity response that was affected by time (Table 1); however, thresholds and response times were variable. The most notable difference among species was in their sensitivity at the lowest salinities; *H. ovalis* was the most sensitive, whereas *Z. muelleri* was the most tolerant of very low (3 and 6 PSU) salinities. Furthermore, *Z. muelleri* increased shoot density by the largest magnitude at low-mid salinities.

More specifically, in *H. ovalis* leaf pair density had declined after a one-week exposure to 3 PSU (Fig 3) and after 2 weeks it was significantly (p<0.01) lower than all other salinity treatments (Table 2). At the same time, leaf pair density showed an initial increase by 24% at 6 PSU after just 1 week, but reduced soon afterwards with significant (p<0.01) reductions at 6 PSU compared to low and mid salinities after 3 weeks. After 6 weeks there were no shoots remaining at 3 PSU and after 10 weeks there were just 3% remaining at 6 PSU. There was a very distinct threshold between 6 and 9 PSU, with leaf pair density increasing relative to starting density and being the highest at salinities ranging from 9 to 15 PSU; however, significant (p<0.01) increases in shoot density occurred only at 12 and 15 PSU. Shoot density increased at 36 and 33 PSU (by 30% and 55%), which was followed by negligible change in density at 30 PSU (2% increase), but density then increased again at lower salinities until reaching mortality thresholds.

For *H. uninervis*, the general trends were similar but the reaction time was slower and was more difficult to detect, as initial shoot density was considerably lower. There were no significant differences among treatments up to and including 3 weeks of hypo-salinity exposure (Fig 4). After 4 weeks, density had significantly declined in 3 PSU relative to salinities of 9 through 18 PSU, and after 6 weeks density was significantly lower at 3 PSU than in all other treatments. After 10 weeks, there were no shoots remaining at 3 PSU. At 6 PSU, *H. uninervis* initially increased by 25% after 3 weeks, but started to decline thereafter, being significantly reduced relative to low/mid-range salinities after 4 weeks and there was 54% loss after 10 weeks. There was a distinct threshold between 6 and 9 PSU, with no net loss of shoots at 9 PSU, which instead showed the greatest increase in density among all salinities of 170% after 10 weeks.

In *Z. muelleri*, hypo-salinity had a significant and positive effect on shoot density at salinities from 3 to 15 PSU (Fig 5). After 2 weeks, shoot density had increased significantly (p<0.01) more at 6 PSU compared to higher salinities, and after 3 weeks, density had increased significantly (p<0.01) at salinities from 3 to 15 PSU relative to higher salinities. After an initial increase of 240% at 3 PSU within 4 weeks, shoot density started to decline, but remained elevated relative to pre-treatment conditions throughout the experiment and was 150% greater than starting density after 10 weeks. The largest increase in shoot density was at 6 PSU, where density was 400% higher than pre-treatment after 10 weeks. It was significantly (p<0.01) higher at 6 PSU than all other treatments (except 9 PSU) after 4weeks and remained significantly (p<0.01) elevated throughout. At 9 PSU, shoot density was significantly (p<0.01) higher than all salinities of 15 PSU and greater after just 3 weeks. Shoot density was significantly (p<0.01) higher at 12 PSU than at 21–36 PSU (except 33 PSU) after 8 and 10 weeks. The smallest change in shoot density occurred at 36 and 27 PSU, with 0 and 1% increase in shoot density, respectively, after 10 weeks.

Leaf area

Foliar surface area (SA), calculated from shoot density as well as shoot size (leaf length, width and leaves per shoot) after 10 weeks exposure to hypo-salinity treatments, followed the same general trends and magnitude of response as for shoot density. For *H. ovalis*, salinity had a significant effect (F = 42.041, MS = 37.236, p<0.001, SqRt transformed) on SA. The largest SA occurred at 15 PSU, where it was significantly (p<0.01) higher than all other treatments except 21 PSU, and was more than double that at 36 PSU (Fig 6A). The lowest SA occurred at 3 and 6 PSU where there were just 0 and 2 shoots remaining, resulting in a significantly (p<0.001) reduced SA compared to all other treatments. For *H. uninervis* salinity also had a significant effect on SA (MS = 9.989, F = 6.839, p<0.001, SqRt transformation), the peak in SA at 9 PSU was significantly (p<0.01) greater than SA at 3-6 PSU, and 18–24 PSU, inclusive (Fig 6B). SA was significantly (p<0.05) lower at 3 PSU than SA at all salinities except 21 and 24 PSU. For *Z. muelleri*, the significant effect of salinity on SA (MS = 25.711, F = 10.182, p<0.001, SqRt transformation) peaked at 6 and 9 PSU, which were 5 times greater than at 36 PSU, and which were both significantly (p<0.05) higher than all other treatments (Fig 5C).

Sexual reproduction (flowering)

Reproductive potential, which is the highest mean recorded in weeks 4–10, increased with salinity, with no structures at 3–9 PSU and the largest number (3.25 pot^{-1}) occurring at 36 PSU (Fig 7A). This is in stark contrast to leaf pair density which was greatest at 12–15 PSU for *H. ovalis*. There was an anomaly of reduced reproductive effort at 30 and 33 PSU compared to higher and lower salinities. When plotted against leaf pair density (Fig 7B), the greatest number of reproductive structures occurred at low to moderate leaf pair densities, and at very high leaf pair densities

Table 3. Summary of the sub-lethal (e.g. growth and photosynthesis or some shoot/biomass loss), and complete mortality thresholds (PSU) for seagrasses in responses to hypo- and hyper-salinity exposure.

Species	Experiment	Experimental salinity range	Optimum salinity	Experimental time	Hypo- Sub-lethal	Hypo- Mortality	Hyper- Sub-lethal	Hyper- Mortality	Sub-lethal measure	Citation
Amphibolis antarctica	A	35–65	42	5 mo	-	-	50–57.5	65	Seedling productivity	[33]
Amphibolis antarctica	IS	35–58.5	42	8–10 d	-	-	49–60	-	Leaf growth and biomass	[34]
Cymodocea nodosa	A	0–72	30–39	10 d	16	0	41	57	Leaf growth	[13]
Cymodocea nodosa	A	37–62	37–44	17 d	-	-	54	NE	Leaf growth, PSII efficiency	[35]
Cymodocea nodosa	A	37–43	37	47 d	-	-	39	NE	Photosynthesis	[36]
Halodule uninervis	**A**	**3–36**	**36**	**10 wk**	**15**	**3**	-	-	**Shoot proliferation, not growth**	**This study**
Halodule wrightii	A	35–70	35–65	1 mo	-	-	70	NE	Leaf growth, osmolality, quantum efficiency	[5]
Halodule wrightii	A	5–45	35	14 d	5	NE	45	NE	Leaf extension	[4]
Halophila johnsonii	A	10–30	20–30	28 d	10	10	-	-	Leaf area, PSII efficiency	[37]
Halophila johnsonii	A	8–30	15–30	1 mo	8–10	NE	-	-	PSII efficiency, leaf osmolality, antioxidant activity of leaves	[12]
Halophila johnsonii	A	0–60	30	15 d	20	0	40	60	Photosynthesis, growth	[38]
Halophila ovalis	**A**	**3–36**	**36**	**10 wk**	**12**	**6**	-	-	**Shoot proliferation, not growth**	**This study**
Halophila ovalis	A	5–45	20–35	6 wk	10	5	40	NE	Biomass	[39]
Posidonia oceanica	A	25–57	25–39	15 d	NE	29	42	50	Leaf growth	[40]
Posidonia oceanica	IS	37–38	37	16 yr	-	-	38–40	NE	Leaf growth, NSC, %N	[7]
Posidonia oceanica	A	37–43	37	47 d	-	-	39	NE	Photosynthesis and leaf growth	[41]
Posidonia oceanica	A	37–43	37	3 mo	-	-	43	NE	Leaf turgor, photosynthesis, growth	[19]
Posidonia oceanica	A	37.5–39	37.5	3 mo	-	-	38.5	NE	Shoot size, leaf growth rate	[30]
Ruppia cirrhosa	IS	0–55	0	5 mo	-	-	15	NE	Natural occurrence	[42]
Ruppia maritima	A	35–70	35–40	1 mo	-	-	55–70	NE	Leaf growth, shoot density, PSII efficiency	[5]
Ruppia maritima	A	0–45	0	150 d	-	-	5	35	Seedling biomass	[31]
Syringodium filiforme	A	5–45	25	14 d	5	NE	45	NE	Leaf extension	[4]
Thalassia testudinum	A	0–70	30–40	14 wk	10	0	50–60	70	Leaf growth, leaf area, photosynthesis, osmolality	[11]
Thalassia testudinum	A	35–65	35–55	2 mo	-	-	65	NE	Leaf growth, shoot density, PSII efficiency	[43]
Thalassia testudinum	A	5–45	40	14 d	5	NE	45	NE	Leaf extension	[4]
Zostera capensis	IS	0–55	30	5 mo	15	0	-	55	Plant mass	[42]
Zostera muelleri	**A**	**3–36**	**36**	**10 wk**	**15**	**NE**	-	-	**Shoot proliferation, not growth**	**This study**

The tested range, time of exposure and the optimum salinity are also given. A = Aquarium based, IS = *in-situ*, NE = No effect, and "-" is not measured. Bold = "This study".

(>165 pairs per pot) there was no reproductive structures, nor at very low densities (<50 pairs), where the plants were generally dying.

Growth

Leaf growth (measured as leaf extension, mm d^{-1}) showed very little response to the salinity treatments. *H. uninervis* was significantly affected by the salinity (MS = 2.203, F = 7.955, $p<0.001$, non-transformed, Fig 8A), but only at 3 and 6 PSU in plants that were essentially dead or almost dead. Growth in 3 PSU was significantly lower than all other treatments, while growth at 6 PSU was significantly lower than 9–18 PSU and 30 PSU and 36 PSU, but not other treatments. Growth in *Z. muelleri* was also significantly affected by salinity (MS = 0.720, F = 2.591 and $p<0.05$, Fig 8B), with growth at 3 PSU being lower than 15 PSU only.

Discussion

These coastal Indo-Pacific seagrasses demonstrated very broad salinity tolerance when gradually exposed to hypo-salinity. Even after 10 weeks exposure, *Zostera muelleri* had survived to salinities as low as 3 PSU, while *Halophila ovalis* and *Halodule uninervis* remained abundant at 9 PSU. However, the plant-scale responses were complex, and the high treatment-resolution enabled us to develop a thorough conceptual model to describe this response (Fig 9). The most distinctive finding was a stress-induced morphometric response (SIMR) [20–22], characterised by shoot proliferation. This corresponded to reduced flowering (for *Halophila ovalis*) indicating a diversion of resources away from sexual reproduction to support the lateral branching. This did not come at the expense of leaf growth, which was largely unaffected by salinity, or shoot development (shoot size), as the foliar surface area mirrored the shoot density response. This shoot proliferation was a 'moderate stress response' and as the hypo-salinity treatments progressed to lower salinities, severe stress resulted in die-off (Fig 9). In this way, the shoot proliferation preceded mortality.

A 'sub-optimal transitional zone' between optimum salinity (36 PSU) and SIMR salinities appeared at 27–33 PSU depending on species, recognizable as a zone with small changes in shoot density (Figs 5 and 6). For example, the smallest change in shoot density occurred at 30 PSU in *H. ovalis* (2%); while at salinities both above (optimum salinity) and below this (stress response) there was shoot proliferation. There was also very low sexual reproduction in this transition zone. Previous studies have reported SIMR responses for other non-seagrass species groups [20–22]; however the proposed sub-optimal transitional zone requires further validation.

The broad salinity tolerance indicates intracellular osmoregulation within the plant tissues. In halophytes, selective ion and solute accumulation enables high intra-cellular osmotic potentials to remain. A number of osmolytes occur in seagrass leaves, including, inorganic ions (Na^+, K^+, Cl^-) soluble sugars and amino acids (in particular proline) [3]. Adjusting osmolyte concentration is energetically costly and slow, and this may partially explain why gradual changes in salinity, rather than sudden changes are associated with broad salinity tolerance [3]. Hypo-salinity can progress quickly: for example, sudden changes might result from heavy rainfall falling directly onto very shallow or even exposed intertidal meadows, or during very sudden and heavy run-off. Under these circumstances, the inability to slowly regulate osmolyte concentrations may cause more cellular damage and result in mortality at higher salinities [3,5,12].

Threshold salinities associated with mortality were different among species with *H. ovalis* being the most sensitive, and *Z. muelleri* the most tolerant of hypo-salinity. We have compared salinity thresholds associated with sub-lethal and lethal impacts from this study with published findings (Table 3). This comparison focuses on mortality or changes in abundance. Since the studies summarized in the comprehensive review by Touchette [3] there has been considerable research effort exploring salinity stress, in particular physiological responses to hyper-salinity stress in *Thalassia testudinum* (e.g. [5,17,28]) and *Posidonia oceanica* (e.g. [18,29,30]). As summarized in Table 3 there are fewer data available on hypo-salinity responses, though where measured, seagrasses do tend to have low hypo-salinity thresholds (Table 3). This detailed experimental design has enabled us to identify salinity thresholds with a high level of precision. A significant outcome from this analysis is the identification of a stress-induced morphometric response indicated in Table 3 as a "sub-lethal" response. Furthermore, our exposure time has exceeded that of many previous studies enabling us to consider sensitivity of seagrasses over 'wet season' time-scales.

The question remains as to why these species tend to be restricted to waters that are predominantly marine when they are clearly tolerant of hypo-salinity. There are a number of possibilities. Firstly, this study was conducted over a 10-week period to represent a hypo-salinity flood event. Exposure to hypo-salinity for longer than 10 weeks could result in higher mortality rates. Secondly, low salinity events tend to coincide with elevated turbidity and nutrients as well as fast water flows. These other environmental impacts, or potentially synergistic impacts (for example, mortality increased with ammonium concentration in *Thalassia testudinum* [11]) could prevent habitation in brackish, riverine environments, rather than salinity itself. Thirdly, reproductive effort was severely impaired at low salinities – although this could only be measured for *H. ovalis*. In some species (e.g. *Ruppia maritima*), seedling germination is enhanced by rapid osmotic shock from hyper to hypo-salinity [31]; however, this study demonstrates that seed production, in these species was inhibited by chronic exposure to hypo-salinity. *Halophila ovalis* is a colonizing species, which is highly dependent on seed production for long-term survival and disruptions to sexual reproduction would probably prevent population survival. Furthermore, if seed production and germination are successful, seedling development is highly sensitive to small changes in salinity [32].

In conclusion, hypo-salinity stress caused a stress-induced morphometric response (SIMR) followed by severe mortality in *H. ovalis* and *H. uninervis* at salinities less than 9 PSU. If observed in natural conditions, a SIMR could suggest that the population is not only healthy, but is in fact in a trajectory of increasing abundance when using traditional monitoring tools, such as shoot density or percent cover. A critical next step is to explore how other interacting factors can affect responses to hypo-salinity.

Acknowledgments

We also thank the staff, in particular B. Lawes and S. Wever at the Marine and Aquaculture Research Facilities Unit at James Cook University for assistance with the experimental set-up. This work was undertaken on plants collected from the Great Barrier Reef World Heritage Area by the James Cook University Authorization MTB41.

Author Contributions

Conceived and designed the experiments: CC CV-R K-jvD MW. Performed the experiments: CV-R CC MT K-jvD. Analyzed the data: CC CV-R K-jvD. Wrote the paper: CC CV-R K-jvD MT MW.

References

1. Les DH, Cleland MA, Waycott M (1997) Phylogenetic studies in Alismatidae, II: Evolution of marine angiosperms (seagrasses) and hydrophily. Syst Bot. 22: 443–463.
2. Janssen T, Bremer K (2004) The age of major monocot groups inferred from 800+rbcL sequences. Bot J Linn Soc. 146: 385–398.
3. Touchette BW (2007) Seagrass-salinity interactions: Physiological mechanisms used by submersed marine angiosperms for a life at sea. J Exp Mar Biol Ecol. 350: 194–215.
4. Lirman D, Cropper WP, Jr. (2003) The influence of salinity on seagrass growth, survivorship, and distribution within Biscayne Bay, Florida: field, experimental, and modeling studies. Estuaries 26: 131–141.
5. Koch MS, Schopmeyer SA, Kyhn-Hansen C, Madden CJ, Peters JS (2007) Tropical seagrass species tolerance to hypersalinity stress. Aquat Bot. 86: 14–24.
6. Walker DI, Kendrick GA, McComb AJ (1988) The distribution of seagrass species in shark bay, Western Australia, with notes on their ecology. Aquat Bot. 30: 305–317.
7. Gacia E, Invers O, Manzanera M, Ballesteros E, Romero J (2007) Impact of the brine from a desalination plant on a shallow seagrass (*Posidonia oceanica*) meadow. Estuarine, Coastal and Shelf Science. 72: 579–590.
8. Furnas M (2003) Catchment and corals: terrestrial runoff to the Great Barrier Reef. Townville Queensland: Australian Institute of Marine Science. 334 p.
9. Preen AR, Long WJL, Coles RG (1995) Flood and cyclone related loss, and partial recovery, of more than 1000 km^2 of seagrass in Hervey Bay, Queensland, Australia. Aquat Bot. 52: 3–17.
10. Campbell SJ, McKenzie LJ (2004) Flood related loss and recovery of intertidal seagrass meadows in southern Queensland, Australia. Estuarine, Coastal and Shelf Science. 60: 477–490.
11. Kahn AE, Durako MJ (2006) *Thalassia testudinum* seedling responses to changes in salinity and nitrogen levels. J Exp Mar Biol Ecol. 335: 1–12.
12. Griffin NE, Durako MJ (2012) The effect of pulsed versus gradual salinity reduction on the physiology and survival of *Halophila johnsonii* Eiseman. Marine Biology. 159: 1439–1447.
13. Fernandez-Torquemada Y, Sanchez-Lizaso JL (2011) Responses of two Meditteranean seagrasses to experimental changes in salinity. Hydrobiologia. 669: 21–33.
14. Chaves MM, Flexas J, Pinheiro C (2009) Photosynthesis under drought and salt stress: regulation mechanisms from whole plant to cell. Annals of Botany. 103: 551–560.
15. Munns R, Tester M (2008) Mechanisms of salinity tolerance. Annu Rev Plant Biol. 59: 651–681.
16. Ralph P (1998) Photosynthetic responses of *Halophila ovalis* (R. Br.) Hook. *f.* to osmotic stress. J Exp Mar Biol Ecol. 227: 203–220.
17. Howarth JF, Durako MJ (2013) Variation in pigment content of *Thalssia testudinum* seedlings in response to changes in salinity and light. Bot Mar. 56: 261–273.
18. Sandoval-Gil J, Marín-Guirao L, Ruiz J (2012) Tolerance of Mediterranean seagrasses (*Posidonia oceanica* and *Cymodocea nodosa*) to hypersaline stress: water relations and osmolyte concentrations. Marine Biology. 159: 1129–1141.
19. Marín-Guirao L, Sandoval-Gil JM, Bernardeau-Esteller J, Ruíz JM, Sánchez-Lizaso JL (2013) Responses of the Mediterranean seagrass *Posidonia oceanica* to hypersaline stress duration and recovery. Mar Environ Res. 84: 60–75.
20. Potters G, Pasternak TP, Guisez Y, Palme KJ, Jansen MAK (2007) Stress-induced morphogenic responses: growing out of trouble? Trends Plant Sci. 12: 98–105.
21. Potters G, Pasternak TP, Guisez Y, Jansen MAK (2009) Different stresses, similar morphogenic responses: integrating a plethora of pathways. Plant, Cell & Environment. 32: 158–169.
22. Zolla G, Heimer YM, Barak S (2010) Mild salinity stimulates a stress-induced morphogenic response in *Arabidopsis thaliana* roots. J Exp Bot. 61: 211–224.
23. Waycott M, McMahon K, Mellors J, Calladine A, Kleine D (2004) A guide to tropical seagrasses of the Indo-West Pacific. Townsville: James Cook University. 72 p.
24. Collier CJ, Waycott M, Giraldo-Ospina A (2012) Responses of four Indo-West Pacific seagrass species to shading. Mar Pollut Bull. 65: 342–354.
25. McMahon KM (2005) Recovery of subtropical seagrasses from natural disturbance. Brisbane: The University of Queensland. 198 p.
26. Short FT, Duarte CM (2001) Methods for the measurement of seagrass growth and production. In: Short FT, Coles R, editors. Global seagrass research methods. Amsterdam: Elsevier Science. pp. 473.
27. Underwood AJ (1997) Experiments in ecology: Their logical design and interpretation using analysis of variance. Cambridge: Cambridge University Press.
28. Jiang Z, Huang X, Zhang J (2013) Effect of nitrate enrichment and salinity reduction on the seagrass *Thalassia hemprichii* previously grown in low light. J Exp Mar Biol Ecol. 443: 114–122.
29. Marín-Guirao L, Ruiz JM, Sandoval-Gil JM, Bernardeau-Esteller J, Stinco CM, et al. (2013) Xanthophyll cycle-related photoprotective mechanism in the Mediterranean seagrasses *Posidonia oceanica* and *Cymodocea nodosa* under normal and stressful hypersaline conditions. Aquat Bot. 109: 14–24.
30. Ruiz JM, Marin-Guirao L, Sandoval-Gil JM (2009) Responses of the Mediterranean seagrass *Posidonia oceanica* to *in situ* simulated salinity increase. Bot Mar. 52: 459–470.
31. Strazisar T, Koch MS, Madden CJ, Filina J, Lara PU, et al. (2013) Salinity effects on *Ruppia maritima* L. seed germination and seedling survival at the Everglades-Florida Bay ecotone. J Exp Mar Biol Ecol. 445: 129–139.
32. Kirkman H, Kuo J (1990) Pattern and process in southern Western Australian seagrasses. Aquatic Botany. 37: 367–382.
33. Walker DI, McComb AJ (1990) Salinity response of the seagrass *Amphibolis antarctica* (Labill.) Sonder et Aschers.: an experimental validation of field results. Aquat Bot. 36: 359–366.
34. Walker DI (1985) Correlations between salinity and growth of the seagrass *Amphibolis antarctica* (labill.) Sonder ex Aschers., In Shark Bay, Western Australia, using a new method for measuring production rate. Aquat Bot. 23: 13–26.
35. Pages JF, Perez M, Romero J (2010) Sensitivity fo the seagrass *Cymodocea nodosa* to hypersaline conditions: a microcosm approach. J Exp Mar Biol Ecol. 386: 34–38.
36. Sandoval-Gil JM, Marín-Guirao L, Ruiz JM (2012) The effect of salinity increase on the photosynthesis, growth and survival of the Mediterranean seagrass *Cymodocea nodosa*. Estuarine, Coastal and Shelf Science. 115: 260–271.
37. Kahn A, Durako MJ (2008) Photophysiological responses of *Halophila johnsonii* to experimental hyposaline and hyper-CDOM conditions. J Exp Mar Biol Ecol. 367: 230–235.
38. Torquemada YF, Durako MJ, Lizaso JLS (2005) Effects of salinity and possible interactions with temperature and pH on growth and photosynthesis of *Halophila johnsonii* Eiseman. Marine Biology. 148: 251–260.
39. Hillman K, McComb AJ, Walker DI (1995) The distribution, biomass and primary production of the seagrass *Halophila ovalis* in the Swan/Canning Estuary, Western Australia. Aquat Bot. 51: 1–54.
40. Fernández-Torquemada Y, Sánchez-Lizaso JL (2005) Effects of salinity on leaf growth and survival of the Mediterranean seagrass *Posidonia oceanica* (L.) Delile. J Exp Mar Biol Ecol. 320: 57–63.
41. Marín-Guirao L, Sandoval-Gil JM, Ruíz JM, Sánchez-Lizaso JL (2011) Photosynthesis, growth and survival of the Mediterranean seagrass *Posidonia oceanica* in response to simulated salinity increases in a laboratory mesocosm system. Estuarine, Coastal and Shelf Science. 92: 286–296.
42. Adams JB, Bate GC (1994) The tolerance to desiccation of the submerged macrophytes *Ruppia cirrhosa* (Petagna) grande and *Zostera capensis* setchell. J Exp Mar Biol Ecol. 183: 53–62.
43. Koch MS, Schopmeyer SA, Holmer M, Madden CJ, Kyhn-Hansen C (2007) *Thalassia testudinum* response to the interactive stressors hypersalinity, sulfide and hypoxia. Aquat Bot. 87: 104–110.

15

Evaluation of Algal Biofilms on Indium Tin Oxide (ITO) for Use in Biophotovoltaic Platforms Based on Photosynthetic Performance

Fong-Lee Ng[1,2], Siew-Moi Phang[1,2]*, Vengadesh Periasamy[3], Kamran Yunus[4], Adrian C. Fisher[4]

1 Institute of Ocean and Earth Sciences, University of Malaya, Kuala Lumpur, Malaysia, **2** Institute of Biological Sciences, Faculty of Science, University of Malaya, Kuala Lumpur, Malaysia, **3** Low Dimensional Materials Research Centre, Department of Physics, Faculty of Science, University of Malaya, Kuala Lumpur, Malaysia, **4** Centre of Research for Electrochemical, Science and Technology (CREST), Department of Chemical Engineering and Biotechnology, University of Cambridge, Cambridge, United Kingdom

Abstract

In photosynthesis, a very small amount of the solar energy absorbed is transformed into chemical energy, while the rest is wasted as heat and fluorescence. This excess energy can be harvested through biophotovoltaic platforms to generate electrical energy. In this study, algal biofilms formed on ITO anodes were investigated for use in the algal biophotovoltaic platforms. Sixteen algal strains, comprising local isolates and two diatoms obtained from the Culture Collection of Marine Phytoplankton (CCMP), USA, were screened and eight were selected based on the growth rate, biochemical composition and photosynthesis performance using suspension cultures. Differences in biofilm formation between the eight algal strains as well as their rapid light curve (RLC) generated using a pulse amplitude modulation (PAM) fluorometer, were examined. The RLC provides detailed information on the saturation characteristics of electron transport and overall photosynthetic performance of the algae. Four algal strains, belonging to the Cyanophyta (Cyanobacteria) *Synechococcus elongatus* (UMACC 105), *Spirulina platensis*. (UMACC 159) and the Chlorophyta *Chlorella vulgaris* (UMACC 051), and *Chlorella* sp. (UMACC 313) were finally selected for investigation using biophotovoltaic platforms. Based on power output per Chl-a content, the algae can be ranked as follows: *Synechococcus elongatus* (UMACC 105) (6.38×10^{-5} Wm^{-2}/μgChl-a)>*Chlorella vulgaris* UMACC 051 (2.24×10^{-5} Wm^{-2}/μgChl-a)>*Chlorella* sp.(UMACC 313) (1.43×10^{-5} Wm^{-2}/μgChl-a)>*Spirulina platensis* (UMACC 159) (4.90×10^{-6} Wm^{-2}/μgChl-a). Our study showed that local algal strains have potential for use in biophotovoltaic platforms due to their high photosynthetic performance, ability to produce biofilm and generation of electrical power.

Editor: Nikolai Lebedev, US Naval Reseach Laboratory, United States of America

Funding: This work was supported by the High Impact Research Grant, University of Malaya Research Grant (J-21002-73823), University of Malaya Postgraduate Research Fund (PG051-2012B), IOES-Cambridge Biophotovoltaic Cell Research Grant (TC004-2010A), Fundamental Research Grant (KPT1059-2012), Malaysian Palm Oil Board (MPOB) Grant (55-02-03-1054) and Knowledge Management Grant (RP0010-13SUS). The funders had no role in study design, data collection and analysis, decision to publish, or preparation of the manuscript.

Competing Interests: The authors have declared that no competing interests exist.

* E-mail: phang@um.edu.my

Introduction

Algae are amongst the most efficient photosynthetic organisms with fast growth rates, diverse products and tolerance to extreme environments. Diatoms, green algae and cyanobacteria (also referred to as the blue-green algae, Cyanophyta) are the major primary producers in the aquatic ecosystem, contributing to carbon dioxide removal, photo-oxygenation and also serving as sources of valuable biochemicals [1]. The biomass productivity of microalgae was estimated to be 50 times higher than switchgrass, which is the fastest growing terrestial plant [2]. With the increased interest in alternative energy sources, algae are being investigated as feedstock for biodiesel, bioethanol, biohydrogen and bioelectricity production [3,4,5]. Microalgae have oil content exceeding 80% DW, grow fast and can be mass cultured using open ponds or enclosed photobioreactors [6]. The filamentous cyanobacterium *Anabaena* was reportedly the first to be used for hydrogen generation [7,8]. In 1997, electricity was generated using 2-hydroxy-1,4-naphthoquinone as an electron shuttle between *Synechococcus* sp. (UTEX 2380) and a carbon-cloth anode [9]. The current intensity increased with increasing cell concentration, reaching 320 μAcm^{-2} [9]. A bioreactor (microbial fuel cell) with an air cathode and a graphite-felt anode coated by a biofilm of bacteria and algae, generated electricity when irradiated. On day 10, the voltage output and current density produced by the reactor were 0.32 V and 8.6 μAcm^{-2} respectively [10].

Biological components have since then been introduced into fuel cells (FCs), giving rise to microbial fuel cells (MFCs). MFCs were designed to operate as a new generation of solar cells called biological photovoltaic devices (Biophotovoltaic, BPV) [11]. Both the MFCs and BPV share similar functions; where MFCs generate electricity from the metabolic process of living microbes, whereas BPV produce electricity from light energy via the light harvesting apparatus of photosynthetic organisms [12]. Previous BPV studies have utilized various exogenous soluble mediator compounds to facilitate electron transfer such as 5 mM ferricyanide as mediator between the biological materials and the anode [11]. In their paper, a new BPV device was fabricated with several advantages over the previous experiments such as multiple microchannels to facilitate multiple simultaneous experiments, removal of the need for external energy supply and a new design to cater for investigation with various types of biological materials instead of solely a single type. Results indicated a direct relationship between effects of cell density, electron mediator concentration and light intensity with the efficiency of the BPV device. The biological materials used in this study were intact *Synechocystis* cells and thylakoid membranes isolated from the cells, which generated total power output of 4.71 and 9.28 nWμmol Chl^{-1} respectively. In another study, McCormick et al. [13] replaced the use of exogenous mediators with the development of biofilms. Several algae were grown directly on an Indium tin oxide- Polyethylene terephthalate (ITO-PET) anode on a sandwich type or an open air design. This work demonstrates the ability to produce simple, portable BPV devices without the need of an artificial electron mediator. The *Synechococcus* biofilm in this work produced a peak power output at 1.03×10^{-2} Wm^{-2} under 10 Wm^{-2} of white light [13]. Biofilm development offers several advantages by increased power output due to direct contact between cell and electrode and reduced internal potential losses. In 1684, Antonie van Leeuwenhoek [14] reported the existence of animalcules on teeth, representing the first scientific report on biofilms. Microorganisms are able to attach to surfaces and form a hydrated polymeric matrix termed "extra-cellular polymeric substances (EPS)" that hold the biofilm together [15]. In an EPS matrix, there are interstitial water channels, which separate bacterial cells from each other, thereby allowing the transportation of nutrients, oxygen and genes [16]. Biofilms composed of microorganisms attached to surfaces, form a hydrated polymeric matrix consisting of polysaccharides, protein and nucleic acids [15]. There is a growing interest in the study of artificial phototrophic biofilms. In biofuel production, the cultivation of algae as biofilms reduces costs due to reduction of water volume and power required for pumping. This increases biomass and also avoids costly harvesting and dewatering technologies [17]. Traditional methods of monitoring changes in biomass of microphytobenthic biofilms are destructive, for example, pigment sampling by using syringe coring method. This method involves pulling the syringe plunger up while pushing the syringe to the sediment to create counter pressure [18]. Fluorescence techniques have been employed to measure photosynthetic performance of biofilms in sediments without interfering with the sediment surface [19]. Radiant energy absorbed by chlorophyll can undergo one of three fates: (i) used for photosynthesis (ii) dissipated as heat or (iii) re-emitted as chlorophyll fluorescence [20]. Hence, by measuring the yield of chlorophyll fluorescence, information about the efficiency of photochemistry and heat dissipation can be generated using a pulse amplitude modulation (PAM) fluorometer (Diving PAM, Walz, Germany) [21]. Photosynthesis parameters such as electron transport rate (ETR) can be calculated to measure the efficiency of photochemistry of Photosystem II (PSII) [22].

In a recent paper, Luimstra et al. [23] produced a cost effective microbial fuel cell design that can be used on algae and cyanobacteria. Several strains of benthic cyanobacteria were screened and displayed electrogenic qualities suitable for microbial fuel cell purpose. The paper also describes a particularly good green alga, *P. pseudovolvox* that demonstrated greater electrogenic activity compared to the others. However, they also highlighted that the mechanisms used in the microorganisms to donate electrons have to be determined at this moment of time. Ng et al. [24] reported that *Chlorella* sp. (UMACC 313) and *Spirulina platensis* (UMACC 159) are able to form biofilms on ITO anode. *Chlorella* and *Spirulina* formed biofilms with coverage and maximum relative electron transport rate (rETRmax) of 99.46% and 140.796 μmol electrons m^{-2} s^{-1} and 80.70% and 153.507 μmol electrons m^{-2} s^{-1} respectively. Results indicate the potential for generating electrical energy from these microalgae using biophotovoltaic platforms.

The objectives of the present study are to (i) establish libraries of 16 strains of algae for selection of most suitable strains for application in biophotovoltaic platforms; (ii) investigate biofilm formation of selected algae on Indium tin oxide (ITO) and glass and determine the photosynthetic efficiency of the biofilms; (iii) Produce and test biofilms on Indium tin oxide (ITO) in a biophotovoltaic (BPV) device for electric power generation. Glass and ITO were chosen as substrates for biofilm formation to compare the adhesion of algae cells to these surfaces. Glass is hydrophilic with high surface energies and ITO is hydrophobic with low surface energy respectively [25]. ITO was an early favorite for hole injection cathode with good transparency and conductivity [26]. ITO was also successfully used to cultivate healthy biofilms for MFCs application [13].

Materials and Methods

Ethics statement

Not relevant. Only microalgae were used in this study.

Algae cultures

Fourteen local tropical algal strains from the University of Malaya Algae Culture Collection (UMACC) [27] and two strains from the Culture Collection of Marine Phytoplankton (CCMP), USA were screened for growth rate, biochemical composition and photosynthetic performance to be used for compiling the algal libraries of potential candidates for BPV platforms. The Cyanobacteria are treated as blue-green algae and will be referred to as Cyanophyta in this paper. All cultures were grown in Bold's Basal Medium for Chlorophyta, the green algae [28], Prov Medium for marine algae [27], Kosaric Medium for Cyanophyta, the blue-green algae [27] and f/2 Medium [29] for marine diatoms (Table 1). An inoculum size of 20%, standardized at an optical density at of 0.2 at 620 nm ($OD_{620\ nm}$) from exponential phase cultures was used. The cultures were grown in 1 L conical flasks in an incubator shaker (120 rpm) at 25°C, with irradiances of 30 μmol photons m^{-2} s^{-1}on a 12:12 light dark cycle. Each microalga was grown in triplicate with a total volume of 500 ml. Growth was monitored based on $OD_{620\ nm}$; which has a high correlation ($r^2 = 0.9$) with chlorophyll a (Chl-a) [30]. In the present study, $OD_{620\ nm}$ was strongly correlated to Chl-a content (r^2 ranged from 0.9102 to 0.9877) for the 16 strains (2 Cyanophytes, 2 diatoms, 14 Chlorophytes) used.

Algal libraries

Growth and biochemical profiling. The specific growth rate (μ, day^{-1}) for all cultures were based $OD_{620\ nm}$ and calculated using the following formula:

$$\mu,day^{-1} = (\ln N_2 - N_1)/(t_2 - t_1)$$

where N_2 is $OD_{620\ nm}$ at t_2; N_1 is $OD_{620\ nm}$ at t_1, and t_2 and t_1 are time periods within the exponential phase [31].

The algal samples were harvested at the end of the experiment (stationary phase) by Millipore filtration using glass fibre filter paper (Whatman GF/C, 0.45 μm) for determination of dry weight and extraction of biochemicals. For dry weight (DW) determina-

Table 1. List of strains used.

Strain	*Habitat	Origin	#Medium
Cyanophyta (Blue-green algae)			
Synechococcus elongatus Nageli UMACC 105	FW	Contaminant of *Spirulina* sp. culture	Kos
Spirulina platensis (Arthrospira)(Gomont)Geitler UMACC 159	FW	Israel	Kos
Chlorophyta (Green algae)			
Chlorella vulgaris Beijerinck UMACC 001	FW	Fish pond at IPSP Farm, University of Malaya	BBM
Scenedesmus sp. Meyen UMACC 036	FW	Fish tank (Tilapia), IPSP Farm, University of Malaya	BBM
Scenedesmus quadricauda (Turpin) Berbisson UMACC 041	FW	Fish tank (Tilapia), IPSP Farm, University of Malaya	BBM
Chlorella vulgaris Beijerinck UMACC 051	FW	POME Aerobic pond, Batang Berjuntai, Selangor, Malaysia	BBM
Scenedesmus sp. Meyen UMACC 068	FW	Fish tank with goat dung, IPSP Farm, University of Malaya	BBM
Oocystis sp. Nageli et A.Braun UMACC 074	FW	Painted Masonry, Science Faculty, University of Malaya	BBM
Chlorococcum oviforme Archibald et Bold UMACC 110	FW	Pond at the IPSP farm, University of Malaya	BBM
Chlorococcum sp. Meneghini UMACC 207	FW	Plastic container, shop houses, Johor Bahru, Malaysia	BBM
Chlorella sp. Beijerinck UMACC 255	M	Sea Bass Pond at Sepang, Selangor, Malaysia	Prov
Chlorella sp. Beijerinck UMACC 256	M	Sea Bass Pond at Sepang, Selangor, Malaysia	Prov
Chlorella sp. Beijerinck UMACC 258	M	Sea Bass Pond at Sepang, Selangor, Malaysia	Prov
Chlorella sp. Beijerinck UMACC 313	FW	POME Anaerobic Pond, Labu Palm Oil Mill, Malaysia	BBM
Bacillariophyta (Diatoms)			
Coscinodiscus granii Gough CCMP 1817	M	University of Rhode Island, Rhode Island USA	f/2
Coscinodiscus wailesii Gran et Angst CCMP 2513	M	Atlantic Ocean (off Georgia Coast, USA)	f/2

*FW: Fresh water; M: Marine.
#Kos: Kosaric Medium (modified after Zarrouk,1966); BBM: Bold's Basal Medium (Nichols and Bold, 1965); Prov: Prov Medium (CCMP, 1996); f/2: f/2 Medium (Guillard and Ryther, 1962).

tion, a known volume of the culture was filtered onto an oven-dried pre-weighed glass fibre filter, which was then dried in an oven at 100°C for 24 h. The DW was calculated as follows:

$$DW(mg.L^{-1}) = \frac{\text{Weight of filter with algae (mg)} - \text{Weight of blank filter (mg)}}{\text{Volume of culture (L)}}$$

The biomass at stationary phase was determined as dry weight (B_{DW}) as well as calculated from the Chl-a content ×67 (B_{CHL}), assuming that Chl-a makes up 1.5% of the cell biomass [32,33]. Protein content of cells was determined by the dye-binding method after extraction in 0.5 N NaOH [34]. Carbohydrates extracted from the cells in 2N HCL were determined using the phenol-sulphuric acid method [35]. Lipids were extracted in MeOH-$CHCl_3$-H_2O (2:1:0.8) and determined by gravimetric method [36]. These biochemicals were expressed as %B_{CHL}. This may avoid errors with the diatoms that contain silicate frustules but since the content of Chl-a may vary with taxa, these values can only be considered as estimates for comparison between strains.

Pulse amplitude modulation (PAM) fluorometer measurement of 16 algal strains. Photosynthetic parameters were measured fluorometrically using a Diving-PAM (Walz, Germany) [37,21,38]. Data provided by the PAM based on fluorescence is useful for assessing the performance of a photosynthetic microbial fuel cell. Inglesby et al. [39] showed that in a study using *Arthrospira maxima*, the use of *in situ* fluorescence detection allowed for a direct correlation between photosynthetic activity and current density. Rapid light curves (RLC) were obtained under software control (Wincontrol, Walz). Red light emitting diodes (LEDs) provided the actinic light used in the RLC at the level of 0, 307, 426, 627, 846, 1267, 1829, 2657 and 4264 μmol photons m^{-2} s^{-1}. The cultures of each strain were dark-adapted for 15 minutes before exposure to each light level for 10 seconds. Maximum quantum efficiency (F_v/F_m), a parameter to indicate the physiological state of phytoplankton was used to indicate if the cells were stressed by the exposure to light: $F_v/F_m = (F_m - F_0)/F_m$ where F_m is the maximum fluorescence and F_0 is the minimum fluorescence resulting in the variable fluorescence F_v. The maximum photosynthetic efficiency was determined from the initial slope (α) of the RLC. The relative electron transport rate (rETR) was calculated by multiplying the irradiance by quantum yield measured at the end of each light interval. The RLC consists of eight consecutive ten-second intervals of actinic light with increasing intensity. The photo-adaptive index (E_k) is obtained from the curve fitting model [40]. The interception point of the alpha (α) value with the maximum photosynthetic rate (rETRmax) is defined as: $E_k = rETRmax/\alpha$. The Non-Photochemical Quenching (NPQ) reflects the ability of a cell to dissipate excess light energy harvested during photosynthesis as heat and is used as an indicator of photoprotection. NPQ is calculated as (F_m'). NPQ=(F_m-F_m')/F_m'. F_m is the maximum fluorescence yield during the saturating flash and F_m' is the maximum fluorescence in the light-adapted state during the saturating flash. All statistical analyses were performed using the Statistica 8 program. Eight strains were selected according to photosynthetic performance based on F_v/F_m, rETRmax, alpha, E_k and NPQ. Different algal types and different habitats are also considered as factors that influence the selection of strain.

Growth and photosynthetic efficiency of algal biofilms

Eight strains namely the Cyanophytes *Synechoccus elongatus* (UMACC 105), *Spirulina platensis* (UMACC 159), the Chlorophytes *Chlorella vulgaris* (UMACC 001), *Chlorella vulgaris* (UMACC 051), *Chlorococcum* sp. (UMACC 207), *Chlorella* sp. (UMACC 256), *Chlorella* sp. (UMACC 313) and the diatom *Coscinodiscus wailesii* (CCMP 2513) were used for the biofilm studies. 100 ml of exponential phase cultures of $OD_{620\ nm}=0.5$ were used. Each culture was placed into a 200 ml autoclaved glass staining jar. ITO coated glass slides (purchased from KINTEC, Hong Kong) and glass slides measuring 20×20 mm were placed in the staining jar with the microalgae culture and transferred into an incubator at 24°C illuminated by cool white fluorescent lamps (30 μmol m^{-2} s^{-1}) on a 12:12 hour light-dark cycle to allow for the algae biofilms to form on the slides. This experiment was conducted in triplicates. The biofilm growth was monitored by photographing the slide surface with a Sony Cyber-Shot DSC-WX30 Camera every three days until the slides were completely covered by the biofilm. The surface area coverage, SAC (%) of the biofilm captured in the photograph was calculated using ImageJ software [41]. At the end of the experiment (day 15), the biofilm thickness of each slide was measured using Elcometer 3230 Wet Film Wheels [42]. The wheel was held using a finger and thumb by its centre and the wheel was placed on the wet film ensuring that it was perpendicular to the algal film. The wheel was rolled across the algal film through an angle of 180° and then removed from the surface. The thickness of the biofilms was recorded based on the scale on the side of the wheel. The biofilms were removed by washing using jets of distilled water from a pipette, into a sterile beaker for extracting biomass for determination of Chl-a content. The microalgae cells were then harvested by millipore filtration using filter paper (Whatman GF/C, 0.45 μm) and the Chl-a of the eight strains were extracted using acetone [31].The absorption of the extract was measured at 665 nm, 645 nm and 630 nm and the Chl-a content calculated using the formula below:

$$\text{Chl-a}(\text{mgm}^{-3}) = (\text{Ca} \times \text{Va})/\text{Vc}$$

Where,

$$\text{Ca} = 11.6 \times OD_{665\ nm} - 1.31 \times OD_{645\ nm} - 0.14 \times OD_{630\ nm}$$

$$\text{Va} = \text{Volume of acetone (mL) used for extraction}$$

$$\text{Vc} = \text{Volume of culture (L)}$$

$$\text{Chl-a}(\text{mg/L}) = \text{Chl-a}(\text{mgm}^{-3})/1000$$

Pulse amplitude modulation (PAM) fluorometer measurement of biofilms

Photosynthetic parameters were measured fluorometrically using a Diving-PAM (Walz, Germany) as described above [21,38]. RLC were obtained under software control (Wincontrol, Walz). Red light emitting diodes (LEDs) provided the actinic light used in the RLC at the level of 0, 127, 205, 307, 426, 627, 846, 1267 and 1829 μmol photons m^{-2} s^{-1}. The biofilm of each ITO slide on day 15 was dark adapted for 15 minutes prior to the exposure to each light level for 10 seconds.

BPV set up and electrical measurement

The BPV devices used in these studies was provided by our collaborators from the University of Cambridge. The closed, single-chamber BPV consisted of a 50×50 mm platinum-coated glass cathode placed in parallel with 35×35 mm ITO coated glass with biofilm grown on the surface (10 mm apart) in a clear Perspex chamber sealed with polydimethylsiloxane (PDMS) and then filled with medium (Figure 1 a–d). The body of the open-air, single-chamber BPV was constructed of clear Perspex [11,13].

Biofilms of the two strains of Cyanophytes, the *Synechoccus elongatus* (UMACC 105) and *Spirulina platensis* (UMACC 159), and the two Chlorophytes *Chlorella vulgaris* (UMACC 051), and *Chlorella* sp. (UMACC 313) grown on ITO, were placed in the devices and the experiment conducted in triplicates. Crocodile clips and copper wire served as connection between anode and cathode to the external circuit. Prior to operation, the chambers were filled with fresh medium (Bold's Basal medium and Kosaric medium) and maintained at 25°C with irradiance of 30 μmol photons m^{-2} s^{-1} for the duration of the experiments. Current outputs were measured using a multimeter (Agilent U1251B). Polarization curves were generated for each strains by applying different resistance (10 MΩ, 5.6 MΩ, 2 MΩ, 560 KΩ, 240 KΩ, 62 KΩ, 22 KΩ, 9.1 KΩ, 3.3 KΩ and 1.1 KΩ) loads to the external circuit. All experiments were conducted in triplicates.

Results

Algae libraries

Growth and biochemical profiling. Table 2 shows the growth rates, biochemical characteristics (protein, lipid, carbohydrate) and PAM data of the 16 strains. Specific growth rate ranged from 0.18 to 0.56 d^{-1}. The fastest growing algae were the Chlorophytes, *Chlorella* sp. (UMACC 258) (0.58 d^{-1}), *Scenedesmus quadricauda* (UMACC 041) (0.37 d^{-1}), *Scenedesmus* sp. (UMACC 068) (0.36 d^{-1}) and the diatom *Coscinodiscus granii* (CCMP 1817) (0.36 d^{-1}). Stationary phase was reached between 9 to 20 days with biomass (B_{CHL}) ranging from 139.70±15.70 μg.ml^{-1} (*Chlorella vulgaris* UMACC 051) to 1158.77±30.28 μg.ml^{-1} (*Spirulina platensis* UMACC 159). The lipid, carbohydrate and protein contents were calculated based on biomass (stationary phase culture) from Chl-a (B_{CHL}). Lipid content ranged from 24.92±6.07 (*Chlorococcum* sp.) to 63.64±2.20% DW (*Chlorella* sp. UMACC 256). Carbohydrate content ranged from 2.36±0.10% (*Spirulina platensis* UMACC 159) to 18.77±1.58% DW (*Chlorococcum oviforme* UMACC 207). Protein ranged from 14.69±1.18 (*Chlorella* sp. UMACC 256) to 53.74±4.28% DW (*Chlorella vulgaris* UMACC 051). In general, the Chlorophytes, especially strains of *Chlorella*, had higher lipid and protein contents than the diatoms and Cyanophytes.

Pulse amplitude modulation (PAM) fluorometer measurement of 16 strains. Maximum relative electron transport rate (rETRmax) ranged from 64.68 μmol electrons m^{-2} s^{-1} to 166.32 μmol electrons m^{-2} s^{-1}. The highest rE-

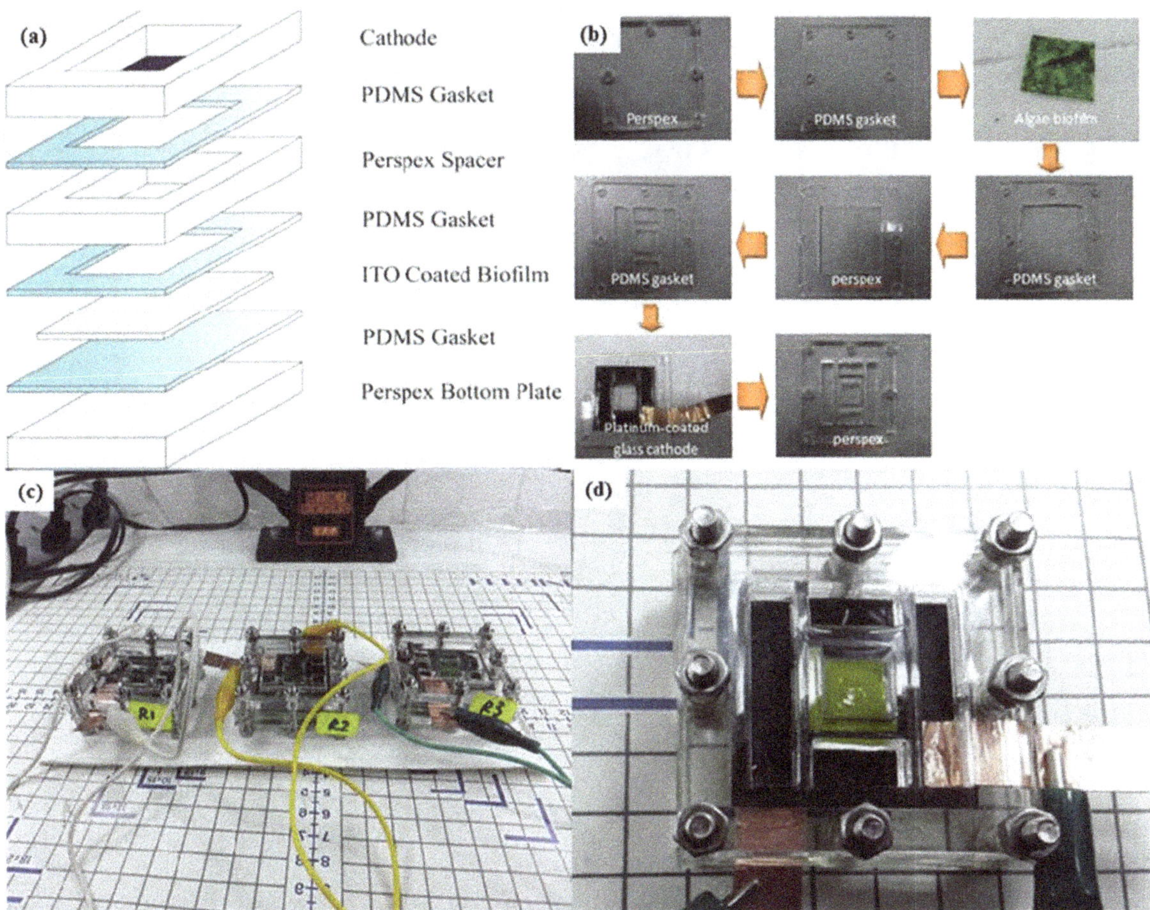

Figure 1. The construction of the Biophotovoltaic (BPV) device (a) Exploded view of a BPV device, (b) Stepwise set-up of the biophotovoltaic device, (c) BPV devices set up in triplicate, (d)Top view of a working BPV device with an algal biofilm formed on an ITO anode.

TRmax were observed from the diatoms *Coscinodiscus wailesii* (CCMP 2513) (166.32 µmol electrons m^{-2} s^{-1}) and *Coscinodiscus granii* (CCMP 1817) (161.60 µmol electrons m^{-2} s^{-1}), followed by the Chlorophyte *Chlorococcum* sp. (UMACC 207) (154.07 µmol electrons m^{-2} s^{-1}) and the two Cyanophytes *Synechococcus elongatus* (UMACC 105) (147.50 µmol electrons m^{-2} s^{-1}) and *Spirulina platensis* (UMACC 159). High rETRmax was correlated to high photoadaptive index (E_k) for all five strains above, but not to the index of light efficiency, α, and the maximum quantum efficiency of PSII (F_v/F_m). The strains that reached stationary phase earlier (Chlorophytes) had higher F_v/F_m values than the Cyanophytes and diatoms that took 14 to 20 days to reach stationary phase. The α values varied between 0.31±0.01 (*Synechococcus elongatus* UMACC 105) to 0.88±0.12 (*Chlorella* sp. UMACC 256). The photoadaptive index, E_k varied from 84.48±1.46 µmol photons m^{-2} s^{-1} (*Scenedesmus* sp. UMACC 036) to 473.86±79.05 µmol photons m^{-2} s^{-1} (*Synechococcus elongatus* UMACC 105). Maximum quantum yields, F_v/F_m varied from 0.43±0.02 (*Spirulina platensis* UMACC 159) to 0.80±0.02 (*Oocystis* sp. UMACC 074).

For all 16 strains, NPQ presented at low irradiances (307 µmol photons m^{-2} s^{-1}) (Figure 2). The NPQ values that indicate photoprotection were very low or not observed in both the Cyanophytes and diatoms. Most strains had NPQ values lower than 0.43 except for four Chlorophyte strains. *Scenedesmus* sp. (UMACC 036), *Scenedesmus* sp. (UMACC 068), *Chlorella vulgaris* (UMACC 001) and *Scenedesmus quadricauda* (UMACC 041) that registered highest NPQ values at 0.877, 0.909, 1.06 and 1.114 respectively. *Chlorella vulgaris* (UMACC 001) reached highest NPQ at 2657 µmol photons m^{-2} s^{-1} and showed significant reduction (0.866) at 4264 µmol photons m^{-2} s^{-1}, indicating that the photoprotective level started to decrease when coping with extremely high actinic light.

Growth and photosynthetic efficiency of biofilms

The biofilms were grown on glass and ITO slides in triplicates. Table 3 shows the surface area coverage (% SAC) of biofilms produced by the eight strains on two different substrates (glass and ITO) on day 3, 6, 9, 12 and 15. The % SAC was monitored every three days. Visible growth of the biofilms was already observed on day 3 except for *Coscinodiscus wailesii* (CCMP 2513) which did not show any biofilm growth on both the glass and ITO slides until day 12 and registered the lowest % SAC among the eight strains. After three days of incubation, the Cyanophytes *Synechococcus elongatus* (UMACC 105) and *Spirulina platensis* (UMACC 159) had formed appreciable biofilms on both glass and ITO. Of the Chlorophytes, the *Chlorella* (UMACC 313) from POME pond and the freshwater *Chlorella* (UMACC 051) had high % SAC as well. All strains reached their maximum % SAC after 15 days of inoculation. *Chlorella* sp. (UMACC 313) showed fastest biofilm coverage, having achieved around 80% SAC on day

Table 2. Statistical comparison of biochemical profile and PAM data of 16 algal strains, data as means ±S.D. (n = 3).

Strain	Specific Growth Rate (μ,day −1)	*Stat. Phase (day)	Biomass at stationary phase based on dry weight (BDW) (μg/ml)	Biomass at stationary phase based on Chl- a content ×67 (B_{CHL}) (μg/ml)	#Lipid (% dry weight)	#δCarbohy. (% dry weight)	#Protein (% dry weight)	rETRmax (μmol photons m^{-2} s^{-1})	Alpha (α)	E_k (μmol photons m^{-2} s^{-1})	F_v/F_m
Synechococcus elongatus UMACC 105	0.25 ± 0.02^{b}	17	$325.82\pm23.04^{d,e}$	208.37 ± 4.39^{a}	26.48 ± 1.68^{f}	$7.44\pm0.39^{a,b}$	44.93 ± 3.40^{a}	$147.50\pm18.42^{a,b,c}$	0.31 ± 0.01^{e}	473.86 ± 79.05^{a}	0.46 ± 0.02^{b}
Spirulina platensis UMACC 159	0.24 ± 0.02^{b}	20	1158.77 ± 30.28^{a}	330.76 ± 2.54^{a}	$53.27\pm7.45^{a,b,c}$	2.36 ± 0.10^{b}	40.44 ± 9.34^{a}	$147.61\pm6.80^{a,b,c}$	0.39 ± 0.01^{e}	$380.77\pm22.74^{a,d}$	0.43 ± 0.02^{b}
Chlorella vulgaris UMACC 001	0.18 ± 0.02^{d}	10	226.69 ± 13.38^{f}	127.52 ± 5.37^{f}	$45.52\pm6.21^{c,d,e}$	$4.30\pm1.59^{a,b}$	$36.87\pm1.46^{a,b,c}$	$68.53\pm0.46^{e,f}$	$0.74\pm0.02^{a,b,c,d}$	93.20 ± 2.33^{f}	0.76 ± 0.02^{a}
Scenedesmus sp. UMACC 036	0.27 ± 0.02^{b}	9	$349.92\pm27.30^{c,d}$	191.40 ± 4.75^{c}	$43.28\pm3.88^{c,d,e}$	$6.32\pm1.70^{a,b}$	$38.98\pm5.43^{b,c,d}$	64.68 ± 11.44^{f}	0.84 ± 0.01^{a}	84.48 ± 1.46^{f}	0.78 ± 0.01^{a}
Scenedesmus quadricauda UMACC 041	0.37 ± 0.03^{c}	10	$270.17\pm16.68^{e,f}$	123.50 ± 7.85^{f}	$42.45\pm6.20^{c,d,e}$	$4.26\pm0.58^{a,b}$	$35.14\pm5.62^{a,b,c}$	$80.44\pm20.78^{d,e,f}$	$0.70\pm0.10^{b,c,d,e}$	$127.89\pm54.32^{d,e,f}$	$0.75\pm0.03^{a,b}$
Chlorella vulgaris UMACC 051	0.25 ± 0.02^{c}	14	139.70 ± 15.07^{g}	127.52 ± 4.94^{f}	$34.09\pm2.47^{e,f}$	$4.31\pm0.59^{a,b}$	53.74 ± 4.28^{a}	$118.21\pm20.51^{b,c,d}$	$0.49\pm0.01^{c,d,e}$	$240.26\pm48.22^{c,d,e}$	$0.60\pm0.00^{a,b}$
Scenedesmus sp. UMACC 068	0.36 ± 0.03^{b}	9	$393.67\pm15.57^{b,c}$	238.74 ± 3.44^{b}	$49.09\pm1.38^{b,c,d}$	$5.92\pm1.82^{a,b}$	$30.45\pm6.60^{a,b}$	$71.16\pm1.98^{e,f}$	$0.76\pm0.01^{a,b,c}$	94.08 ± 3.84^{f}	0.77 ± 0.01^{a}
Oocystis sp. UMACC 074	0.34 ± 0.02^{c}	9	439.61 ± 18.66^{b}	190.95 ± 5.84^{c}	$41.84\pm5.11^{c,d,e}$	$7.01\pm1.73^{a,b}$	$36.87\pm2.96^{a,b,c}$	$97.91\pm17.23^{d,e,f}$	0.85 ± 0.04^{a}	$115.81\pm27.06^{d,e,f}$	0.80 ± 0.02^{a}
Chlorococcum oviforme UMACC 110	0.34 ± 0.02^{b}	10	257.78 ± 19.24^{f}	$136.90\pm7.23^{e,f}$	$43.11\pm4.22^{c,d,e}$	$6.76\pm1.64^{a,b}$	$38.72\pm1.88^{a,b}$	$84.82\pm7.58^{d,e,f}$	0.82 ± 0.04^{a}	104.47 ± 13.80^{f}	0.74 ± 0.00^{a}
Chlorococcum sp. UMACC 207	0.23 ± 0.02^{c}	15	240.49 ± 14.18^{f}	148.96 ± 4.85^{e}	24.92 ± 6.07^{f}	18.77 ± 1.58^{a}	48.92 ± 2.94^{a}	$154.07\pm22.00^{a,b,c}$	$0.51\pm0.05^{b,c,d,e}$	$302.60\pm48.08^{b,c}$	$0.57\pm0.02^{a,b}$
Chlorella sp. UMACC 255	0.33 ± 0.02^{b}	9	$362.31\pm23.59^{c,d}$	119.26 ± 8.23^{f}	$51.65\pm1.05^{a,b,c}$	$5.56\pm0.53^{a,b}$	$26.56\pm2.68^{b,c}$	$86.60\pm11.88^{d,e,f}$	$0.78\pm0.03^{a,b,c}$	$111.69\pm20.62^{d,e,f}$	0.76 ± 0.01^{a}
Chlorella sp. UMACC 256	0.23 ± 0.02^{a}	9	$352.00\pm37.28^{c,d}$	149.86 ± 5.62^{e}	63.64 ± 2.20^{a}	$6.62\pm0.88^{a,b}$	$15.12\pm3.28^{b,c}$	$94.06\pm10.87^{d,e,f}$	0.88 ± 0.12^{a}	$108.58\pm22.07^{e,f}$	0.77 ± 0.04^{a}
Chlorella sp. UMACC 258	0.56 ± 0.02^{c}	9	$337.88\pm27.85^{c,d}$	173.08 ± 9.24^{d}	$58.98\pm5.04^{a,b,c}$	3.92 ± 0.20^{b}	14.69 ± 1.18^{c}	$85.51\pm13.87^{d,e,f}$	$0.80\pm0.04^{a,b}$	$107.52\pm23.23^{e,f}$	0.76 ± 0.01^{a}
Chlorella sp. UMACC 313	0.33 ± 0.02^{b}	14	$349.29\pm15.07^{c,d}$	194.52 ± 2.05^{c}	$50.65\pm4.06^{a,b,c}$	2.68 ± 1.15^{b}	$33.17\pm2.24^{a,b,c}$	$109.98\pm9.23^{c,d,e}$	$0.45\pm0.03^{d,e}$	$245.55\pm31.79^{c,d}$	0.47 ± 0.04^{b}
Coscinodiscus granii CCMP 1817	0.36 ± 0.01^{b}	14	$338.71\pm18.42^{c,d}$	78.84 ± 7.35^{g}	$36.76\pm0.91^{d,e,f}$	16.79 ± 2.82^{a}	40.87 ± 1.23^{a}	$161.60\pm23.41^{a,b}$	0.41 ± 0.04^{e}	$403.25\pm93.92^{a,b}$	0.45 ± 0.03^{b}
Coscinodiscus wailesii CCMP 2513	0.33 ± 0.01^{c}	15	$372.10\pm12.83^{c,d}$	94.02 ± 3.93^{g}	$46.07\pm2.97^{b,c,d,e}$	17.35 ± 2.87^{a}	$31.22\pm3.24^{a,b,c}$	166.32 ± 13.57^{a}	0.40 ± 0.03^{e}	$425.18\pm71.67^{a,d}$	0.45 ± 0.02^{b}

Differences between alphabets indicate significant difference between different strains. (ANOVA, Turkey HSD test, $p<0.05$). Alpha is photosynthetic efficiency and indicates the amount of ETR per photon. E_k is the photoadaptive index and indicates how well cells are adapted to their light environment.

*Stat.: Stationary;

δCarbohy.;

#Values based on biomass (B_{CHL}) calculated from (Chl-a ×67).

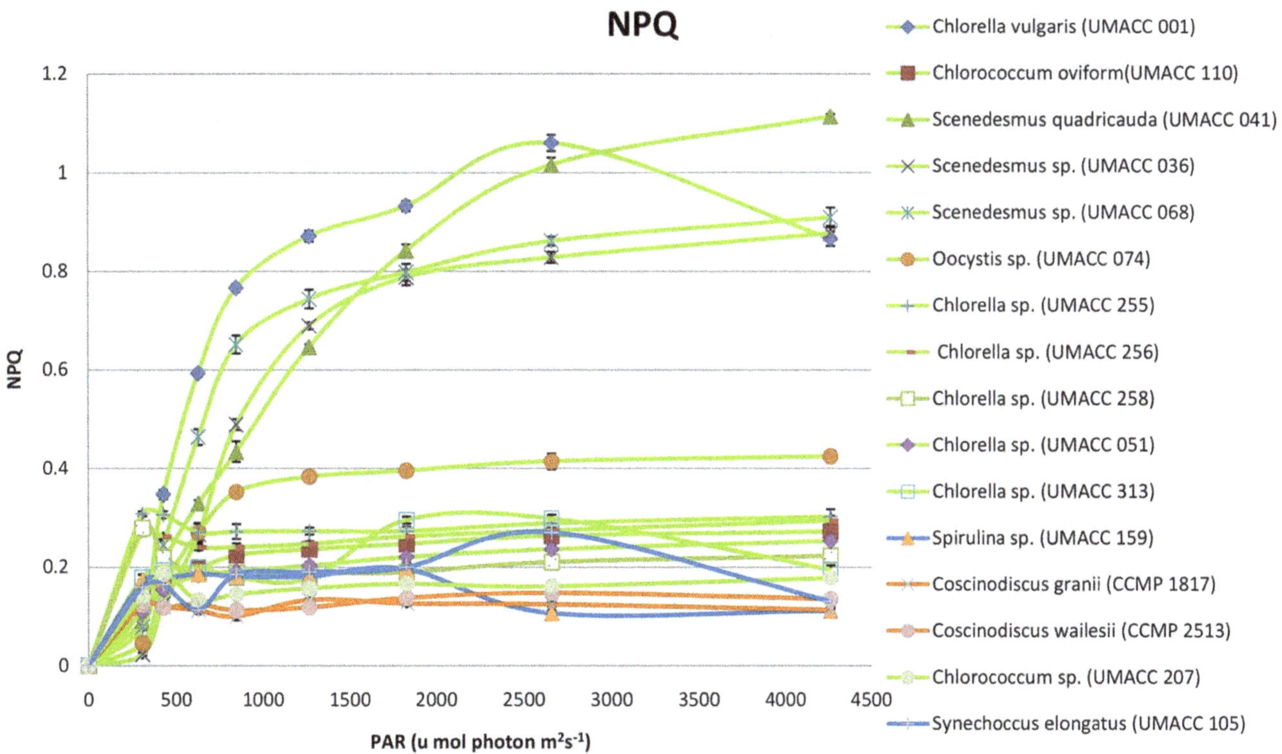

Figure 2. Non-photochemical quenching of sixteen microalgae strains (n = 3).

6 with both glass and ITO. It achieved highest biofilm coverage with 98.18% SAC on glass and 98.69% SAC on ITO on day 15.

Table 4 gives details of the biofilms and the PAM data on day 15 of the study. The fluorescence characteristics of the biofilms from the eight strains exposed to a similar range of irradiance were

Table 3. Statistical comparison of surface are coverage (% of eight strains on ITO and glass slides on day 3, 6, 9, 12 and 15, data as means ±S.D. (n = 3).

Strain	Substrate Material	Surface Area Coverage (%)				
		Day 3	Day 6	Day 9	Day 12	Day 15
Synechococcus elongatus	Glass	$58.74 \pm 7.08^{g,h,i}$	$59.70 \pm 1.18^{g,h,i}$	$76.25 \pm 0.90^{e,f,g}$	$85.90 \pm 6.25^{b,c,d}$	84.68 ± 4.16^{b}
UMACC 105	ITO	$65.95 \pm 6.56^{g,h}$	$77.38 \pm 2.82^{d,e,f}$	$81.9 \pm 2.12^{d,e,f}$	$82.66 \pm 2.78^{d,e,f}$	96.07 ± 1.69^{a}
Spirulina platensis	Glass	$23.59 \pm 3.13^{p,q}$	$34.26 \pm 2.39^{m,n}$	$53.97 \pm 7.74^{i,j,k}$	$51.51 \pm 3.93^{i,j,k}$	75.56 ± 6.45^{c}
UMACC 159	ITO	$38.90 \pm 1.47^{l,m,n}$	$78.83 \pm 1.38^{d,e,f}$	$84.95 \pm 2.00^{c,d,e}$	$88.40 \pm 1.21^{b,c,d}$	97.57 ± 0.77^{a}
Chlorella vulgaris	Glass	$14.74 \pm 0.79^{p,q,r}$	$29.36 \pm 1.77^{o,p}$	$40.79 \pm 0.97^{l,m,n}$	$48.77 \pm 1.41^{i,j,k}$	$55.83 \pm 1.597^{d,e}$
UMACC 001	ITO	$13.58 \pm 0.91^{q,r}$	$32.90 \pm 1.83^{m,n,o}$	$43.56 \pm 1.31^{k,l,m}$	$51.68 \pm 0.58^{j,k,l}$	56.57 ± 2.11^{d}
Chlorella vulgaris	Glass	$44.90 \pm 2.12^{j,k,l}$	$63.38 \pm 6.19^{g,h}$	$70.65 \pm 3.06^{f,g}$	$77.69 \pm 1.69^{d,e,f}$	95.37 ± 1.36^{a}
UMACC 051	ITO	$44.58 \pm 1.19^{k,l,m}$	$62.35 \pm 1.83^{g,h,i}$	$70.57 \pm 1.45^{f,g}$	$79.21 \pm 1.93^{d,e,f}$	93.32 ± 1.33^{a}
Chlorococcum sp.	Glass	9.56±0.76q,r	$18.42 \pm 1.48^{p,q,r}$	$24.61 \pm 2.49^{o,p}$	$36.30 \pm 1.31^{l,m,n}$	45.98 ± 1.50^{f}
UMACC 207	ITO	$9.06 \pm 0.32^{r,s}$	$18.19 \pm 1.16^{p,q,r}$	$22.41 \pm 1.34^{o,p,q}$	$34.05 \pm 1.72^{m,n}$	46.71 ± 1.12^{f}
Chlorella sp.	Glass	$30.78 \pm 1.07^{o,p}$	$46.17 \pm 10.80^{j,k,l}$	$41.22 \pm 1.00^{l,m}$	$46.42 \pm 0.58^{j,k,l}$	$49 \pm 1.56^{e,f}$
UMACC 256	ITO	$31.59 \pm 2.51^{m,n,o}$	$39.36 \pm 0.55^{l,m,n}$	$40.66 \pm 0.78^{l,m,n}$	$46.06 \pm 0.67^{j,k,l}$	62.79 ± 2.18^{d}
Chlorella sp.	Glass	$66.08 \pm 1.51^{g,h}$	$80.28 \pm 0.38^{d,e,f}$	$88.19 \pm 1.08^{b,c,d}$	$93.71 \pm 1.16^{a,b,c}$	98.18 ± 0.52^{a}
UMACC 313	ITO	$67.24 \pm 0.78^{g,h}$	$80.25 \pm 1.41^{d,e,f}$	$89.69 \pm 1.00^{a,b,c}$	$94.88 \pm 0.75^{a,b}$	98.69 ± 0.09^{a}
Coscinodiscus wailesii	Glass	0^{s}	0^{s}	0^{s}	$4.83 \pm 0.38^{r,s}$	10.84 ± 1.08^{g}
CCMP 2513	ITO	0^{s}	0^{s}	0^{s}	$5.37 \pm 0.90^{r,s}$	13.22 ± 1.30^{g}

Differences between alphabets indicate significant difference between different strains. (ANOVA, Turkey HSD test, $p<0.05$).

different. The photosynthetic performance was strain dependent. Biomass (B_{DW}) ranged from 11.07±2.89 μg/ml (*Coscinodiscus wailesii* CCMP 2513 on ITO) to 180.00±10.00 μg/ml (*Chlorella* sp. UMACC 313 on ITO). The Chl-a content of the biofilms ranged from 0.037±0.006 μg/ml (*Coscinodiscus wailesii* CCMP 2513 on ITO) to 2.533±0.311 μg/ml (*Chlorella* sp. UMACC 313 on ITO). Biofilm thickness ranged from 7±1 μm (*Coscinodiscus wailesii* CCMP 2513 on glass) to 96±2 μm (*Spirulina platensis* UMACC 159 on ITO). The rETRmax varied from 370.334±2.004 to 508.399±17.077 μmol electrons m^{-2} s^{-1}. The α values ranged from 0.485±0.001 (*Chlorella* sp. UMACC 313 on glass) to 0.750±0.050 (*Chlorella vulgaris* UMACC 001 on ITO). E_k values were all relatively high and ranged from 582.724±12.559 μmol photons m^{-2} s^{-1} (*Chlorella vulgaris* UMACC 001 on ITO) to 761.731±0.716 μmol photons m^{-2} s^{-1} (*Chlorella* sp. UMACC 313 on glass). F_v/F_m ranged from 0.799±0.011 (*Synechococcus elongatus* UMACC 105 on ITO) to 0.857±0.006 (*Chlorococcum* sp. UMACC 207 on glass).

NPQ was active in most of the strains except for the Cyanophyte *Synechococcus elongatus* (UMACC 105) which failed to produce any NPQ during the RLC (Figure 3). NPQ increased markedly and was almost linear following the increase of irradiance. Most of the strains started to produce NPQ at 127 μmol photons m^{-2} s^{-1}, with values ranging from 0.004 to 0.156. Strains such as *Chlorella* sp. (UMACC 256), *Chlorella vulgaris* (UMACC 051), *Chlorella vulgaris* (UMACC 001) and *Chlorococcum* sp. (UMACC 207) developed moderate to high levels of NPQ. *Chlorella* sp. (UMACC 256) biofilm grown on glass produced the highest NPQ value of 0.294 while the diatom *Coscinodiscus wailesii* (CCMP 2513) developed the lowest NPQ (<0.01).

BPV set up and electrical measurements

A study was carried out to correlate the %SAC of biofilm on ITO with the amount of photocurrent being generated as well as the overall performance of the device. Figure 4 shows a polarization curve observed for *Chlorella vulgaris* (UMACC 051), *Chlorella* sp. (UMACC 313), and the Cyanophytes *Spirulina platensis* (UMACC 159) and *Synechococcus elongatus* (UMACC 105). Significant power outputs were seen from all strains. The Cyanophytes *Synechococcus elongatus* UMACC105 (461 mV) and *Spirulina platensis* UMACC 159 (327 mV) produced highest cell voltage at open circuit, followed by the Chlorophytes *Chlorella vulgaris* UMACC 051 (249 mV) and *Chlorella* sp. UMACC 313 (239 mV). Meanwhile, the value of maximum power density in descending order was *Synechococcus elongatus* UMACC 105 (3.13×10^{-4} Wm^{-2})> *Chlorella* sp. UMACC 313 (1.24×10^{-4} Wm^{-2})>*Spirulina platensis* UMACC 159 (1.21×10^{-4} Wm^{-2})>*Chlorella vulgaris* UMACC 051 (1.12×10^{-4} Wm^{-2}). As for the value of maximum current density, a descending order, *Synechococcus elongatus* UMACC 105 (2.93×10^{-3} Am^{-2})>*Chlorella* sp. UMACC 313 (2.83×10^{-3} Am^{-2})>*Chlorella vulgaris* UMACC 051 (2.02×10^{-3} Am^{-2})>*Spirulina platensis* UMACC 159 (1.72×10^{-3} Am^{-2}) was observed.

Discussion

The preliminary screening of 16 strains, comprising 14 local strains belonging to the blue-green and green algae, and two diatoms from the CCMP, allowed the selection of eight out of the 16 strains that were characterized in terms of growth rate, biochemical composition (protein, lipid and carbohydrate contents) and photosynthetic efficiency, for further studies on biofilm formation. The 16 strains were cultured till stationary phase to observe the batch culture characteristics. At stationary phase, the lipid contents are expected to be higher than at exponential phase of culture, as lipids are accumulated upon reaching stationary phase, as opposed to protein [30]. This was clearly observed with most of the strains except for the Cyanophyta (*Synechoccus elongatus* UMACC 105) and two Chlorophytes (*Chlorella vulgaris* UMACC 051 and *Chlorococcum* UMACC 207), where protein was highest at day 15. In general, the Chlorophytes contained higher lipid than the other strains, with *Chlorella* sp. (UMACC 256) having the highest lipid (63.64%DW) contents on day 15.

Eight out of the 16 strains, namely two Cyanophytes, five Chlorophytes and one diatom, were selected for the biofilm studies. The Cyanophytes were selected because blue-green algae produce abundant mucilage and are expected to form biofilms easily. The blue-green algae has higher potential to form a hydrated polymeric matrix termed extracellular polymeric substances (EPS) which encourages the biofilm formation [15]. Also from the preliminary screening, the *Synechococcus elongatus* (UMACC 105) and *Spirulina platensis* (UMACC 159) had high rETRmax values, indicating efficient photosynthesis. *Chlorella vulgaris* (UMACC 001) has been used for many studies in our laboratory and has shown ability to grow in wastewaters [43,44,45] and has potential for biofuel production [30]. In 2007, Wong and co-workers [46] compared the tolerance of Antarctic, tropical and temperate microalgae to ultraviolet radiation (UVR) stress. When the *Chlorella vulgaris* (UMACC 001) was exposed to ambient light, the specific growth rate was 0.16 d^{-1} but decreased to 0.04 d^{-1} when exposed to ultraviolet radiation (UVR) treatment. Vejesri et al. [30] reported that *Chlorella vulgaris* (UMACC 001) was a potential feedstock for biodiesel production due to high specific growth rate ($\mu = 0.42$ d^{-1}) and high saturated fatty acid (SFA) content (68.2%DW). Chu and co-workers [43] reported that immobilized cultures of *Chlorella vulgaris* (UMACC 001) in alginate removed 48.9% of color from textile wastewater. *Chlorella vulgaris* (UMACC 001) showed high NPQ and strong photoprotection with higher capacity to cope with high irradiance. *Chlorococcum* (UMACC 207) had high rETRm as well as high protein (30.29%DW) content. *Chlorella* sp. (UMACC 313) and *Chlorella vulgaris* (UMACC 051) isolated from the aerobic pond for palm oil mill effluent (POME) treatment were selected because they formed biofilms on the base and sides of the culture flasks, and had very high lipid content. One marine alga *Chorella* sp. (UMACC 256) was included because it had the highest rETRmax value among the three marine strains. It was also shown that this strain had high specific growth rate (0.75 d^{-1}) and high saturated fatty acid (SFA) content (53.8%DW) in our previous study [30]. One strain from the CCMP culture collection was also selected for comparison with the UMACC cultures. The strain, *Coscinodiscus wailesii* (CCMP 2513) was a centric diatom with high rETRmax (166.32 μmol electrons m^{-2} s^{-1}) and higher lipid content of the two diatoms. In addition to the photosynthetic parameter of rETRmax, consideration was also given to the biofuel potential of the strains, in terms of lipid content.

In the biofilm studies, strains like *Synechococcus elongatus* (UMACC105) and *Chlorella* (UMACC 313) were observed to have formed appreciable biofilms after only three days. This may be based on the high production of EPS by *Synechococcus elongatus* (UMACC 105) and the high growth rate of *Chlorella* (UMACC 313). Application of PAM fluorometry in the study of algae biofilms has provided valuable information on the operation of algal BPV platforms. One of the most important information is the non-destructive generation of light-response curves of photosynthetic activity [47]. The distance between the optical fiber-optics and the sample surface was set at 2 mm [48]. Eight selected strains

Table 4. Statistical comparison of biomass and PAM data of eight strains, on day 15; data as means ±S.D. (n = 3).

Strain	Substrate	Biomass (μg/ml)	Chl-a (μg/ml)	Biofilm Thickness (μm)	rETRmax (μmol electrons $m^{-2}s^{-1}$)	Alpha (α)	E_k (μmol photons $m^{-2}s^{-1}$)	F_v/F_m
Synechococcus elongatus	Glass	$55.00 \pm 10.00^{d,e}$	0.191 ± 0.009^{e}	62 ± 8^{c}	427.243 ± 14.846^{d}	0.569 ± 0.0210^{d}	$750.481 \pm 2.329^{a,b,c}$	$0.810 \pm 0.012^{b,c,d}$
UMACC 105	ITO	86.67 ± 7.64^{c}	0.490 ± 0.054^{d}	80 ± 5^{b}	$417.775 \pm 5.148^{d,e}$	$0.553 \pm 0.009^{d,e,f}$	$755.498 \pm 3.135^{a,b}$	0.799 ± 0.011^{d}
Spirulina platensis	Glass	$70.00 \pm 10.00^{c,d}$	$0.301 \pm 0.062^{d,e}$	87 ± 3^{b}	$466.672 \pm 4.530^{a,b,c,d}$	$0.623 \pm 0.006^{b,c,d}$	$749.0923 \pm 6.839^{a,b,c,d}$	$0.845 \pm 0.016^{a,b,c}$
UMACC 159	ITO	130.00 ± 10.00^{b}	0.845 ± 0.028^{c}	96 ± 2^{a}	$452.767 \pm 16.133^{b,c,d}$	$0.605 \pm 0.024^{c,d}$	$748.029 \pm 2.763^{b,c,d}$	$0.828 \pm 0.019^{a,b,c,d}$
Chlorella vulgaris	Glass	$23.33 \pm 2.89^{g,h}$	0.048 ± 0.003^{e}	$17 \pm 1^{e,f}$	430.730 ± 32.745^{d}	0.732 ± 0.054^{a}	588.41 ± 6.445^{f}	$0.826 \pm 0.002^{a,b,c,d}$
UMACC 001	ITO	$35.00 \pm 5.00^{e,f,g}$	0.058 ± 0.006^{e}	$20 \pm 2^{e,f}$	437.444 ± 38.471^{d}	0.750 ± 0.050^{a}	582.724 ± 12.559^{f}	$0.831 \pm 0.012^{a,b,c,d}$
Chlorella vulgaris	Glass	$50.00 \pm 5.00^{d,e}$	$0.302 \pm 0.205^{d,e}$	24 ± 1^{e}	$420.458 \pm 32.686^{d,e}$	$0.562 \pm 0.047^{d,e}$	$748.377 \pm 4.590^{b,c,d}$	$0.824 \pm 0.019^{a,b,c,d}$
UMACC 051	ITO	151.67 ± 10.41^{b}	0.501 ± 0.044^{d}	38 ± 3^{d}	429.798 ± 1.716^{d}	0.573 ± 0.004^{d}	$750.528 \pm 1.612^{a,b,c}$	$0.851 \pm 0.004^{a,b}$
Chlorococcum sp.	Glass	$26.67 \pm 2.89^{f,g,h}$	0.075 ± 0.011^{e}	$12 \pm 1^{f,g,h}$	$457.495 \pm 10.938^{a,b,c,d}$	$0.620 \pm 0.015^{b,c,d}$	$737.498 \pm 1.133^{c,d,e}$	0.857 ± 0.006^{a}
UMACC 207	ITO	$26.67 \pm 7.64^{f,g,h}$	0.087 ± 0.007^{e}	$15 \pm 1^{e,f,g,h}$	$446.736 \pm 15.241^{c,d}$	$0.606 \pm 0.021^{c,d}$	$736.798 \pm 0.770^{d,e}$	$0.840 \pm 0.013^{a,b,c,d}$
Chlorella sp.	Glass	$46.67 \pm 5.77^{e,f}$	0.184 ± 0.008^{e}	$15 \pm 1^{f,g,h}$	$498.801 \pm 14.443^{a,b,c}$	$0.684 \pm 0.021^{a,b}$	729.266 ± 1.226^{e}	$0.802 \pm 0.021^{c,d}$
UMACC 256	ITO	$53.33 \pm 5.77^{d,e}$	0.192 ± 0.006^{e}	$16 \pm 1^{e,f,g}$	508.399 ± 17.077^{a}	$0.694 \pm 0.021^{a,b}$	732.519 ± 2.837^{e}	$0.825 \pm 0.018^{a,b,c,d}$
Chlorella sp.	Glass	146.67 ± 7.64^{b}	1.364 ± 0.056^{b}	82 ± 3^{b}	369.186 ± 0.407^{e}	0.485 ± 0.001^{f}	761.731 ± 0.716^{a}	$0.819 \pm 0.006^{a,b,c,d}$
UMACC 313	ITO	180.00 ± 10.00^{a}	2.533 ± 0.311^{a}	87 ± 3^{b}	370.334 ± 2.004^{e}	$0.487 \pm 0.004^{e,f}$	$760.447 \pm 1.825^{a,b}$	$0.827 \pm 0.006^{a,b,c,d}$
Coscinodiscus wailesii	Glass	$13.33 \pm 2.89^{g,h}$	0.037 ± 0.006^{e}	7 ± 1^{h}	$501.960 \pm 0.480^{a,b}$	$0.681 \pm 0.001^{a,b,c}$	$737.454 \pm 0.805^{c,d,e}$	$0.854 \pm 0.030^{a,b}$
CCMP 2513	ITO	11.67 ± 2.89^{h}	0.037 ± 0.006^{e}	$8 \pm 0^{g,h}$	$502.239 \pm 0.421^{a,b}$	$0.681 \pm 0.001^{a,b,c}$	$737.503 \pm 0.828^{c,d,e}$	$0.834 \pm 0.004^{a,b,c,d}$

Differences between alphabets indicate significant difference between different strains. (ANOVA, Turkey HSD test, $p<0.05$). Alpha is photosynthetic efficiency and indicates the amount of ETR per photon. E_k is the photoadaptive index and indicates how well cells are adapted to their light environment.

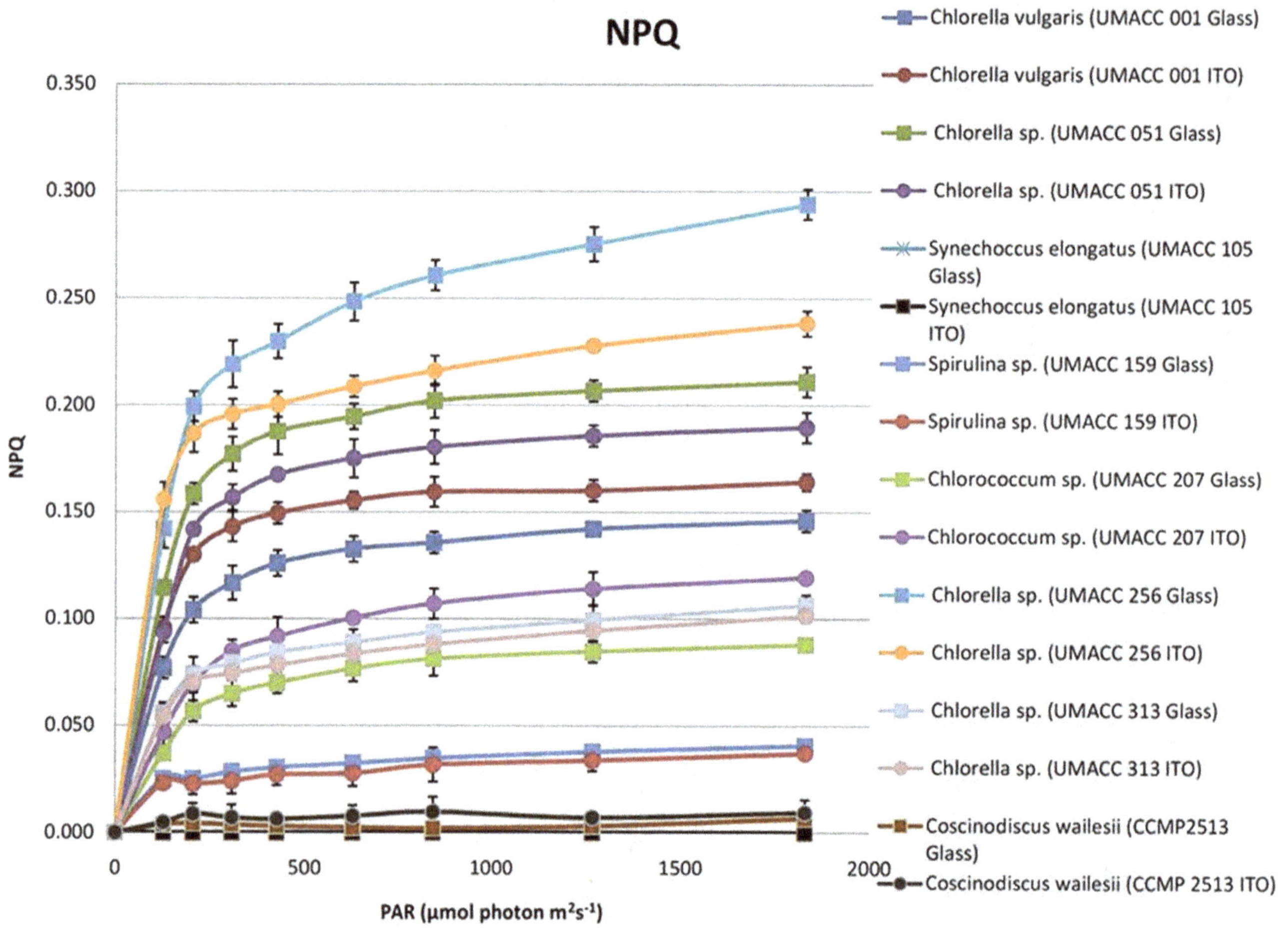

Figure 3. Non-photochemical quenching of biofilms formed by eight strains grown on ITO and glass slides on day 15 (n = 3).

starting with the same concentration of pure cultures at $OD_{620\ nm}$ at 0.5. All cultures were transferred into the same incubator at 24°C illuminated by cool white fluorescent lamps (30 μmol $m^{-2}\ s^{-1}$) on a 12:12 hour light-dark cycle to ensure that the only variable was the strain of algae in the biofilm development. E_k values were all relatively high and ranged from 582.724 to 761.731 μmol photons $m^{-2}\ s^{-1}$ indicating that all biofilms were able to adapt to the high light intensities (307–1829 μmol photons $m^{-2}\ s^{-1}$). Under high irradiances, an inverse relationship between NPQ and rETRmax was observed in all strains; higher NPQ values being matched by lower rETRmax values. The Chl-a content of the eight strains increased with the thickness of biofilms. The *Chlorella* species (UMACC 313 and UMACC 051) isolated from the aerobic pond used in POME treatment formed significant biofilms on day 3 on both substrate materials. However, the % SAC of the biofilm from the diatom *Coscinodiscus wailesii* (CCMP 2513) was low compared to other strains probably due to the need for longer time to form hydrated polymeric matrix. Even in the suspension culture during the preliminary screening, this diatom took 15 days to reach stationary phase (Table 2). The substrate effect was minimal as there was no significant difference of % SAC of all strains on both glass and ITO slides, except for the filamentous blue-green alga *Spirulina platensis* (UMACC 159) which grew better on ITO slides compared to glass slides with higher % SAC, Chl-a content and biofilm thickness. From this study, four strains, the Chlorophytes *Chlorella* (UMACC 051, UMACC 313) and the Cyanophytes *Synechococcus elongatus* (UMACC 105) and *Spirulina platensis* (UMACC 159) were identified to produce the most stable biofilms on ITO. The growth and viability of these microorganisms in a working BPV device would have a significant impact on the photocurrent that would be generated [13].

The biofilm samples were dark adapted for at least 15 minutes before PAM measurement. When cultures were placed in the dark, minimum (F_o) and maximum (F_m) fluorescence both increased, as did the maximum PSII quantum efficiency (F_v/F_m). After dark adaption, the reaction centres were in a relaxed stage, and ready to receive light for photosynthesis [19]. The maximum quantum yield F_v/F_m was obtained when all reaction centres were opened and was proportional to the fraction of reaction centres capable of converting absorbed light to photochemical energy [49]. F_v/F_m is often used as an indicator of photosynthetic capacity. Various studies have reported values of F_v/F_m ranging between 0.1 to 0.65 for natural populations of microalgae [50]. Results in the present study indicated that most samples were in a healthy condition, as the F_v/F_m values were all above 0.800 except for *Synechococcus elongatus* UMACC 105 with a F_v/F_m value of 0.799. Serôdio et al. [51] reported rapid light-response curves of chlorophyll fluorescence in benthic microalgae from a mesotidal estuary located in the west coast of Portugal. The values of photosynthetic parameters were as follows: alpha (α): 0.249–0.473, rETRmax (μmol electrons $m^{-2}\ s^{-1}$): 118.9 to 424.0, E_k (μmol photons $m^{-2}\ s^{-1}$): 474.1 to 821.4. A similar study carried out by Cartaxana et al. [52] on the same area reported

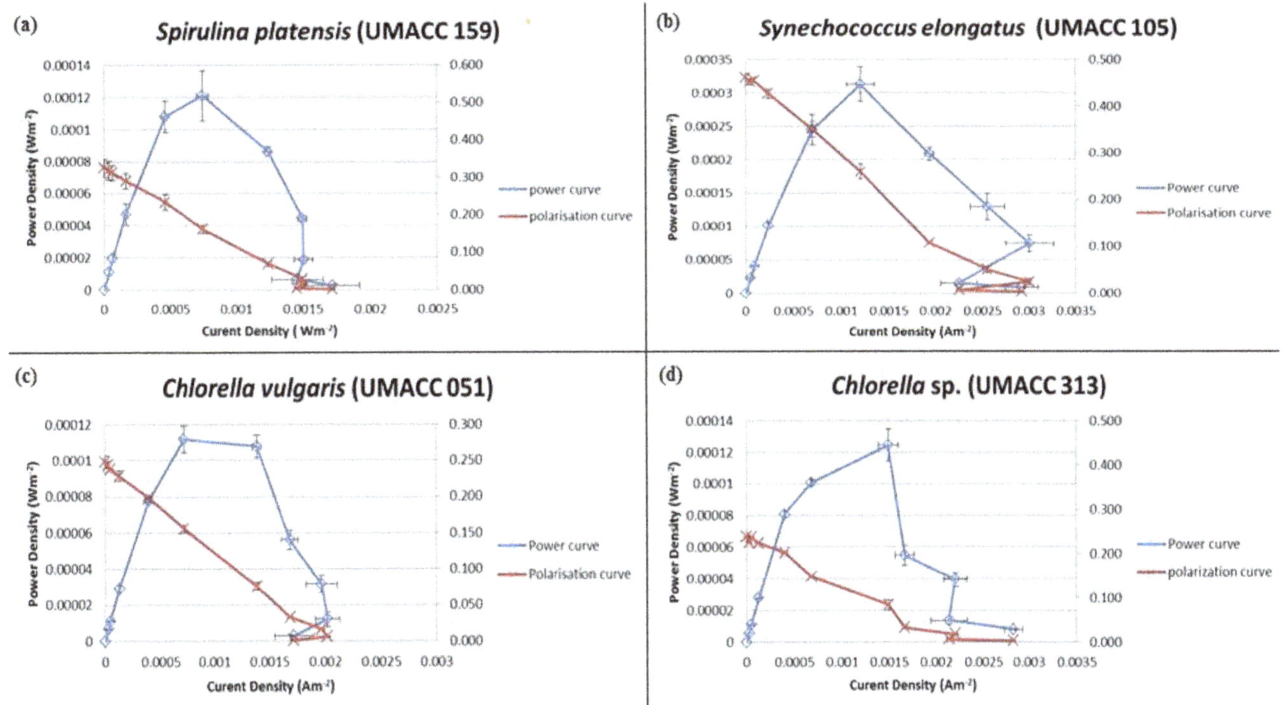

Figure 4. Polarization curves observed for (a) Spirulina platensis (UMACC 159), (b) Synechococcus elongatus (UMACC 105), (c) Chlorella vulgaris (UMACC 051), and (d) Chlorella sp.(UMACC 313).

PAM fluorometry as a tool to assess photophysiology of intertidal microphytobenthic biofilms. The microphytobenthos communities were dominated by diatoms, resulting in the following photosynthetic parameters; alpha (α): 0.657, rETRmax (μmol electrons m^{-2} s^{-1}): 331, E_k (μmol photons m^{-2} s^{-1}): 582.72–761.73. In our study, the range of photosynthetic parameters for suspension cultures of the 16 strains were: alpha (α): 0.31–0.91, rETRmax (μmol electrons m^{-2} s^{-1}): 64.68–166.32, E_k (μmol photons m^{-2} s^{-1}): 72.73–473.85 and F_v/F_m: 0.43–0.80. For biofilms of the eight strains, the range of photosynthetic parameters were: alpha (α): 0.48–0.75, rETRmax (μmol electrons m^{-2} s^{-1}): 369.19–508.40, E_k (μmol photons m^{-2} s^{-1}): 582.72–760.45 and F_v/F_m: 0.80–0.85. Our results showed that algal biofilms have better photosynthetic performance than suspension cultures. Results showed that the algal strains used in this study have high photosynthetic efficiency indicating their potential for use in BPV platforms. In addition, high rETRmax and E_k values obtained from the present study reflected an adequate adaptation to high irradiance [53], and was observed in the two Cyanophytes (*Synechococcus elongatus* UMACC 105) and the diatom (*Coscinodiscus wailesii* CCMP 2513). The NPQ values for each strain were compared and an increase in NPQ with changing light intensities was observed in all cases of the biofilms (Figure 3). The change in NPQ is expected due to photoprotection of photosystem II to avoid light induced damage [54]. In the case of the Cyanobacteria (Cyanophyta), the resulting changes in NPQ have been reported to be triggered by the light activation of orange carotenoid protein found in phycobilisomes [55]. In comparison, the NPQ in diatoms is mediated by the light-harvesting complex stress-related (LHCSR) protein and the conversion of diadinoxanthin (Ddx) to diatoxanthin (Dtx) [56]. In green algae (Chlorophyta), the controlled change in pH and its effect on the xanthophyll cycle has a predominated effect on NPQ [54]. In the present study, the Chlorophyte strains were observed to have higher ability (higher NPQ values) to photoprotect themselves than the Cyanophytes and diatoms. The Cyanophytes (blue-green algae) were observed to have the lowest value or in some cases, no NPQ, indicating that they have the lowest capacity for photoprotection. There was no clear correlation between the F_v/F_m values and the NPQ values of the 16 strains in the screening experiment as well as the eight strains used for the biofilm study, except where the lowest F_v/F_m values of the two Cyanophytes also corresponded to the lowest NPQ values (Tables 2 & 4 and Figures 2 & 3). Interestingly the low NPQ values of the two Cyanophytes and two diatoms corresponded to higher E_k values, and may indicate that these strains may in fact be tolerant of higher irradiance. This however, has to be confirmed by detailed studies on the relationship between the various parameters. In this study, the four strains, Cyanophytes *Synechococcus elongatus* UMACC 105, *Spirulina platensis* UMACC 159 and Chlorophytes *Chlorella* UMACC 051 and UMACC 313, were selected for the BPV study based on strong ability to form biofilms; however their NPQ ranged from low to high values, to allow the correlation of NPQ to photo current production in a working BPV device. The diatoms had high rETRmax but very low NPQ, and were not selected due to poor ability to form biofilms. On analysis of data from the BPV studies, it was apparent that the Cyanophyte *Synechococcus elongatus* (UMACC 105), was able to give highest performance in the BPV device but it had the lowest NPQ value. The maximum power density was similar for the other Cyanophyte *Spirulina* sp. (UMACC159) and two Chlorophytes *Chlorella* (UMACC 051 and 313), which had higher NPQ. Further studies would need to be carried out to correlate the NPQ of the strains with the ability to extract charge from biofilms in a working BPV device.

All four strains, the Cyanophytes *Synechoccus elongatus* (UMACC 105) and *Spirulina platensis* (UMACC 159), and the Chlorophytes

Chlorella vulgaris (UMACC 051) and *Chlorella* sp. (UMACC 313), showed ability to produce electrical power outputs (Figure 4 a–d). *Synechococcus elongatus* (UMACC 105) showed higher power outputs compared to other strains. This may be due to the fact that *Synechococcus elongatus* (UMACC 105) registered a high rETRmax values (147.50 μmol electrons m^{-2} s^{-1}), high E_k value and readily produced biofilms (84.68% SAC on ITO) which were able to accumulate a high level of charge. The *Synechococcus* produces unicells of small dimensions and with production of EPS, would form compact biofilms on the anode, compared to the filamentous *Spirulina* which like the *Arthrospira* used in the study by Inglesby et al. [39], formed porous biofilms. The maximum power density obtained in this study ranged from 1.12×10^{-4} to 3.13×10^{-4} Wm^{-2}. Inglesby et al. [39] reported very low power densities from 6.7 to a maximum of 24.8 μW m^{-2} with a similar BPV device using biofilm of the Cyanophyte *Arthrospira maxima* on ITO as well, considering that the theoretical maximum power density achievable when using microalgae in microbial fuel cells is 2.8 Wm^{-2}. The low power density obtained in the present study reflects an unoptimised system with the results aimed only for comparison between the four strains. Optimisation of the process in terms of biofilm thickness and irradiance properties may improve the power output. Optimisation of temperature and light intensity in the BPV device may enhance power output, as was shown by Inglesby et al. [39], where an increase of temperature from 25 to 35°C increased power output from 9.9 to 24.8 μW m^{-2} while there may be a limitation to how high the light intensity can be, due to photoinhibition. In the present study. Based on power output per Chl-a content, the four strains can be ranked as follows: *Synechococcus elongatus* (UMACC 105) (6.38×10^{-5} Wm^{-2}/μgChl-a)>*Chlorella vulgaris* (UMACC 051) (2.24×10^{-5} Wm^{-2}/μgChl-a)>*Chlorella* sp. (UMACC 313) (1.43×10^{-5} Wm^{-2}/μgChl-a)>*Spirulina platensis* (UMACC 159) (4.90×10^{-6} Wm^{-2}/μgChl-a). These values were obtained by dividing the maximum power density by the total Chl-a extracted from the whole algal biofilm on the ITO slides. However no correlation ($P<0.05$) was observed between power output and Chl-a content. Of the four strains, *Chlorella vulgaris* (UMACC 051) had highest lipid (31.09% DW) and crude protein (46.70% DW) contents, indicating additional advantages over the other strains. Although it is premature to speculate on the long-term usage of algal cultures in BPV devices, on a large scale, biomass may be generated continuously in the BPV. The biomass may be harvested on a regular basis to prevent over-buildup of the biofilm, which results in reduced bioactivity at the surface of the biofilm furthest from the anode. Photosynthetically efficient strains with high lipid and protein productivities, may have an added advantage over other strains, providing a valuable biomass that may enter an alternative energy-producing system like biodiesel production. This may overcome the lower energy output from algal BPV systems compared to commercially available photovoltaics [39]. However the additional cost of continuous nutrient supply to the biofilms on the anodes and cost of removal of surplus biomass from the biofilms may necessitate innovative design and operation of the BPV devices.

Conclusions

The present study indicated that local algal strains were good candidates for utilization in BPV platforms in future. According to our screening results, all eight strains were able to form biofilms on ITO anode surfaces with good photosynthetic performance. From this, four strains were selected for the BPV studies, based on their high photosynthetic performance and ability to produce biofilms. The Chlorophytes *Chlorella* species (UMACC 051, UMACC 313) and the Cyanophytes, *Spirulina platensis* (UMACC 159) and *Synechococcus elongatus* (UMACC 105), demonstrated exoelectrogenic activity and showed their capacity to produce significant electrical power outputs without the requirement of additional organic fuel. More work is required to further understand the mechanisms of harnessing light energy and converting them to electricity as well as to investigate the correlation between PAM data and the BPV power output.

Acknowledgments

Special thanks to Vejeysri Vello, Chan Zhijian, Muhammad Musoddiq bin Jaafar, Ng Poh Kheng, Dr. Song Sze Looi, Dr. Low Van Lun, Dr. Tan Ji, Tan Cheng Yao, Dr. Emienour Muzalina binti Mustafa for technical support and advice.

Author Contributions

Conceived and designed the experiments: FN SP VP KY. Performed the experiments: FN. Analyzed the data: FN SP. Contributed reagents/materials/analysis tools: FN SP VP KY AF. Wrote the paper: FN SP.

References

1. Roeselers G, Barbara Z, Staal M, van Loosdrecht M, Muyzer G (2006) On the reproducibility of microcosm experiments – different community composition in parallel phototrophic biofilm microcosms. FEMS Microbiol Ecol 58(2): 169–178.
2. Li Y, Horsman M, Wu N, Lan CQ, Calero ND (2008) Articles: Biocatalysis and bioreactor design. Biotechnol Prog 24: 815–820.
3. Yazdi HR, Carver SM, Christy AD, Tuovinen OH (2008) Cathodic limitations in microbial fuel cells: An overview. J Power Sources 180: 683–694.
4. Daroch M, Geng S, Wang G (2013) Recent advances in liquid biofuel production from algae feedstocks. Applied Energy 102: 1371–1381.
5. Blatti JL, Beld J, Behnke CA, Mendez M, Mayfiels SP (2012) Manipulating fatty acid biosynthesis in microalgae for biofuel through protein-protein interactions. Plos One 7(9): e42949.doi:10.1317/journal.pone.0042949.
6. Amaro HM, Guedes AC, Malcata FX (2011) Advances and perspectives in using microalgae to produce biodiesel. Applied Energy 88: 3402–3410.
7. Jackson DD, Ellms JW (1896) On odors and tastes of surface waters with special reference to *Anabaena*, a microscopical organism found in certain water supplies of Massachusetts. Rep. Mass State Board Health 1896: 410–420.
8. Prince RC, Kheshgi HS (2005) The photobiological production of hydrogen: potential efficiency and effectiveness as a renewable fuel. Crit Rev Microbiol 31: 19–31.
9. Yagishita T, Sawayama S, Tsukahara K, Ogi T (1997) Effects of intensity of incident light and concentrations of *Synechococcus* sp. and 2-hydroxy-1, 4-naphthoquinone on the current output of photosynthetic electrochemical cell. Solar Energy 61: 347–353.
10. Koichi N, Kazuhito H, Kazuya W (2010) Light/electricity conversion by a self-organized photosynthetic biofilm in a single-chamber reactor. Appl Microbiol Biotechnol 86: 957–964.
11. Bombelli P, McCormick A, Bradley R, Yunus K, Philips J, et al. (2011) Harnessing solar energy by bio-photovoltaic (BPV) devices. Commum Agric Biol Sci 76(2): 89–91.
12. Mao L, Verwoerd WS (2013) Selection of organisms for systems biology study of microbial electricity generation: a review. Journal of Energy and Environmental Engineering 4(17): 1–18.
13. McCormick AJ, Bombelli P, Scott AM, Philips AJ, et al. (2011) Photosynthetic biofilms in pure culture harness solar energy in a mediatorless bio-photovoltaic cell (BPV) system. Energy Environ Sci 4: 4699–4709.
14. van Leeuwenhoek A (1684). Some microscopical observations about animals in the scurf of the teeth. Phil Trans, 14, 568–574.
15. Sauer K, Rickard AH, Davies DG (2007) Biofilms and Biocomplexity. Microbe 2: 347–353.
16. Declerck P (2010) Biofilms: the environmental playground of *Legionella pneumophila*. Environmental Microbiology 12(3): 557–566.
17. Ozkan A, Berberoglu H (2012) Reduction of water and energy requirement of algae cultivation using an algae biofilm photobioreactor. Bioresource Technol 114: 542–548.

18. Wiltshire KH, Blackburn J, Paterson DM (1997) The cryolander: a new method for fine-scale in situ sampling of intertidal surface sediments. J Sediment Res 67: 977–981.
19. Consalvey M, Jesus B, Perkins RG, Brotas V, Underwood GJC, et al. (2004) Monitoring migration and measuring biomass in benthic biofilms: the effects of dark/far-red adaption and vertical migration on fluorescence measurements. Photosynthesis Research 81: 91–101.
20. Maxwell K, Johnson N (2000) Chlorophyll fluorescence- a practical guide. Journal of Experimental Botany 51(345): 659–668.
21. McMinn A, Martin A, Ryan K (2010) Phytoplankton and sea ice algal biomass and physiology during the transition between winter and spring (McMurdo Sound, Antarctica). Polar Biol 33: 1547–1556.
22. Perkins RG, Underwood GJC, Brotas V, Snow GC, Jesus B, et al. (2001) Responses of microphytobenthos to light: primary production and carbohydrate allocation over an emersion period. Mar Ecol Prog Ser 223: 101–112.
23. Luimstra VM, Kennedy SJ, Güttler J, Wood SA, Williams DE, et al. (2013) A cost-effective microbial fuel cell to detect and select for photosynthetic electrogenic activity in algae and cyanobacteria. J Appl Phycol 26: 15–23. doi: 10.1007/s10811-013-0051-2.
24. Ng FL, Phang SM, Vengadesh P, Yunus K and Fisher AC (2013) Algae Biofilm on Indium Tin Oxide Electrode for Use in Biophotovoltaic Platforms. Advanced Materials Research 895: 116–121.
25. Ozkan A, Berberoglu H (2013) Adhesion of algal cells to surfaces. The Journal of Bioadhesion and Biofilm Research 29(4): 469–482.
26. Hwang J, Amy F, Kahn A (2006) Spectroscopic study on sputtered PEDOT-PSS: Role of surface Pss layer. Organic Electronics 7: 387–396.
27. Phang SM, Chu WL (1999) Catalogue of Strains, University of Malaya Algae Culture Collection (UMACC). Institute of Postgraduate Studies and Research, University of Malaya, Kuala Lumpur.
28. Nichols HW & Bold HC (1965) *Trichorsarcina polymorpha* gen. et sp. nov. J Phycol 1: 34–38.
29. Guillard RRL, Ryther JH (1962) Studies of marine planktonic diatoms. I. *Cyclotella nana* Hustedt and *Detomula confervacea* Cleve. Can J Microbiol 8: 229–239.
30. Vejeysri V, Phang SM, Chu WL, Majid NA, Lim PE, et al. (2013) Lipid productivity and fatty acid composition-guided selection of *Chlorella* strains isolated from Malaysia for biodiesel production. J Appl Phycol doi: 10.1007/s10811-013-0160.
31. Strickland JDH, Parsons TR (1968) A Practical Handbook of Seawater Analysis. Bull. Fish Res Bd Can 167: 311.
32. APHA (American Public Health Association, American Waterworks Association, Water Pollution Control Federation (1998) Standard methods for the examination of water and wastewater, 18th Edition. Diaz de Santos, Madrid.
33. Phang SM, Ong KC (1988) Algal Biomass Production in digested palm oil mill effluent. Biological Wastes 25: 171–193.
34. Bradford MM (1976) A rapid and sensitive method for the quantitation of microgram quantities of protein utilizing the principle of dye-binding. Analyt. Biochem 72: 248–254.
35. Kochert AG (1978) Carbohydrate determination by the phenol–sulfuric acid method. In: Hellebust JA, Craigie JS (eds), Handbook of Phycological Methods: Physiological and Biochemical Methods. Cambridge University Press, Cambridge. 95–97 p.
36. Bligh EG, Dyer WJ (1959) A rapid method of total lipid extraction and purification. Can. J. Biochem. Physiol 37: 911–917.
37. Keng FSL, Phang SM, Rahman NAR, Leedham EC, Hughes C, et al. (2013) Volatile halocarbon emissions by three tropical brown seaweeds under different irradiances. J Applied Phycol 25: 1377–1386.
38. Pankowski A, McMinn A (2009) Iron availability regulates growth, photosynthesis, and production of ferredoxin and flavodoxin in Antartic Sea ice diatoms. Aquatic Biology 4: 274–288.
39. Inglesby AE, Yunus K, Fisher AC (2013) In situ fluorescence and electrochemical monitoring of a photosynthetic microbial fuel cell. Phys Chem Chem Phys 15: 6903–6911.
40. Platt T, Callegos CL, Harrison WG (1980) Photoinhibition of photosynthesis in natural assemblages of marine phytoplankton. J Mar Res 38: 687–701.
41. Ferreira TA, Rasband W (2010) Image J User Guide. http://imagej.nih.gov/ij/docs/user-Guide.pdf. Accessed 25 September 2012.
42. Elcometer wet film combs and wheel operating instructions. Available: http://www.elcometer.com/images/stories/PDFs/InstructionBooks/112_115_154_3230_3236_3238.pdf. Accessed 1 September 2012.
43. Chu WL, See YC, Phang SM (2008) Use of immobilised *Chlorella vulgaris* for the removal of colour from textile dyes. J Appl Phycol 21: 641–648.
44. Lim SL, Chu WL and Phang SM (2010) Use of *Chlorella vulgaris* for bioremediation of textile wastewater. Bioresour Technol101: 7314–7322.
45. Emienour Muzalina Mustafa, Phang SM, Chu WL (2012). Use of an algal consortium of five algae in the treatment of landfill leachate using the high-rate algal pond system. J Appl Phycol 24:953–963.
46. Wong CY, Chu WL, Marchant H, Phang SM (2007) Comparing the response of Antarctic, tropical and temperate microalgae to ultraviolet radiation (UVR) stress. J Appl Phycol 19: 689–699.
47. Vieira S, Calado R, Coelho H, Serôdio J (2009) Effect of light exposure on the retention of kleptoplastic photosynthetic activity in the sacoglossan mollusc *Elysia viridis*. Mar Biol 156: 1007–1020.
48. Bonnineau C, Sague IG, Urrea G, Guasch H (2012) Light history modulates antioxidant and photosynthetic responses of biofilms to both natural (light) and chemical (herbicides). Ecotoxicology 21: 1208–1224.
49. Jordan L, McMinn A, Thompson P (2009) Diurnal changes of photoadaptive pigments in microphytobenthos. Journal of the Marine Biological Association of the United Kingdom 1–8.
50. Reeves S, McMinn A, Martin A (2011) The effect of prolonged darkness on the growth, recovery and survival of Antarctic sea ice diatoms. Polar Biol 34: 1019–1032.
51. Serôdio J, Vieira S, Cruz S, Coelho H (2006) Rapid light-response curves of chlorophyll fluorescence in microalgae: relationship to steady-state light curves and non-photochemical quenching in benthic diatom-dominated assemblages. Photosynth Res 90: 29–43.
52. Cartaxana P, Serôdio (2008) Inhibiting diatom motility: a new tool for the study of the photophysiology of intertidal microphytobenthic biofilms. Limnology and Oceanography Methods 6: 466–476.
53. McMinn A, Hattori H (2006) Effect of time of day on the recovery from light exposure in ice algae from Saroma Ko lagoon, Hokkaido. Polar Biosci 20: 30–36.
54. Jahns P, Holzwarth AR (2012) The role of xanthophyll cycle and of lutein in photoprotection of photosystem II. Biochimica et Biophysica Acta 1817: 182–193.
55. Boulay C, Abasova L, Six C, Vass I, Kirilovsky D (2008) Occurrence and function of the orange carotenoid protein in photoprotective mechanisms in various cyanobacteria. Biochimica et Biophysica Acta 1777: 1344–1354.
56. Lavaud J, Strzepek RF, Kroth PG (2007) Photoprotection capacity differs among diatoms: Possible consequences on the spatial distribution of diatoms related to fluctuations in the underwater light climate. Limnol Oceanogr 52(3): 1188–1194.

16

Plastic Traits of an Exotic Grass Contribute to Its Abundance but Are Not Always Favourable

Jennifer Firn[1]*, Suzanne M. Prober[2], Yvonne M. Buckley[3,4]

1 School of Earth, Environment and Biological Sciences, Queensland University of Technology, Brisbane, Queensland, Australia, **2** Ecosystem Sciences, CSIRO, Wembley, Western Australia, Australia, **3** School of Biological Sciences, The University of Queensland, St. Lucia, Queensland, Australia, **4** Ecosystem Sciences CSIRO, Dutton Park, Queensland, Australia

Abstract

In herbaceous ecosystems worldwide, biodiversity has been negatively impacted by changed grazing regimes and nutrient enrichment. Altered disturbance regimes are thought to favour invasive species that have a high phenotypic plasticity, although most studies measure plasticity under controlled conditions in the greenhouse and then assume plasticity is an advantage in the field. Here, we compare trait plasticity between three co-occurring, C_4 perennial grass species, an invader *Eragrostis curvula*, and natives *Eragrostis sororia* and *Aristida personata* to grazing and fertilizer in a three-year field trial. We measured abundances and several leaf traits known to correlate with strategies used by plants to fix carbon and acquire resources, i.e. specific leaf area (SLA), leaf dry matter content (LDMC), leaf nutrient concentrations (N, C:N, P), assimilation rates (*Amax*) and photosynthetic nitrogen use efficiency (PNUE). In the control treatment (grazed only), trait values for SLA, leaf C:N ratios, *Amax* and PNUE differed significantly between the three grass species. When trait values were compared across treatments, *E. curvula* showed higher trait plasticity than the native grasses, and this correlated with an increase in abundance across all but the grazed/fertilized treatment. The native grasses showed little trait plasticity in response to the treatments. *Aristida personata* decreased significantly in the treatments where *E. curvula* increased, and *E. sororia* abundance increased possibly due to increased rainfall and not in response to treatments or invader abundance. Overall, we found that plasticity did not favour an increase in abundance of *E. curvula* under the grazed/fertilized treatment likely because leaf nutrient contents increased and subsequently its' palatability to consumers. *E. curvula* also displayed a higher resource use efficiency than the native grasses. These findings suggest resource conditions and disturbance regimes can be manipulated to disadvantage the success of even plastic exotic species.

Editor: Justin Wright, Duke University, United States of America

Funding: Thank you to CSIRO Ecosystem Sciences and Australian Research Council (YB, DP0771387) for funding. The funders had no role in study design, data collection and analysis, decision to publish, or preparation of the manuscript.

Competing Interests: The authors have declared that no competing interests exist.

* E-mail: jennifer.firn@qut.edu.au

Introduction

Exotic plant species can establish and dominate sites despite lacking evolutionary familiarity with local conditions and having small founder populations (i.e. the invasion paradox [1,2]). Substantial evidence suggests disturbances such as changed grazing regimes and nutrient addition increase opportunities for invasive species to establish [3,4,5,6,7,8], particularly if disturbances are novel to an ecosystem [9,10]. Disturbance favours the growth and survival of some species over others depending on the characteristics of the disturbance itself including the type, frequency, duration and intensity [11], but also on the traits of species present [12,13]. Despite extensive research, evidence for a generic set of traits that favour exotic over native species remains inconclusive [14,15,16,17].

Evidence suggests invasive species tend to display traits of fast growing species that are resource acquisition specialists and native species tend to display traits of slow-growing species that are conservation specialists [18,19,20,21,22]. The leaf economic spectrum proposes a fundamental trade-off in the traits held by fast- and slow-growing plant species [23,24,25]. Fast growing species, better at resource capture, tend to dominate disturbed ecosystems where resource availability is not limited. These fast growing species have generally higher specific leaf area (SLA, mm^2/mg, fresh leaf area/oven-dry mass), lower leaf dry matter content (LDMC, mg/g, oven dry mass/water-saturated fresh mass), higher nutrient contents and higher rates of assimilation (*Amax*) [23,24,25]. Slower growing plant species generally occupying low resource and less disturbed sites are better at resource conservation and to tend to hold opposite traits—lower SLAs, higher LDMCs, lower nutrient contents and lower rates of *Amax* [23,24,25]. Studies comparing the leaf traits of exotics and natives have consistently found evidence for this trade-off, with exotics showing better resource acquisition strategies and natives better resource conservation strategies [26]. However, recent findings by Leishman et al. [27] suggest that exotic and native plant species can hold similar strategies for capturing resources, with exotic and native species at disturbed sites possessing similar traits, but different traits to natives at pristine sites.

To date, most studies investigating plant traits focused on differences between species (interspecific variability) and across sites affected by different disturbances and environmental conditions, but recent research has highlighted the importance of intraspecific variability in traits or phenotypic plasticity [28,29].

Table 1. General characteristics of the invasive exotic lovegrass, and the native grasses purple wiregrass and woodlands lovegrass.

Characteristics	*Eragrostis curvula* **lovegrass, Exotic grass**	*Aristida personata* **purple wiregrass Native grass**	*Eragrostis sororia* **woodlands lovegrass Native grass**
Mean abundance at site (± S.E.) at time 0	47.56%±3.98	22.66±5.12	1.61%±0.67
Growth Habit	Tufted perennial	Tufted perennial	Tufted perennial
Photosynthetic Pathway	C_4	C_4	C_4
Height	Up to 120 cm	Up to 120 cm	Up to 70 cm
Growth season	Summer	Summer	Summer
Flowering time	Spring to Autumn	Summer to Autumn	Summer
Palatability to livestock	Low	Low	Moderate
Native continental distribution	Africa	Australia	Australia

[61,62,63,64,65].

Evidence suggests that invasive exotic species display higher phenotypic plasticity than natives—the potential of each individual genotype to produce different traits/phenotypes in response to disturbance and fluctuating environmental conditions [14,30,31]. This capacity to change morphological or physiological traits may allow genotypes of a species to thrive across a wider range of environmental conditions (genotype-level plasticity), and/or allow individuals within a population to thrive at sites during and after disturbance or resource pulses (species-level plasticity) [14,17,30,32,33].

Invasive species have shown higher trait plasticity in response to increased resources, e.g soil nutrients and water, in comparison to phylogenetically-related non-invasive species from high resource environments [34,35], and phylogenetically-related native species from low resource environments [33]. Although a recent greenhouse study comparing 20 phylogenetically-related invasive and native trees and shrubs found similar levels of trait plasticity in response to nutrient and light treatments, but enhanced performance by invasive species measured as mean trait values [36]. Studies have also shown individuals of the same species sampled from both introduced and native sites have a higher trait plasticity at introduced sites [37]. However, studies measuring trait plasticity have generally grown species over short-periods of time under controlled greenhouse conditions. Adults growing in the field may display different morphological and physiological traits when subjected to a wider range of resource conditions and biotic interactions in comparison to controlled greenhouse experiments [38].

Here, we use a factorial field trial to compare trait plasticity between an invasive exotic grass (*Eragrostis curvula* (Schrad.) Nees, hereafter lovegrass), and two native grasses (*Aristida personata* Henrard, hereafter purple wiregrass, and *Eragrostis sororia* Domin, hereafter woodlands lovegrass). Our study is unique as we measure how traits of key species in a community change in response to treatments, and measure these changes under 'realistic conditions' to increase the reliability of the results for explaining invasion success. We measured how grazing and fertilizer addition treatments altered abundances and several leaf traits. We hypothesised that under the existing site conditions (grazing) that the invader would display traits consistent with faster growth than the natives. We also hypothesised that under different experimental treatments the invader would exhibit higher phenotypic plasticity than the natives, evidenced by predictable changes in traits based on trends identified in the leaf economic spectrum. We then relate these results to differences in abundance of all three species between the treatments. This invasion scenario is a model system to compare plasticity because these species share life-history traits, co-exist at the same site, and native woodlands lovegrass is a congener of the invader lovegrass (Table 1).

Results

After three years of treatments, the abundance of all species was significantly correlated with abundance prior to the start of the treatments, time 0 (Table 2). Lovegrass (exotic) abundance was best explained by the additive effects of grazing and fertilizer, but not the interaction (Table 2 a). Lovegrass abundance increased across all treatments except the grazed/fertilized treatment where its abundance decreased (Fig. 1 a, 53.11%±27.80 reduction in comparison to time 0). Grazing treatments had the strongest effect ($F_{1,\ 2}$ = 139.91, P<0.008); but fertilizer treatments also had a significant effect ($F_{1,\ 54}$ = 6.14, P<0.02). Purple wiregrass abundance was best explained by the effect of the grazing treatment ($F_{1,\ 2}$ = 44.51, P<0.025). After three years, the abundance of purple wiregrass was reduced across all treatments, but most significantly in the grazing exclusion treatments (>35% decrease, Fig. 1 b), which was also the treatment where lovegrass abundance increased the most (>20% increase Fig. 1a). Woodlands lovegrass was low in abundance pre-treatment (Table 1), and increased across all treatments when compared to its abundance at year 0. It increased in abundance in the grazing exclusion treatments to more than 2% and in the grazed treatments to more than 10% (Fig. 1 c), but the difference between treatments was not significant.

In year 3, the availability of soil nutrients also varied significantly depending on the treatments (Fig. S1 and Table S1). Soil nitrate (NO_3) levels varied marginally by the interaction of grazing and fertilizer treatments ($F_{1,\ 58}$ = 3.00, P<0.09, Fig. 1 a), although overall nitrate levels were higher in the grazing exclusion treatments and highest in the grazing exclusion and unfertilized treatment. This decreasing trend in soil nitrate levels, despite the application of fertilizer, is likely reflective of increased leaching and/or use by plants and soil fauna. Soil ammonium (NH_4) levels did not vary significantly between the treatments (Fig. S1b and Table S1b). Soil phosphate (PO4) levels increased significantly with the grazing treatments ($F_{1,\ 2}$ = 30.53, P<0.03, Fig. 1 c) and the fertilizer treatments ($F_{1,\ 58}$ = 11.67, P<0.001, Fig. 1 c), but not the interaction.

Traits differed between species in the control treatment

In the control treatment (grazed/no fertilizer), mean LDMC values did not vary significantly between species ($F_{2,\ 60}$ = 0.89,

Table 2. Results from an ANOVA conducted to assess the significance of the fixed effects for LMEMs of abundance (arc-sine transformed) in year 3, with a fixed effects structure of grazing and fertilizer treatments and a co-variate of abundance in time 0, and a random effects structure of block/plot.

abundance$_{time\ 3}$	**Fixed effects**	**F values (df as subscript), *P* value**
a) Lovegrass	**grazing**	**$F_{1,\ 2}=139.91$, $P<0.008$**
	fertilizer	**$F_{1,\ 54}=6.14$, $P<0.02$**
	abundance$_{time\ 0}$	**$F_{1,\ 54}=48.79$, $P<0.0002$**
	grazing × fertilizer	$F_{1,\ 54}=1.96$, $P<0.20$
	grazing × abundance$_{time\ 0}$	$F_{1,\ 54}=1.98$, $P<0.20$
	fertilizer × abundance$_{time\ 0}$	$F_{1,\ 54}=0.38$, $P<0.60$
	grazing × fertilizer × abundance$_{time\ 0}$	$F_{1,\ 54}=0.26$, $P<0.65$
b) Purple wiregrass	**grazing**	**$F_{1,\ 2}=44.51$, $P<0.025$**
	fertilizer	$F_{1,\ 54}=0.25$, $P<0.70$
	abundance$_{time\ 0}$	**$F_{1,\ 54}=33.25$, $P<0.002$**
	grazing × fertilizer	$F_{1,\ 54}=0.02$, $P<0.90$
	grazing × abundance$_{time\ 0}$	$F_{1,\ 54}=0.16$, $P<0.70$
	fertilizer × abundance$_{time\ 0}$	$F_{1,\ 54}=1.16$, $P<0.30$
	grazing × fertilizer × abundance$_{time\ 0}$	$F_{1,\ 54}=0.31$, $P<0.60$
c) Woodlands lovegrass	grazing	$F_{1,\ 2}=4.18$, $P<0.20$
	fertilizer	$F_{1,\ 32}=4.18$, $P<0.20$
	abundance$_{time\ 0}$	**$F_{1,\ 32}=8.00$, $P<0.04$**
	grazing × fertilizer	$F_{1,\ 32}=1.18$, $P<0.30$
	grazing × abundance$_{time\ 0}$	$F_{1,\ 32}=2.29$, $P<0.20$
	fertilizer × abundance$_{time\ 0}$	$F_{1,\ 32}=0.91$, $P<0.40$
	grazing × fertilizer × abundance$_{time\ 0}$	$F_{1,\ 32}=0.13$, $P<0.80$

$P<0.50$, Fig. 2 a). Mean SLA values differed marginally between species, but contrary to expectations, with lovegrass showing a lower mean SLA value than Purple wiregrass ($F_{2,\ 60}=2.69$, $P<0.08$, Fig. 2 b), a trait indicative of a slower growing species. In agreement with expectations that lovegrass would display characteristics of a faster growing species under the grazing treatment (control), lovegrass had a significantly higher assimilation rate (*Amax*; $F_{2,\ 30}=10.1$, $P<0.002$, Fig. S2 "grazing") and photosynthetic nitrogen use efficiency (PNUE = *Amax*/leaf nitrogen; $F_{2,\ 30}=8.78$, $P<0.001$, Fig. 3 "grazing") than the two native grasses. Leaf nutrient concentrations in the control treatment did not vary significantly between species, except in the case of Leaf C:N ratios where lovegrass showed a significantly higher ratio than woodlands lovegrass ($F_{2,\ 30}=3.13$, $P<0.05$, Fig. 4 b), again contrary to expectations as this is a trait indicative of a slower growing species.

Trait plasticity in response to the treatments differed amongst species

The traits of lovegrass varied predictably with the treatments, with adult individuals showing significant differences in LDMC, SLA, PNUE, *Amax* and leaf nutrients. The traits of purple wiregrass also changed with the nutrient treatments, but woodlands lovegrass showed little change (Fig. 3–5, Fig. S2, Table 3–5 and Table S1). Differences in LDMC and SLA values for lovegrass were best explained by the interaction of grazing and fertilizer treatments (LDMC: $F_{1,\ 58}=10.10$, $P<0.002$ and SLA: $F_{1,\ 58}=3.89$, $P<0.05$, Table 3). In agreement with expectations, LDMC decreased and SLA increased for lovegrass with increasing amounts of disturbance from grazing exclusion treatments to the grazed/fertilized treatments (Fig. 5 a and b). The highest LDMC and lowest SLA values were found in both grazing exclusion treatments, whereas the lowest LDMC and highest SLA values were shown in the grazed/fertilized treatments. For purple wiregrass, differences in LDMC were not explained by the grazing or fertilizer treatments, whereas differences in SLA were explained by fertilizer treatments ($F_{1,\ 58}=4.25$, $P<0.05$, Table 3).

PNUE varied depending on the interaction between species and treatments ($F_{6,\ 42}=2.38$, $P<0.05$). Lovegrass showed a three-fold increase in PNUE between the grazed and exclusion treatments (Fig. 3). Woodlands lovegrass overall had a lower PNUE than the other grasses, but rates did not vary between treatments (Fig. 3). Purple wiregrass had a higher PNUE rate than woodlands lovegrass, but did not show a significant difference between treatments (Fig. 3). *Amax* varied similarly to PNUE depending on the interaction between species and treatments ($F_{6,\ 42}=2.84$, $P<0.02$, Fig. S2).

Differences in total nitrogen concentration were marginally significant and C:N ratio were significant for lovegrass leaves collected from fertilized and unfertilized treatments (Fig. 4 a and b, Table 4 a and Table S2 a). Lovegrass leaves showed a marginally significant increase in leaf nitrogen concentration in the fertilized plots, with the highest increase occurring in the grazed/fertilized treatment (Fig. 4 a). Consistent with this increase in N, leaf C:N ratios for lovegrass decreased when fertilizer was added (Fig. 4 b and Table S2 a). The total phosphorus concentration of lovegrass leaves varied significantly with the grazing treatment, with the highest phosphorus concentration occurring in treatments where grazing was maintained (Fig. 4 c, Table 4 a and Table S2 a). Total

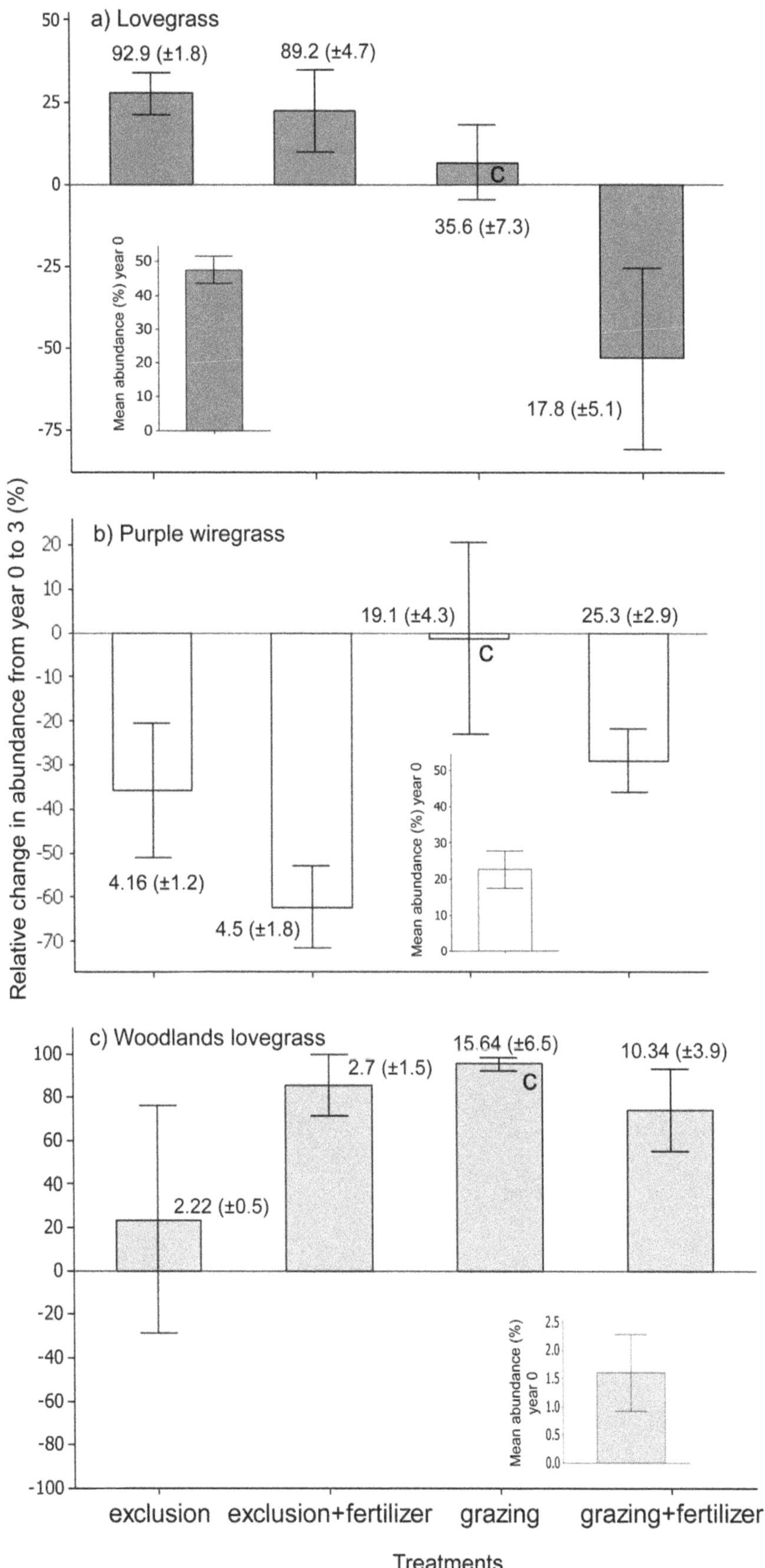

Figure 1. Relative change in abundance of each species from measurements taken prior to the start of the field trial and measurements taken again after three years of treatments (calculated as ((abundance$_{T3}$-abundance$_{T0}$)/abundance$_{T0}$)×100). The

insets show the mean abundance values (± S.E.) at time 0 and the values shown next to each bar are the mean abundance values (± S.E.) after three years. **C** indicates the control treatment grazing/no fertilizer.

leaf phosphorus for purple wiregrass did not vary significantly between treatments, but total nitrogen and C:N ratios differed marginally with the grazing treatment (Table 4 b and Table S2 b). The total nitrogen concentration of purple wiregrass leaves was lower in the grazed treatment, and C:N ratios higher in the grazed treatment for purple wiregrass. The nitrogen concentration of woodlands lovegrass leaves did not vary significantly, but did vary depending on the grazing treatments for both C:N ratios and total phosphorus concentration (Table 4 c and Table S2 c). In both cases, woodlands lovegrass leaves, collected from the exclusion treatment, had the highest C:N ratios and the highest total phosphorus concentration (Fig. 4 c).

Table 5 summarises the response of each trait to the grazing and fertilizer treatments for each of the three grass species and indicates whether the change followed or was contrary to expectations.

Discussion

Overall we found the invasive exotic grass displayed higher trait plasticity in response to the treatments than the two native grasses (Table 5). Lovegrass changed its traits according to predictions based on trends from the Leaf Economic Spectrum for all six traits, compared with only one trait for the native grasses [23]. A recent meta-analysis comparing 75 invasive/non-invasive pairs of plant species found invaders were more plastic in their response to increased resource availability than non-invaders, but plasticity was only a fitness advantage for the invasive species when resource conditions were high [39]. Increased resource availability is widely

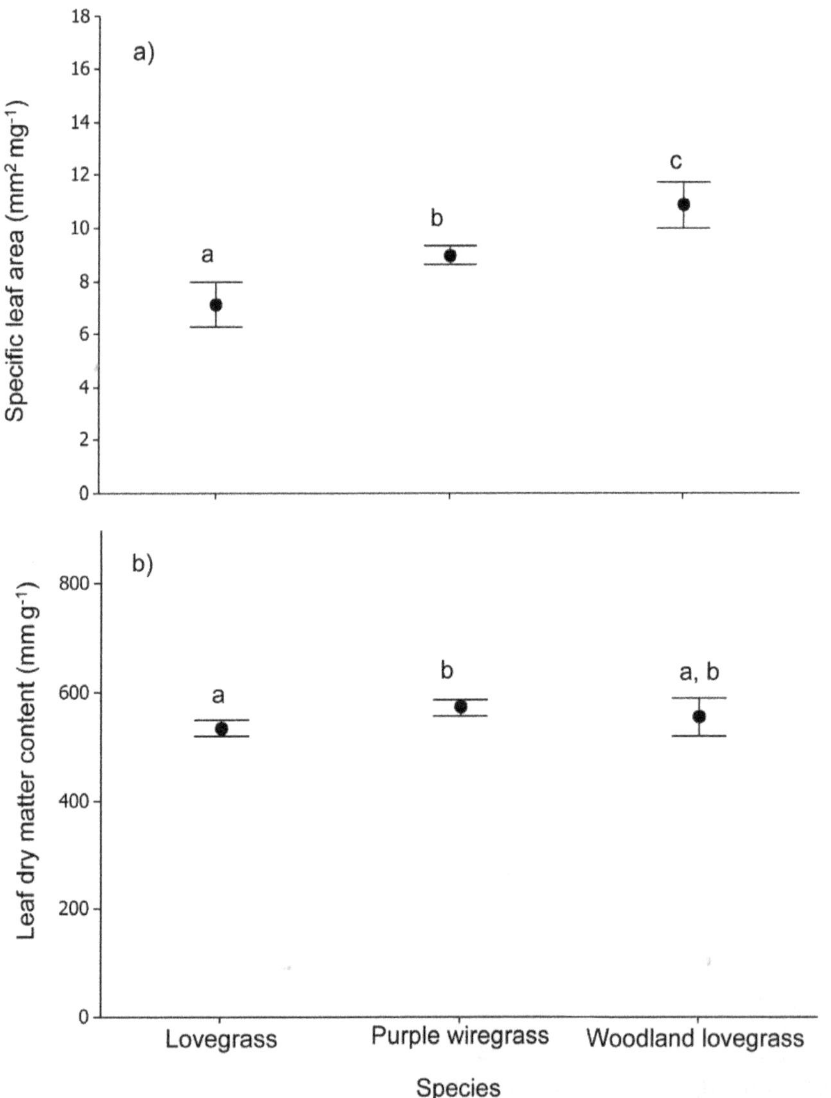

Figure 2. Comparison of mean trait values between species at the site level. Mean LDMC and SLA values (± S.E.) for each species for the grazing only treatment, which was the original disturbance at this site and therefore represents a control. Different letters indicate means are significantly different at $p<0.05$.

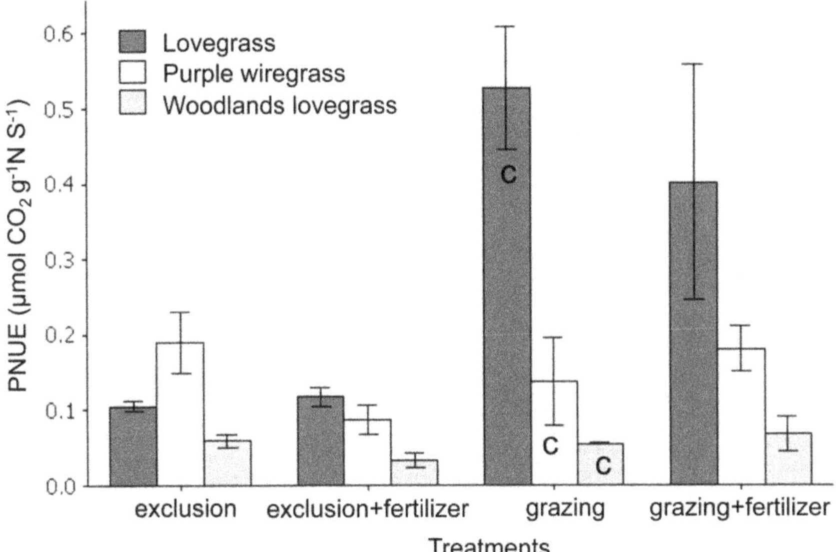

Figure 3. Mean photosynthetic nitrogen use efficiency (± SE) for each species depending on the grazing and fertilizer treatments. C indicates the control treatment grazing/no fertilizer.

agreed to promote invasion [40,41,42]. Because most studies in this meta-analysis were pot trials growing plants in the absence of competition and other biotic interactions such as grazing, it is difficult to extrapolate these findings to field conditions [38]. We found the plastic response of Lovegrass was not an advantage in the field when resources are high as increased soil nutrients, led to increased resource uptake by the exotic but also increased selective grazing pressure.

Using a three year field study, we found increased nutrients coupled with grazing decreased the abundance of the exotic, a trend also measured in the first two years of the study and published in Firn et al. [43]. Under these conditions lovegrass leaves increased in SLA, decreased in LDMC and increased in leaf total nitrogen and phosphorus concentration in response to fertilizer, but this response likely also increased its palatability to grazing livestock. An alternative explanation for these results to trait plasticity may be increased genetic diversity prior to the start of the experiment within the lovegrass population, and the treatments led to differential survival or 'filtering' of phenotypes better adapted to the different experimental conditions. This explanation is, however, unlikely as the grasses are long-lived perennials and we were careful to measure traits from large mature tussocks. Also, if genetic diversity were the explanation, genotypes in the experimental treatments would likely be subsets of those in the control; therefore, we would have expected higher trait variation in the control treatment (grazing only).

The leaf traits of lovegrass also changed in the exclusion treatments showing lower SLA and higher LDMC suggesting it has at least a comparable capacity to conserve resources as the native grasses. Using a greenhouse study, Funk [33] compared the response of several related exotic and native species from resource limiting environments and also found exotics were equally or more efficient at resource conservation. Lovegrass and both native species showed similar mean traits under the control treatment of grazing only, except the exotic had a lower SLA (indicative of a slower growing species) than purple wiregrass and a higher PNUE and *Amax* than both native species. This result suggests lovegrass is more efficient at resource capture than the native species. A study comparing traits of exotics to native species in the same region of Australia, found several C_4 exotic grass species (including lovegrass), had higher LDMC than native species [44], similarly suggesting successful exotic grasses in this region may be resource conservation specialists [44].

We also found evidence that lovegrass has a higher resource use efficiency (RUE, carbon assimilation per unit of resource, measured as PNUE) than the native grasses. Funk and Vitousek [45] compared RUE between related and co-occurring exotic and native species within Hawaii, and also found exotics had a higher RUE. PNUE increased more than three-fold in the grazed versus exclusion treatments. Leaf C:N ratios decreased in the treatments that were fertilized, but this same response was not shown by the native grasses. Higher RUE would be an advantage at the field site, as rainfall is highly variable and soil nutrients low.

Lovegrass was the dominant species at the site in year 0 and displayed the highest plasticity in response to grazing and fertilizer after three years of treatment, and this plasticity correlated with changes in its abundance in the short-term. Grime's mass ratio hypothesis [46] describes dominant species as having the highest impact on ecosystem functions. Dominant species may then be the most plastic in response to changed conditions. While intermediate/subordinate species abundance may be most influenced by the abundance of the dominant species and transient species (a species whose abundance fluctuates depending on resources) abundance responsive to environmental fluctuations [46].

Purple wiregrass, a subordinate species, was reduced in abundance across the treatments with the highest reductions occurring where grazing was excluded. Purple wiregrass did show some trait plasticity. Although it was the least disturbed treatments where purple wiregrass showed some plasticity, including an increased SLA in exclusion/fertilised treatment and an increased leaf nitrogen concentration in the exclusion treatments. In accordance with Grime's mass ratio hypothesis, increased abundance of lovegrass (>85%) in the exclusion treatments may account for the significant reduction of purple wiregrass abundance (reduced by >35%).

Woodlands lovegrass increased in abundance across the treatments. Although related to lovegrass, woodlands lovegrass did not show similar trait plasticity. This finding suggests that the

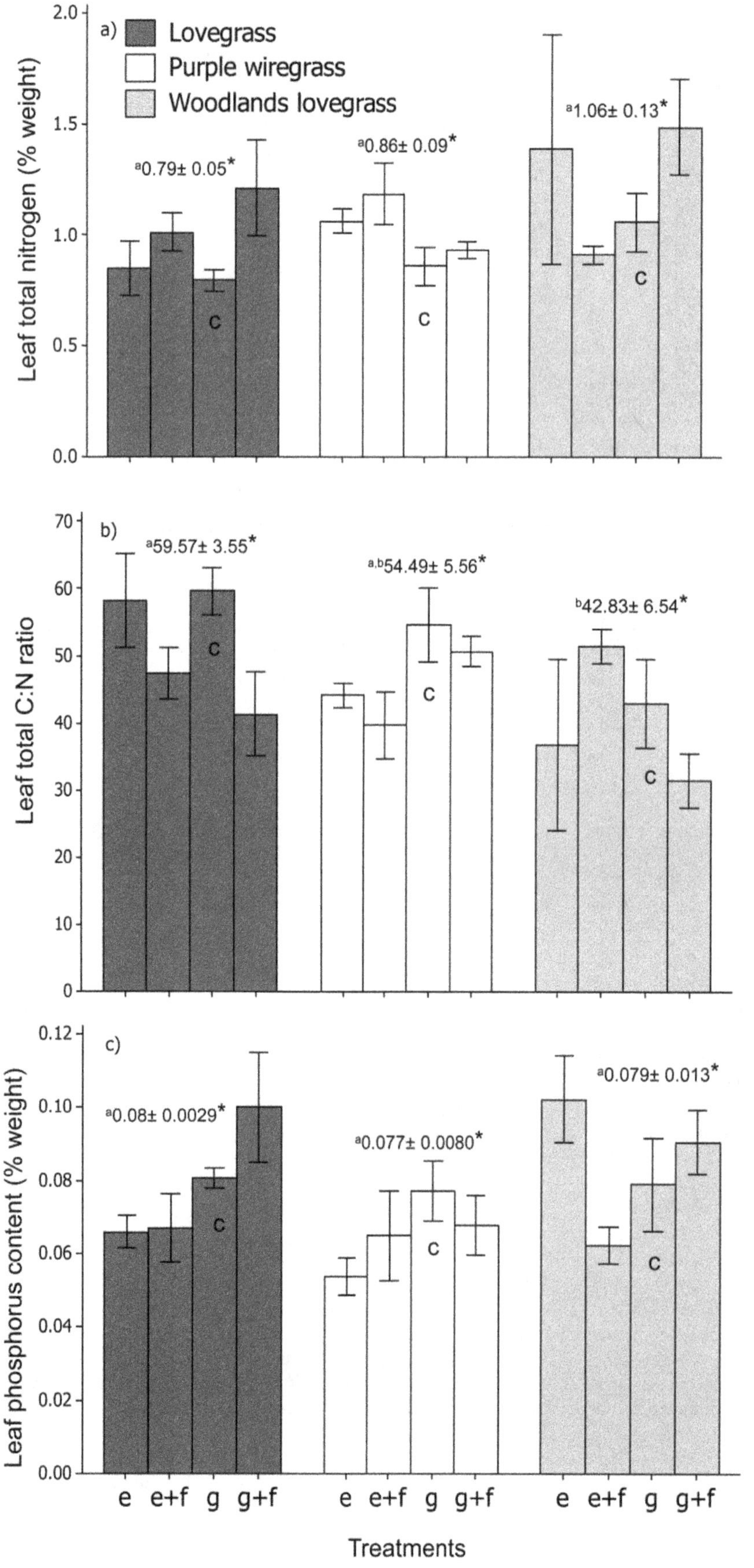
a)
Lovegrass
Purple wiregrass
Woodlands lovegrass
Leaf total nitrogen (% weight)
a0.79± 0.05*
a0.86± 0.09*
a1.06± 0.13*
c
b)
Leaf total C:N ratio
a59.57± 3.55*
a,b54.49± 5.56*
b42.83± 6.54*
c)
Leaf phosphorus content (% weight)
a0.08± 0.0029*
a0.077± 0.0080*
a0.079± 0.013*
e e+f g g+f
Treatments

Figure 4. Mean leaf nutrient concentrations (± SE) for each species depending on the grazing and fertilizer treatments. Panel a) shows leaf total nitrogen concentration (% weight), b) leaf carbon to nitrogen ratios and c) leaf phosphorus concentration (% weight). Values shown in each panel are the mean leaf nutrient concentrations (± SE) for each species at the site regardless of treatment. Different letters indicate means are significantly different at $p<0.05$. **C** indicates control treatment, grazing/no fertilizer.

increased abundance of woodlands lovegrass may be driven by other factors such as increased rainfall in year 3 of the study. Mean rainfall in year 0 was 215 mm, which was lower than the local 20 year average of 600 mm [47]; while mean rainfall in year 3 was higher than the local average at 652 mm.

At disturbed sites, invasive exotic species may be successful because of traits that allow quick growth in response to increased resource conditions [40,41], but in generally low resource environment these species likely also need traits that temper growth to survive lulls between resource pulses [48]. Pursuit of a tangible set of generic traits that distinguish exotics from natives may not be plausible or meaningful [49]; instead, we suggest the pursuit should focus on plasticity, as this may be the trait that leads to characteristically dominant plant species whether native or exotic. Lovegrass may have replaced a more plastic and characteristically dominant native species, and future studies should compare invasive and native species that are all generally considered to hold a similar hierarchical role in a community (i.e.

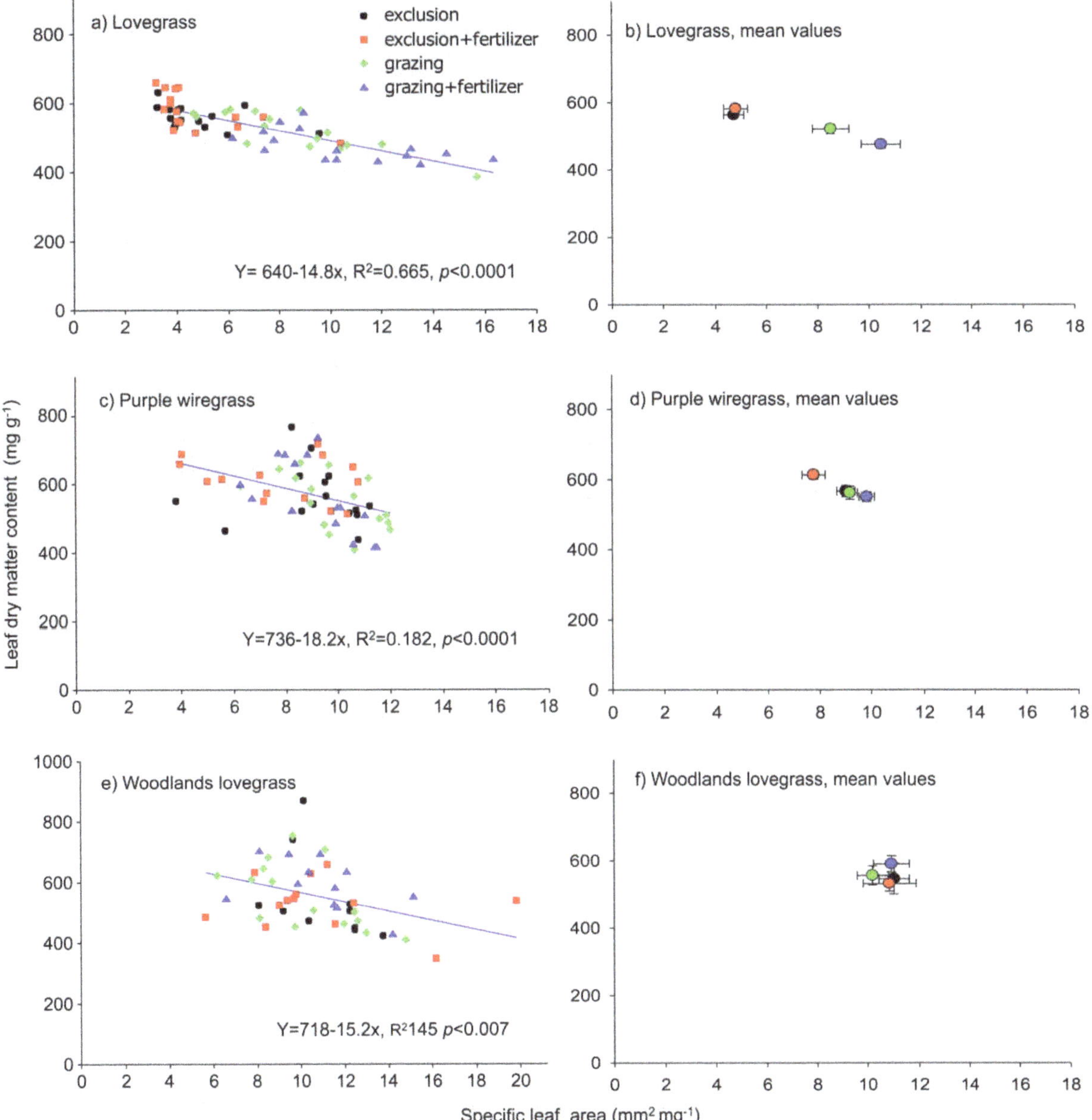

Figure 5. Correlations between LDMC and SLA values for each species depending on the four grazing and fertilizer treatments. Panel a), c) and e) show the mean LDMC and SLA values for each species collected from each plot and b), d), f) show the mean values for LDMC and SLA (±) for each treatment.

Table 3. Results from an ANOVA of LMEMs of leaf dry matter content (LDMC) and specific leaf area (SLA) for each of the grass species, with a fixed effects structure of grazing and fertilizer treatments and a random effects structure of block/plot.

Species & response variable	Fixed effects	*F* values (dfs as subscript), *P* values
a) Lovegrass LDMC (mg g^{-1})	grazing	$F_{1,2}=5.60$, $P<0.20$
	fertilizer	$F_{1,58}=2.58$, $P<0.10$
	grazing×fertilizer	$\mathbf{F_{1,58}=10.10}$, $\mathbf{P<0.002}$
SLA (mm^2 mg^{-1})	grazing	$F_{1,2}=8.85$, $P<0.10$
	fertilizer	$\mathbf{F_{1,58}=4.55}$, $\mathbf{P<0.04}$
	grazing×fertilizer	$\mathbf{F_{1,58}=3.89}$, $\mathbf{P<0.05}$
b) Purple wiregrass LDMC (mg g^{-1})	grazing	$F_{1,2}=0.40$, $P<0.60$
	fertilizer	$F_{1,58}=1.91$, $P<0.20$
	grazing×fertilizer	$F_{1,58}=1.00$, $P<0.30$
SLA (mm^2 mg^{-1})	grazing	$F_{1,2}=0.64$, $P<0.50$
	fertilizer	$\mathbf{F_{1,58}=4.25}$, $\mathbf{P<0.05}$
	grazing×fertilizer	$F_{1,58}=0.28$, $P<0.60$
c) Woodlands lovegrass LDMC (mg g^{-1})	grazing	$F_{1,2}=0.08$, $P<0.80$
	fertilizer	$F_{1,58}=0.17$, $P<0.70$
	grazing×fertilizer	$F_{1,58}=0.51$, $P<0.50$
SLA (mm^2 mg^{-1})	grazing	$F_{1,2}=0.10$, $P<0.80$
	fertilizer	$F_{1,58}=0.54$, $P<0.50$
	grazing×fertilizer	$F_{1,58}=0.35$, $P<0.60$

compare dominants to dominants, subordinates to subordinates and rare to rare).

Overall, our results show that plasticity at the species level, however, does not necessarily equate to a 'super invader'; instead, if plasticity of an undesirable species is understood, biotic interactions and resource availability can be manipulated to limit abundance. Exotic species with high species-level phenotypic plasticity may then be vulnerable to changed resource conditions as a direct result of this plasticity.

Materials and Methods

Study species

Lovegrass was introduced into Australia in the early 1900s for pasture improvement and soil conservation [50], and is now found in every Australian state, spreading into many regions where it was never intentionally introduced [51]. The increased dominance of lovegrass poses a significant threat to native biodiversity because of its ability to dominant communities, and the sustainability of production in farming communities because it is not palatable (low nutrients and crude protein content) to grazing livestock in the low

Table 4. For each of the three grass species, treatments that significantly predicted differences in total leaf nitrogen concentration, leaf carbon to nitrogen ratio and total leaf phosphorus concentration.

	Predictor variables	*F* values (df as subscript), *P* value
a) Lovegrass leaves		
Total nitrogen (% weight)	Fertilizer treatment	$F_1=4.67$, $P<0.06$
C:N ratio	Fertilizer treatment	$F_1=7.26$, $P<0.02$
Total phosphorus (% weight)	Grazing treatment	$F_1=6.16$, $P<0.03$
b) Purple wiregrass leaves		
Total nitrogen (% weight)	Grazing treatment	$F_1=5.33$, $P<0.08$
C:N ratio	Grazing treatment	$F_1=7.17$, $P<0.06$
Total phosphorus (% weight)	NS	
d) Woodlands lovegrass leaves		
Total nitrogen (% weight)	NS	
C:N ratio	Grazing treatment	$F_1=7.82$, $P<0.06$
Total phosphorus (% weight)	Grazing treatment	$F_1=9.33$, $P<0.02$

Table 5. Summary of the traits of each grass species that showed a plastic response to the treatments according to expectations "√", contrary to expectations "X" or traits that did not change in response to the treatments "-".

Trait	*Eragrostis curvula* **lovegrass, Exotic grass**	*Aristida personata* **purple wiregrass Native grass**	*Eragrostis sororia* **woodlands lovegrass, Native grass**
LDMC (mg g^{-1})	✓ (grazing & fertilizer)	-	-
SLA (mm^2 mg^{-1})	✓ (grazing & fertilizer)	✓ (fertilizer)	-
Amax (μmol co_2 g^{-1} s^{-1})	✓ (grazing)	-	-
PNUE (μmol co_2 g^{-1}N s^{-1})	✓ (grazing)	-	-
Leaf total Nitrogen (%)	✓ (fertilizer)	X (grazing)	-
Leaf total C:N	✓ (fertilizer)	X (grazing)	X (grazing)
Leaf Phosphorus (%)	✓ (grazing)	X (grazing)	X (grazing)

productivity regions where it is spreading, and difficult to control [51].

In June 2006, we established a large field trial on a private cattle grazing property in the Millmerran region of south-western Queensland, Australia [43]. The field site had been grazed by cattle with a low stocking rate since 1980 and has never been cultivated or fertilized. Lovegrass was first identified on the property in 1998 by the landholder. Average rainfall of this area is 600 mm p.a. [47] with two-thirds of the rain occurring during the summer months from October to April. The soil is a yellow sodosol derived from sandstone, which is characteristically low in nutrients, slightly acidic (ranging from pH 4.8 to 5.9), low in water holding capacity, and highly susceptible to compaction [47].

Data collection and sampling design

In this experiment, we measured traits and abundances from a subset of treatments and plots from a larger field trial with a randomized split-plot design [43]. In June 2006, four large blocks (35×40 m) were established randomly in a pasture dominated by lovegrass. Two blocks were fenced to exclude grazing by cattle and limit access by other native and exotic gazers (e.g. kangaroos, wallabies, hares and rabbits). The other two blocks were left open to grazing. In each block, we established 48 plots (each was 9 m^2 in size with an additional 4 m^2 buffer between each plot). In this study, we sampled 16 plots in each block, 8 fertilized plots and 8 unfertilized plots. We applied a slow-release fertilizer to half of the plots in a pellet form (N 21.6%, P 1.1%, K 4.1%) at a low application rate of 2 kg/ha at the start of each growing season from 2006 to 2009, which can begin anytime between October and December depending on rainfall. In this experiment, the grazed/-unfertilized treatment is considered the control treatment, because this was the disturbance acting on the site prior to the start of the experiment. Firn *et al.* (2010) contains species abundance results from 2006 to 2008.

In December 2009, prior to applying the fourth year of treatments, we measured species abundance and leaf traits (specific leaf area, leaf dry matter content, *Amax* and leaf nitrogen, phosphorus, and carbon to nitrogen levels) from the grazing and fertilizer treatment combinations. The abundances of all species were recorded within each plot in the central 9 m^2 section using the point-intercept method (modified from [52]). A 4 mm dowel was placed vertically on set points along a grid of 100 points. Relative abundance was measured by identifying and counting each leaf, stem and inflorescence that touched the dowel at each point along the grid.

To measure SLA, LDMC, LN traits, we collected five young but fully expanded leaves from three mature individuals of each species within each of the 64 plots, using the standardised protocols detailed by Cornelissen et al. [53] and the rehydration methods proposed by Garnier et al. [54]. Because of the low abundance of woodlands lovegrass, we did not find individuals of this species in all plots, but we were able to collect samples from 42 plots. The leaves collected from each species in each plot were combined (leaves of most species were small), weighed and scanned for area, using a flat bed scanner (Epson perfection V300) and image analysis software (ImageJ, [55]). Leaf samples were then dried in an oven for 72 hours at 60°C and re-weighed.

These leaf samples were then bulk sampled by species and treatment and analysed for total nitrogen and carbon concentration using a LECO CNS 2000 combustion analyser set at 1100°C. Total leaf phosphorus concentration was measured using a Varian Vista Pro ICPOES on samples digested in 5:1 nitric:perchloric acid (six samples were analysed per species per treatment) [56]. We measured leaf nitrogen and phosphorus concentration because extensive research has shown a stronger relationship between these nutrients within the leaf economic spectrum than other nutrients [57]. Six soil samples (core radius of 5 cm, 10 cm deep) were collected from each plot at the same time as the botanical surveys. Available soil nitrate, ammonium and phosphorus were analysed with colorimetric methods using a SEAL AQ2 [58]. Soil and leaf nutrient analyses were conducted by the Analytical Services Unit, School of Land and Food Sciences, the University of Queensland.

In September 2010, we measured assimilation rates (A_{max}) of eight individuals of lovegrass and purple wiregrass within four fertilised plots and four unfertilized plots in each of the four blocks. We were very careful with our leaf selection, choosing young, intact leaves with a healthy appearance and growing in full sun. We took measurements between 6:00 am and 10:00 am over five days to standardize measurements. Because of the smaller size of woodlands lovegrass leaves and its low abundance, we were only able to measure four individuals within each of the treatments. We used a LI-COR LI-6400 photosynthesis system and the narrow leaf chamber LI-COR LI-6400-11. For assimilation rates ambient CO_2 conditions were maintained at 400 μmol L^{-1}, relative humidity at 40–50%, leaf temperature at 22–23°C and PAR at 1300–1530 μL L^{-1}. To fill the chamber, multiple leaves growing in full sun were selected from each individual and leaf area was measured with a LI-COR, LI-3000c Portable Area Meter. We calculated photosynthetic nitrogen use efficiency (PNUE) as the ratio of *Amax* to leaf nitrogen.

Data analysis

To analyse the effects of different treatments on species abundance and leaf traits, we developed Linear Mixed Effects Models (hereafter LMEM), using R 2.12.1 (R Foundation for Statistical Computing©) and the nlme package. We modelled the abundance (arc-sine transformed) of each study species in time 3 as a function of grazing and fertilizer treatments and abundance at time 0 as a covariate with a nested random effects structure of block/plot. We also modelled each of the leaf traits and soil nutrient levels as a function of the grazing and fertilizer treatments with a nested random effects structure of block/plot. Maximum likelihood was used when comparing nested models to simplify the model for fixed effects [59,60]. We used diagnostic plots to check model assumptions [59]; there was no evidence of correlation of observations within groups and we assumed that within group errors were normally distributed. Finally, we used ANOVAs to assess the significance of the fixed effects within the LMEMs (Pinheiro and Bates, 2000). To analyse leaf nutrient concentrations, we used ANOVAs as opposed to LMEMs because these values were measured from bulked samples.

Supporting Information

Table S1 Effect of treatments on Soil NO_3, NH_4 and PO_4 levels in year 4. Results of an ANOVA conducted to assess the significance of the fixed effects for LMEMs of soil nutrient levels, with a fixed effects structure of grazing and fertilizer treatments, and a random effects structure of block/plot.

Table S2 Effect of treatments on leaf nitrogen concentration, leaf carbon to nitrogen ratio and total leaf phosphorus concentration. Results of an ANOVA conducted for each of the three grass species.

Figure S1 Soil nitrate, ammonium and phosphate levels taken across the treatments in year 3 of the field trial. e = grazing exclusion treatment, e+f = grazing exclusion and fertilized treatment, g = grazing treatment, and g+f = grazing and fertilized treatment.

Figure S2 Mean assimilation rates (± SE) for each species depending on the grazing and fertilizer treatments. **C** indicates the control treatment grazing/no fertilizer.

Acknowledgments

Thank you to the land-owner for generously permitting us to set-up this experiment on the property and Katrina Cousins, Karri Hartley and Richard Unwin for help in the field and laboratory. Thank you to Dr. S. McIntyre and Dr. H. Murphy from CSIRO Ecosystem Sciences for commenting on an earlier draft of this manuscript.

Author Contributions

Conceived and designed the experiments: JF SMP YMB. Performed the experiments: JF. Analyzed the data: JF. Wrote the paper: JF SMP YMB.

References

1. Sax DF, Brown JH (2000) The paradox of invasion. Global Ecology and Biogeography 9: 363–371.
2. Rout ME, Callaway RM (2009) An Invasive Plant Paradox. Science 324: 734–735.
3. MacDougall AS, Turkington R (2005) Are invasive species the drivers or passengers of change in degraded ecosystems? Ecology 86: 42–55.
4. Hobbs RJ, Atkins L (1988) Effect of disturbance and nutrient addition on native and introduced annuals in plant communities in the Western Australian wheatbelt. Australian Journal of Ecology 13: 171–179.
5. HilleRisLambers J, Yelenik SG, Colman BP, Levine JM (2010) California annual grass invaders: the drivers or passengers of change? Journal of Ecology 98: 1147–1156.
6. Seabloom EW, Harpole WS, Reichman OJ, Tilman D (2003) Invasion, competitive dominance, and resource use by exotic and native California grassland species. Proceedings of the National Academy of Science 100: 13384–13389.
7. Suding KN, Goldberg D (2001) Do disturbances alter competitive hierarchies? Mechanisms of change following gap creation. Ecology 82: 2133–2149.
8. Firn J, Rout T, Possingham HP, Buckley YM (2008) Managing beyond the invader: manipulating disturbance of natives simplifies control efforts. Journal of Applied Ecology 45: 1143–1151.
9. Hobbs RJ, Arico S, Aronson J, Baron JS, Bridgewater P, et al. (2006) Novel Ecosystems: theoretical and management aspects of the new ecological world order. Global Ecology and Biogeography 15: 1–7.
10. Zavaleta ES, Hobbs RJ, Mooney HA (2001) Viewing invasive species removal in a whole-ecosystem context. Trends in Ecology and Evolution 16: 454–459.
11. Pickett STA, Cadenasso ML, Meiners SJ (2008) Ever since Clements: from succession to vegetation dynamics and understanding intervention. Applied Vegetation Science 12: 9–21.
12. McIntyre S, Lavorel S (2006) A conceptual model of land use effects on the structure and function of herbaceous vegetation. Agriculture, Ecosystems and Environment 2007: 11–21.
13. Suding KN, Lavorel S, Chapin FS, Cornelissen JHC, Diaz S, et al. (2008) Scaling environmental change through the community-level: a trait-based response-and-effect framework for plants. Global Change Biology 14: 1125–1140.
14. Daehler CC (2003) Performance comparisons of co-occurring native and alien invasive plants: implications for conservation and restoration. Annual Review of Ecology and Systematics 34: 183–211.
15. Bazzaz FA (1986) Life history of colonizing plants: some demographic, genetic and physiological features. In: Mooney HA, Drake JA, eds. Ecology of Biological Invasion of North America and Hawaii. New York: Springer-Verlag. pp 96–110.
16. Rejmanek M, Richardson DM (1996) What attributes make some plants species more invasive? Ecology 77: 1655–1661.
17. Baker HG (1964) Characteristics and modes of origin of weeds. In: Baker HG, Ledyard Stebbins G, eds. The Genetics of Colonizing Species. New York: Academic Press Inc. pp 147–172.
18. Leishman MR, Haslehurst T, Ares A, Baruch Z (2007) Leaf trait relationships of native and invasive plants: community- and global-scale comparisons. New Phytologist 176: 635–643.
19. Baruch Z, Goldstein G (1999) Leaf contruction cost, nutrient concentration, and net CO_2 assimilation of native and invasive species in Hawaii. Oecologia 121: 183–192.
20. Durand LZ, Goldstein G (2001) Growth, leaf characteristics and spore production in native and invasive tree ferns in Hawaii. American Fern Society 91: 25–35.
21. Gulias J, Flexas J, Mus M, Cifre J, Lefi E, et al. (2003) Relationship between maximum leaf photosynthesis, nitrogen content and specific leaf area in Balearic endemic and non-endemic Mediterranean species. Annals of Botany 92: 215–222.
22. Grotkopp E, Rejmanek M, Rost TL (2002) Toward a causal explanation of plant invasiveness: seedling growth and life-history strategies of 29 pine (*Pinus*) species. American Naturalist 159: 396–419.
23. Wright IJ, Reich PB, Westoby M, Ackerly DD, Baruch Z, et al. (2004) The worldwide leaf economics spectrum. Nature 428: 821–827.
24. Westoby M, Wright IJ (2006) Land-plant ecology on the basis of functional traits. Trends in Ecology & Evolution 21: 261–268.
25. Diaz S, Hodgson JG, Thompson K, Cabido M, Cornelissen JHC, et al. (2004) The plant traits that drive ecosystems: Evidence from three continents. Journal of Vegetation Science 15: 295–304.
26. Moles AT, Gruber MAM, Bonser S (2008) A new framework for predicting invasive plant species. Journal of Ecology 96: 13–17.
27. Leishman MR, Thomson VP, Cooke J (2010) Native and exotic invasive plants have fundamentally similar carbon capture strategies. Journal of Ecology 98: 28–42.
28. Albert CH, Grassein F, Schurr FM, Vieilledent G, Violle C (2011) When and how should intraspecific variability be considered in trait-based plant ecology? Perspectives in Plant Ecology, Evolution and Systematics doi:10.1016/j.ppees.2011.04.003. pp 1–9.
29. Albert CH, Thuiller W, Yoccoz NG, Soudant A, Boucher F, et al. (2010) Intraspecific functional variabillity: extent, structure and sources of variation. Journal of Ecology 98: 604–613.
30. Richards CL, Bossdorf O, Muth NZ, Gurevitch J, Pigliucci M (2006) Jack of all trades, master of some? On the role of phenotypic plasticity in plant invasions. Ecology Letters 9: 981–993.

31. DeWitt TJ, Sih A, Wilson DS (1998) Costs and limits of phenotypic plasticity. Trends in Ecology and Evolution 13: 77–81.
32. Elton CS (1958) The Ecology of Invasions by Animals and Plants. London: Methuen and CO Ltd.
33. Funk JL (2008) Differences in plasticity between invasive and native plants from a low resource environment. Journal of Ecology 96: 1162–1173.
34. Burns JH, Winn AA (2006) A comparison of plastic responses to competition by invasive and non-invasive congeners in Commelinaceae. Biological Invasions 8: 797–807.
35. Muth NZ, Pigliucci M (2007) Implementation of a novel framework for assessing species plasticity in biological invasions: responses of Centaurea and Crepis to phosphorus and water availability. Journal of Ecology 95: 1001–1013.
36. Godoy O, Valladares F, Castro-Diez P (2011) Multispecies comparison reveals that invasive and native plants differ in their traits but not in their plasticity. Functional Ecology 25: 1428–1259.
37. Zou J, Rogers WE, Siemann E (2007) Differences in morphological and physiological traits between native and invasive populations of *Sapium sebiferum*. Functional Ecology 21: 721–730.
38. Hulme PE (2008) Phenotypic plasticity and plant invasions: is it all Jack? Functional Ecology 22: 3–7.
39. Davidson AM, Jennions M, Nicotra AB (2011) Do invasive species show higher phenotypic plasticity than native species and, if so, is it adaptive? A meta-analyses. Ecology Letters 14: 419–431.
40. Melbourne BA, Cornell HV, Davies KF, Dugaw CJ, Elmendorf S, et al. (2007) Invasion in a heterogeneous world: resistance, coexistence or hostile takeover? Ecology Letters 10: 77–94.
41. Davis MA, Grime JP, Thompson K (2000) Fluctuating resources in plant communities: a general theory of invasibility. Journal of Ecology 88: 528–534.
42. Pysek P, Richardson DM (2007) Traits associated with invasiveness in alien plants: where do we stand? In: Nentwig W, ed. Biological Invasions. Berlin Heidelberg: Springer-Verlag.
43. Firn J, House APN, Buckley YM (2010) Alternative states models provide an effective framework for invasive species control and restoration of native communities. Journal of Applied Ecology 47: 96–105.
44. McIntyre S, Martin TG, Heard KM, Kinloch J (2005) Plant traits predict impact of invading species: an analyses of herbaceous vegetation in the subtropics. Australian Journal of Botany 53: 757–770.
45. Funk JL, Vitousek PM (2007) Resource-use efficiency and plant invasion in low-resource systems. Nature 446: 1079–1081.
46. Grime JP (1998) Benefits of plant diversity to ecosystems: immediate, filter and founder effects. Journal of Ecology 86: 902–910.
47. Biggs A, Coutts A, Harris PS, eds. (1999) Central Darling Down Land Management Manual: Department of Primary Industries.
48. Chesson P, Gebauer RLE, Schwinning S, Huntly N, Wiegand K, et al. (2004) Resource pulses, species interactions, and diversity maintenance in arid and semi-arid environments. Oecologia 141: 236–253.
49. Thompson K, Davis MA (2011) Why research on traits of invasive plants tells us very little. Trends in Ecology and Evolution 26: 155–156.
50. Leigh JH, Davidson RL (1968) *Eragrostis curvula* (Schrad.) Nees and some other African lovegrasses. Plant Introduction Review 5: 21–46.
51. Firn J (2009) African lovegrass in Australia: a valuable pasture species or embarrassing invader. Tropical Grasslands 43: 86–97.
52. Everson CS (1987) A comparison of six methods of botanical analysis in the montane grasslands of Natal. Vegetatio 73: 47–51.
53. Cornelissen JHC, Lavorel S, Garnier E, Diaz S, Buchmann N, et al. (2003) A handbook of protocols for standardised and easy measurement of plant functional traits worldwide. Australian Journal of Botany 51: 335–380.
54. Garnier E, Shipley B, Roumet C, Laurent G (2001) A standardized protocol for the determination of specific leaf area and leaf dry matter content. Functional Ecology 15: 688–695.
55. Rasband WS (1997–2011) ImageJ website, U.S. National Institutes of Health, Betheseda Maryland, U.S.A. http://imagejnihgov/ij/. Accessed 2012 March 29.
56. Reuter DJ, Robinson JB (1986) Plant Analysis and Interpretation Manual. Melbourne, Australia: Inkata Press.
57. Wright IJ, Reich PB, Cornelissen JHC, Falster DS, Garnier E, et al. (2005) Assessing the generality of global leaf trait relationships. New Phytologist 166: 485–496.
58. Raymont GE, Higginson FR (1992) The Australian Handbook of Soil and Water Chemical Methods. Melbourne: Inkata Press.
59. Pinheiro JC, Bates DM (2000) Mixed-Effects Models in S and S-Plus; Chambers J, Eddy W, Hardle W, Sheather S, Tierney L, eds. New York: Springer Verlag.
60. Ives AR, Zhu J (2006) Statistics for correlated data: phylogenies, space, and time. Ecological Applications 16: 20–32.
61. Voight PW, Kneebone WR, McIlvain EH, Shoop MC, Webster JE (1968) Palatability, chemical composition, and animal gains from selection pf Weeping Lovegrass, Eragrostis curvula (Schrad.) Nees. Agronomy Journal 62: 673–676.
62. Mitchell M (2002) Native Grasses: an identification handbook for Temperate Australia. Collingwood: Landlinks Press 42 pages p.
63. Anderson E (2003) Plants of Central Queensland. Brisbane: Department of Primary Industries. 271 p.
64. Henry DR, Hall TJ, Jordan DJ, Milson JA, Schefe CM, et al. (1995) Pasture Plants of Southern Inland Queensland; Pavasaris S, ed. Brisbane: Queensland Department of Primary Industries.
65. Sharp D, Simon BK (2002) AusGrass: Grasses of Australia Australian Biological Resources Study, Canberra and Environmental Protection Agency, Queensland.

17

Photosynthesis in *Chromera velia* Represents a Simple System with High Efficiency

Antonietta Quigg[1,2*], Eva Kotabová[3,4], Jana Jarešová[3,4], Radek Kaňa[3,4], Jiří Šetlík[3], Barbora Šedivá[3], Ondřej Komárek[3], Ondřej Prášil[3,4]

1 Department of Marine Biology, Texas A&M University at Galveston, Galveston, Texas, United States of America, **2** Department of Oceanography, Texas A&M University, College Station, Texas, United States of America, **3** Institute of Microbiology, Academy of Sciences of the Czech Republic, Třeboň, Czech Republic, **4** Faculty of Sciences, University of South Bohemia, České Budějovice, Czech Republic

Abstract

Chromera velia (Alveolata) is a close relative to apicomplexan parasites with a functional photosynthetic plastid. Even though *C. velia* has a primitive complement of pigments (lacks chlorophyll *c*) and uses an ancient type II form of RuBISCO, we found that its photosynthesis is very efficient with the ability to acclimate to a wide range of irradiances. *C. velia* maintain similar maximal photosynthetic rates when grown under continual light-limited (low light) or light-saturated (high light) conditions. This flexible acclimation to continuous light is provided by an increase of the chlorophyll content and photosystem II connectivity under light limited conditions and by an increase in the content of protective carotenoids together with stimulation of effective non-photochemical quenching under high light. *C. velia* is able to significantly increase photosynthetic rates when grown under a light-dark cycle with sinusoidal changes in light intensity. Photosynthetic activities were nonlinearly related to light intensity, with maximum performance measured at mid-morning. *C. velia* efficiently acclimates to changing irradiance by stimulation of photorespiration and non-photochemical quenching, thus avoiding any measurable photoinhibition. We suggest that the very high CO_2 assimilation rates under sinusoidal light regime are allowed by activation of the oxygen consuming process (possibly chlororespiration) that maintains high efficiency of RuBISCO (type II). Despite the overall simplicity of the *C. velia* photosynthetic system, it operates with great efficiency.

Editor: Senjie Lin, University of Connecticut, United States of America

Funding: This research has been supported by the Grant Agency of the Czech Academy of Sciences, grant GAAV IAA601410907 and by Grant Agency of the Czech Republic, grant GACR P501/12/G055. International Research Travel Awards from Texas A&M Universities International programs office to A.Q. supported several visits to the Institute of Microbiology in Třeboň, Czech Republic were the work was conducted. The Institute of Microbiology is funded by the Czech Academy of Sciences (contract RVO 61388971). The funders had no role in study design, data collection and analysis, decision to publish, or preparation of the manuscript.

Competing Interests: The authors have declared that no competing interests exist.

* E-mail: quigga@tamug.edu

These authors contributed equally to this work.

Introduction

Most of all the diverse assemblage of eukaryotic oxygenic photosynthetic autotrophs present today belong to either the green (chlorophyll *b*-containing) or red (chlorophyll *c*-containing) plastid lineages [1,2,3]. There is however a subgroup of non-photosynthetic relatives, thought to have lost their plastids secondarily. The apicomplexans, which are non-photosynthetic sporozoan parasites (e.g., the malaria organism, *Plasmodium falciparum*), have a relic unpigmented plastid (apicoplast) indicating that the ancestors of these organisms were once photosynthetic, and that part of the plastid metabolic machinery is indispensable to the present organism. This may include the fatty acid synthesis enzymes [4] and isoprenoid biosynthesis [5]. Recently, two distinct families were described - *Chromeraceae* and *Vitrellaceae* [6] which include *Chromera velia* and *Vitrella brassicaformis* respectively. While apicomplexans are not currently placed in the polyphyletic group 'algae' by taxonomists, their algal roots have been long acknowledged [7,8,9].

C. velia (Chromerida, Alveolata) associated with the scleractinian coral *Leptastrea purpurea* was isolated in 2001 by Moore et al. [10] from Sydney Harbour (Australia). This is the first extant relative of apicomplexan parasites discovered to have a heritable functional photosynthetic plastid. *C. velia* plastid shares an origin with the apicoplasts, is surrounded by four membranes, pigmented with chlorophyll (chl) *a* and various carotenoids. Gene phylogenies relate the apicoplasts to the chloroplasts of peridinin-containing dinoflagellates [3,11,12]. Yet, *C. velia* plastids differ from those in dinoflagellates: they lack the accessory pigment chl *c* [10] and operate modified heterotrophic heme synthesis pathway [13]. It has been suggested that the ancestor of peridinin dinoflagellates and apicomplexans possessed a photosynthetic chromalveolate plastid containing chl *a* and *c* [7,8]. Other dinoflagellates, in a series of complex events not discussed herein [see 7,11,12], now have fucoxanthin-containing plastids or green plastids, and up to 50% of others lost (and did not replace) their chromalveolate plastid resulting in heterotrophy [7,8]. Nonetheless, many dinoflagellates today can still switch between autotrophy, mixotrophy and heterotrophy depending on environmental conditions.

Zooxanthellae, a group of symbiotic dinoflagellates, have important relationships with corals and other invertebrates [14,15,16,17]. Like many zooxanthellae, *C. velia* can live independently from its host and is culturable. With the discovery of *C. velia*, we now have a model organism to study apicomplexan evolution and zooxanthellae photosynthesis.

There has been a flurry of publications since the pivotal paper of Moore et al. [10] announcing the discovery of this unique phototroph. Keeling [9], Oborník et al. [18] and Janouškovec et al. [19] have re-evaluated and revised current views on plastid distribution in the red lineage with *C. velia* now providing an important "missing link". Keeling [9] reported that the appearance of Chromera has, along with other cryptic plastids, provided important support for the chromalveolate hypothesis proposed by Cavalier-Smith some 10 years earlier (see 7) and more importantly, transformed the view of plastid distribution in the red lineage. Janouškovec et al. [19] conducted a careful phylogenetic analysis of plastid genomes to find extant plastids of apicomplexans and dinoflagellates were inherited by linear descent from a common red algal endosymbiont. It appears that plastids of heterokont algae and apicomplexa all derive from the same endosymbiosis. Okamoto and McFadden [20] concluded that the discovery of *C. velia* ends the debate on the origin of apicoplasts, providing strong evidence for origins with a red algal endosymbiont. Oborník et al. [18] instead focused on the pathway by which apicomplexa evolved from free-living heterotrophs through phototrophs to being the omnipresent obligatory intracellular parasite. More recently, Oborník et al. [6,21] presented a careful examination of the morphology and ultra structure of multiple life cycle stages; Weatherby et al. [22] provide details of the cell surface and flagella morphology of the motile form of *C. velia*; Sutak et al. [23] provide details of a nonreductive iron uptake mechanism while Guo et al. [24] reported that both nutrient concentrations and salinity are important in regulating the transformation of immotitle-motile *C. velia*. Kořený at al. [13] found that unlike other eukaryotic phototrophs, *C. velia* synthesizes chl from glycine and succinyl-CoA, Kotabová et al. [25] found that fast de-epoxidation of violaxanthin in *C. velia* enables highly efficient non-photochemical fluorescence quenching (NPQ) and Pan et al. [26] published a detailed phylogenetic analysis of the light-harvesting antennae of *C. velia*. Recently, Botté et al. [27] identified plant-like galactolipids and Leblond et al. [28] determined sterols in a Chromera. Given the potential for *C. velia* in the screening of anti-apicoplast drugs for the treatment of malaria (*Plasmodium* sp.) and diseases caused by related parasites (e.g., *Toxoplasma*), Okamoto and McFadden [20] concluded that "the little alga from the bottom of Sydney Harbour" may eventually be enlisted in developing new treatments for these diseases using herbicides which will attack the photosynthetic apparatus.

However, to date, there is no information on the properties and efficiency of photosynthesis in *C. velia* which displays simple pigmentation (only chl *a* together with violaxanthin, ß,ß-carotene and a novel isofucoxanthin-like carotenoid but without any accessory pigments like chl *c* [10]) and utilizes the primitive form (type II) of RuBisCO [19]. Herein, we investigated photosynthesis in *C. velia* in detail (electron transport and O_2 evolution in Photosystem (PS) II, ^{14}C fixation rates in Calvin-Benson cycle, pigment composition) together with its ability for photoacclimation to both continuous "low" and "high" light (15 and 200 µmol photons m^{-2} s^{-1}) as well as its response to "natural" changes in irradiance provided with a sinusoidal light:dark regime. We found that photosynthesis in *C. velia* represents a simple system with surprisingly high efficiency. *C. velia* protects itself against photoinhibition at high irradiance by utilizing NPQ, energy spillover and photorespiration. At low irradiances *C. velia* maximizes its performance by reorganizing its antennae to ensure a constant light-dependent rate of photosynthesis across all growth environments.

Methods

Organism and culture conditions

C. velia (strain RM12) was maintained in f/2 culture medium and 28°C. For low light (LL) and high light (HL) experiments, cells were kept in semi-continuous batch growth with 24 h continuous light of 15 and 200 µmol photons m^{-2} s^{-1} respectively. For the sinusoidal light:dark cycle experiments, cells were grown under a 12:12 h light:dark cycle. Light intensity was controlled by computer [29] with a midday peak of 500 µmol photons m^{-2} s^{-1} (dashed curves e.g. in Fig. 1). Nutrient concentrations were saturating, pH buffered at 8.2, bubbling ensured CO_2 supply and mixing. Cells were counted daily and size determined with a calibrated Coulter Counter (Beckman Mulitsizer III) equipped with a 70 µm aperture; cell densities were maintained between $1.0–2.0\times10^6$ cells ml^{-1} by periodic dilutions with fresh f/2 medium. Specific growth rates (µ; day^{-1}) were determined from $\mu = (\ln c - \ln c_0)/t - t_0$ where c is the cell concentration and t is measured in days once cells had acclimated to the respective light treatment.

Cell composition

Cultures (n≥3) were harvested onto precombusted (400°C, 4 hrs) GF/F filters and frozen until analysis of cellular carbon and nitrogen. Samples for pigment extractions were, in the same way as for fluorescence measurements, dark acclimated for 20 min prior to collection on GF/F filters and frozen immediately. Thawed filters were soaked in 100% methanol at −20°C and subsequently disrupted using a mechanical tissue grinder. Centrifugation (12000 g, 15 min) immediately before HPLC analysis on an Agilent 1200 chromatography system equipped with the DAD detector removed debris. Pigments were separated using a Phenomenex column (Luna 3µ C8, size 100×4.60 mm) at 35°C by applying a 0.028 M ammonium acetate/MeOH gradient (20/80) with a flow rate of 0.8 ml/min [30]. Eluted pigments were quantified by their absorption at 440 nm with consideration of their different extinction coefficients. For chl quantification, we used the same extract, but measured the sample on a UV/VIS spectrophotometer (Unicam UV 550, Thermo Spectronic, UK). Chl *a* concentration was calculated according to et al. [31].

Fluorescence emission spectra

Room temperature fluorescence emission spectra were measured in cuvette with a SM-9000 spectrophotometer (Photon Systems Instruments, Czech Republic) for blue light excitation (λ = 464 nm) with a dark acclimated sample in the F_m (maximum fluorescence) state induced by a saturating pulse according to Kaňa et al. [32]. 77 K Chl fluorescence emission spectra were measured using the Aminco Bowman series 2 spectrofluorometer (Thermo Fisher Scientific, USA). The excitation was at 435 nm and 4 nm slit width. The emission spectra were recorded in 0.4 nm steps from 600 to 800 nm, with 1 nm slit width. The instrument function was corrected by dividing raw emission spectra by simultaneously recorded signal from the reference diode. Spectra were normalized to 690 nm. Fluorescence nomenclature is summarized in Table 1.

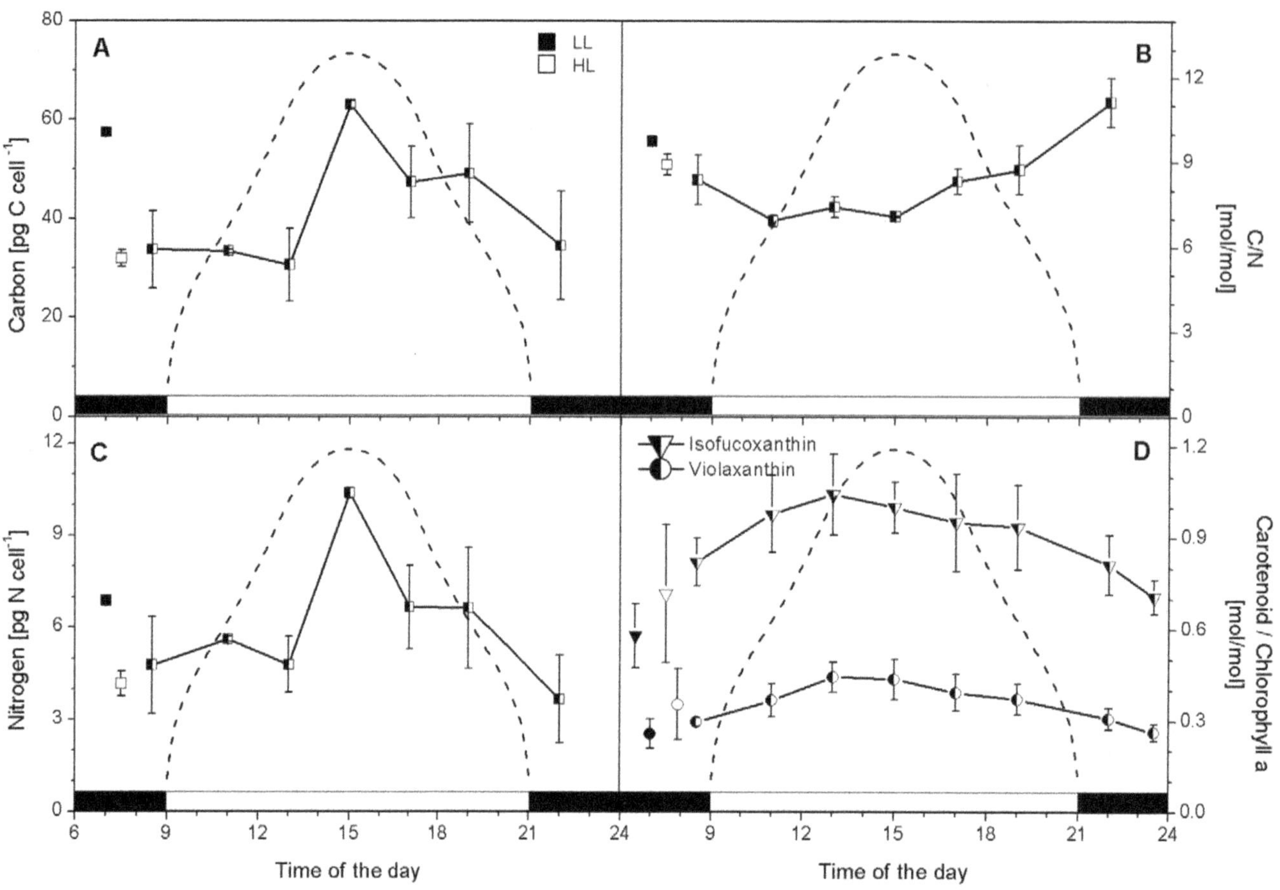

Figure 1. Changes in *C. velia* C, N, C:N and carotenoids during a sinusoidal light:dark cycle. Changes in C (pg C cell^{-1}, panel A), the ratio of C:N (mol:mol, panel B), N quotas(pg N cell^{-1}, panel C) and the major carotenoids relative to chl *a* (presented as ratio per chl *a*, panel D) are shown. Error bars are calculated as standard deviations of n≥3 replicated. Average values (plus error bars) measured for LL (■) and HL (□) grown *C. velia* are included for comparison. The dashed line shows the light intensity during the light part of the cycle, with a midday peak of 500 μmol photons m^{-2} s^{-1}.

Variable fluorescence measurements

Chlorophyll fluorescence was measured using a double-modulation fluorometer FL-3000 (Photon System Instruments, Czech Republic). Before measurements started, cells were dark acclimated for 20 min to oxidize the electron transport chain. A multiple turnover saturating flash was applied to measure the maximum quantum yield of photochemistry of PS II (F_v/F_m) according to $(F_m-F_o)/F_m$ where the difference between the maximum (F_m) and minimal fluorescence (F_o) is used to calculate the variable fluorescence (F_v) [33]. Cells were then illuminated with an orange actinic light (625 nm; 480 μmol photons m^{-2} s^{-1}). After 2 min, another saturating flash was applied and NPQ calculated as $(F_m-F_m')/F_m'$ in which case F_m' is the maximum fluorescence measured in the light. Photochemical quenching (qP) was calculated as $(F_m'-F_t)/(F_m'-F_o')$. The effective quantum yield of PSII photochemistry (Genty's parameter, Φ_{PSII}) was calculated as $(F_m'-F_t)/F_m'$, where F_t was the actual fluorescence level at given time excited by the actinic light.

Fast rate repetition fluorescence was measured using specially designed FM 3500 fluorometer (Photon Systems Instruments, Czech Republic). After 20 min dark acclimation, a series of 100 blue (463 nm) single turnover (1 μs) saturating flashes for sequential PSII closure were applied. This was done for eleven levels of blue (463 nm) actinic light intensities (0–1650 μmol photons m^{-2} s^{-1}). The data were fitted to model of Kolber et al. [33] including parameters such as maximum and minimal fluorescence, effective PSII cross-section (σ_{PSII}) and connectivity (p). These parameters were used for calculation of the electron transport rate ETR_{PSII} as $\sigma_{PSII}\ (F_q'/F_v')/(F_v/F_m)\ E$, where F_q' is $(F_m'-F')$ and E (light intensity) according to Suggett et al. [34]. The specific absorption of PSII (a^*_{PSII}) was calculated as $(\sigma_{PSII}\ (RC_{PSII}/chl\ a))/(F_v'/F_m')$, where $RC_{PSII}/chl\ a$ is equal to 0.002 [34].

Gas exchange measurements

Photosynthesis and dark respiration rates at 28°C were measured using a Clark-type oxygen electrode (Theta 90, Czech Republic) in the presence of 1 mM sodium bicarbonate. Light intensity in the electrode chamber was measured using a calibrated microspherical quantum sensor US-SQS/A (Walz, Germany) and light meter LI-250 (Li-Cor, USA). Gross photosynthesis (A_g) was calculated from the slope of O_2 evolution at a saturating irradiance plus the slope of respiratory O_2 utilization measured in the dark after the light exposure. O_2 evolution rates were normalized to chl *a*.

^{14}C fixation

The relationship between photosynthesis and irradiance was determined using the small volume ^{14}C incubation method of Lewis and Smith [35]. Cultures were kept in the dark for 20 mins before starting by spiking an aliquot (25 ml) of culture with ^{14}C-

Table 1. Abbreviations, equations and units.

α	photosynthetic efficiency; measured from the initial slope of a PI curve	mgC mg chl a^{-1} s^{-1}/µmol photons m^{-2} s^{-1}
a^*_{PSII}	chl a-specific PSII absorption coefficient	m^2 [mg chl a]$^{-1}$
Ag	gross photosynthetic rate	µmol O_2 mg chl a^{-1} h^{-1}
c	cell concentration	cells mL^{-1}
Chl *a*	chlorophyll *a*	pg $cell^{-1}$
E_k	index of light saturation = P_m/α	µmol m^{-2} s^{-1}
ETR_{PSII}	Electron transport rate	µmol electrons mg chl a^{-1} h^{-1}
F_o and $F_{o'}$	minimal fluorescence yield in the dark and in the light respectively	relative units
F_m and $F_{m'}$	maximum fluorescence yield in the dark and in the light respectively	relative units
F_v	variable fluorescence yield = $F_m - F_o$	relative units
F_v/F_m	maximum quantum yield of photochemistry = $(F_m - F_o)/F_m$	relative units
F_q	Difference between fluorescence yields = $F_m' - F_t$	relative units
F_t	actual fluorescence level at a given time excited by the actinic light	relative units
ΦPSII	effective quantum yield of PSII photochemistry (Genty's parameter) = $(F_{m'}-F_t)/F_{m'}$	relative units
HL	high light	
LL	low light	
n	number of replicates performed for a given experiment	
NPQ	non-photochemical quenching = $(F_m-F_{m'})/F_{m'}$	relative units
p	Connectivity factor	
P^{max}	maximum chl-specific carbon fixation rate	mgC mg chl a^{-1} h^{-1}
PS I	photosystem one	
PS II	photosystem two	
PQ	Photosynthetic quotient	
PTOX	plastid-localized terminal oxidase enzyme	
qP	photochemical quenching	relative units
RC_{PSII}/Chl *a*	photosynthetic unit size	m^2 (mol Chl $a)^{-1-1}$
σ_{PSII}	effective absorption cross section of PSII under dark acclimation	A^2 PSII
t	time	day
µ	specific growth rate = $(\ln c - \ln c_0)/(t - t_0)$	day^{-1}

sodium bicarbonate (MP Biochemicals, USA; final concentration of 1 µCi ml^{-1}) and incubating for 40 mins at 28°C and a range of light intensities from 5 to 1500 µmol photons m^{-2} s^{-1}. Triplicate samples for background counts (with 100 µL of buffered formalin) and total counts (with 250 µL of phenethylamine and 5 ml of Ecolume scintillation cocktail) were prepared at the start. Buffered formalin (100 µL) terminated the reactions; samples were acidified with 50% HCl (1 mL) were left overnight to purge off unincorporated label before disintegrations per minute were counted on a calibrated Tricarb 1500 Scintillation Counter. Dissolved inorganic carbon concentrations were determined in a cell-free medium by the Gran titration technique described by Butler [36]. Photosynthesis-irradiance curves were fitted using $P = P_{max} \times \tanh(\alpha \times E/P_{max})$ according to Jassby and Platt [37] where the maximum chl-specific carbon fixation rate (P_{max} = mg C mg chl^{-1} h^{-1}) and the initial slope of the curve (α) = mg C mg chl^{-1} h^{-1} (µmole photons m^{-2} $s^{-1})^{-1}$ were estimated from measurements of photosynthesis (P) and irradiance (E). The index of light saturation (E_k; µmol m^{-2} s^{-1}) was calculated as P_{max}/α.

Results

All findings are presented as averages of n≥3 cultures plus/minus standard deviations.

Basic physiological parameters

Growth rates were determined once cultures of *C. velia* had acclimated to the respective irradiance (Table 2). While LL cells grew faster (0.21±0.02day^{-1}) than those at HL (0.16±0.01 day^{-1}), cells growing on the sinusoidal light:dark cycle had the fastest overall growth rate of 0.37±0.01 day^{-1}. Cell size was dependent on the light intensity for growth and light regime (Table 2), *C. velia* cell diameter decreased in the following order: LL (6.87±0.09 µm), sinusoidal cycle (6.07±0.63 µm) and HL (5.68±0.31 µm).

Cellular C and N concentrations for *C. velia* were dependent of the irradiance for growth as well as the time of day (Table 2; Fig. 1A, B, C). The average cellular C quota was significantly higher in LL (57 pg C $cell^{-1}$±1) than in HL (32 pg C $cell^{-1}$±2) ($p = 0.002$; n = 2) cells but this was not the case for N quotas (6.88 pg N $cell^{-1}$±0.14 and 4.18 pg N $cell^{-1}$±0.40 respectively, $p = 0.093$; n = 2). However, given the different cell sizes, the average cellular densities of C and N were almost identical for both treatments (0.33–0.34 pg C $µm^{-3}$ and 0.041–0.044 pg N $µm^{-3}$, respectively). In the sinusoidal grown *C. velia* (Fig. 1 A, C), C quotas increased throughout the light period from predawn values of 34 pg C $cell^{-1}$ (±8) to 63 pg C $cell^{-1}$(±1) in the middle of the day corresponding to a 85% increase (paralleled by a 116%

Table 2. Summary of cellular responses in *C. velia* grown under three different photon treatments.

	LL	HL	Sinusoidal light:dark cycle	Response in sinusoidal cultures
Irradiance μmol photons m-2 s-1	15	200	Max. 500	Sinusoidal function
Irradiance regime	24 h continuous	24 h continuous	12 h:12 h light:dark	
Photon dose per day mol photons m-2 day-1	1.3	17.3	13.2	
Growth rate d-1	0.21±0.02	0.16±0.01	0.37±0.01	
Cell size μm	6.87±0.09	5.68±0.31	6.07±0.63	no change
C quota pg cell-1	57±1	32±2	53±12	See Fig. 1A
C:N mol:mol	9.8±0.1	8.9±0.4	8.4±0.7	See Fig. 1B
N quota pg cell-1	6.88±0.14	4.18±0.40	6.4±1.7	See Fig. 1C
Chl a pg cell-1	0.60±0.08	0.21±0.05	0.45±0.04	no change
violaxanthin/Chl *a* mol:mol	0.26±0.05	0.35±0.12	0.36±0.07	See Fig. 1D
isofucoxanthin/Chl *a* mol:mol	0.58±0.11	0.72±0.23	0.91±0.11	See Fig. 1D
ß,ß-carotene/Chl *a* mol:mol	0.030±0.006	0.034±0.009	0.045±0.007	no change
total carotenoids/Chl *a* mol:mol	0.87±0.15	1.11±0.31	1.31±0.18	See Fig. 1D
Chl *a* specific absorption of PSII a*PSII	0.0071±0.0002	0.0101±0.0003	0.0109±0.0003	Not shown

increase in N quotas). Within an hour of lights off, C and N quotas were back down to predawn levels. The average cellular density of C was comparable to LL and HL cultures (0.36 pg C μm^{-3}) but the cellular density of N was higher (0.052 pg N μm^{-3}) in the sinusoidal cultures. Molar C:N ratios changed throughout the light photoperiod (Fig. 1B).

Pigment composition

Chl *a* concentrations responded to the irradiance for growth (Table 2). In *C. velia*, cellular chl *a* concentrations were three times higher in LL cells compared to those growing at HL (0.60 ± 0.08 pg chl *a* $cell^{-1}$ and 0.21 ± 0.05 pg chl *a* $cell^{-1}$respectively). Chl *a* concentrations did not vary significantly ($p = 0.236$; n = 21) throughout the sinusoidal light:dark cycle (0.45± 0.04 chl *a* pg $cell^{-1}$). Despite growing at very different irradiances, the cellular density of pigments was comparable for LL (3.5 fg chl *a* μm^{-3}) and sinusoidal grown *C. velia* (3.8 fg chl *a* μm^{-3}) but were the lowest when cells were grown at HL (2.2 fg chl *a* μm^{-3}).

Unlike its closest extant photosynthetic relatives, *C. velia* lacks chl *c* and only expresses a limited set of carotenoids [10]. Changes in the fraction of violaxanthin, isofucoxanthin, ß,ß-carotene and total carotenoids were examined relative to chl *a* (Table 2). In HL relative to LL grown cells, there was in total, 28% more carotenoids (1.11 ± 0.31 and 0.87 ± 0.15 respectively), specifically 35% more violaxanthin (0.35 ±0.12 versus 0.26±0.05 respectively) and 24% more isofucoxanthin (0.72±0.23 versus 0.58±0.11 respectively). ß,ß-carotene was a minor component of the accessory pigment pool, and did not change significantly between HL and LL grown *C. velia* (0.034±0.009 and 0.030±0.006 respectively; $p = 0.237$; n = 3). In *C. velia* cells growing on a sinusoidal regime, predawn total carotenoids: chl *a* ratios were similar to those of HL grown cells (1.17 ± 0.09), increasing to 1.55 (± 0.17) before noon, then decreasing to 1.00 (± 0.07) several hours post illumination (Fig. 1D). ß,ß-carotene also changed significantly ($p<0.001$; n = 23) throughout the sinusoidal cycle (not shown), following a similar trend to total carotenoids. At midday, there was 47% more violaxanthin and 22% more isofucoxanthin present in *C. velia* than before dawn. Ratios of accessory pigments to chl *a* at midday were similar, albeit higher, than those measured in HL growing *C. velia*. When translated to cellular densities, then the density of violaxanthin was similar in LL and HL cells (1.22 and 1.15 fg μm^{-3}, respectively, while it almost doubled in the sinusoidal cells (2.54 fg μm^{-3}). The cellular density of light harvesting isofucoxanthin was also the highest in the sinusoidal grown cells (5.8 fg μm^{-3}), than in LL cells (3.08 fg μm^{-3}) and HL cells (2.36 fg μm^{-3}).

Rearrangement of light harvesting complexes

The changes in arrangement of light harvesting complexes during acclimation to HL and LL were deduced from fluorescence emission spectroscopy. Spectra were measured at room and at low (77 K) temperatures. The room temperature (RT) fluorescence emission spectra of *C. velia* detected a clear red-shift of the PSII maximum between HL and LL grown *C. velia* (from 685 nm for HL to 688 nm for LL, Fig. 2A). As changes in intensities of both emission bands observed at RT (685 nm and 688 nm) showed similar kinetics during fluorescence induction with continuous light (data not shown), we attribute them to the fluorescence emission of PSII core proteins. In LL grown *C. velia*, an additional red-shifted fluorescence band at 710 nm can be seen at RT (Fig. 2A). Since PSI is not known to emit at RT and the intensity of the 710 nm fluorescence band at RT showed similar variability as the PSII emission band at 685 nm (data not shown), this indicates its origin in some red-shifted antennae of PSII.

At 77 K, a major emission band with maximum at 690 nm was observed (Fig. 2B). Under LL conditions, an additional red-shifted emission maximum at 77 K was observed as a shoulder at 712 nm (Fig. 2B). These results indicate that acclimation to different light intensities induces antennae reorganization. Since there is no distinct band that can be attributed to PSI fluorescence in the 77 K emission spectra, we cannot determine whether light acclimation also affected the stoichiometry of PSI and II. The emission spectra of cells during the sinusoidal light:dark cycle in *C. velia* were identical to HL grown *C. velia* (not shown).

Variable fluorescence parameters

The maximal efficiency of PSII photochemistry, F_v/F_m, was 0.61 (±0.01) for LL grown *C. velia* (Table 3). By contrast, HL

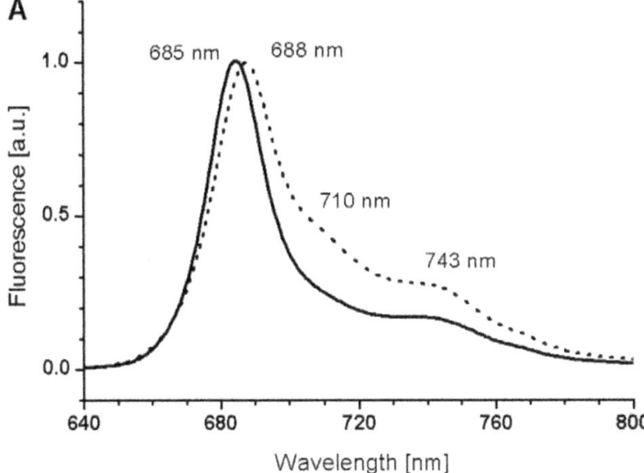

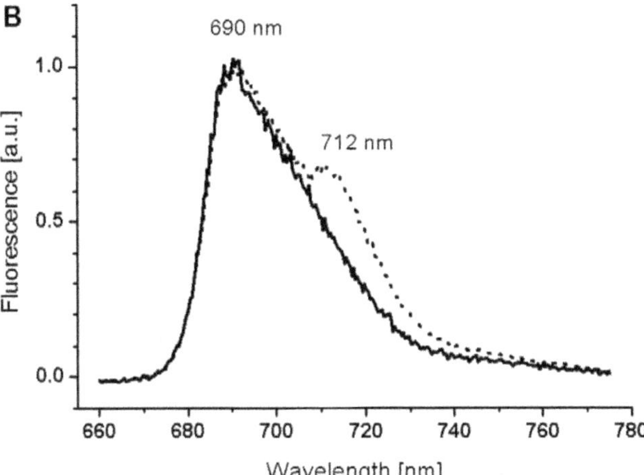

Figure 2. Room temperature and 77K fluorescence emission spectra for *C. velia* grown under LL and HL. Spectra were normalized at the chlorophyll a fluorescence emission maxima. Given the fluorescence spectra was identical for HL and sinusoidal light:dark cycle grown *C. velia*, we only show the results for the HL grown cells.

grown *C. velia* had an F_v/F_m ratio of 0.52 (±0.01) (Table 3) indicating a decrease in the maximal efficiency of PSII photochemistry. Similarly, during the sinusoidal cycle, F_v/F_m ratios declined from 0.56 (±0.01) at the beginning of the light period to 0.51 (±0.01) by the evening and then started to increase again after the light was turned-off (Fig. 3A).

The efficiency of PSII photochemistry in the light, Φ_{PSII} (also known as the Genty parameter), was slightly greater when *C. velia* was grown at HL (0.14±0.01) than at LL (0.11±0.02) (Table 3). In *C. velia* cells grown under a sinusoidal light:dark cycle, the predawn value of Φ_{PSII} was similar to LL grown cultures (0.11±0.01), then increasing rapidly after onset of light and reaching the highest values before midday (0.28±0.01). Subsequently Φ_{PSII} started to decline towards the dark period to value of 0.14 (±0.01) (Fig. 3B).

qP reflects the number of open PSII reaction centers and denotes the proportion of excitation energy trapped by them – therefore the higher qP, the more efficient it is in utilization of incident light. We have found that HL relative to LL grown *C. velia* had higher qP values (0.54±0.02 and 0.34±0.04 respectively) (Table 3). In the sinusoidal *C. velia* cultures, we observed a gradual increase in qP with increasing light intensity during light period (Fig. 3C). Average predawn qP values were 0.41 (±0.01), while those at noon were almost double (0.78±0.01). qP started to decline already pre dusk to 0.52 (±0.03) after the onset of night, (Fig. 3C). All these results showed ramping up of the photosynthetic apparatus during mid-morning (qP increase, Fig. 3C) and the ability of *C. velia* to use most of the incident light for photosynthesis during maximal irradiance (no decay of qP at noon, Fig. 3C).

Given that NPQ is proportional to heat-dissipation of excitation energy in the antenna system in the dark acclimated state [see 25], or more simply, the amount of energy not used in photochemistry, changes in NPQ in the sinusoidal cultures reflect the cells dynamically responding to changes in irradiance for growth. NPQ was highest predawn (1.26±0.07) and declined rapidly after onset of light to a minimum 0.35(±0.03) before midday. From midday NPQ showed a continual increase (Fig. 3D) through to the dark period. NPQ was higher (in fact doubled) in HL grown *C. velia* (1.42±0.29) than LL cultures (0.73±0.08) (Table 3). These data show that *C. velia* can cope efficiently with increasing light intensity during diel cycle as NPQ values were minimal and relatively stable during the first half of light period (up to 15 h, see Fig. 3D), and the non-radiative energy dissipation was stimulated only afterwards.

O_2 evolution, ^{14}C fixation and the Photosynthetic quotient

The ability of *C. velia* to efficiently acclimate photosynthesis to wide range of constant irradiance was reflected in the values of the gross rate of O_2 evolution. Ag was comparable for HL and LL *C. velia* when expressed on a per chl *a* basis (Table 4; Fig. 4A). This was also the case for ^{14}C fixation, found to be 3.67±0.07 and 2.97±0.07 mg C mg chl^{-1} h^{-1}for HL and LL *C. velia* respectively (Table 4; Fig. 4C).

We observed a pronounced 'hysteresis effect' in photosynthetic parameters of *C. velia* grown under the sinusoidal light:dark cycle. The hysteresis effect represents an asymmetric response of photosynthesis to the same irradiance in the morning versus the afternoon [38,39,40] and references therein. We observed maximum O_2 evolution rates before noon with reduced O_2 evolution in the afternoon at the same light intensity (Fig. 4A). In the same way, the maximum ^{14}C fixation rate was measured before noon was greater than that measured after (Fig. 4C). Thus, both dark and light photosynthetic reactions show the mid-morning maximum. However, while ^{14}C fixation rates gradually decreased after the mid-morning peak, O_2 evolution rates declined rapidly to ca. 60% at midday and then remained constant until late evening (Fig. 4A, C).

Changes in photosynthetic efficiency did not follow those for the maximum photosynthetic rate in the sinusoidal cultures (Fig. 4B); α increased steadily throughout the day to ca. 0.18±0.01 mgC mg chl^{-1} h^{-1} (μmole photons m^{-2} s^{-1})$^{-1}$ with a midday depression. In both HL and LL grown *C. velia*, α values were similar to predawn values measured of the sinusoidal cycle (Fig. 4B), (0.06±0.00 and 0.09±0.01 mgC mg chl^{-1} h^{-1} (μmole photons m^{-2} s^{-1})$^{-1}$ respectively).

We found the saturation intensity for carbon fixation (E_k) was higher, in fact doubled, in HL grown relative to LL grown *C. velia* (63±3 μmole photons m^{-2} s^{-1} and 33±3 μmole photons m^{-2} s^{-1} respectively) (Table 4) but still below the growth irradiance for HL cultures. In sinusoidal cells of *C. velia*, E_k was much higher than in cultures growing under continuous light and followed the diel changes in irradiance, that is, E_k increased from the predawn value of 82±5 μmole photons m^{-2} s^{-1} to 129±9 μmole photons m^{-2} s^{-1} at midday and dropped back down to 58±3 μmole photons m^{-2} s^{-1} after the period of light

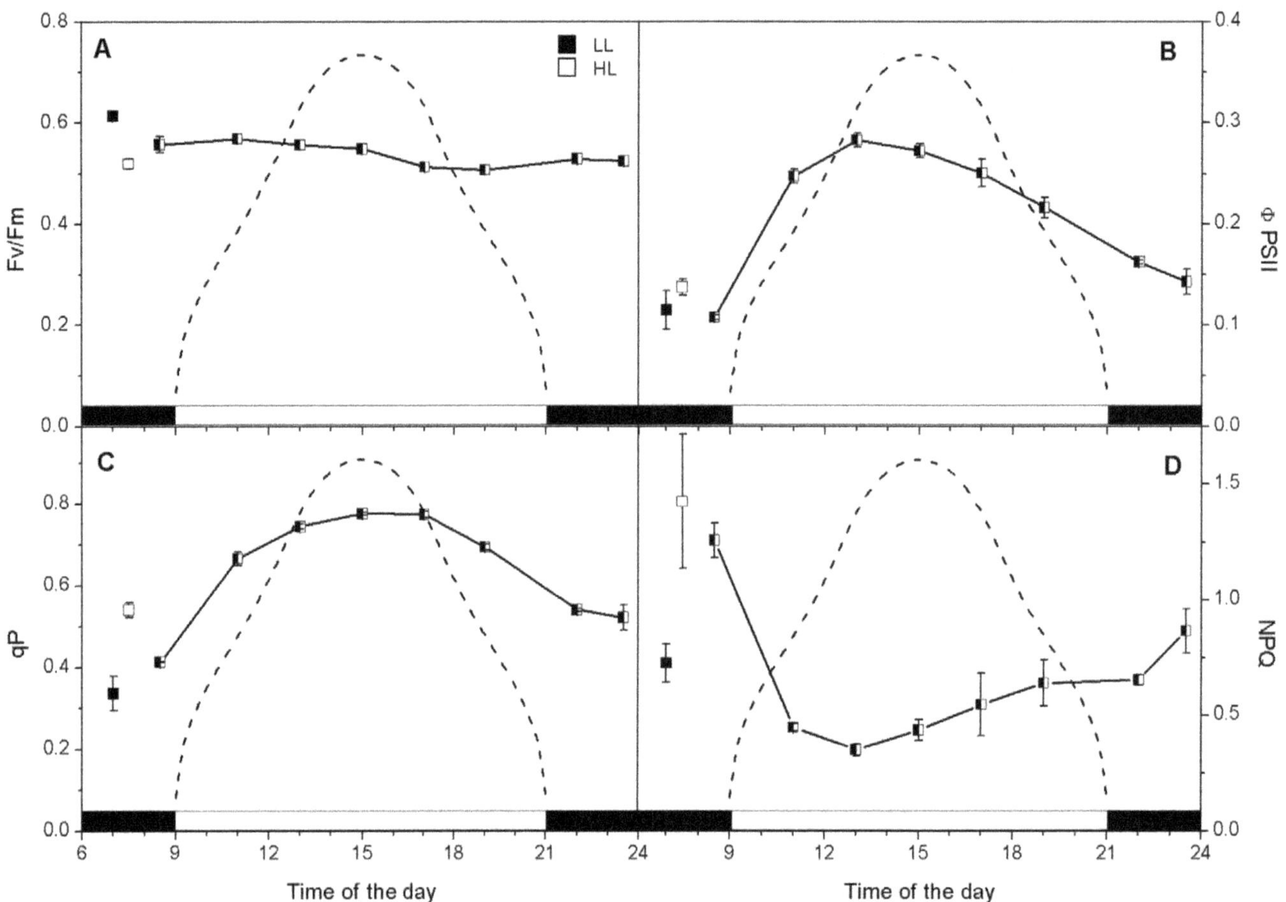

Figure 3. Changes in *C. velia* major fluorescence parameters during a sinusoidal light:dark cycle. Changes in the F_v/F_m (panel A), Φ_{PSII} (panel B), qP (panel C) and NPQ (panel D) are shown. Error bars are calculated as standard deviations of n≥3 replicated. Average values (plus error bars) measured for LL (■) and HL (□) grown *C. velia* are included for comparison. The dashed line shows the light intensity during the light part of the cycle, with a midday peak of 500 μmol photons m^{-2} s^{-1}.

(not shown). This indicates that it is possible for C. *velia* to attain a high E_k, but not in the HL cells grown on continuous light.

The photosynthetic quotient (PQ) is defined by the molar ratio of the rate of oxygen production relative to carbon dioxide assimilated. We found the PQ ratio to be ~1.3 when examining *C. velia* grown in continuous light, both in the HL and LL grown cultures (Table 4). *C. velia* cells growing on the sinusoidal light:dark cycle modulated their photosynthetic quotient in response to the changing irradiance during the light period; with the midday minimum 0.76. While the photosynthetic quotient was still close to 1 an hour after dark, the predawn value was doubled (PQ≈2) (Fig. 4D).

Table 3. Summary of physiological responses measured using fluorescence techniques in *C. velia* grown under three different photon treatments.

	LL	HL	Sinusoidal light:dark cycle	Response in sinusoidal cultures
F_v/F_m	0.61±0.01	0.52±0.01	0.54±0.02	Fig. 3A
Φ_{PSII}	0.11±0.02	0.14±0.01	0.21±0.06	Fig. 3B
qP	0.34±0.04	0.54±0.02	0.64±0.13	Fig. 3C
NPQ	0.73±0.08	1.42±0.29	0.65±0.29	Fig. 3D
1-qP	0.66±0.04	0.46±0.02	0.36±0.13	Not shown
σ_{PSII}	307±8	380±6	444±8	Not shown
p	0.37±0.01	0.22±0.01	0.26±0.04	Not shown

(Note: *p* in this table refers to the connectivity factor).

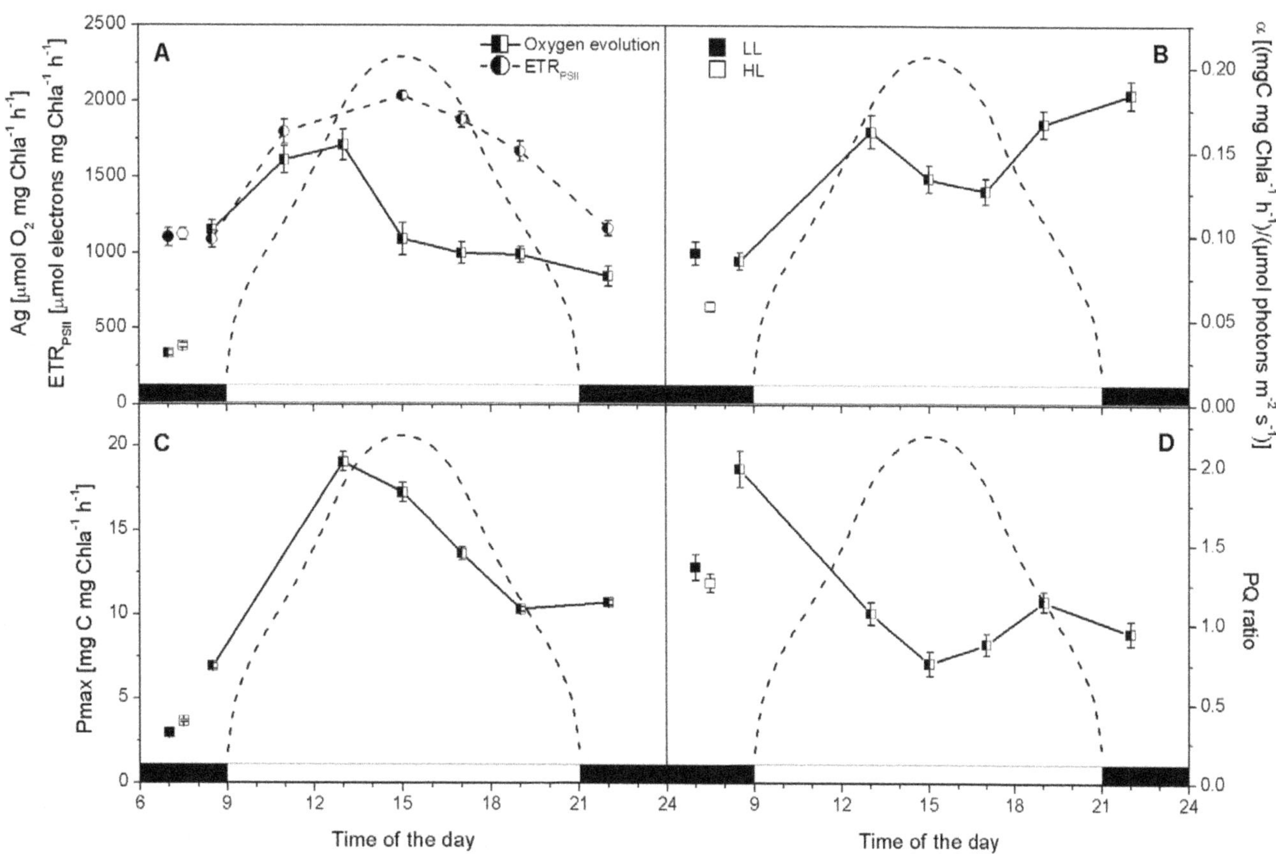

Figure 4. Changes in *C. velia* light and dark reactions during a sinusoidal light:dark cycle. Changes in O_2 evolution and ETR_{PSII} (panel A), alpha (photosynthetic efficiency, panel B), C fixation rates (panel C) and the photosynthetic quotient (PQ, panel D) are shown. Error bars are calculated as standard deviations of n≥2 replicated. Average values (plus error bars) measured for LL (■) and HL (□) grown *C. velia* are included for comparison. The dashed line shows the light intensity during the light part of the cycle, with a midday peak of 500 μmol photons m^{-2} s^{-1}.

Discussion

We have measured the photosynthetic activities and photo-acclimation strategies of the coral associated alveolate alga *Chromera velia*, the closest photosynthetic relative to apicomplexan parasites and dinoflagellate algae [10,19]. The majority of reef building scleractinian corals contain endosymbiotic dinoflagellate algae (zooxanthellae) of the genus *Symbiodinium* [14,15,16,17]. The photosynthetic performance of zooxanthellae is affected by environmental factors including ambient light (quantity and quality), temperature, CO_2, and nutrient availability. The response to light is arguably the most important factor controlling productivity, physiology and ecology of corals [16,40,41,42,43,44,45]. Many studies have investigated photo-acclimation strategies in scleractinian and other corals e.g., [14,15,16,17,43,44,45,46,47,48] but studies of the free-living zooxanthellae are less common e.g., [14,45,50]. Studies have shown that corals and their symbiotic algae may be vulnerable to

Table 4. Summary of physiological responses measured by oxygen evolution and ^{14}C fixation in *C. velia* grown under three different photon treatments.

	LL	HL	Sinusoidal light:dark cycle	Response in sinusoidal cultures
A_g μmol O_2 mg chla^{-1} h^{-1}	338±19	386±16	1200±329	Fig. 4A
P_{max} μmolC mg chla^{-1} h^{-1}	248±5	306±6	1084±381	Not shown
P_{max} mgC mg chla^{-1} h^{-1}	2.97±0.07	3.67±0.07	13.0±4.6	Fig. 4C
ETR_{PSII} μmole electrons mg chl a^{-1} h^{-1}	1102±59	1120±38	1605±391	Fig. 4A
PQ	1.36±0.08	1.26±0.06	1.13±0.44	Fig. 4D
α mgC mg chla^{-1} s^{-1}/μmol photons m^{-2} s^{-1}	0.09±0.01	0.06±0.00	0.14±0.04	Fig. 4B
E_k μmol photons m^{-2} s^{-1}	33±3	63±3	90±33	Not shown

very high irradiances, expressed in the natural environment as a localized solar bleaching response (e.g., reduction in numbers of symbiotic dinoflagellates, loss of photosynthetic pigments, photo-inhibition of photosynthesis, or a combination of these) [45,48,49,51,52].

LL and HL acclimation strategies

To understand the basic physiological behavior and acclimation strategies we cultivated *C. velia* at continuous irradiance under two extreme light intensities: low (15 µmol m^{-2} s^{-1}) and high (200 µmol m^{-2} s^{-1}). LL *C. velia* had greater C, N and chl *a* quotas, higher F_v/F_m in comparison to HL grown cells which in turn had more photoprotective carotenoids. The LL grown *C. velia* accumulated both C and N reserves and increased their light harvesting potential (package effect aside) relative to those grown at HL (Table 2). The acclimation to LL conditions also required substantial antennae reorganization, as indicated by appearance of an extra emission band (710 nm at RT, 712 nm at 77 K; see Fig. 2). Moreover, PSII reactions centers of LL grown *C. velia* are probably better interconnected as deduced from the 78% higher value of the connectivity parameter compared to HL grown cells (Table 3). On the other hand, there is probably also a redistribution of some PSII antennae towards PSI in LL grown cells. This is also observed in the reduction of chl *a* specific absorption of PSII by 29% (a^*_{PSII}; see Table 2) as well as the functional absorption cross section of PSII by 19% (σ_{PSII}; see Table 3) in spite of the three times higher chl *a* concentrations. We suggest that the PSI abundance increases PSI activity that could be used for optimization of photosynthesis under light limited conditions.

The acclimation to HL conditions considerably decreased the chl content and increased carotenoid:Chl *a* ratio (Table 2). It was recently shown that both major carotenoids, isofucoxanthin as well as violaxanthin, contribute to light-harvesting in *C. velia* [25]. The increase in pigment content was much more pronounced for violaxanthin (35%) than for isofucoxanthin (24%); this is related to their photoprotective roles [25]. Violaxanthin in *C. velia* undergoes very fast de-epoxidation to zeaxanthin under the excessive irradiance that enables efficient photoprotection by NPQ [25]. In line with that, HL grown cells, containing appreciably more violaxanthin had NPQ values twice those measured in LL cells (Table 3). The NPQ reflects non-photochemical energy dissipation and reduces the excitation pressure over PSII in HL grown *C. velia* (see 1-qP lower by 30% in comparison with LL in Table 3). This reveals the ability of *C. velia* to protect itself under high-light growth conditions (200 µmol m^2 s^{-1}). As a result, in spite of the reduction of photosynthetic efficiencies (see the lower of F_v/F_m and α in Tables 3 and 4), *C. velia* can maintain the rates of photosynthesis for both, light (O_2 evolution, A_g) and dark (C fixation, P_{max}) reactions during growth under HL conditions (see Table 4).

Growth rates, C:N ratios, Chl *a* and photosynthetic rates for *C. velia* are similar to those reported for *Symbiodinium* spp. e.g. [14,45,47] and other eukaryotic algae grown under continuous light e.g. [38,53]. The low values of Φ_{PSII} measured in *C. velia* are similar to those measured in coral residing in the shallowest habitats [54]. This is usually interpreted as suggesting low efficiency of PSII but that cannot be the case in *C. velia* as we measured comparatively high rates of oxygen evolution. Alternatively, it can indicate redistribution (spillover) of excitation between PSII and PSI. Other experimental data (fluorescence recovery after photobleaching, R.Kaňa, unpublished) also suggest high mobility of *C. velia* antennae and thus support the spillover hypothesis. In the case of spillover the F_m fluorescence of PSII could be lowered by non-fluorescent PSI. Such limitation of excitation energy transfer to PSII at high irradiances allows *C. velia* to limit photodamage to the D1 protein in PSII. It seems that *C. velia* possess a unique organization of the antennae system where absorbed light is distributed among both photosystems to increase their efficiency.

Photosynthesis under sinusoidal light:dark cycle

The efficient photosynthesis in *C. velia* was further stimulated under sinusoidal light:dark cycle that better simulates physiological conditions in nature. In this case, we measured very high rates of O_2 evolution (up to 1708 µmol O_2 mg chl^{-1} h^{-1}) and ^{14}C fixation (up to 19 mg C mg chl^{-1} h^{-1}) during the light phase of the sinusoidal light:dark cycle, some 4–5 times higher than those measured in cultures receiving continuous light. In addition, the sinusoidal light regime allowed greater flexibility and dynamics of the photosynthesis apparatus with more than 60% of NPQ (depending the time of day) recovered within 3 minutes of sampling, while only 40% recovered in cultures grown under continuous irradiance (data not shown). Hence, the diel periodicity in irradiance during the growth cycle is crucial to obtaining maximal photosynthetic rates (P_{max}, A_g) and consistent with the need of *C. velia* to maintain energetic balance within the primary photosynthetic reactions.

We also found that while the actual PSII photochemistry in the light (Φ_{PSII}, Fig. 3B) tracked changes in ^{14}C fixation and O_2 evolution rates (mid-morning maximum and afternoon depression, see Fig. 4A, C), the maximal quantum yield of PSII (F_v/F_m) and qP were maintained until the late afternoon (Fig. 3A, C). This afternoon difference between the maximal capacity of PSII photochemistry (represented by F_v/F_m, Fig. 3A) together with qP (Fig. 3C) and actual efficiency of PSII in light (Φ_{PSII}, Fig. 3B) correlated with gradual increase of NPQ during the afternoon (Fig. 3D). Stimulation of NPQ (Fig. 3D) could be the consequence of slower CO_2 assimilation during the afternoon (Fig. 4C) resulting in slower ATP regeneration that would cause higher lumen acidification [55], a main stimulus of NPQ increase in *C. velia* [25]. Our results therefore suggest that optimization of light reactions during day period proceeds on the level of regulation of light-harvesting antennae (by NPQ, see Fig. 3 D), rather than through photoinhibitory destruction of core proteins of PSII. Further support for this comes from the lack of a significant mid-day depression in F_v/F_m, a sign of photoinhibition [56,57]. Also, the excitation pressure over PSII (1-qP) during the highest irradiances was minimal (around 0.22); this excludes an overexcitation of PSII and therefore photoinhibition.

Hysteresis effects were observed during the sinusoidal light cycle when examining O_2 evolution rates and ^{14}C fixation rates (Fig. 4A, C). Indeed, *C. velia* appeared to be taking an 'afternoon nap'. What we found interesting is that the afternoon depression varied between the light and dark reactions of photosynthesis. While ^{14}C fixation rates were gradually decreasing after the mid-morning peak (Fig. 4C), O_2 evolution rates declined rapidly to ca. 60% at midday and then remained constant until e in the evening (Fig. 4A). Given the culture conditions, the hysteresis we observed were not related to changes in nutrient and CO_2 availability. Additional support for this comes from the C and N quotas which were not that different from Redfield (Fig. 1) and pH values which were close to 8.2 (not shown) throughout the sinusoidal light:dark cycle. The decoupling between O_2 evolution and ^{14}C fixation rates during the midday and early afternoon maybe related to changes in photorespiration, the process whereby phototrophes fix O_2 and liberate CO_2 [58]. Photorespiration has been previously reported in zooxanthellae [see 16,59], but rates were generally found to be low, presumably because the type II RuBisCO was found exclusively in the pyrenoid [60] thus limiting exposure to O_2 and/or due to the presence of a carbon concentrating mechanism [61].

Burris [62] reported that photosynthetic quotients could be even as low as 0.1 when photorespiration is the dominant process while little or no photorespiration takes place when PQs are > than 1. As we measured PQ's of 0.76 at midday (Fig. 4D), we suggest the presence of photorespiration in *C. velia*, but only with rather low rates, similarly as was observed for zooxanthellae. This would imply that *C. velia* switches from primarily production of glycolate and its excretion (the greater the glycolate, the lower the quotient) during the day to glycerol synthesis at night (PQ≈2, see Fig. 4D) [57]. The presence of photorespiration in *C. velia* suggests that it has a similar lifestyle to symbiotic dinoflagellates. The released carbon compounds (referred to as 'junk food' by Falkowski et al. [41]) are used by the coral for respiratory energy generation via the synthesis of energy-rich storage products such as lipids and starch [16,41]. The high carbon fixation rates in *C. velia* may be associated with supplying carbon to this cycle.

Photorespiration is thought to be an evolutionary relic as primitive photosynthesis originated from the early atmosphere with very little oxygen and thus the early RuBisCO lacked discrimination between O_2 and CO_2 [58]. *C. velia* possess the primitive form of RuBisCO (type II) homologous to that of dinoflagellates that was acquired during evolution by the horizontal gene transfer from a proteobacterium [19]. The type II RuBisCO comprises only large subunits, has a high K_c but has a poor affinity for CO_2 and discriminates CO_2 from O_2 less well than type I RuBisCO [58,61,63,64]. We suggest that type II RuBisCO could be a reason for such a high ^{14}C fixation rates observed in *C. velia*. However, type II RuBisCO is very sensitive to presence of O_2 that could be used for its oxygenase activity and thus reduce carboxylation activity. Therefore, there must be a mechanism to reduce O_2 accessibility to RuBisCO. One of the most known mechanisms that operates to suppress the oxygenase activity of RuBisCO is the carbon-concentrating mechanism (CCM) that accumulates C_i and elevates [CO_2] around RuBisCO. *Symbiodinum* spp. fix inorganic carbon efficiently because they use CCMs [59,61]. Nonetheless, even in zooxanthellae, CCM activity has not been found to be as high as might be expected [61]. Another possibility is the presence of special organelles (pyrenoids) that elevate [CO_2] within chloroplasts. Pyrenoids are often present in dinoflagellates [65]; their formation appears to be correlated with CO_2 fixation rates in *Gonyaulax* sp. [60] which also utilizes type II RuBisCO. Currently, we lack direct evidence of CCMs and pyrenoids in *C. velia* but given the very high carbon fixation rates parallel the highest O_2 evolution rates mid-morning, it appears that *C. velia* is somehow able to build a sufficient carbon pool around RuBisCO so that it functions predominately as a carboxylase.

An alternative possibility explaining high carbon fixation rates in high oxygen concentrations could be the occurrence of some alterative, oxygen consuming, electron flow, e.g. Mehler reaction or chlororespiration. During the Mehler reaction, the O_2 produced by PSII is reduced again by PSI thereby decreasing its concentration [66]. Chlororespiration is defined as a respiratory-like electron transport activity from NAD(P)H to O_2, catalysed by NADH dehydrogenase and the plastid-localized terminal oxidase (PTOX) enzyme [58,67]. Even though PTOX has not yet been found in dinoflagellates nor their closest relatives (ciliates, perkinsea and apicomplexa; see Peltier et al. [68]), constitutive chlororespiration has already been proposed for a HL acclimated clade A *Symbiodinium* [69]. These alternative electron sinks would not only reduce the O_2 concentrations in chloroplast, but also generate large trans-thylakoid ΔpH, provide the extra ATP and enhance photoprotection by NPQ see [70] for review. We did observe an uncoupling of ETR_{PSII} from the rate of gross oxygen production by PSII from the midday until late afternoon (see Fig. 4A) suggesting the presence of some alternative electron flow to oxygen. Moreover, we have observed a temporary fast post-illumination decrease in O_2 concentration (Fig. 5) at the time of maximal oxygen evolution that may play a role in oxygen consumption. Interestingly, this respiration occurs only for sinusoidal grown cells with maximal rate of CO_2 fixation and was absent in HL and LL grown cells. We hypothesize that the observed respiration may represent a mechanism by which the O_2 concentration is reduced and thus the oxygenase activity of RuBisCO type II is minimized so that RuBisCO is turned towards higher carboxylation. This oxygen consuming process could play the role of an "optimizer" reducing O_2 concentrations at RuBisCO. To resolve the origin of this respiration, more experiments are necessary that will also show if it proceeds in the whole thylakoid or if it is present only in some parts of membrane, where RuBisCO is located.

Conclusions

C. velia is effectively a mixture of different organisms: heme synthesis as observed in Apicomplexans, simple pigmentation as in Eustigmatophyceae, primitive type II RuBisCO as in Dinoflagellates, and antenna organized as observed in Bacillariophyceae (diatoms). We have shown for the first time that this simple photosynthetic system is surprisingly efficient in photosynthetic carbon assimilation. Uniquely to *C. velia*, we propose that members of this new family of Chromeraceae use photorespiration, together with the thermal energy dissipation via NPQ, as a mechanism in their photoacclimation strategy. We propose that future studies examine the role of photorespiration in this apicomplexan. Rather than considering it a wasteful processes (compared to photosynthesis due to its high consumption of NADPH and ATP), it should

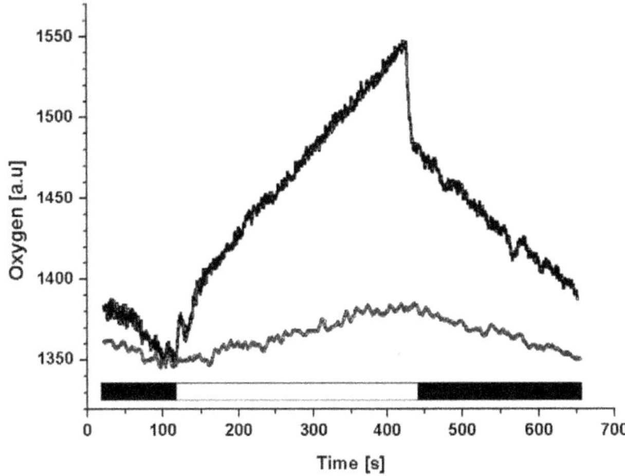

Figure 5. Respiratory and photosynthetic activities of *C. velia*. Representative data of low (grown under HL conditions; grey line) and high (mid-morning maximum under sinusoidal light:dark cycle; black line) photosynthetic rates are shown. Oxygen concentration was continuously monitored using Clark-type electrode in the dark (black banner) or in the light (white banner). For better comparison, curves were artificially shifted to the same initial value at the time 115 s. Data were normalized to chl *a* concentration of samples.

be considered as mechanism for energy dissipation as well as producer of additional glycolate to its host. If photorespiration is acting in photoprotection, then corals which harbor *C. velia* (and symbiotic dinoflagellates such as *Symbiodinium*) may benefit from this particular symbiont(s).

Acknowledgments

We thank the scientists, staff and students at the Institute of Microbiology Třeboň, Czech Republic for their support during this study. We thank the associate editor and review process for improving the manuscript.

Author Contributions

Conceived and designed the experiments: AQ EK OP. Performed the experiments: AQ EK JJ RK JŠ BŠ OK OP. Analyzed the data: AQ EK RK OP. Contributed reagents/materials/analysis tools: AQ EK JJ RK JŠ BŠ OK OP. Wrote the paper: AQ EK OP.

References

1. Quigg A, Finkel ZV, Irwin AJ, Reinfelder JR, Rosenthal Y, et al. (2003) The evolutionary inheritance of elemental stoichiometry in marine phytoplankton. Nature 425: 291–294.
2. Quigg A, Irwin AJ, Finkel ZV (2011) Evolutionary imprint of endosymbiosis of elemental stoichiometry: testing inheritance hypotheses. Proc R Soc Lond: B 278: 526–534.
3. Falkowski PG, Katz ME, Knoll AH, Quigg A, Raven JA, et al 2004. The evolution of modern eukaryotic phytoplankton. Science 305: 354–360.
4. McFadden GI, Waller RE, Reith ME, Langunnasch N (1997) Plastids in apicomplexan parasites. Plant System Evol 11:261–287.
5. Ralph SA, van Dooren GG, Waller RF, Crawford MJ, Fraunholz MJ, et al. (2004) Tropical infectious diseases: metabolic maps and functions of the *Plasmodium falciparum* apicoplast. Nat Rev Microbiol 2: 203–216.
6. Oborník M, Modrý D, Lukeš M, Černotíková-Stříbrná E, Cihlář J, et al. (2012) Morphology, ultrastructure and life cycle of *Vitrella brassicaformis* n. sp., n. gen., a novel chromerid from the Great Barrier Reef. Protist 163: 306–323
7. Cavalier-Smith T (1999) Principles of protein and lipid targeting in secondary symbiogenesis: Euglenoid, dinoflagellate, and sporozoan plastid origins and theeukaryote family tree. J Eukaryot Microbiol 46:347–366.
8. Cavalier-Smith T (2004) Chromalveolate diversity and cell megaevolution: interplay of membranes, genomes and cytoskeleton. In: Hirt RP, Horner D, editors. Organelles, genomes and eukaryotic evolution. Taylor and Francis, London. pp. 71–103.
9. Keeling PJ (2009) Chromalveolates and the evolution of plastids by secondary endosymbiosis. J Eukaryot Microbiol 56: 1–8.
10. Moore RB, Oborník M, Janouškovec J, Chrudimský T, Vancová M, et al. (2008) A photosynthetic alveolate closely related to apicomplexan parasites. Nature 451: 959–963.
11. Zhang Z, Green BR, Cavalier-Smith T (2000) Phylogeny of ultra-rapidly evolving dinoflagellate chloroplast genes: A possible common origin for sporozoan and dinoflagellate plastids. J Mol Evol 51: 26–40.
12. Fast NM, Kissinger JC, Roos DS, Keeling PJ (2001)Nuclear-encoded, plastid targeted genes suggest a single common origin for apicomplexan and dinoflagellate plastids. Mol Biol Evol 18: 418–426.
13. Kořený L, Sobotka R, Janouškovec J, Keeling PJ, Oborník M (2011) Tetrapyrrole synthesis of photosynthetic chromerids is likely homologous to the unusual pathway of apicomplexan parasites. Plant Cell23: 3454–3462.
14. Chang SS, Prézelin BB, Trench RK (1983) Mechanisms of photoadaptation in three strains of the symbiotic dinoflagellate *Symbiodinium microadriaticum*. Mar Biol 76: 219–229.
15. Muscatine L, Falkowski PG, Porter JW, Dubinsky Z (1984) Fate of photosynthetic fixed carbon in light-adapted and shade-adapted colonies of the symbiotic coral *Stylophora pistillata*. Proc R Soc Lond B 222:181–202.
16. Yellowlees D, Warner M (2003) Photosynthesis in symbiotic algae. In: Larkum AWD, Douglas SE, Raven JA, editors. Photosynthesis in Algae. Advances in Photosynthesis and Respiration. Springer. pp. 437–455.
17. Sheppard CRC, Davy SK, Pilling GM (2009) The Biology of Coral Reefs. The Biology of Habitats Series, Oxford University Press. 352 p.
18. Oborník M, Janouskovec J, Chrudimský T, Lukes J (2009) Evolution of the apicoplast and its hosts: From heterotrophy to autotrophy and back again. Int J Parasitol 39: 1–12.
19. Janouškovec J, Horák A, Oborník M, Lukes J, Keeling PJ (2010) A common red algal origin of the apicomplexan, dinoflagellate, and heterokont plastids. Proc Natl Acad Sci U.S.A. 107: 10949–10954.
20. Okamoto N, McFadden GI (2008) The mother of all parasites. Future Microbiol.3: 391–395.
21. Oborník M, Vancová M, Lai DH, Janouškovec J, Keeling PJ, et al. (2011) Morphology and ultrastructure of multiple life cycle stages of the photosynthetic relative of apicomplexa, *Chromera velia*. Protist 162:115–130.
22. Weatherby K, Murray S, Carter DJ, Šlapeta J (2011) Surface and flagella morphology of the motile form of *Chromera velia* revealed by field-emission scanning electron microscopy. Protist 162: 142–153.
23. Sutak R, Šlapeta J, San Roman M, Camadro J-M, Lesuisse E (2010) Nonreductive iron uptake mechanism in the marine alveolate *Chromera velia*. Plant Physiol154: 991–1000.
24. Guo JT, Weatherby K, Carter D, Šlapeta J (2010) Effect of nutrient concentration and salinity on immotile–motile transformation of *Chromera velia*. J Eukaryot Microbiol 57: 444–446.
25. Kotabová E, Kaňa R, Jarešová J, Prášil O (2011) Non-photochemical fluorescence quenching in *Chromera velia* is enabled by fast violaxanthin de-epoxidation. Febs Letters 585: 1941–1945.
26. Pan H, Šlapeta J, Carter D, Chen M (2012) Phylogenetic analysis of the light-harvesting system in *Chromera velia*. Photosyn Res111: 19–28.
27. Botté CY, Yamaryo-Botté Y, Janouškovec J, Rupasinghe T, Keeling PJ, et al. (2011) Identification of plant-like galactolipids in *Chromera velia*, a photosynthetic relative of malaria parasites. J Biol Chem 286: 29893–29903.
28. Leblond JD, Dodson J, Khadka M, Holder S, Seipelt RL (2012) Sterol Composition and Biosynthetic Genes of the Recently Discovered Photosynthetic Alveolate, *Chromera velia* (Chromerida), a Close Relative of Apicomplexans. J. Eukaryot. Microbiol., 0(0), pp. 1–7, DOI: 10.1111/j.1550-7408.2012.00611.x
29. Havelková-Doušová H, Prášil O, Behrenfeld MJ (2004) Photoacclimation of *Dunaliella tertiolecta* (Chlorophyceae) under fluctuating irradiance. Photosynthetica 42: 273–281.
30. Jeffrey SW, Vesk M (1997) Introduction to marine phytoplankton and their pigment signatures. In: Jeffrey SW, Mantoura RFC, Wright SW, editors. Phytoplankton pigments in oceanography. Paris: UNESCO Publishing. pp 37–84.
31. Porra RJ, Thompson WA, Kriedemann PE (1989) determination of accurate extinction coefficients and simultaneous equations for assaying chlorophylls a and b extracted with four different solvents: Verification of the concentration of chlorophyll standards by atomic absorption spectrometry. Biochim Biophys Acta 975: 384–394.
32. Kaňa R, Kotabová E, Komárek O, Papageorgiou GC, Govindjee, et al. (2012a) Slow S to M fluorescence rise in cyanobacteria is due to a state 2 to state 1 transition. Biochim Biophys Acta, in press.
33. Kolber ZS, PrasilO, Falkowski PG (1998)Measurements of variable chlorophyll fluorescence using fast repetition rate techniques: defining methodology and experimental protocols. Biochim Biophys Acta1367: 88–106.
34. Suggett DJ, Moore CM, Geider RJ (2010) Estimating Aquatic Productivity from Active Fluorescence Measurement. In: Suggett DJ, Prasil O, Borowitzka M.A, editors. Chlorophyll a Fluorescence in Aquatic Sciences: Methods and Applications. Springer, Dordrecht, pp 103–127.
35. Lewis MR, Smith JC (1983) A small volume, short-incubation-time method for measurement of photosynthesis as a function of incident irradiance. Mar Ecol Prog Ser 13:99–102.
36. Butler NB (1982) Carbon dioxide equilibria and their applications. New York: Addison-Wesley Publishing Company Inc. 259 p.
37. Jassby AD, Platt T (1976) Mathematical formulation of the relationship between photosynthesis and light for phytoplankton. Limnol Oceanogr 21: 540–547.
38. Falkowski PG, Dubinsky Z, Wyman K (1985a) Growth-irradiance relationships in phytoplankton. Limnol Oceanogr 30: 311–321.
39. Falkowski PG, Dubinsky Z, Santostefano G (1985b) Light-enhanced dark respiration in phytoplankton. Verh Internat Verein Limnol 22: 2830–2833.
40. Levy O, Dubinsky Z, Schneider K, Achituv Y, Zakai D, et al. (2004) Diurnal hysteresis in coral photosynthesis. Mar Ecol Prog Ser 268: 105–117.
41. Falkowski PG, Dubinsky Z, Muscatine L, Porter JW (1984) Light and bioenergetics of symbiotic coral. BioScience 34: 705–709.
42. Porter JW, Muscatine L, Dubinsky Z, Falkowski PG (1984) Primary production and photoadaptation in light-adapted and shade-adapted colonies of the symbiotic coral, *Stylophora pistillata*. Proc R Soc Lond B 222: 161–180.
43. Gorbunov MY, Falkowski PG, Kolber ZS (2000) Measurement of photosynthetic parameters in benthic organisms in situ using a SCUBA-based fast repetition rate fluorometer. Limnol Oceanogr 45:242–245.
44. Gorbunov MY, Kolber ZS, Lesser MP, Falkowski PG (2001) Photosynthesis and photoprotection in symbiotic corals. Limnol Oceanogr 46:75–85.
45. Hennige SJ, Suggett DJ, Warner ME, McDougall KE, Smith DJ (2009) Photobiology of *Symbiodinium* revisited: bio-physical and bio-optical signatures. Coral Reefs 28: 179–195.
46. Hoegh-Guldberg O, Jones R (1999) Photoinhibition and photoprotection in symbiotic dinoflagellates from reef-building corals. Mar Ecol Prog Ser 183:73–86.

47. Jones RJ, Hoegh-Guldberg O (2001) Diurnal changes in the photochemical efficiency of the symbiotic dinoflagellates (Dinophyceae) of corals: photoprotection, photoinactivation and the relationship to coral bleaching. Plant Cell Environ 24: 89–99.
48. Ralph PJ, Gademann R, Larkum AWD, Kahl M (2002) Spatial heterogeneity in active chlorophyll fluorescence and PSII activity of coral tissues. Mar Biol 141: 639–646.
49. Ragni M, Airs RL, Hennige SJ, Suggett DJ, Warner ME, et al. (2010) PSII photoinhibition and photorepair of *Symbiodinium* (Pyrrophyta) differs between thermally tolerant and sensitive phylotypes. Mar Ecol Prog Ser 406: 57–70.
50. Suggett DJ, Warner M, Smith DJ, Davey P, Hennige S, et al. (2008) Photosynthesis and production of hydrogen peroxide by symbiotic dinoflagellates during short-term heat stress. J Phycol44: 948–956.
51. Iglesias-Prieto R, Beltran V, La Jeunesse T, Reyes-Bonilla H, Thome P (2004) Different algal symbionts explain the vertical distribution of dominant reef corals in the eastern Pacific. Proc R Soc Lond B 271: 1757–1763.
52. Hoegh-Guldberg O, Mumby PJ, Hooten AJ (2007) Coral reefs under rapid climate change and ocean acidification. Science318:739–742.
53. Quigg A, Beardall J (2003) Protein turnover in relation to maintenance metabolism at low photon flux in two marine microalgae. Plant, Cell Environ 26: 1–10.
54. Warner ME, Lesser MP, Ralph PJ (2010) Chlorophyll fluorescence in reef building corals. In: Suggett DJ, Prášil O, Borowitzka M, editors. Chlorophyll Fluorescence in Aquatic Sciences: Methods and Applications. Springer. pp. 209–222.
55. Kaňa R, Kotabová E, Sobotka R, Prášil O (2012b) Non-photochemical quenching in cryptophyte alga *Rhodomona ssalina* is located in chlorophyll a/c antennae. PLoS One. 7: e29700.
56. Prášil O, Adir N, Ohad I (1992) Dynamics of Photosystem II: mechanism of photoinhibition and recovery processes. In: Barber J, editor. The Photosystems: Structure, Function and Molecular Biology. Elsevier Science, Oxford. pp. 295–348.
57. Kaňa R, Lazar D, Prášil O, Naus J (2002)Experimental and theoretical studies on the excess capacity of Photosystem II. Photosyn Res72: 271–284.
58. Beardall J, Quigg A, Raven JA (2003) Oxygen consumption: Photorespiration and chlororespiration. In: Larkum AWD, Douglas SE, Raven JA, editors. Photosynthesis in Algae. Advances in Photosynthesis and Respiration. Springer. pp. 157–181.
59. Crawley A, Kline DI, Dunn S, Anthony K, Dove S (2010) The effect of ocean acidification on symbiont photorespiration and productivity in *Acropora formosa*. Global Change Biol16: 851–863.
60. Nassoury N, Fritz L, Morse D (2001) Circadian changes in ribulose-1,5-bisphosphate carboxylase/oxygenase distribution inside individual chloroplasts can account for the rhythm in dinoflagellate carbonfixation. Plant Cell 13: 923–934.
61. Leggat W, Badger M, Yellowlees D (1999) Evidence for an inorganic carbon-concentrating mechanism in the symbiotic dinoflagellate *Symbiodinium* sp. Plant Physiol 121: 1247–1255.
62. Burris JE (1981) Effects of oxygen and inorganic carbon concentrations on the photosynthetic quotients of marine algae. Mar Biol 65: 215–219.
63. Whitney SM, Andrews TJ (1998) The CO_2/O_2 specificity of single-subunit ribulose-bisphosphate carboxylase from the dinoflagellate, *Amphidinium carterae*. Aus J Plant Physiol25: 131–138.
64. Badger MR, Bek EJ (2008) Multiple Rubisco forms in proteobacteria: their functional significance in relation to CO_2 acquisition by the CBB cycle. J Exp Bot 59:1525–1541.
65. Schnepf E, Elbrachter M (1999) Dinophyte chloroplasts and phylogeny - A review. GRANA 38: 81–97.
66. Mehler AH (1957) Studies on reactions of illuminated chloroplasts, I: Mechanism of the reduction of oxygen and other Hill reagents. Arch BiochemBiophys33:65–77.
67. Peltier G, Cournac L (2002) Chlororespiration. Ann Rev Plant Physiol Plant Mol Biol 53: 523–550.
68. Peltier G, Tolleter D, Billon E, Cournac L (2010) Auxiliary electron transport pathways in chloroplasts of microalgae. Photosyn Res106: 19–31.
69. Reynolds JM, Bruns BU, Fitt WK, Schmidt GW (2008) Enhanced photoprotection pathways in symbiotic dinoflagellates of shallow-water corals and other cnidarians. Proc Natl Acad Sci U.S.A. 105: 13674–13678.
70. Cardol P, Forti G, Finazzi G (2011) Regulation of electron transport in microalgae. Biochim Biophys Acta, 1807: 912–918.

18

Endophyte-Mediated Effects on the Growth and Physiology of *Achnatherum sibiricum* Are Conditional on Both N and P Availability

Xia Li, Anzhi Ren*, Rong Han, Lijia Yin, Maoying Wei, Yubao Gao

College of Life Sciences, Nankai University, Tianjin, P. R. China

Abstract

The interaction of endophyte–grass associations are conditional on nitrogen (N) availability, but the reported responses of these associations to N are inconsistent. We hypothesized that this inconsistency is caused, at least in part, by phosphorus (P) availability. In this experiment, we compared the performance of endophyte-infected (EI) and endophyte-free (EF) *Achnatherum sibiricum* subjected to four treatments comprising a factorial combination of two levels of N (N+ vs. N−, i.e. N supply vs. N deficiency) and two levels of P (P+ vs. P−, i.e. P supply vs. P deficiency) availability. The results showed that *A. sibiricum–Neotyphodium* associations were conditional on both N and P availability, but more conditional on N than P. Under N+P− conditions, endophyte infection significantly improved acid phosphatase activity of EI plants, such that the biomass of EI plants was not affected by P deficiency (i.e. similar growth to N+P+ conditions), and resulted in more biomass in EI than EF plants. Under N−P+ conditions, biomass of both EI and EF decreased compared with N+P+; however, EI biomass decreased slowly by decreasing leaf N concentration more rapidly but allocating higher fractions of N to photosynthetic machinery compared with EF plants. This change of N allocation not only improved photosynthetic ability of EI plants but also significantly increased their biomass. Under N−P− conditions, EI plants allocated higher fractions of N to photosynthesis and had greater P concentrations in roots, but there was no significant difference in biomass between EI and EF plants. Our results support the hypothesis that endophyte–grass interactions are dependent on both N and P availability. However, we did not find a clear cost of endophyte infection in *A. sibiricum*.

Editor: Mark van Kleunen, University of Konstanz, Germany

Funding: This research was funded by the National Natural Science Foundation (30970460) and the Scientific Research Foundation for Returned Overseas Chinese Scholars, State Education Ministry (2009–2011). The funders had no role in study design, data collection and analysis, decision to publish, or preparation of the manuscript.

Competing Interests: The authors have declared that no competing interests exist.

* E-mail: renanzhi@nankai.edu.cn

Introduction

Many grasses are infected by clavicipitaceous fungal endophytes that occur in aboveground plant tissues. Asexual endophytes live asymptomatically within the host tissues, receiving protection and nutrients, and are vertically transmitted to the next plant generation via host seeds. Based on numerous studies using tall fescue and perennial ryegrass in agronomic, fertilized soils, this symbiosis has been considered strongly mutualistic – mainly because endophyte infection may improve herbivore resistance of the host grasses due to production of alkaloids [1], and increase plant vigor and tolerance to a wide range of abiotic environmental conditions (e.g. drought) [2–3]. There is increasing evidence that the benefits from endophyte infection depend largely on the availability of other resources, in particular nutrients [4]. Resource limitation can increase the cost of supporting some endophytes [5–6], potentially changing the interaction from mutualism to parasitism or commensalism [7]. In fact, many of the studies that have found improved growth in endophyte-infected (EI) grasses were done under benign conditions of moderate to high soil nutrient availability [8–11].

Studies on endophyte-related responses of grasses to nutrient acquisition have focused on the influence of nitrogen (N), since this element is not only a constituent of alkaloids in infected plants but also one of the most important limiting resources for plant growth in nature. In the plant the photosynthetic apparatus is the largest sink of N [12]. Photosynthetic capacity and photosynthetic N use efficiency (PNUE) correlates strongly with N allocation to the photosynthetic machinery [13]. Small changes in N allocation can greatly influence light-saturated photosynthetic rate (P_{max}) and PNUE, and therefore plant performance [14–16]. Consequently, leaf N allocation to photosynthesis is an important factor explaining differences in P_{max} and PNUE [14]. Published reports of the effects of endophyte infection on N use efficiency of grass-endophyte associations are inconsistent. Arachavaleta et al. [17] found beneficial effects of endophyte infection in tall fescue only at high N concentrations, and this result was further supported by our previous study in perennial ryegrass [11]. In contrast, Ravel et al. [18] found an advantage of EI plants over EF (endophyte-free) plants was greater at low N levels. It has been documented that increased N availability may also change the relative availability of other nutrients such as phosphorus (P) [19–20]. Therefore, we asked whether the inconsistent results are caused, in part, by other nutrients such as P.

Similar to N nutrition, P availability also influences ergot alkaloid production in EI grasses [21]. However, published reports of the effects of endophyte infection on P use efficiency of grass-endophyte associations are limited [22]. Malinowski et al. [23] found that EI tall fescue expressed an increased root absorption area through reduced root diameter and increased root hair length compared with the EF counterpart. The Fe^{3+} reducing activity on the root surface and total phenolic concentration in roots also increased dramatically in response to endophyte infection [24]. N addition may change the relative availability of P. On one hand, N addition could stimulate phosphatase activity of the root [25], which could potentially promote P uptake from bound-P. In fact, the production and excretion of acid phosphatase is considered to be one component of a plant phosphate-starvation rescue system [26]. On the other hand, a high N:P supply ratio could result in P starvation in the plant [27]. Populations previously limited by N can switch to limitation by P after receiving high N [19–20]. N:P stoichiometry in plant tissues, especially leaves, is related to growth strategy and can be an indicator of vegetation composition, functioning and nutrient limitation at the community level [28]. Until now, however, the effect of both N and P availability on grass–endophyte associations has received little attention.

Endophytic fungi not only occur in agronomic grasses but also in almost all habitats where grasses are common [29]. In our previous survey in the permanent grasslands of northern China, 25 of 41 species of grasses surveyed (61%) were infected by *Neotyphodium* endophytes [30]. However, most of the work for endophyte-plant interactions has been based upon endophyte-plant studies of two, economically important, artificially selected and non-native grass species [31–32]. Few studies exist to predict how wild plant–endophyte symbioses will respond to N and/or P availability, especially when the two elements are considered simultaneously. If infection competes with other plant functions for limiting nutrients, then infection may be more advantageous in environments with high soil nutrients [8,33]. Alternatively, if systemic endophytes enhance nutrient uptake by the host [34], then infected plants may outcompete uninfected plants when nutrients are limited. Therefore, endophyte infection may have strong influences on plant community composition by altering the performance of host grasses relative to other species present in the community in response to different nutrient availability.

Achnatherum sibiricum (L.) Keng is a caespitose perennial grass that is widely distributed in northern China. After five years of continuous survey, Wei et al. [30] found that *A. sibiricum* was highly infected by *Neotyphodium* fungi, and there was little difference in infection rates among different geographic populations. Within the genus *Achnatherum* there are five sections, and *A. sibiricum* belongs to section Achnatheropsis (Tzvel.) N. S. Probatova. There are nine species in this section, including seven Asian and two American species [35]. Except for *A. sibiricum*, only two species, *A. inebrians* (Hance) Keng ex Tzvelev and *A. robustum* (Vasey) Barkworth, are reported to be infected by *Neotyphodium* endophytes. Both are notorious for their narcotic effects on livestock, and hence are known as 'drunken horse grass' and 'sleepy grass', respectively [36–37]. In contrast to *A. inebrians* and *A. robustum*, *A. sibiricum* has no obvious herbivore deterrence according to local records and our own observations. Here, we investigated whether the responses of *A. sibiricum* to endophyte infection depended on N and/or P availability. Specifically, we addressed the following questions: (1) does the endophyte improve performance of the native grass host? (2) does N and/or P availability affect the symbiosis-dependent benefits? If this is the case, (3) how does the nutrient availability affect the symbiosis-dependent benefits?

Materials and Methods

Ethics statement

No specific permissions were required since in this study we only collected a limited amount of seeds from a native grassland, and this grassland is not privately-owned or protected in any way. Our field study did not involve any endangered or protected species.

Study System

Achnatherum sibiricum is a perennial, sparse bunch grass that is native to the Inner Mongolia Steppe of China. It is usually a companion species in the grassland and can sometimes become a dominant species. High incidences of *Neotyphodium* endophyte infection (86–100%) in *A. sibiricum* were recorded in seven native populations in our previous study [30]. In the present study, seeds of *A. sibiricum* were collected from natural population in Hailar in Northeast China (119.67°E, 49.10°N), where the annual mean temperature is around −2°C and annual precipitation about 367 mm. This meadow steppe belongs to a transitional type of habitat between forest and steppe. *Achnatherum sibiricum* within this area is less preferred by mammalian herbivores compared to other dominant species in the community [38]. In the sampled area the dominant species included *Stipa baicalensis* Roshev. and *Leymus chinensis* L., with *A. sibiricum* and *Koeleria cristata* (Linn.) Pers. as common species. Within this population, we collected seeds in August 2008 and stored them at 4°C.

Detection of endophytes using the aniline blue staining method [39] showed that endophyte infection frequency of the Hailar population was 100%. To eliminate the endophyte, we heat treated a subset of randomly chosen seeds in a convection drying oven according to Kannadan and Rudgers [40]. Because disinfection procedures have not been established for this species, we initially treated seeds for 0, 5, 10, 15, 20, 25 or 30 d at 60°C to determine the optimal treatment time. Then all treated seeds were planted in plastic pots filled with vermiculite in November 2008. To assess treatment effectiveness, we examined three leaf peels from each plant under a microscope [39]. In addition, we assessed potential effects of the heat treatment on seed germination and seedling growth. After a 30-d heat treatment, none of the seedlings were infected. Moreover, high temperature treatment had no significant effect on germination rate, germination potential and germination index [41].

Experimental Design

The plants used in this experiment were cloned from 100 plants grown from seeds that were not heat treated (endophyte-infected, EI) and 100 plants from seeds that were heat treated for 30 d (endophyte-free, EF), multiplied and selected for uniformity in spring of 2009 and 2010. During this period, the plants were clipped repeatedly and kept in vegetative growth. This procedure allowed the subsequent assessment of plant performance to be separated from the initial heat treatment by a round of vegetative reproduction, and is commonly used in endophyte studies [40,42]. On June 2010, we randomly chose 100 EI and 100 EF tillers (one tiller from each plant), of approximately equal size, and transplanted them evenly into 40 white plastic pots (20 EI and 20 EF pots, five tillers per pot). One pot was 23 cm in diameter and 25 cm in depth and filled with 5 kg of sand. The design of the experiment was completely randomized and a 2×2×2 factorial, with infection status (EI vs. EF), N availability (N+ vs. N−, i.e. supply vs. deficiency), and P availability (P+ vs. P−, i.e. supply vs. deficiency) as the variables. There were five replicates per treatment group. The experiment lasted 49 d, from 5 August to 23 September 2010, and was carried out at the campus

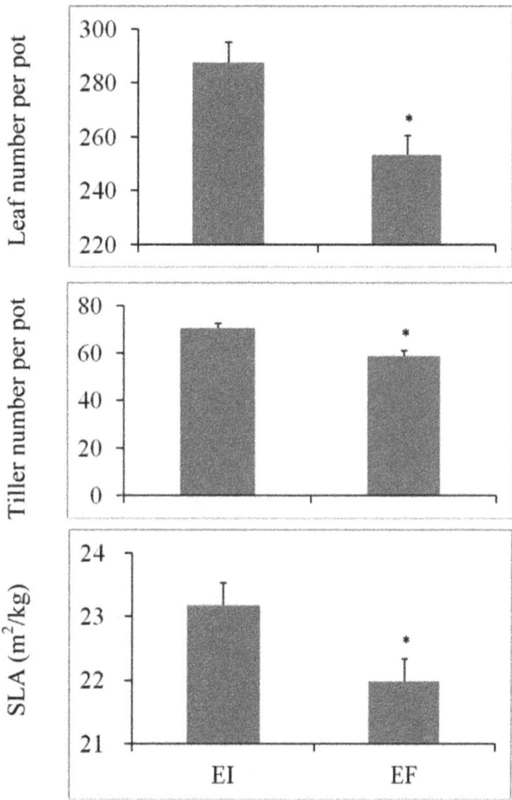

Figure 1. Leaf number, tiller number, and specific leaf area (SLA) of endophyte-infected (EI) or endophyte-free (EF) *Achnatherum sibiricum*. Bars are means+1 SE. Means are data averaged across treatments. An asterisk denotes significance at $P<0.05$.

experimental field at Nankai University, Tianjin, China. Each ramet (from a single tiller) was examined for endophyte status following staining with lactophenol aniline blue [39] at the end of the experiment.

Nutrient Treatment

We established four treatments in which nutrient availability was varied, i.e. N+P+, N+P−, N−P+ and N−P−. Ramets from each EI and EF group were grown under all combinations of nutrient availability. The nutrients were supplied by the addition of complete Hoagland nutrient solution. The composition of the nutrient solution was 5.0 mM $Ca(NO_3)_2$, 5.0 mM KNO_3, 2.5 mM $MgSO_4 \cdot 7H_2O$, 2.0 mM KH_2PO_4, 29 μM Na_2-EDTA, 20 μM $FeSO_4 \cdot 7H_2O$, 45 μM H_3BO_3, 6.6 μM $MnSO_4$, 0.8 μM $ZnSO_4 \cdot 7H_2O$, 0.6 μM H_2MoO_4, 0.4 μM $CuSO_4 \cdot 5H_2O$ and pH 6.0±0.1. For N− treatment, 5.0 mM $CaCl_2$ and 5.0 mM KCl were added instead of $Ca(NO_3)_2$ and KNO_3. For P− treatment, 2.0 mM KCl was added instead of KH_2PO_4. The pH was adjusted to 6.0±0.1. During the experiment, 0.8 L of nutrient solution was added twice a week per pot, and 15 times in total. Plants were subjected to ambient light and temperature regimes. The positions of the pots were randomly rotated each week to minimize location effects.

Growth and Biomass

Measurements of tiller number, leaf number and shoot height of the longest tiller were made on all ramets at the beginning and end of the experiment. At the end of the experiment, leaves, sheaths

Table 1. Three-way ANOVA for vegetative growth of endophyte-infected (EI) or endophyte-free (EF) *Achnatherum sibiricum*.

		Tiller No.			Leaf No.			SLA			Shoot biomass			Root biomass			Total biomass		
	df	MS	F	P	MS	F	P	MS	F	P	MS	F	P	MS	F	P	MS	F	P
Endophyte (E)	1	1334	13.39	<0.01	11730	10.99	<0.01	53.94	5.426	0.026	28.80	26.58	<0.01	9.120	21.21	<0.01	70.33	31.04	<0.01
Nitrogen (N)	1	8208	82.41	<0.01	61701	57.83	<0.01	88.66	8.917	<0.01	383.2	353.6	<0.01	9.351	21.75	<0.01	512.2	226.1	<0.01
Phosphorus(P)	1	570.0	5.723	0.023	1918	1.798	0.189	13.17	1.324	0.258	87.38	80.65	<0.01	6.956	16.18	<0.01	143.6	63.40	<0.01
E×N	1	7.225	0.073	0.789	24.03	0.023	0.882	0.390	0.039	0.844	0.445	0.411	0.526	0.004	0.008	0.928	0.529	0.233	0.632
E×P	1	42.03	0.422	0.521	1113	1.043	0.315	10.44	1.050	0.313	0.702	0.648	0.427	0.180	0.418	0.523	1.592	0.703	0.408
N×P	1	207.0	2.079	0.159	235.2	0.220	0.642	4.323	0.435	0.514	3.091	2.853	0.101	1.616	3.759	0.061	9.178	4.051	0.053
E×N×P	1	286.2	2.874	0.100	60.03	0.056	0.814	0.206	0.021	0.886	11.13	10.27	<0.01	1.552	3.611	0.066	20.99	9.267	<0.01
Residual	32	99.60			1067			9.942			1.083			0.430			2.266		

Table 2. Biomass allocation of endophyte-infected (EI) or endophyte-free (EF) *Achnatherum sibiricum* under various conditions of N and P availability.

Treatment			Shoot biomass (g)	Root biomass(g)	Total biomass(g)	Root: Shoot
P+	N+	EI	14.15±1.299a	5.14±0.846a	19.30±1.999a	0.36±0.042c
		EF	13.56±1.501a	4.70±0.691a	18.26±1.929a	0.35±0.049c
	N−	EI	9.36±0.775b	4.95±0.428a	14.32±1.088b	0.53±0.042 b
		EF	7.09±0.579c	3.76±0.437b	10.85±0.790c	0.53±0.069 b
P−	N+	EI	13.07±1.271a	5.24±0.547a	18.31±1.748a	0.40±0.024c
		EF	9.84±1.247b	3.74±0.915b	13.58±1.897b	0.38±0.078c
	N−	EI	5.06±0.424d	3.46±0.494bc	8.52±0.446d	0.69±0.142a
		EF	4.37±0.672d	2.78±0.701c	7.15±1.332d	0.63±0.088a

Note. Values are means ± SE. Significant differences ($P<0.05$) for each variable are indicated by lowercase letters for variables where N, P availability and endophyte infection were analyzed together.

and roots were harvested separately. Ten fully expanded leaves growing on vegetative tillers per pot were chosen to measure the area and were weighed separately for determination of specific leaf area (SLA). Roots were washed free of soil. Then all plant parts, including leaf blades, sheaths, roots and senesced leaves were separately oven-dried at 60°C.

Gas Exchange

At the end of the treatments, gas exchange measurements (see below) were made on the youngest fully expanded attached leaf in a pot with a LI-COR 6400 infrared gas analyzer (LI-Cor, Lincoln, NE, USA). The same leaf was also used for measurements of SLA and N content. In this way, differences among leaves of the same plant could be avoided when the relationships among the variables were analyzed.

Photosynthesis-light responses of plants were assessed under 400 μmol mol^{-1} CO_2. Net photosynthetic rate (Pn) was measured at 1500, 1200, 1000, 800, 500, 300, 200, 150, 100, 50, 20 and 0 μmol m^{-2} s^{-1} PPFD (photosynthetic photon flux density). From the Pn-PPFD curve, Pmax and staturation PPFD were determined.

Photosynthesis-CO_2 responses of plants were assessed under saturation PPFD, 1200 μmol m^{-2} s^{-1}. P_n was measured at 1500, 1200, 1000, 800, 600, 400, 300, 200, 150, 100 and 50 μmol mol^{-1} CO_2 in the reference chamber. The leaf temperature was held constant at 25°C by the equipment. From the P_n-C_i (internal CO_2 concentration) curve, the parameters needed to calculate the fraction of leaf N allocated to the photosynthetic machinery were determined. The calculation details are as follows.

The P_n-C_i curve was fitted with a linear equation ($P_n = kC_i+i$) within 50–200 μmol mol^{-1} C_i [43]. Maximum carboxylation rate (V_{cmax}) and dark respiration rate (R_d) were calculated according to Farquhar and Sharkey [44] as follows:

$$V_{c\,\max} = k\left[C_i + K_c\left(1 + {}^{O}\!/_{K_0}\right)\right]^2 \Big/ \left[\Gamma^* + K_c\left(1 + {}^{O}\!/_{K_0}\right)\right]$$

$$Rd = {}^{V_{c\,\max}(C_i - \Gamma^*)}\Big/\left[C_i + K_c\left(1 + {}^{O}\!/_{K_c}\right)\right]^{-(kC_i+i)}$$

where K_c and K_o are the Michaelis–Menten constants of Rubisco for carboxylation and oxidation, respectively, and calculated according to Niinemets and Tenhunen [45]. Γ^* is the CO_2

Table 3. Three-way ANOVA for photosynthetic parameters, N allocation and acid phosphatase activity of endophyte-infected (EI) or uninfected (EF) ramets of *Achnatherum sibiricum*.

		N_A			P_{max}			PNUE			P_T			Acid phosphatase activity		
	df	MS	*F*	*P*	MS	*F*	*P*	MS	*F*	*P*	MS	*F*	*P*	MS	*F*	*P*
Endophyte (E)	1	0.567	90.06	<0.01	31.91	17.67	<0.01	895.2	320.7	<0.01	1.108	83.08	<0.01	0.003	18.35	<0.01
Nitrogen (N)	1	4.332	687.6	<0.01	215.1	119.1	<0.01	771.8	276.5	<0.01	1.396	104.7	<0.01	0.029	185.1	<0.01
Phosphorus(P)	1	0.050	7.945	<0.01	3.869	2.142	0.156	18.99	6.803	0.015	0.035	2.628	0.125	0.005	30.73	<0.01
E×N	1	0.001	0.080	0.780	5.946	3.292	0.082	209.3	74.97	<0.01	0.434	32.56	<0.01	0.000	1.587	0.217
E×P	1	0.045	7.112	0.012	4.095	2.267	0.145	19.26	6.901	0.015	0.000	0.015	0.905	0.000	1.587	0.217
N×P	1	0.017	2.646	0.114	0.581	0.322	0.576	21.24	7.609	0.011	0.118	8.816	<0.01	0.021	134.3	<0.01
E×N×P	1	0.030	4.688	0.038	5.115	2.832	0.105	0.647	0.232	0.635	0.027	1.996	0.177	0.000	1.587	0.217
Residual	32	0.006			1.806			2.791			0.013			0.000		

Note. N_A, total leaf nitrogen content; P_{max}, maximum net photosynthetic rate; PNUE, photosynthetic nitrogen use efficiency; P_T the fraction of leaf nitrogen allocated to all components of the photosynthetic machinery.

Table 4. N allocation and maximum photosynthetic rate of endophyte-infected (EI) or endophyte-free (EF) *Achnatherum sibiricum* under various conditions of N and P availability.

Treatment			N_A	P_T	P_{max}
P+	N+	EI	0.92±0.084c	0.719±0.059c	15.58±1.323a
		EF	1.28±0.100a	0.474±0.034d	12.80±1.068b
	N−	EI	0.35±0.036f	1.240±0.227b	10.60±0.708c
		EF	0.61±0.081d	0.614±0.124cd	7.95±0.844d
P−	N+	EI	1.02±0.114c	0.571±0.037cd	15.23±1.223a
		EF	1.13±0.080b	0.471±0.051d	12.29±2.773bc
	N−	EI	0.25±0.075f	1.529±0.116a	8.12±0.463d
		EF	0.49±0.023e	0.758±0.139c	8.50±0.972d

Note. N_A, total leaf nitrogen content in g m^{-2}; P_T, the fraction of leaf nitrogen allocated to all components of the photosynthetic machinery in g g^{-1}; P_{max}, maximum net photosynthetic rate in μmol m^{-2} s^{-1}. Values are means ± SE. Significant differences ($P<0.05$) for each variable are indicated by lowercase letters for variables N, P availability and endophyte infection were analyzed together.

compensation point and O is the intercellular oxygen concentration (close to 210 mmol mol^{-1}).

Maximum electron transport rate (J_{max}) was calculated according to Loustau et al. [46] as follows:

$$J_{max} = \left[4\left(P_{max}{}' + R_d\right)\left(C_i + 2\right)\right] / \left(C_i - \Gamma^*\right)$$

where $P_{max}{}'$ was determined under saturation PPFD and CO_2 concentration.

The fractions of total leaf N allocated to carboxylation (P_C), bioenergetics (P_B) and light-harvesting (P_L) of the photosynthetic apparatus were calculated according to Niinemets and Tenhunen [45] as:

$$P_C = V_{c\max} / (6.25 \times V_{cr} \times N_A)$$

$$P_B = J_{\max} / (8.06 \times J_{mc} \times N_A)$$

$$P_L = C_C / (N_M \times C_B)$$

where V_{cr} and J_{mc} are the specific activities of Rubisco and cytochrome f, respectively. N_A is total leaf N content, C_C is leaf chlorophyll concentration, N_M is mass-based leaf N content and C_B is ratio of leaf chlorophyll to leaf N in light-harvesting components. The fraction of leaf N allocated to all components of the photosynthetic machinery (P_T) was calculated as the sum of P_C, P_B and P_L. Photosynthetic nitrogen use efficiency (PNUE) was calculated as the ratio of P_{max} to area-based leaf N concentration.

Other Response Variables

The youngest fully expanded leaves were collected for measuring photosynthetic pigment content [47], N and P concentrations. Dried roots were sampled for measuring N and P concentrations. N concentrations of the plant were analyzed using the Kjeldahl method, and P concentrations were measured by molybdenum–antimony colorimetric method [48].

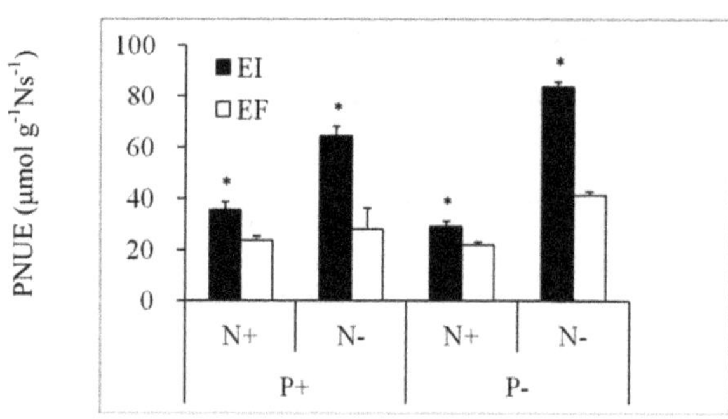

Figure 2. Photosynthetic nitrogen use efficiency (PNUE) of endophyte-infected (EI) or endophyte-free (EF) *Achnatherum sibiricum* under various conditions of N and P availability. Bars are means+1 SE. An asterisk denotes significance at P<0.05.

The acid phosphatase secreted by the roots was measured according to the method of Mclanchlan [49]. At harvest, the sand was washed from the roots, water was removed with tissue paper, and 2.0 g of fresh roots (representative sub-sample) were added to sodium acetate-acetic acid buffer with para- nitrophenyl phosphate (PNPP). The concentration of para-nitrophenol in the solution was determined in a spectrophotometer by measuring the absorbance at 405 nm. Phosphatase activity was calculated as the amount of para-nitrophenol produced per g fresh root mass and per hour.

Statistical Analyses

All statistical analyses were performed with SPSS 10.0 (SPSS, Chicago). For some variables (tiller number, leaf number and biomass allocation), natural log transformation was used to homogenize variance and to obtain a normal distribution of residuals. Effects of N availability, P availability and endophyte infection were analyzed using a three-way analysis of variance (ANOVA). Differences between the means of different treatments and endophyte infection were compared using Duncan's multiple-range tests at $P<0.05$.

Results

Shoot Growth and Biomass Allocation

At the beginning of the experiment, there were no significant differences between the EI and EF plants in tiller number ($F=0.073$, $P=0.999$), leaf number ($F=0.279$, $P=0.958$) and

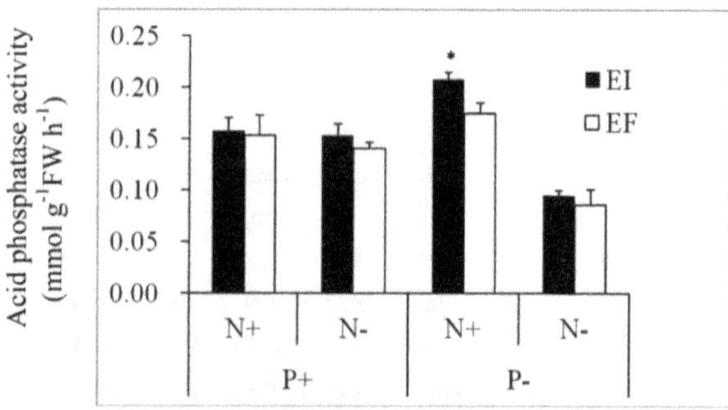

Figure 3. Acid phosphatase activity of endophyte-infected (EI) or endophyte-free (EF) *Achnatherum sibiricum* under various conditions of N and P availability. Bars are means+1 SE. An asterisk denotes significance at P<0.05.

Table 5. Three-way ANOVA for ecological stoichiometry of endophyte-infected (EI) or endophyte-free (EF) ramets of *Achnatherum sibiricum*.

		N concentration									**P concentration**								
		Leaf			**Root**			**Total**			**Leaf**			**Root**			**Total**		
	df	**MS**	*F*	*P*	**MS**	*F*	*P*	**MS**	*F*	*P*	**MS**	*F*	*P*	**MS**	*F*	*P*	**MS**	*F*	*P*
Endophyte (E)	1	196.7	141.6	<0.01	13.47	8.609	<0.01	34.56	68.19	<0.01	0.456	3.617	0.066	7.639	165.8	<0.01	1.845	31.37	<0.01
Nitrogen (N)	1	2563	1845	<0.01	579.5	370.4	<0.01	1688	3331	<0.01	2.793	22.16	<0.01	2.266	49.18	<0.01	1.914	32.55	<0.01
Phosphorus(P)	1	33.62	24.20	<0.01	14.92	9.538	<0.01	41.01	80.92	<0.01	7.048	55.92	<0.01	0.471	10.22	<0.01	3.209	54.57	<0.01
E×N	1	2.084	1.500	0.230	0.034	0.022	0.883	0.600	1.184	0.285	0.006	0.048	0.829	3.919	85.05	<0.01	0.460	7.823	<0.01
E×P	1	13.70	9.864	<0.01	27.41	17.52	<0.01	1.406	2.775	0.106	0.215	1.703	0.201	0.119	2.579	0.118	0.065	1.102	0.302
N×P	1	8.363	6.021	0.020	2.475	1.582	0.218	0.204	0.404	0.530	5.206	41.31	<0.01	0.870	18.89	<0.01	1.853	31.51	<0.01
E×N×P	1	12.92	9.300	<0.01	42.42	27.11	<0.01	0.001	0.002	0.961	0.047	0.372	0.546	0.004	0.096	0.759	0.044	0.752	0.392
Residual	32	1.389			1.564			0.507			0.126			0.046			0.059		

shoot height of the longest tiller ($F = 0.266$, $P = 0.963$). Endophyte presence significantly increased tiller number, leaf number and SLA of *A. sibiricum* irrespective of N or P availability (Table 1, Fig. 1). Total biomass was significantly affected by main effects of endophyte status, N and P availability, and the interaction of endophyte ×N×P (Tables 1 and 2). Under N+P− and N−P+ conditions, EI plants had significantly higher shoot, root and total biomass than EF plants. Under N+P+ and N−P− conditions, however, there were no significant differences in biomass between EI and EF plants. Under P+ condition, the biomass of both EI and EF plants decreased with N deficiency; however, the degree of decrease was lower for EI than EF plants. At the same time, both EI and EF plants allocated more resources to roots and thus the root:shoot ratio increased with N deficiency. Under N+ condition, when compared with P supply, the total biomass of EI plants was maintained with P deficiency; for EF populations, however, the biomass decreased significantly with P deficiency (Table 2).

N Allocation and Photosynthesis

Area-based leaf N content (N_A) was significantly affected by endophyte infection, N and P availability as well as their interaction (Table 3). In all treatments, N_A of EI was lower than that of EF plants (Table 4). N_A of EI was significantly affected by N supply but not by P supply. For N_A of EF, however, it was significantly affected by both N and P supply. When N allocation was considered, there were differences between EI and EF plants and/or among different treatments. With N supply, EI plants had similar or slightly higher N fractions allocated to the photosynthetic machinery (P_T) when compared with their EF counterparts. With N deficiency, however, the above N fraction in EI plants was significantly higher compared to their EF counterparts.

The maximum net photosynthetic rate for EI tended to be higher than that of EF, but there was a significant difference only in N+P− and N−P+ treatments (Table 4). When PNUE was considered, it was significantly higher for EI compared to EF plants in all treatments (Fig. 2).

Acid Phosphatase Activity

Acid phosphatase activity was significantly affected by main effects of endophyte status, N and P availability and the interaction N×P (Table 3). In all treatments, the acid phosphatase activity of EI plants tended to be higher than that of EF plants, but the difference was significant only in N+P− treatment (Fig. 3).

Plant N and P Concentrations

Both leaf and root N and P concentrations of *A. sibiricum* were significantly affected by N and P availability as well as endophyte infection (Table 5). Leaf N concentration was significantly lower for EI compared to EF plants. Under N+ conditions, EI leaf N concentration was not affected by P deficiency, while EF leaf N decreased significantly with P deficiency. Under P+ conditions, both EI and EF leaf P concentrations decreased significantly with N deficiency. Endophyte infection had no effect on leaf P concentration but significantly increased root P concentration. Total N concentration (N concentration of the whole plant) was significantly decreased, while total P concentration (P concentration of the whole plant) was significantly increased by endophyte infection, and thus N:P ratio was significantly lower for EI compared to EF plants (Table 6).

Discussion

The present study demonstrated that a beneficial interaction between the native grass *A. sibiricum*, and its associated fungal endophyte (*Neotyphodium* sp.) depended on both N and P availability. When only N or P was limited, EI plants accumulated significantly more aboveground biomass and total biomass than EF plants. When both N and P were limited, however, the benefits of endophyte infection declined. These findings are in agreement with reports on the response of perennial ryegrass to N deficiency [18] and tall fescue to P deficiency [24], in which only one element (N or P) was deficient; and also in agreement with previous research that EI plants have no advantage over EF plants at low nutrient availabilities [5,8,50]. We did not find a significant advantage of EI plants over their EF counterparts when N and P were supplied – this has been reported previously in tall fescue [8]. A possible explanation for this difference is that the nutrients in the medium where *A. sibiricum* grew were not sufficiently high. McCormick et al. [4] also found EI *Danthonia spicata* did not have a performance advantage relative to EF plants under fertilized conditions, in which the medium where *D. spicata* grew was extremely nutrient poor even in the fertilized treatment. Overall, our results suggested that the benefits from endophyte infection depended largely on the supply of N and/or P. When both N and P were limited simultaneously, the benefits from endophyte infection disappeared.

Table 6. N and P concentration of endophyte-infected (EI) or endophyte-free (EF) *Achnatherum sibiricum* under various conditions of N and P availability.

Treatment			N concentration (g/kg)			P concentration (g/kg)			N:P ratio		
			Leaf	Root	Total	Leaf	Root	Total	Leaf	Root	Total
P+	N+	EI	22.61±0.325c	18.55±1.255a	20.61±0.331b	4.40±0.304a	2.28±0.145cd	3.61±0.203a	5.16±0.334c	8.15±0.845a	5.72±0.331c
		EF	28.89±1.559a	13.62±1.712c	23.08±0.688a	3.99±0.259a	2.12±0.265d	3.25±0.168b	7.27±0.778b	6.52±1.364b	7.11±0.452b
	N−	EI	8.19±0.717f	8.32±1.240d	8.00±0.412e	3.10±0.433b	2.70±0.125ab	2.89±0.193c	2.71±0.665ef	3.09±0.522d	2.77±0.216ef
		EF	13.12±1.215d	7.63±1.410de	10.00±0.913d	2.79±0.426b	1.34±0.190e	2.24±0.374ef	4.77±0.630cd	5.82±1.439b	4.58±0.920d
P−	N+	EI	24.00±1.585c	13.11±0.780c	19.10±0.922c	2.62±0.237b	2.88±0.205a	2.47±0.091de	9.19±0.715a	4.57±0.454c	7.74±0.408b
		EF	25.66±1.338b	15.62±0.833b	20.84±0.805b	2.65±0.518b	2.55±0.391bc	2.40±0.403ef	9.97±1.830a	6.24±0.950b	8.86±1.371a
	N−	EI	5.48±1.409g	8.00±1.258de	6.21±0.713f	2.91±0.229b	2.76±0.132ab	2.74±0.139cd	1.88±0.450f	2.89±0.370d	2.26±0.161f
		EF	10.33±0.531e	6.50±1.258e	7.44±0.685e	2.750.313b	1.13±0.125e	2.12±0.187f	3.80±0.470de	5.73±0.602bc	3.53±0.290e

Note. Values are means ± SE. Significant differences ($P<0.05$) for each variable are indicated by lowercase letters for variables N, P availability and endophyte infection were analyzed together.

When N was supplied, EI plants had similar N concentration and total biomass regardless of P status, while EF plants had significantly lower N and total biomass in P deficiency compared to P supply. There were no significant differences in P concentration between shoots of EI and EF plants in all treatments. This suggests that the beneficial effect of endophyte infection on biomass production of the host plants was more strongly regulated by the availability of N rather than P [51]. With N supply, endophyte infection may help the host grass in maintaining biomass regardless of P status. With N deficiency, even with P supply, the biomass of both EI and EF plants decreased; however, EI biomass decreased slowly.

Current knowledge suggests that leaf N content is correlated with photosynthetic capacity [52]. In the present experiment, N concentration was lower for EI compared to EF plants; however, EI plants allocated significantly higher fractions of N to photosynthetic machinery with N deficiency. EI plants had significantly lower leaf N concentration but significantly higher maximum photosynthetic rate, PNUE and total biomass than did EF plants in the N−P+ treatment. It has been reported that organisms with a greater growth advantage in nutrient-poor environments are those able to modify their body nutrient content and increase efficiency of nutrient use without major decreases in their growth rates [53–54]. Under N−P+ conditions, EI plants grew better than EF plants by lowering their N concentration while increasing their N allocation to photosynthetic machinery. Therefore, it is N allocation to photosynthetic machinery instead of leaf N concentration itself that was more highly correlated with plant growth [55–56].

In N+P− treatment, P concentration in the shoot of EI and EF plants was similar but EI roots had significantly higher P concentration than EF roots, and similar results were reported by Zabalgogeazcoa et al. [57] on the response of *Festuca rubra* grown in low nutrient soil. Higher root P concentration here was attributed to higher acid phosphatase activity of EI roots. Phosphatase is an enzyme excreted by plant roots, fungi and bacteria and may contribute to as much as 65% of the annual P uptake of grasses [58]. A series of studies have shown that phosphatase activity was increased by AM fungal colonization [59]. Thus, was high acid phosphatase activity of EI roots related to AM colonization? Endophytes in grasses have been reported to reduce mycorrhizal colonization of host roots as well as spore densities in the soil [60–61]. In our sampled area in the Inner Mongolia Steppe, Bao [62] found that AM infected over 80% of Gramineae; however, the average infection rate was relatively low (i.e. about 28%). AM infection was not found in the *Achnatherum* genus. In the present study, although we did not measure AM colonization of the roots, the plants were grown from seeds collected in the natural grassland where AM colonization was not found in the *Achnatherum* genus, so it is reasonable to assume they were not colonized by mycorrhizae. Therefore, in the N+P− treatment in the present study, it is endophyte infection that significantly improved acid phosphatase activity of the host grass, which led to higher root P concentration and further higher total biomass in EI compared to EF plants.

The results presented here agreed with the initial prediction that beneficial interaction between the native grass *A. sibiricum* and its associated fungal endophyte depended on both N and P availability. The results further suggested that the beneficial effect of endophyte infection was more conditional on N than P. Under N+P− conditions, endophyte infection significantly improved acid phosphatase activity of EI plants, and so biomass of EI plants was not affected by P deficiency, and resulted in a greater P concentration and more biomass in EI than EF plants. Under

N−P+ conditions, both EI and EF biomass decreased compared with N+P+ conditions. EI plants had decreased leaf N concentration but allocated higher fractions of N to photosynthetic machinery compared to EF plants, which resulted in a slow decrease of EI growth– thus EI plants had significantly more biomass than EF plants. Under N−P− conditions, EI plants allocated higher fractions of N to photosynthesis and had a greater P concentration in roots, but there was no significant difference in biomass between EI and EF plants. Additionally, we did not find a clear cost of endophyte infection even in the N−P− treatment. Admittedly, the duration of the field pot experiment was short in comparison with the natural life span of the grass host and our results should be interpreted with caution. We propose that future studies should examine a wider range of native grass-endophyte systems in long-term field studies to better understand the general role of defensive mutualism in endophyte-plant interactions.

Author Contributions

Conceived and designed the experiments: AR. Performed the experiments: XL RH LY MW. Analyzed the data: XL AR. Contributed reagents/materials/analysis tools: YG. Wrote the paper: AR.

References

1. Cheplick GP, Clay K (1988) Acquired chemical defences in grasses: the role of fungal endophytes. Oikos 52:309–318.
2. Elmi AA, West CP (1995) Endophyte infection effects on stomatal conductance, osmotic adjustment and drought recovery of tall fescue. New Phytol 131: 61–67.
3. Hesse U, Schoberlein W, Wittenmayer L, Forster K, Warnstorff K, et al. (2003) Effects of *Neotyphodium* endophytes on growth, reproduction and drought-stress tolerance of three *Lolium perenne* L. genotypes. Grass Forage Sci 58: 407–415.
4. McCormick MK, Gross KL, Smith RA (2001) *Danthonia spicata* (Poaceae) and *Atkinsonella hypoxylon* (Balansiae): environmental dependence of a symbiosis. Am J Bot 88: 903–909.
5. Cheplick GP (2007) Costs of fungal endophyte infection in *Lolium perenne* genotypes from Eurasia and North Africa under extreme resource limitation. Environ Exp Bot 60:202–210.
6. Müller CB, Krauss J (2005) Symbiosis between grasses and asexual fungal endophytes. Curr Opin Plant Biol 8: 450–456.
7. Marks S, Clay K (2007) Low resource availability differentially affects the growth of host grasses infected by fungal endophytes. Int J Plant Sci 168:1269–1277.
8. Cheplick GP, Clay K, Marks S (1989) Interactions between infection by endophytic fungi and nutrient limitation in the grasses *Lolium perenne* and *Festuca arundinacea*. New Phytol 111: 89–97.
9. Ravel C, Balfourier F, Guillaumin JJ (1999) Enhancement of yield and persistence of perennial ryegrass inoculated with one endophyte isolate in France. Agronomie 19:635–644.
10. Lewis GC (2004): Effects of biotic and abiotic stress on the growth of three genotypes of *Lolium perenne* with and without infection by the fungal endophyte *Neotyphodium lolii*. Ann Appl Biol 144: 53–63.
11. Ren AZ, Gao YB, Wang W, Wang JL, Zhao NX (2009) Influence of nitrogen fertilizer and endophyte infection on ecophysiological parameters and mineral element content of perennial ryegrass. J Integr Plant Biol 51: 75–83.
12. Poorter H, Evans JR (1998) Photosynthetic nitrogen-use efficiency of species that differ inherently in specific leaf area. Oecologia 116: 26–37.
13. NiinemetsÜ, Valladares F, Ceulemans R (2003) Leaf-level phenotypic variability and plasticity of invasive *Rhododendron ponticum* and non-invasive *Ilex aquifolium* co-occurring at two contrasting European sites. Plant Cell Environ 26:941–956.
14. Onoda Y, Hikosaka K, Hirose T (2004) Allocation of nitrogen to cell walls decreases photosynthetic nitrogen-use efficiency. Funct Ecol 18:419–425.
15. Feng YL, Fu GL, Zheng YL (2008) Specific leaf area relates to the differences in leaf construction cost, photosynthesis, nitrogen allocation and use efficiencies between invasive and noninvasive alien congeners. Planta 228:383–390.
16. FengYL, Lei YB, Wang RF, Callaway RM, Valiente-Banuet A, et al. (2009) Evolutionary tradeoffs for nitrogen allocation to photosynthesis versus cell walls in an invasive plant. PNAS 106:853–1856.
17. Arachevaleta M, Bacon CW, Hoveland CS, Radcliffe E. (1989) Effect of the tall fescue endophyte on plant response to environmental stress. Agron J 81: 83–90.
18. Ravel C, Courty C, Coudret A, Charmet G (1997). Beneficial effects of *Neotyphodium lolii* on the growth and the water status in perennial ryegrass cultivated under nitrogen deficiency or drought stress. Agronomie 17: 173–181.
19. Mohren GMJ, Van Den Burg J, Burger FW (1986) Phosphorus deficiency induced by nitrogen input in Douglas fir in the Netherlands. Plant Soil 95:191–200.
20. Van Der Woude BJ, Pegtel DM, Bakker JP (1994) Nutrient limitation after long-term nitrogen-fertilizer application in cut grasslands. J Appl Ecol 31: 405–412.
21. Malinowski DP, Belesky DP, Baligar VC, Fedders JM (1998b) Influence of phosphorus on the growth and ergot alkaloid content of *Neotyphodium coenophialum*-infected tall fescue (*Festuca arundinacea*). Plant Soil 198: 53–61.
22. Ren AZ, Gao YB, Zhou F (2007) Response of *Neotyphodium lolii*-infected perennial ryegrass to phosphorus deficiency. Plant Soil Environ 53: 113–119.
23. Malinowski DP, Brauer DK, Belesky DP (1999) *Neotyphodium coenophialum*-endophyte affects root morphology of tall fescue grown under phosphorus deficiency. J Agron Crop Sci 183:53–60.
24. Malinowski DP, Alloush GA, Belesky DP (1998) Evidence for chemical changes on the root surface of tall fescue in response to infection with the fungal endophyte *Neotyphodium coenophialum*. Plant Soil 205:1–12.
25. Fujita Y, Robroek BJM, de Ruiter PC, Heil GW, Wassen MJ (2010) Increased N affects P uptake of eight grassland species: the role of root surface phosphatase activity. Oikos 119:1665–1673.
26. Wasaki J, Yamamura T, Shinana T, Osaki M (2003): Secreted acid phosphatase is expressed in cluster roots of lupin in response to phosphorus deficiency. Plant Soil, 248: 129–136.
27. Duff SMG, Sarath G, Plaxton WC (1994) The role of acid phosphatases in plant phosphorus metabolism. Physiol Plant 90: 791–800.
28. Sterner RW, Elser JJ (2002). Ecological Stoichiometry: The Biology of Elements from Molecules to the Biosphere. Princeton, NJ, , USA: Princeton University Press.
29. Faeth SH (2002) Are endophytic fungi defensive plant mutualists? Oikos 98: 25–36.
30. Wei YK, Gao YB, Xu H, Su D, Zhang X, et al. (2006) Occurrence of endophytes in grasses native to northern China. Grass Forage Sci 61: 422–429.
31. Saikkonen L, Helander K, Faeth SH (2006) Model systems in ecology: dissecting the endophyte-grass literature. Trends Plant Sci 11: 428–433.
32. Saikkonen K, Saari S, Helander M (2010) Defensive mutualism between plants and endophytic fungi? Fungal Diversity 41: 101–113.
33. Marks S, Clay K, Cheplick GP (1991) Effects of fungal endophytes on interspecific and intraspecific competition in the grasses *Festuca arundinacea* and *Lolium perenne*. J Appl Ecol 28:194–204.
34. Malinowski D, Belesky DP (2000) Adaptations of endophyte-infected cool-season grasses to environmental stresses: mechanisms of drought and mineral stress tolerance. Crop Sci 40: 923–940.
35. Wu ZL, Lu SL (1995) On geographical distribution of *Achnatherum* beauv. (Gramineae). Acta Phytotaxonomica Sinica 34: 152–161.
36. Petroski RJ, Powell RG, Clay K (1992) Alkaloids of *Stipa robusta* (sleepygrass) infected with an *Acremonium* endophyte. Nat Toxins 1: 84–88.
37. Bruehl GW, Kaiser WJ, Klenin RE (1994) An endophyte of *Achnatherum inebrians*, an intoxicating grass of northwest China. Mycologia 86:773–776.
38. Jin XM, Han GD (2010) Effects of grazing intensity on species diversity and structure of meadow steppe community. Pratacultural Sci (in Chinese) 27:7–10.
39. Latch GCM, Christensen MJ, Samuels GJ (1984) Five endophytes of *Lolium* and *Festuca* in New Zealand. Mycotaxon 20:535–550.
40. Kannadan S, Rudgers JA (2008) Endophyte symbiosis benefits a rare grass under low water availability. Funct Ecol 22: 706–713.
41. Li X, Han R, Ren AZ, Gao YB (2010) Using high-temperature treatment to construct endophyte-free *Achnatherum sibiricum*. Microbiology China 37: 1395–1400.
42. Morse LJ, Faeth SH, Day TA (2007) *Neotyphodium* interactions with a wild grass are driven mainly by endophyte haplotype. Funct Ecol, 21: 813–822.
43. Laisk A (1977) Kinetics photosynthesis and photorespiration in C_3 plants. Nauka, Moscow.
44. Farquhar GD, Sharkey TD (1982) Stomatal conductance and photosynthesis. Annu Rev Plant Physiol 11: 191–210.
45. Niinemets Ü, Tenhunen JD (1997) A model separating leaf structural and physiological effects on carbon gain along light gradients for the shade-tolerant species *Acer saccharum*. Plant Cell Environ 20: 845–866.
46. Loustau D, Beahim M, Gaudillère JP, Dreyer E (1999) Photosynthetic responses to phosphorous nutrition in two-year-old maritime pine seedlings. Tree Physiol 19: 707–715.
47. Lin FP, Chen ZH, Chen ZP, Zhang DM (1999) Physiological and biochemical responses of the seedlings of four legume tree species to high CO_2 concentration. Chin J Plant Ecol 23: 220–22.
48. Bao SD (2000) Agrochemical Analysis of Soil. Chinese Agricultural Press, P44–49.
49. Mclanchlan KD (1980) Acid phosphatase activity of intact roots and phosphorus nutrition in plants (I): Assay conditions and phosphatase activity. Aust J Agric Res 31: 429–440.
50. Ahlholm JU, Helander M, Lehtimäki S, Wäli P, Saikkonen K (2002) Vertically transmitted fungal endophytes: different responses of host-parasite systems to environmental conditions. Oikos 99:173–183.
51. Bai YF, Wu JG, Clark CM, Naeem S, Pan QM, et al. (2010). Tradeoffs and thresholds in the effects of nitrogen addition on biodiversity and ecosystem functioning: evidence from inner Mongolia Grasslands. Global Change Biol 16: 358–372.

52. Hikosaka K (2004) Interspecific difference in the photosynthesis-nitrogen relationship: patterns, physiological causes, and ecological importance. J Plant Res 117: 481–494.
53. Elser JJ, Acharya K, Kyle M, Cotner J, Makino W, et al. (2003) Growth rate-stoichiometry couplings in diverse biota. Ecol Lett 6: 936–943.
54. Mulder K, Bowden WB (2007) Organismal stoichiometry and the adaptive advantage of variable nutrient use and production efficiency in *Daphnia*. Ecol Modell 202: 427–440.
55. González AL, Kominoski JS, Danger M, Ishida S, Iwai N, et al. (2010) Can ecological stoichiometry help explain patterns of biological invasions? Oikos 119: 779–790.
56. Jeyasingh PD, Weider LJ, Sterner RW (2009) Genetically-based tradeoff s in response to stoichiometric food quality influence competition in a keystone aquatic herbivore. Ecol Lett 12:1229–1237.
57. Zabalgogeazcoa I, Ciudad AG, Vázquez de Aldana BR, Criado BG (2006) Effects of the infection of the fungal endophyte *Epichloë* festucae on the growth and nutrient content of *Festuca rubra*. Eur J Agron 24: 374–384.
58. Kroehler CJ, Linkins AE (1988) The root surface phosphatases of *Eriophorum vaginatum*: effects of temperature, pH, substrate concentration and inorganic phosphorus. Plant Soil 105: 3–10.
59. Allen EB, Allen MF, Helm DJ, Trappe JM, Moliva R, et al. (1995) Patterens and regulation of mycorrhizal and fungal diversity. Plant Soil 170: 47–62.
60. Chu-chou M, Guo B, An ZQ, Hendrix JW, Ferriss RS, et al. (1992) Suppression of mycorrhizal fungi in fescue by the *Acremonium coenophialum* endophyte. Soil Biol Biochem. 24: 633_637.
61. Mack KML, Rudgers JA (2008) Balancing multiple mutualists: asymmetric interactions among plants, arbuscular mycorrhizal fungi, and fungal endophytes. Oikos 117: 310–320.
62. Bao YY (2004) Diversity and ecological distribution of arbuscular mycorrhizal association in the grassland and desert of Inner Mongolia. Ph.D thesis, Inner Mongolia Agricultural University, China.

19

Can Physiological Endpoints Improve the Sensitivity of Assays with Plants in the Risk Assessment of Contaminated Soils?

Ana Gavina[1,2], Sara C. Antunes[2,3], Glória Pinto[1,2], Maria Teresa Claro[1,2], Conceição Santos[1,2], Fernando Gonçalves[1,2], Ruth Pereira[3,2]*

1 Departamento de Biologia, Universidade de Aveiro, Campus de Santiago, Aveiro, Portugal, **2** CESAM - Centro de Estudos do Ambiente e do Mar, Universidade de Aveiro, Campus de Santiago, Aveiro, Portugal, **3** Departamento de Biologia da Faculdade de Ciências, Universidade do Porto, Rua do Campo Alegre, Porto, Portugal

Abstract

Site-specific risk assessment of contaminated areas indicates prior areas for intervention, and provides helpful information for risk managers. This study was conducted in the Ervedosa mine area (Bragança, Portugal), where both underground and open pit exploration of tin and arsenic minerals were performed for about one century (1857 - 1969). We aimed at obtaining ecotoxicological information with terrestrial and aquatic plant species to integrate in the risk assessment of this mine area. Further we also intended to evaluate if the assessment of other parameters, in standard assays with terrestrial plants, can improve the identification of phytotoxic soils. For this purpose, soil samples were collected on 16 sampling sites distributed along four transects, defined within the mine area, and in one reference site. General soil physical and chemical parameters, total and extractable metal contents were analyzed. Assays were performed for soil elutriates and for the whole soil matrix following standard guidelines for growth inhibition assay with *Lemna minor* and emergence and seedling growth assay with *Zea mays*. At the end of the *Z. mays* assay, relative water content, membrane permeability, leaf area, content of photosynthetic pigments (chlorophylls and carotenoids), malondialdehyde levels, proline content, and chlorophyll fluorescence (F_v/F_m and Φ_{PSII}) parameters were evaluated. In general, the soils near the exploration area revealed high levels of Al, Mn, Fe and Cu. Almost all the soils from transepts C, D and F presented total concentrations of arsenic well above soils screening benchmark values available. Elutriates of several soils from sampling sites near the exploration and ore treatment areas were toxic to *L. minor*, suggesting that the retention function of these soils was seriously compromised. In *Z. mays* assay, plant performance parameters (other than those recommended by standard protocols), allowed the identification of more phytotoxic soils. The results suggest that these parameters could improve the sensitivity of the standard assays.

Editor: Vishal Shah, Dowling College, United States of America

Funding: The authors have no support or funding to report.

Competing Interests: The authors have declared that no competing interests exist.

* E-mail: ruthp@ua.pt

Introduction

Plants are essential components of ecosystems as they are primary producers of organic matter and oxygen, and a food source for heterotrophic organisms, humans included. They are considered versatile tools to monitor the presence and the effects of pollutants in soil, for they are in close contact with the soil matrix and with soil pore water, absorbing both nutrients and pollutants and responding to changes in soil properties [1,2,3,4]. Several are the reasons why plants have been widely used in assays, to evaluate soil quality and risk assessment of phytotoxic compounds: i) they have a sedentary existence, so they can be continuously exposed to a source of pollution throughout their life cycle; ii) seeds are relatively inexpensive and plants are easily cultured in laboratory; iii) their biological responses can be evaluated in a short period of time and, iv) their condition/performance can be monitored in different ways, from physical observations to spectroscopic methods [5,6,7]. In order to ensure comparability of results across studies and laboratories, there is a list of standardized plant species that can be used in toxicity tests [8,9].

As far as tests with terrestrial plants are considered, the standardized protocols suggest that parameters such as seed germination, growth above soil and/or root growth have to be evaluated [8,10]. As with other tests, these can be considered acute when they evaluate potential immediate effects, as inhibition of seed germination, inhibition of seedling growth and biomass production, and chronic when evaluating long-term effects involving those occurring in the life cycle of the plant [11]. However, there are several other ecophysiological parameters that can be evaluated in plants, which can potentially be more sensitive and indicative of stress conditions. These parameters are usually not considered in plant tests because they are not previewed in standard protocols. However, besides the standard parameters, the evaluation of other physiological (e.g. chlorophyll fluorescence, pigments content) and biochemical (e.g. content of malondialdehyde and proline, enzymes activity) parameters may also be important [2,6], as they can help finding out potential false negative results.

Photosynthesis is a core function in the physiology of plants, during which light is captured by chlorophyll molecules and by two photosystems (PSI and PSII) in the membrane of thylakoids and then used to remove electrons from water molecules. Such electrons are transported through an electron transport system and finally accepted by $NADP^+$ molecules. Meanwhile, the transportation of electrons occurs in close association with the passive movement of protons to the lumen of thylakoids. The energy of this gradient is used for the phosphorylation of ADP. Both ATP and NADPH molecules are key products for CO_2 fixation and the production of sugars in the dark step of the process (Calvin Cycle) [12].

The photosynthetic system of higher plants has been shown to be sensitive, reacting to different kinds of stress agents like drought [13], salinity [14], metals [15,16,17] and herbicides [18], in shorter periods of time. During stress conditions plants lose their ability to use light energy and dissipation mechanisms are triggered to protect the plant from photoinhibition and photoxidation [19]. The excess of light energy can be dissipated as heat or as chlorophyll fluorescence [20]. Hence, impairments in the photosynthetic activity can be evaluated measuring chlorophyll fluorescence parameters like Φ_{PSII}, which measures the efficiency of the PSII photochemistry (i.e. the proportion of light absorbed by chlorophyll molecules used in photochemistry reactions) and F_v/F_m the maximum efficiency of PSII (the efficiency of the PSII when all the reactive centres are open) [20]. The evaluation of these parameters has been facilitated by the marketing of user friendly and portable devices, which makes routine evaluations possible. Further, these measurements have the great advantage of being non-destructive allowing multiple evaluations throughout plant exposures to stressful conditions.

Additionally, when the rate of excitation of chlorophyll molecules exceeds the conversion of energy in the reaction centres of PSII, excited chlorophyll molecules can generate singlet oxygen molecules, which can promote photoxidation. At this stage, carotenoids, which are also components of PSII, take action, as non-enzymatic antioxidants, scavenging excited chlorophyll molecules and dissipating energy as heat [14,19]. However, not only singlet oxygen species but also other reactive oxygen species (ROS), generated by different stress agents, may induce oxidative damage to pigments, impairing overall photosynthetic activity (photoinhibition). The aminoacids metabolism has being shown as crucial in the response of plants to oxidative stress agents because aminoacids like proline, amongst other functions, may act as hydroxyl radical scavengers [21].

Having in mind all of these mechanisms involved in plants response to toxicants, the aim of the present study was to evaluate the ability of new endpoints to increase the sensitivity of plant assays, to identify natural soils, seriously contaminated with metals, based on their phytotoxicity. To attain this purpose, both the whole soil matrix and soil elutriates, for a set of soil samples from an abandoned mine area, were assessed through seed germination and growth assay with *Zea mays* and a growth inhibition assay with *Lemna minor*, respectively. The assays were performed according to, standard protocols. Further, other plant physiological parameters as water content, chlorophylls (*a* and *b*) and carotenoids content, chlorophyll fluorescence (F_v/F_m and Φ_{PSII}), membrane permeability and oxidative stress parameters (proline and MDA content) were assessed in *Zea mays* at the end of the assay. Here, we hypothesized that more soils will be identified as phytotoxic, if more plant performance parameters are measured. Here we hypothesized that the more plant performance parameters are measured, the more soils will be identified as phytotoxic.

Materials and Methods

No specific permisiions were required for these locations activities. We confirm that the location is not privately-owned or protected in any way and we confirm that the field studies did not involve endangered or protected species.

Study site and soil sampling

The Ervedosa Mine is located in Vinhais, district of Bragança, in northeast Portugal. In this mine arsenic (As) and tin (Sn) were explored for about one hundred years (1857–1969) (figure 1) deeply changing the overall landscape [22]. Environmental contamination of local soils by metals was evaluated and reported by Novais [23]. The levels of metals detected in soils, of this area, have raised concerns about the potential risks to local natural communities. Some soils have also shown to be highly toxic for species like *Eisenia andrei*, *Folsomia candida*, *Pseudokirchneriella subcapitata*, *Daphnia magna* and *Vibrio fischeri* (unpublished data) confirming their hazard for edaphic species.

In the mine area four transects (C, D, E and F) were considered with four sampling points each, set apart from each other for about 50 m (figure 1). Additionally, a reference site was selected, 3 km away from the mine area. Transect C began in the ore treatment area and extended north. Transect D extended from the mining area to the river Tuela. Transect E started in the ore exploration area and extended south, to the Ervedosa village. Further, transect F was set parallell to the river Tuela and crossed the area where the ore was treated to extract metals of interest.

Surface soil samples (0–20 cm) were collected in the seventeen sampling points and brought to the laboratory where they were left to dry at room temperature. Thereafter the samples were sieved and the <4 mm fraction was stored for physical and chemical characterization and for plant assays.

General physical and chemical characterization of soil samples

Soil conductivity was measured in a soil-water suspension according to the method described by FAOUN [24]. For this purpose 10 g of soil, were mechanically shaken with 50 mL of distilled water during 15 min. The suspension was left to rest overnight and conductivity was measured using a pre-calibrated LF330/SET conductivity meter. Soil pH_{KCl} was measured in a suspension of soil, prepared with a solution of KCl 1M, according to ISO 10390 [25].

Water holding capacity (WHC) of soils was measured according to the procedure described in the ISO 10390 guidelines [25]. Soil samples were placed in polypropylene flasks, with the bottom replaced by filter paper and immersed in water for 3 h. After this period flasks were placed on absorbent paper for 2 h to reject the excess of water that could not be retained by soil. The WHC was then determined by weighting each replicate before and after drying at 105°C until weight stabilization [25].

Soil water content (moisture) was determined by weight loss, at 105°C, for 24 h. The organic matter content (OM) was determined by weight loss on ignition at 450°C, during 8 h, according to SPAC [26]. All parameters described above were measured in three soil replicates.

Soil metal content: total and extractable concentrations

The content in metals of soil samples was determined by two extraction methods: a strong one with *aqua regia* and a mild extraction with calcium chloride 0.01 M [27]. For the *aqua regia* extraction, 1 g of each soil replicate was digested with 3 mL of 37% hydrochloric acid (*pro analysis*, Panreac) and 1 mL of 65%

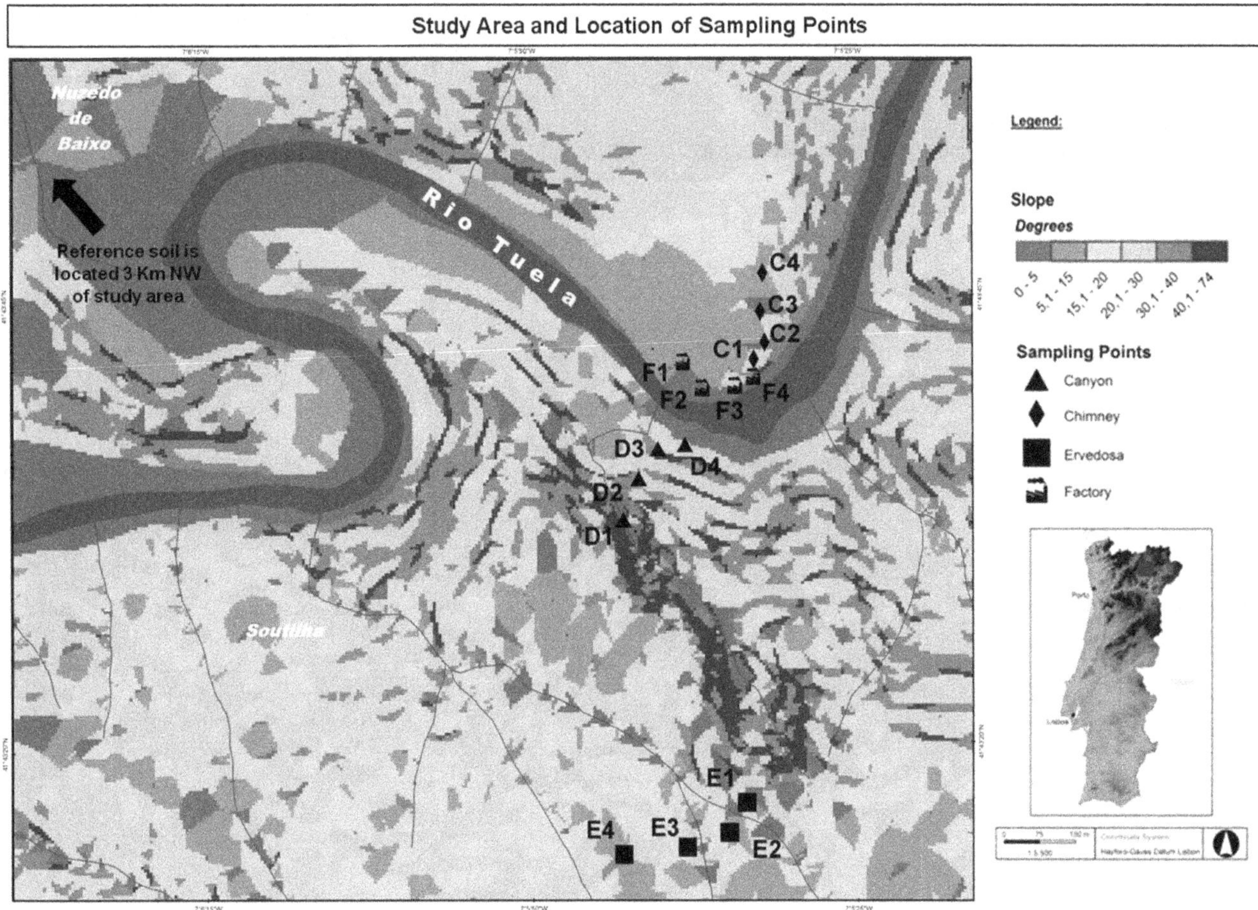

Figure 1. Study area and location of transects and sampling points (adapted from Carvalho et al. [22]).

nitric acid (Suprapur, Merck), in closed Teflon flasks. The flasks were heated on a sand bath at 100°C for 5 h. After this period, 10 mL of HNO_3 (4N) were added to the flasks and the solution was filtered, through 0.2 μm FT30/0.2CA-S filters, to remove all coarser particles, and transferred to polypropylene volumetric tubes. At the end of acid digestion the volume of each extract was adjusted with distilled water until a final volume of 25 mL was attained. For quality control of the extraction procedure, the same process was carried out using the same reagents but without the soil sample and three blank samples were prepared and sent for analysis. For the extraction with calcium chloride 0.01 M, suspensions of soil in the $CaCl_2$ solution (0.01 M) (1:10 m/v)) [27] were prepared for all the soil replicates. The soil suspensions were shaken mechanically for 2 h, at 20±2°C. After this, the suspensions were centrifuged at 4000 rpm and stored (acidified to pH<2 with HNO_3) for quantification of metals. Total and extractable concentrations of Al, Pb, P, V, Mn, Fe, Cu, Zn, As, Sb, Ba and Sn were analyzed by ICP-MS (Thermo X-Series quadrupole ICP-MS, Thermo Scientific).

Lemna minor assay

L. minor was obtained from laboratorial cultures reared under controlled conditions (temperature 20±2°C; photoperiod 16 h^L:8 h^D; illuminance: 10000 lux) in Steinberg medium according to the guideline OECD 221 [28]. The tests were performed with soil elutriates obtained from suspensions of soil samples in Steinberg medium (1:4 m/v). These suspensions were mechanically shaken overnight and then left to stand for 12 h for sedimentation. After this period, suspensions/elutriates were decanted and the supernatant portion was collected. *L. minor* was exposed, in three replicates, to a range of elutriate dilutions (100 mL/replicate). The assay was started placing nine fronds of *L. minor*, per vessel, under the controlled conditions described above. In the control replicates *L. minor* fronds were exposed only to the Steinberg medium. After 7 days of exposure, the fronds of each replicate were collected, dried at 70°C, till weight stabilization, and weighted. Growth rate was quantified according to the equation: $GR = (Ln(W_f)-Ln(W_i))/7$ (W_f and W_i are final and initial weights, respectively) [28]. IC_{50} values and corresponding 95% confidence limits, for each elutriate, were determined by nonlinear regression analysis, fitting a logistic equation to the data using technique of least squares. The software Statistica 10.0 was used for this purpose.

Zea mays seed germination and growth assay

Seed germination and growth assay with *Z. mays* were performed according to the ISO 11269-2 guideline [8]. Seeds were purchased from a local supplier and the damaged ones were discarded after visual inspection. Assays were performed in plastic pots, which were filled with 200 g of soil (four replicates per soil). Control was conducted with OECD standard soil [29]. Twenty seeds were added to each pot. In the beginning of the assay, a commercial solution of nutrients (Substral™ 10%) was added to each pot. Pots were maintained at controlled temperature

(20±2°C), photoperiod (16 h^L: 8 h^D) and illuminance (about 25000 lux). During daily observations, the number of emerged seeds was recorded and the water content of the pots was checked and adjusted. Only the first five emerged seeds were left to growth, the remaining ones were counted and harvested. The assay was validated and started after 50% of the seeds from the control pots emerged. Fourteen days later, the assay finished. Chlorophyll fluorescence measures were taken in the adaxial side of leaves of two plants from each soil replicate. The biomass above soil was harvested (only for four plants per pot) and wet weight was immediately determined. Dry biomass was weighted after drying at 70°C.

The leaves collected to measure water content and membrane permeability of plant cells were immediately processed, while the leaves for the quantification of chlorophylls, carotenoids, proline and malondialdehyde content were immediately frozen in liquid nitrogen and stored in a deep freezer for further analysis. These parameters were measured in plant leaves that were collected from one plant per replicate.

Plant performance parameters in *Zea mays* assay

Specific leaf area. The leaves harvested to determine specific leaf area (SLA) were placed on graph paper (used as scale) and then photographed. Afterwards the leaf area was determined with the ImageJ 1.43 µ software (Internet free). The leaves were weighed on an analytical balance and were dried at 60°C, until stabilization, and the dry weight was determined. SLA was then calculated as the ratio of leaf area (cm^2) to leaf dry weight (g).

Photosynthetic pigments (chlorophyll and carotenoid contents). Chlorophylls (chl *a* and chl *b*) and carotenoids were determined spectrophotometrically according to the method described by Sims and Gamon [30]. Pigments were extracted from leaves samples of about 0.5 g, and were homogenized in 2 mL of cold acetone (99% Cleanse®)/Tris buffer 50 mM (99.8% Merck®) (80:20, v/v). Then the extracts were transferred to centrifuge tubes, homogenized in vortex for about 30 s and centrifuged for 5 min, at 4000 rpm and 4°C. The supernatant was transferred to new tubes which were stored in ice and in the dark. The extraction procedure was repeated by adding more 1.5 mL of the same extraction solution to the pellet. The resulting supernatant was collected into former tubes, kept in the dark, and once again the extraction solution was added till a final volume of 6 mL was attained.

The quantification of chlorophyll (*a* and *b*) and carotenoid contents was achieved by spectrophotometry, measuring absorbance of the extracts at 470, 537, 647 and 663 nm in a Thermo Scientific Vis Spectrophotometer 10S TM. The extraction solution was used as blank for zeroing the absorbance.

Malondialdehyde content. The content of malondialdehyde (MDA) in samples of plant tissue was determined by the thiobarbituric acid method as described by Elkahoui et al. [31]. The MDA is an end product of lipid peroxidation in plant cells. Hence samples of leaves, of about 0.5 g, from one plant per replicate, were homogenized with 5 mL of 0.1% trichloroacetic acid (TCA) (Riedel-de Haën). The homogenates were centrifuged for 5 min, at 4000 rpm and at 4°C. Then, aliquots of 1 mL of the supernatant were transferred to falcon tubes and 4 mL of 20% TCA solution containing 0.5% of thiobarbituric acid (TBA) (≥98%, Sigma-Aldrich) were added to the tubes. The tubes were placed in a water bath, at 95°C, for 30 min. After cooling in ice, the tubes were centrifuged for 10 min at 4000 rpm and at 4 °C. The specific and the non-specific absorbance of the supernatant were measured at 532 and 600 nm, respectively. Distilled water was used as blank for zeroing the absorbance of the spectophotometer Thermo Scientific TM 10S Vis. The MDA content was calculated subtracting the non-specific absorbance at 600 nm and using the molar extinction coefficient $\varepsilon = 155\ mM^{-1}\ cm^{-1}$.

Proline content. The proline content of plant leaves was determined according to the method described by Khedr et al. [32]. From each plant (one plant per replicate) about 100 mg of leaves were homogenized in 1.5 mL of 3% sulfosalicylic acid (≥99%, Sigma). After centrifugation of the extracts at 4000 rpm, 100 µL of the supernatant were transferred to new tubes and mixed with 2 mL of glacial acetic acid (pro analysis, Panreac) and 2 mL of ninhydrin (Riedel-de Haën). The mixture was incubated in a water bathat 100°C, for 1 h. After this period, the tubes were placed in ice, and 1 mL of toluene (99.9%, Merck) was added to cooled tubes, in a hote. Absorbance of the chromophore solution was measured at 520 nm in a Thermo Scientific TM 10S Vis spectrophotometer [32]. The content of proline in samples was then extrapolated from a calibration line obtained measuring the absorbances of proline solutions of known concentration (0.2, 0.1, 0.05, 0.025, 0.0125 mg mL^{-1}).

Relative water content. For the evaluation of this parameter each leaf was weighed on an analytical balance (F_W), and then placed in a Falcon tube completelly filled with distilled water. The tubes were left in the dark, at 4°C, for 12 h. After this periodthe leaves were removed from water and placed on an absorbent paper to remove the excess of water, and the turgid weight (T_w) was determined. Afterwards, leaves were dried at 60°C, until stabilization and the dry weight was determined (D_w). The relative water content of plant leaves (RWC) was calculated using the following equation, and expressed as a percentage:

$$RWC(\%) = [(F_W\text{-}D_W)/(T_w\text{-}F_W)] \times 100$$

Chlorophyll fluorescence (Fv/Fm and ΦPSII). Chlorophyll fluorescence measurements were performed on the same expanded leaves of each plant using a portable fluorometer (Minipan Photosynthesis Yield Analyser, Walz, Effeltrich, Germany). Light exclusion clips were placed on the adaxial side of the leaves for 30 min and the following chlorophyll (chl) fluorescence measurements were taken [20]: minimum chl fluorescence in the dark adapted state (F_0), when all the reaction centres of PSII are opened; maximum chl fluorescence in the dark adapted state (F_m), after a pulse of actinic light (0.8 s to 8000 micromol $m^{-2}\ s^{-1}$) has closed all the reaction centres of PSII; the steady state chl fluorescence in the light adapted state (F_t); and the maximum chl fluorescence in the light adapted state (F'_m) after the same pulse of actinic light has been applied. With these measurements the efficiency of photosystem II (quantum yield) (Φ_{PSII}) and the maximum quantum yield or the maximum photosynthetic efficiency of photosystem II (F_v/F_m) were calculated based on the following equations:

$$\Phi_{PSII} = (F'_m - F_t) / F'_m \text{and} F_v / F_m = (F_m - F_0) / F_m$$

Membrane permeability. Membrane stability was estimated indirectly through quantification of electrolyte leakage according to, the method described by Lutts et al. [33]. One leaf from each replicate was weighed, washed with Milli Q water and then placed in falcon tubes filled with Milli Q water. The ratio mass/volume was the same in all the tubes. The tubes were shaken mechanically for 12 h in an orbital shaker. At the end of this period, the conductivity of the solution was measured with a conductivimeter (CONSORT C830 - Multi-parameter analyzer) ($C_{inicial}$). Then the vials were placed in the autoclave for 10 min, at 121°C. After cooling, the conductivity of the solution (C_{final}) was measured again. The membrane permeability and the ratio of conductivities $C_{inicial}/C_{final}$ were calculated and expressed as a percentage.

Stastical analysis

To test for significant differences in the parameters measured in plants, exposed to different mine soils, one-way analysis of variance (ANOVA) was performed, after the Levene's test for checking homogeneity of variances. When significant differences were recorded by the one-way analysis of variances, a two-tailed Dunnet or Games-Howel test (GHT) (when the assumption of equal variances was not accomplished) was perfomed to compare each soil with the REF soil, in terms of the paremeter under evaluation. The authors chose parametric tests, instead of non-parametric tests, even when the assumptions were not met, because one-way ANOVA has proved to be robust even when some deviations from requirements occur [34].

Results and Discussion

Soil contamination is considered one of the main causes of soil degradation worldwide and in Europe in particular [35]. After the recognition of the high rate of the verified soil loss, the European Union has developed new legal documents to protect the soils within the European territory. Within this scenario a soil framework directive was proposed and has been under discussion, since 2006 [36]. Amongst other aspects, this directive states that each member state should provide a list of the contaminated sites within their territory [36]. Such requirement will lead all the member states to enforce the application of environmental risk assessment (ERA) frameworks. Phytotoxic tests are required by ERA frameworks [37] to assess soil habitat, retention (aquatic species) and production functions. Bearing this idea in mind, this work was developed to assess the phytotoxicity of soils collected in the Ervedosa mine (north of Portugal) explored in the past for tin and arsenic. Further, we have hypothesized that we can improve the sensitivity of the standard phytotoxic assays evaluating other plant physiological, biochemical and chlorophyll fluorescence parameters, based on the assumption that these parameters will be able to detect stress before visible signs have evolved.

Soils physical and chemical characterization

The average values recorded for the different physical and chemical parameters measured in each soil sample collected in the Ervedosa mine area are described in table 1. In general, the soils had low pH (below 4.6±0.02 recorded in the REF soil) as well as low conductivity values. Soil F1 displayed the lowest pH_{KCl} (3.4) and the highest conductivity value (290.33 µS cm^{-1}). Regarding the content of organic matter (OM), and according to USEPA [38] classification, soils were grouped into: i) low content (<2%) – soils D2, E1and F3; ii) medium content (2%≤OM<6%) – soils REF, C1, C3, D1, D3, D4, E2, F1 and F2 soils and iii) high content (≥6%) – soils C2, C4, E3, E4 and F4. Soil E4 presented the highest organic matter content (19.7%) as well as the highest water holding capacity (84.6%).

Total and extractable metal concentrations

Generally, the highest total concentrations of metals were recorded in soils from transects D, E and F. Except for Al, Mn and Fe, which were recorded in high concentrations in all the soils,

Table 1. General physical and chemical parameters measured in soil samples collected in the Ervedosa mine area (average ± STDEV): pH_{KCl}, conductivity, MO -organic matter (%) and WHC_{max} – maximum water holding capacity (%).

	pH_{KCl}	Conductivity(µS cm^{-1})	OM (%)	WRC (%)
REF	4.6±0.02	51.1±0.76	3.8±0.4	22.5±0.1
C1	3.8±0.02	35.7±0.41	4.4±0.4	38.6±0.6
C2	4.6±0.66	36.7±0.5	9.8±0.3	57.7±0.5
C3	4.0±0.01	25.8±1.34	5.6±0.3	40.1±2.3
C4	4.1±0.01	35.1±0.31	7.7±0.2	52.8±0.9
D1	4.0±0.01	34.0±4.40	4.4±0.0	62.5±12.5
D2	4.4±0.01	19.5±3.64	1.3±0.1	8.9±0.1
D3	4.0±0.02	41.0±2.77	2.4±0.3	35.4±0.3
D4	4.4±0.02	12.2±0.41	4.9±0.1	23.3±0.1
E1	4.3 ±0.01	7.5±0.07	1.6±0.5	34.3±2.4
E2	4.3±0.00	20.0±0.29	5.4±0.3	40.8±1.8
E3	3.8±0.04	22.8±0.98	10.9±0.4	66.3±0.9
E4	3.8±0.01	40.0±7.76	19.7±0.3	84.6±2.1
F1	3.4±0.04	290.3±7.75	2.0±0.1	27.9±1.0
F2	3.6±0.05	54.3±12.81	3.0±0.3	33.1±3.9
F3	4.0±0.01	16.3±0.33	1.5±0.1	31.8±0.4
F4	4.1±0.01	32.3±0.67	6.7±0.2	46.8±3.4

Highest values recorded for each parameter were highlighted with bold letter.

including the REF soil. When compared with some soil benchmark values available, almost all the soils from transepts C, D and F presented concentrations of arsenic well above the EPA ECO-SSL (18 mg kg^{-1}) (a plants soil screening benchmark) (http://rais.ornl.gov/) as well as above the HC_5 value proposed for this metalloid (5.63 mg kg^{-1}) by Jänsch et al. [39]. The HC_5 values proposed by these authors were calculated based on EC_{50} values obtained for different species of animals, plants and microbial processes in chronic tests. These results represent the concentrations of the metals below which no more than 5% of the species and/or microbial processes will show a detrimental effect of 50%. This observation creates suspicions about the potential phytoxicity of almost all the soils analyzed in this study, since benchmark values for As were clearly surpassed, as previously mentioned. However, it seemed that this element was particularly available for plants especially in the soil F1, which showed the highest concentration of As in calcium chloride extracts (Table 2). The same calcium chloride extract obtained for soil D1, showed the highest concentrations of P, Mn, Fe and Cu. In fact all the soils from transect D and soil E1 had total concentrations of Cu, well above the EPA ECO-SSL (70 mg kg^{-1})) and the C_5 value (55 mg kg^{-1}) proposed by Jänsch et al. [39]. Soils D2 and F2 had total concentrations of lead also above the soil screening benchmarks mentioned (EPA Eco-SSL: 120 mg kg^{-1}; HC_5: 163.5 mg kg^{-1}), and the same was observed in terms of the total concentration of zinc in soils D1, E1 and E4. Nevertheless, in all the other soils (except D1 and F1), the extractable concentrations of metals were not meaningful, except for Al. The lack of correlation between soil total metal contents and the levels bioaccumulated by plants has been pointed out by several authors. Subsequently, the use of neutral salt solutions has been recommended based on the assumption that the cations provided by these salts are able to displace metals located on mineral surfaces, to the aqueous phase [40], mimicking processes occurring in rhizosphere microenvironment. In fact plants can make metal ions more available in the rhizosphere, both increasing acidity with the support of proton pumps localized in their plasma membranes and through the active secretion of low-molecular mass compounds that function as metal chelators [41]. The negative potential of plasma membranes, the existence of Fe^{2+}, Ca^{2+}, and Zn^{2+} transporter channels of low specificity and of intracellular binding sites for metals are additional driving forces for metals uptake [41] and together they could explain the toxicity of soils other than those with high extractable concentrations of metals.

Lemna minor assay

IC_{50} values and corresponding 95% confidence limits for the growth of *L. minor*, recorded after the exposure to the elutriates of the different mine soils are described in Table 3. Only elutriates of D1, D2, D3, E1, F1, F2 and F3 soils have significantly inhibited the growth of this aquatic plant species. Nevertheless, soil elutriates have displayed quite different toxicities with the lowest IC_{50} values recorded in the first samples of transects D, E and F, which were those collected near the mining and ore treatment area. The high availability of As and Cu, was probably responsible for the high toxicity of the elutriate from soil F1 and D1 to *L. minor*. The high phytotoxicity of As results from its ability to mimic phosphorus,

Table 2. Average concentrations (± STDEV) of metals in soils samples collected in the Ervedosa mine area, after calcium chloride (0.01 M) and aqua regia extraction (total metal contents).

	Al	Pb	P	V	Mn	Fe	Cu	Zn	As	Sb	Ba	Sn
Aqua regia extraction (µg g^{-1})												
REF	14886.3	27.6	208.6	24.3	132.8	13722.7	26.0	37.4	61.3	0.3	35.8	1.8
C1	9700.0	26.3	199.8	13.5	114.8	22884.8	14.2	46.2	158.7	3.8	19.7	2.3
C2	1963.1	2.5	30.9	bdl	12.9	1152.7	1.2	bdl	0.1	0.1	4.2	0.7
C3	6943.8	21.4	304.0	8.6	53.6	18795.6	11.3	17.6	33.9	2.4	23.2	1.2
C4	865.2	1.2	16.2	bdl	5.8	628.8	1.1	bdl	bdl	bdl	1.7	0.4
D1	4281.7	34.8	4604.0	0.8	1448.3	33536.0	604.8	243.1	3163.5	2.5	13.7	2.5
D2	5056.5	215.7	1344.2	42.5	92.6	16047.8	140.2	68.6	867.6	2.5	31.2	5.5
D3	6916.2	50.2	753.0	36.1	199.6	17220.8	67.0	52.6	323.5	2.6	12.6	2.9
D4	17825.2	30.1	438.8	31.4	538.5	29942.0	128.8	93.5	255.2	2.6	30.6	3.3
E1	12777.5	20.4	223.4	47.6	735.5	25414.9	477.7	232.5	439.4	0.8	27.7	1.5
E2	13752.6	33.9	194.5	22.3	139.6	22676.5	28.6	58.1	68.9	0.8	32.2	3.3
E3	764.1	1.5	31.9	bdl	5.4	1334.4	1.1	bdl	0.3	bdl	1.8	bdl
E4	9675.9	82.7	1227.8	17.2	333.3	45944.2	44.3	461.5	367.4	7.4	97.6	16.2
F1	1084.7	79.2	323.5	bdl	18.0	8348.6	7.9	4.7	15251.8	66.0	61.2	69.7
F2	4044.7	144.7	267.8	17.1	37.2	22555.8	23.0	19.0	13742.9	195.4	46.8	82.1
F3	2099.4	56.0	299.1	1.7	17.7	9634.3	45.2	27.0	7969.1	32.4	30.8	32.7
F4	8754.2	27.4	231.3	12.9	208.7	18965.6	18.2	37.0	47.3	0.8	24.4	1.7
$CaCl_2$ extraction (mg L^{-1})												
REF	0.70	0.00	0.01	bdl	0.15	0.06	0.00	0.04	0.00	bdl	0.17	bdl
C1	2.55	0.01	0.02	bdl	0.73	0.12	0.00	0.06	0.01	bdl	0.03	bdl
C2	1.63	0.01	0.01	bdl	3.07	0.12	0.00	0.09	0.00	bdl	0.14	bdl
C3	2.49	0.01	0.01	bdl	0.26	0.08	0.00	0.03	0.00	bdl	0.06	bdl
C4	3.19	0.01	0.02	bdl	0.66	0.22	0.01	0.03	0.00	bdl	0.09	bdl
D1	1.38	0.00	0.20	bdl	3.53	0.60	3.11	1.10	0.09	bdl	0.01	bdl
D2	1.50	0.00	0.04	bdl	0.44	0.09	0.11	0.07	0.02	bdl	0.01	bdl
D3	2.72	0.00	0.02	bdl	0.50	0.18	0.08	0.15	0.01	bdl	0.01	bdl
D4	2.10	0.00	0.05	bdl	0.60	0.33	0.18	0.19	0.01	bdl	0.06	bdl
E1	2.63	0.00	BDL	bdl	0.24	0.04	0.55	0.29	0.00	bdl	0.06	bdl
E2	1.32	0.01	0.01	bdl	0.47	0.13	0.00	0.04	0.00	bdl	0.06	bdl
E3	1.46	0.01	BDL	bdl	1.14	0.15	0.01	0.04	0.00	bdl	0.06	bdl
E4	1.06	0.00	BDL	bdl	0.59	0.25	0.00	0.08	0.00	bdl	0.11	bdl
F1	0.81	0.01	0.06	bdl	0.04	0.07	0.01	0.01	77.49	bdl	0.25	bdl
F2	1.51	0.00	BDL	bdl	0.16	0.14	0.01	0.03	0.32	bdl	0.03	bdl
F3	0.64	0.00	0.01	bdl	0.11	0.10	0.01	0.05	0.27	bdl	0.01	bdl
F4	0.63	0.01	0.17	bdl	0.60	0.42	0.01	1.89	0.03	bdl	0.26	bdl

Highest concentrations were highlighted with bold letter. BDL stands for below detection limit.

causing negative effects in plants metabolic activity [42]. In fact, the concentration of As in the elutriate of the F1 soil was similar to the EC_{50} value reported by Duester et al. [43] for As (V) and for *L. minor* growth (82 mg L^{-1}: 95%CI = 76–87). Even though we have not determined As speciation in our elutriates, this form of arsenic is expected to occur at high concentrations, since As (III) tends to oxidize to As (V) in aqueous suspensions. Copper is also a metal very toxic to *L. minor*. Teisseire et al. [44] determined an IC_{50} of 0.16 mg L^{-1}, which was well below the extractable concentration of Cu found for soils D1 and E1. As far as elutriates of soils D2, D3, F2 and F3 are considered, their toxicity was probably related with aluminum because it was the metal present at highest concentration in the calcium chloride extracts. Nevertheless, Radić et al. [45] have shown the ability of *L. minor* to tolerate concentrations of Al up to 8.09 mg L^{-1} due to their great ability to up-regulate anti-oxidant defenses. Further, we cannot forget the possible differences between metal concentrations extracted with calcium chloride and those extracted with Steinberg medium that were used to produce the soil elutriates tested with *L. minor*. Complexation with organic components of the medium may have promoted a greater availability of metals to the macrophyte. Further, potential synergistic effects between all the metals in elutriates, even at lower concentrations could not be ignored. The known tolerance of *L. minor* and its ability to accumulate metals [45] has supported the suggested use for remediation purposes. However, in this study, *L. minor* was sensitive to different soil elutriates, even with low concentration of metals.

More concerning in terms of risk assessment, was the inhibitory effect on the growth of *L. minor*, recorded for soil elutriates 2 and 3 of the segments D and F. These segments are those extending from the mining area to the River Tuela and parallel to the same river, respectively. The results obtained suggest that there may be a poor retention of the soil near the stream, which contributes for the mobilization of a mixture of metals to the soil aqueous phase and then to the aquatic ecosystem, with potential impact on its biological populations. This suspicion justifies a more detailed evaluation of this water stream, since a contamination, especially with As, may be occurring, with potential risks to natural communities and humans.

Table 3. IC_{50} values and corresponding 95% confidence limits for growth inhibition of *L. minor* exposed to elutriates of different soil samples collected in the Ervedosa mine area. NT stands for no toxicity.

	IC_{50}
REF	NT
C1	NT
C2	NT
C3	NT
C4	NT
D1	2.77<4.28<5.79
D2	37.92<57.03<76.15
D3	60.21<87.36<114.5
D4	NT
E1	17.12<24.66<32.19
E2	NT
E3	NT
E4	NT
F1	0.19<0.22<0.25
F2	23.75<55.67<87.58
F3	31.93<45.18<58.43
F4	NT

Zea mays assay

The assay was validated, since more than 50% of the seeds have emerged in the OECD soil (control), as stated by the standard protocol [25]. After confirming this, the natural REF soil was used as control in the assay, and all the statistical comparisons were made in order to minimize the influence of soil properties in the physiological parameters evaluated. In fact no significant statistical differences were recorded between the REF and the OECD soil for almost all the parameters evaluated (except for fluorescence parameters). No seed germination was recorded in the F soil replicates, since the data available for all the other parameters are unavailable for this soil.

In terms of the parameters recommended by the ISO 11269-2 protocol [8], no significant differences were recorded between the REF and all the other soils, in the average number of emerged seeds ($F = 1.38$; d.f. = 55, 76; $p = 0.185$). The average number of emerged seeds varied between 42.5 and 80% (except for soil F1). The lack of sensitivity of this parameter to soil contamination with metals has already been reported by several authors [3,46,47]. Such fact results from the protection given to embryos by seed coverage. However, this fact is species and metal dependent [48]. In this study it was possible to perceive, once more, that seed germination was inhibited only when extremely high concentrations of metals/metalloids (As in particular) had the potential to mobilize to the soil aqueous phase, becoming bioavailable. This reinforces the usefulness of this parameter only to identify worst-case scenarios of contamination, and probably more important to worst-case scenarios of metals bioavailability.

Concerning the average fresh and dry biomass above soil, significant differences among plants exposed to the different mine soils were recorded ($F = 2.097$; d.f. = 59,43; $p = 0.029$ and $F = 7.722$; d.f. = 55, 38, $p = 0.000$, respectively). A significant reduction in fresh weight was recorded only for plants exposed to soil D1 (GHT: $p = 0.025$) when compared with the REF soil (Figure 2). Plants from soils C4 and F3 (GHT: $p = 0.017$) displayed a significant lower dry weight (Figure 2). The opposite was recorded for plants exposed to soils C2 (GHT: $p = 0.001$) and C3 (GHT: $p \leq 0.001$), which have displayed a substantial high dry biomass when compared to plants exposed to the REF soil.

As far as parameters related with plants growth and development, other than those included in the ISO protocol, are considered, significant differences in the specific leaf area (SLA) were recorded amongst plants exposed to the different mine soils ($F = 3.263$; d.f. = 47, 34; $p = 0.003$). A significant increment in this parameter was recorded for soils E2 (GHT: $p \leq 0.001$) and E3 (GHT: $p = 0.028$) (Figure 3). In terms of biochemical parameters, at the end of the assay, significant differences in total chlorophyll *a+b* and carotenoids contents were recorded in leaves of plants exposed to different mine soils ($F = 6.576$; d.f. = 50, 34; $p \leq 0.001$ and $F = 5.217$; d.f. = 50, 34; $p \leq 0.001$, respectively) (Figure 4). Plants from soils C2 (GHT: $p = 0.002$), C3 (GHT: $p = 0.001$) and F3 (GHT: $p = 0.032$) have displayed a significant higher content of chlorophylls *a+b*, while the same soils plus soils D1 (GHT: $p = 0.019$), F2 (GHT: $p = 0.005$) and F4 (GHT: $p = 0.023$) have induced a significant increment in the content of carotenoids of maize plants. Due to their different physical and chemical properties, metals have three different mechanisms of toxicity:

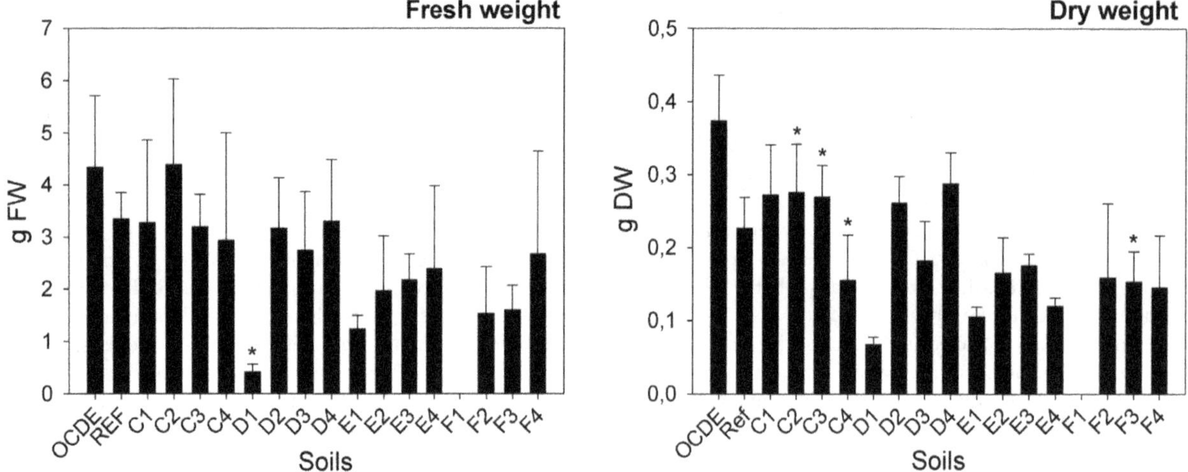

Figure 2. Average fresh and dry weight. Average fresh and dry weight of plants (g FW and gDW) exposed to different soils collected in Ervedosa mine area and to REF and OECD artificial soil. The error bars represent the standard deviation.

production of reactive oxygen species (ROS), blocking of functional groups of enzymes and displacement of metal ions from biomolecules [16,41]. In turn, ROS induce oxidative damage to pigments, proteins and lipids in the thylakoid membranes, compromising the overall photosynthetic activity [19]. Consequently, metals toxicity usually activates anti-oxidant defenses [49]. Carotenoids are non-enzymatic antioxidants that protect plants against photoxidation, protecting chlorophyll molecules from oxidative damages [14,50]. Hence, the increment in the production of carotenoids content may express a response of plants to counteract the toxic effect of metals. The same occurrence was reported for other plants species, exposed to different metals (e.g. 51) or to wastes rich in metals [52]. However, in plants exposed to at least one Ervedosa mine soil (F3) such response was probably insufficient since a significant reduction in biomass still occurred. However, no significant lipid peroxidation was recorded. In fact, despite the slight increase in the MDA content in tissues of plants exposed to soils from transect C and also in some soils from transect E and F, a significant increment was observed only in plants exposed to soil D3 (Figure 5). Such observations indicate that only in these plants were the physiological mechanisms not efficient in counteracting the oxidative stress. As far as total chlorophyll contents are again considered, our results do not comply with the findings of other authors, reporting a decrease in total chlorophyll content caused by metals stress [53,54], at least for soil F3. Plants exposed to this soil have shown a significant increment in total chlorophyll content despite the significant reduction in their dry biomass. Nevertheless, this could have been a punctual situation in which the plants have tried to adapt to metal exposure by increasing chlorophyll synthesis. Nevertheless, this is only an explanative hypothesis, requiring further confirmation.

Several studies, reporting the physiological responses of plants to metals stress, have shown that the amino acid proline usually accumulates in response to metal/metalloid (As included) exposures [21,53,54]. In fact proline has a central role in the ability of plants to react to abiotic stress [21], since it acts as a mediator in osmotic balance, protects macromolecules during dehydration and acts as a hydroxyl radical scavenger [14]. In this study no changes were recorded in this parameter, except for soil C1 (Figure 5). Plants exposed to this soil showed a significant reduction in their proline content ($F = 3.895$; d.f. = 62, 46; $p \leq 0.001$; Dunnet: $p \leq 0.001$). Although contradicting general findings, Pavlík et al. [21] have suggested that under As stress the biosynthesis of proline could be inhibited, due to a preferred utilization of glutamate, which in turns leads to the synthesis of phytochelatins. It was shown that the synthesis of phytochelatins, also called class III metallothioneins, is activated in plant cells after exposure to different metals, as part of another important detoxification mechanism [49].

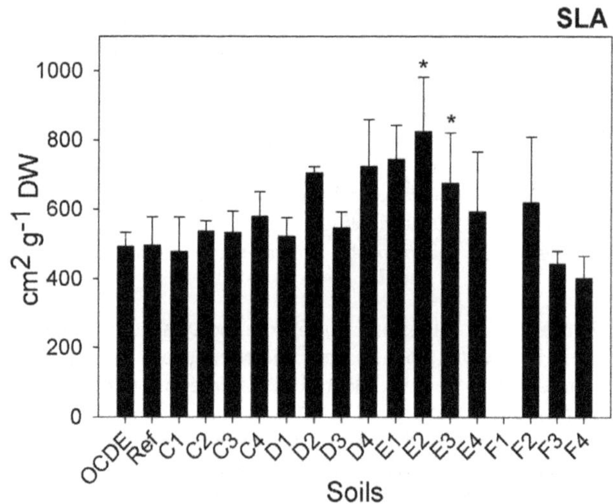

Figure 3. Specific leaf area. Specific leaf area (SLA) of plants ($cm^2\ g^{-1}$ DW) exposed to different soils collected in Ervedosa mine area and to REF and OECD artificial soil. The error bars represent the standard deviation and * correspond to significant differences towards the REF soil.

Plants exposed to soils C3, D1 and F2 have also shown a significant reduction in leaf water content ($F = 3.282$; d.f. = 63, 47; $p = 0.001$; GHT: $p = 0.003$) (Figure 6). Since no significant differences in terms of cells membrane permeability was observed ($F = 1.558$; d.f. = 46, 30; $p = 0.147$) between plants exposed to the different mine soils (Figure 6), we can suggest that the reduction in

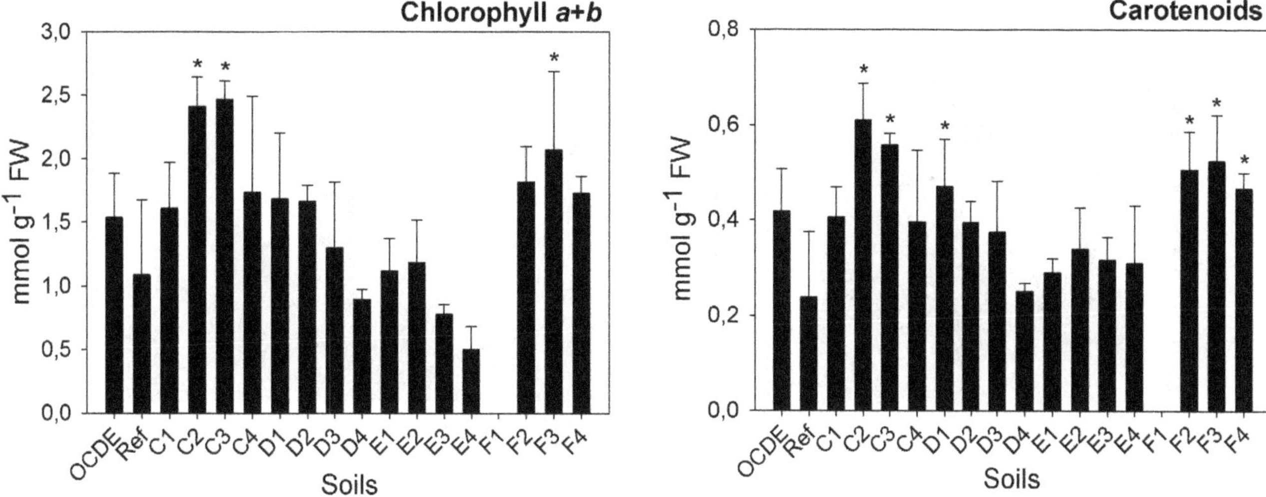

Figure 4. Chlorophyll *a+b* and carotenoids. Chlorophyll *a+b* (μmol g^{-1} FW) and caroptenoids in plants (mmol g^{-1} FW) exposed to different soils collected in Ervedosa mine area and to REF and OECD artificial soil. The error bars represent the standard deviation and * correspond to significant differences towards the REF soil.

water content in plants exposed to soil F3, was probably due to an inhibition in root growth with subsequent reduction in water uptake. Different authors [55,56] have reported the inhibition of roots growth caused by metals/metalloids like As and Cu, in *Triticum aestivum* and *Helianthus annuus*, respectively. Nevertheless, this parameter was not assessed in this study.

An efficient photosynthesis is crucial for plant survival and fitness [19], and chlorophyll fluorescence can give information about the state of the photosynthetic apparatus and, of photosystem II [20] in particular, which is considered to be the most vulnerable component. In terms of chlorophyll fluorescence parameters measured in this study, namely F_v/F_m ratio ($F = 5.058$; d.f. = 121, 105; $p \leq 0.001$) and Φ_{PSII} ($F = 4.335$; d.f. = 122, 106; $p \leq 0.001$) significant differences among the plants exposed to the different mine soils were recorded for both parameters (Figure 7). The F_v/F_m ratio, which measures the photochemical efficiency of photosystem II (PSII) in the dark-adapted state, was significantly reduced in plants exposed to soil C3 (GHT: $p = 0.006$), D4 ($p = 0.009$), E1 (GHT: $p \leq 0.001$), E3 (GHT: $p = 0.011$) and F2 ($p \leq 0.001$), when compared to the reference soil (Figure 7). In all these soils the plants have displayed average F_v/F_m ratios below 0.80. According to Björkman and Demming [57] the F_v/F_m ratio is almost constant for different plant species, under non-stressed conditions and, usually varies between 0.80–0.86. Values below this range suggest impairments in the photosynthetic apparatus. This possibility of damages was further reinforced for soils D4 (GHT: $p \leq 0.001$), E1 (GHT: $p \leq 0.001$), F2 (GHT: $p = 0.020$), which has also shown a significant reduction in Φ_{PSII} values (Figure 7). As it was demonstrated by Küpper et al. [58], different metallic cations may replace the central magnesium ion of the chlorophyll molecules, resulting in "*heavy metals substituted chlorophylls (hm-chls)*"

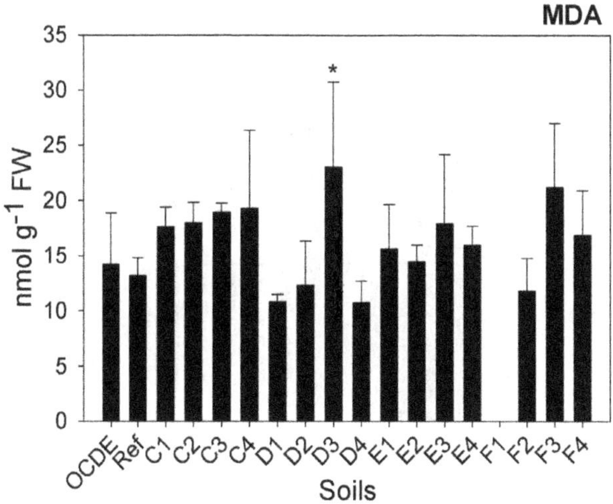

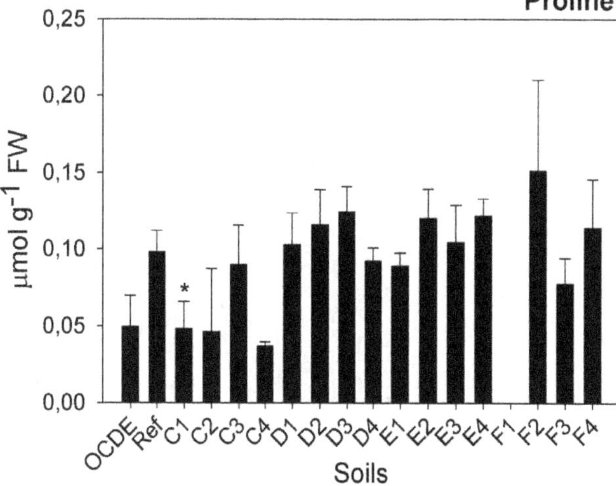

Figure 5. Malondialdehyde and proline content. Malondialdehyde (MDA) (nmol g^{-1} FW) and proline content in plants (μmol g^{-1} FW) exposed to different soils collected in Ervedosa mine area and to REF and OECD artificial soil. The error bars represent the standard deviation and * correspond to significant differences towards the REF soil.

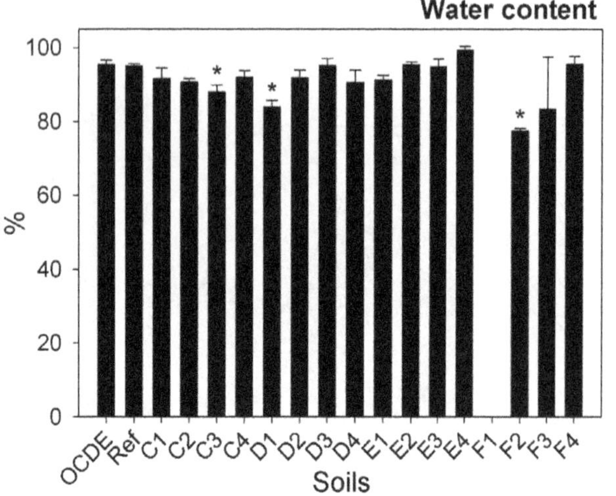

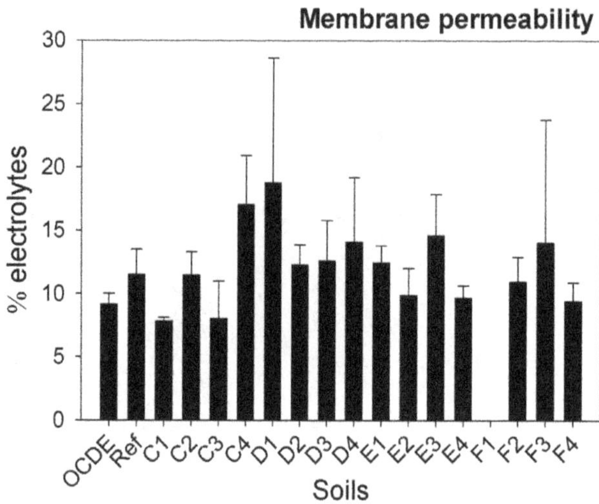

Figure 6. Water content and membrane permeability. Water content (%) and membrane permeability (% electrolytes) in plants exposed to different soils collected in Ervedosa mine area and to REF and OECD artificial soil. The error bars represent the standard deviation and * correspond to significant differences towards the REF soil.

reducing light harvesting by these molecules and, subsequently, reducing their fluorescence yields and compromising photosynthesis. Further, these authors have proved that the rate of substitution reactions varies with light intensity. At lower intensities *hm-chls* are more stable, and plants could appear vital, even when dead. Such fact could explain why a slight decrease in fluorescence parameters was recorded, at least for plants exposed to some soils, even without a concomitant reduction in the total chlorophyll content. Although the light intensity, to which plants were exposed during the assay, was within the range recommended by the standard protocol, the levels were lower than those recorded under a normal sunny day, in temperate latitudes.

The germination and early growth of plants are parameters that cannot be neglected in the evaluation of soils phytotoxicity since they integrate the overall effects of stress [46]. However, some authors suggested the evaluation of other parameters, at lower levels of organization, which may be more sensitive to the impact of chemicals, allowing both the early detection of physiological effects and the comprehension of their mechanisms of action [59,60,61,62]. In this study, the key biomarkers evaluated in the *Zea mays* seedlings were parameters related with plant development, photosynthetic activity, water balance, the synthesis of secondary metabolites, oxidative stress, and detoxification mechanisms. Table 4 summarizes the results, presenting the significant effects detected for each parameter evaluated in *Z. mays* plants exposed to the different soils. As it was possible to perceive by grey columns, five additional soils (C1, D2, D4, E1, F2) induced stress on *Z. mays* with the evaluation of other plant performance parameters. Fluorescence parameters were the more sensitive and those with a greater contribution to detect false negative results in terms of phytotoxicity. In addition, elutriates of three of these soils (D2, E1 and F2) have also proved to be toxic to *L. minor*. Hence,

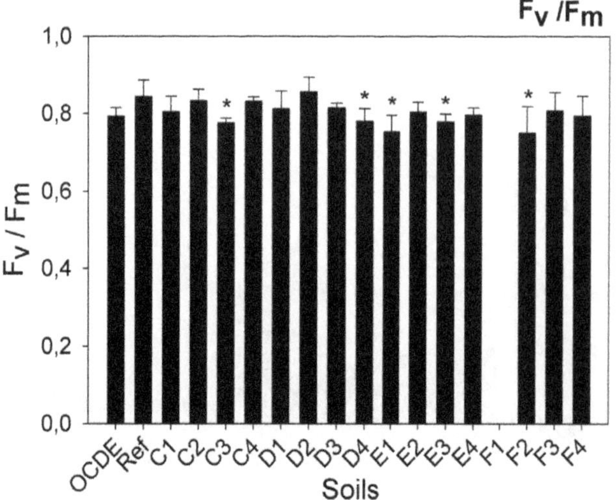

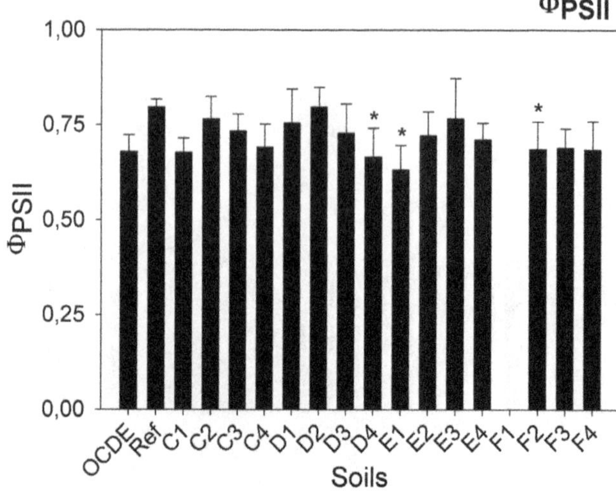

Figure 7. Maximum quantum yield and efficiency of photosystem II. Maximum quantum yield or the maximum photosynthetic efficiency of photosystem II (F_v/F_m) and efficiency of photosystem II (Φ_{PSII}) in plants exposed to different soils collected in Ervedosa mine area and to REF and OECD artificial soil. The error bars represent the standard deviation and * correspond to significant differences towards the REF soil.

Table 4. Summary of the significant effects recorded for all the parameters measured in the *Zea mays* assay, in plants exposed to the different mine soils (arrows point out for significant increments in comparison with REF plants).

Parameters	C1	C2	C3	C4	D1	D2	D3	D4	E1	E2	E3	E4	F1	F2	F3	F4
Seed germination													X			
Fresh biomass					X											
Dry biomass		X (↑)	X (↑)	X	X										X	
Specific Leaf Area										X (↑)	X (↑)					
Carotenoids content		X (↑)	X (↑)		X (↑)									X (↑)	X (↑)	X (↑)
Chlorophylls $a+b$		X (↑)	X (↑)												X (↑)	
MDA content						X (↑)										
Proline content	X															
Membrane permeability (% of electrolytes)																
Water content															X	
F_v/F_m			X					X	X		X			X		
Φ_{PSII}	X			X				X	X					X	X	

this new evaluation of phytotoxicity contributes to increase the evidence of risks posed by these soils. A great number of phytotoxic soils conform to previsions based on comparisons of soils total metal contents with soil benchmark values.

In summary, we can conclude that the inclusion of other physiological (chlorophyll fluorescence and/or stress oxidative parameters) in standard protocols for assays with terrestrial plants can improve their sensitivity, contributing for a more accurate evaluation of risks posed by contaminated soils. Chlorophyll fluorescence parameters, in particular, are non destructive and their measurement does not require specialized skills. However, a similar evaluation should be made, previously, for soils with different kinds of contamination.

Author Contributions

Conceived and designed the experiments: RP SCA AG. Performed the experiments: AG MTC SCA GP. Analyzed the data: AG RP. Contributed reagents/materials/analysis tools: FG CS. Wrote the paper: AG RP.

References

1. Gong P, Wilke BM, Strozzi E, Fleischmann S (2001) Evaluation and refinement of a continuous seed germination and early seedling growth test for the use in ecotoxicological assessment of soils. Chemosphere 44: 491–500.
2. Loureiro S, Santos C, Pinto G, Costa A, Monteiro M, et al. (2006) Toxicity assessment of two soils from Jales mine (Portugal) using plants: growth and biochemical parameters. Arch Environ Contam Toxicol 50: 182–190.
3. Pereira R, Marques CR, Silva Ferreira MJ, Neves MFJV, Caetano AL, et al. (2009) Phytotoxicity and genotoxicity of soils from an abandoned uranium mine area. Appl Soil Ecol 42: 209–220.
4. Lima MPR, Soares AMVM, Loureiro S (2011) Combined effects of soil moisture and carbaryl to earthworms and plants: simulation of flood and drought scenarios. Environ Pollut 159: 1844–1851.
5. Pfleeger TG, Ratsch HC, Shimabuku RA (1993) A review of terrestrial plants as biomonitors. Environmental Toxicology and Risk Assessment: 2nd Volume. ASTM STP 1216.
6. Verkleij JAC (1994) Effects of heavy metals, organic substances, and pesticides on higher plants. Ecotoxicology of soil organisms. SETAC Press. pp. 139–161.
7. Krugh BW, Miles D (1996) Monitoring the effects of five "nonherbicidal" pesticide chemicals on terrestrial plants using chlorophyll fluorescence. Environ Toxicol Chem 4 (15): 495–500.
8. ISO (1995) Soil quality—Determination of the effects of pollutants on soil flora—Part 2: Effects of chemicals on the emergence of higher plants. ISO—The International Organization for Standardization, Geneve. ISO 11269-2 : 7.
9. OECD (2006b) Terrestrial plant test: Seedling emergence and seedling growth test. OECD - Organization for Economic Cooperation and Development, Paris. 208: 6.
10. CARACAS (1998) Concerted action on risk assessment for contaminated sites in the European Union 1996–1998 – Risk assessment for contaminated sites in Europe. Nottingham, UK.
11. Van Assche F, Alonso JL, Kapustka LA, Petrie R, Stephenson GL, et al. (2002) Terrestrial plant toxicity tests. Test methods to determine hazards of sparingly soluble metal compounds in soil. SETAC Press. 128: 59–77.
12. Azevedo H, Pinto C, Fernandes J, Loureiro S, Santos C (2005) Cadmium effects on sunflower growth and photosynthesis. J Plant Nutr 28 (12): 2211–2220.
13. Rong-hua L, Pei-guo G, Baum M, Grando S, Ceccarelli S (2006) Evaluation of chlorophyll content and fluorescence parameters as indicators of drought tolerance in barley. Agriculture Sciences in China 5: 751–757.
14. Li G, Wan S, Zhou J, Yang Z, Qin P (2010) Leaf chlorophyll fluorescence, hyperspectral reflectance, pigments content, malondialdehyde and proline accumulation responses of castor bean (*Ricinus communis* L.) seedlings to salt stress levels. Ind Crops Prod 31:13–19.
15. Stoeva N, Berova M, Zlatev Z (2003) Physiological response of maize to arsenic contamination. Biol Plant 47(3): 449–452.
16. Vernay P, Gauthier-Moussard C, Hitmi A (2007) Interaction of bioaccumulation of heavy metal chromium with water relation, mineral nutrition and photosynthesis in developed leaves of *Lolium perenne* L. Chemosphere 68: 1563–1575.
17. Cherif J, Derbel N, Nakkach M, von Bergmann H, Jemal F, et al. (2010) Analysis of in vivo chlorophyll fluorescence spectra to monitor physiological state of tomato plants growing under zinc stress. J Photochem Photobiol B 101: 332–339.
18. Gao Y, Fang J, Zhang J, Ren L, Mao Y, et al. (2011) The impact of the herbicide atrazine on growth and photosynthesis of seagrass, *Zostera marina* (L.), seedlings. Mar Pollut Bull 62: 1628–1631.
19. Szabó I, Bergantino E, Giacometti GM (2005) Light and oxygenic photosynthesis: energy dissipation as a protection mechanism against photo-oxidation. EMBO Rep 6 (7): 629–634.
20. Maxwell K, Johnson GN (2000) Chlorophyll fluorescence – a practical guide. J Exp Bot 51(345): 659–668.
21. Pavlík M, Pavlíková D, Staszková L, Neuberg M, Kaliszová R, et al. (2010) The effect of arsenic contamination on aminoacids metabolism in *Spinacia oleracea* L. Ecotoxicol Environ Saf 73: 1309–1313.
22. Carvalho J, Gavina A, Cruz T, Gonçalves F, Pereira R, et al. (2010) Minas da Ervedosa: no passado uma mina, hoje um laboratório natural. CAPTAR 2 (2): 29–45.
23. Novais HJGSQ (2006) Avaliação da qualidade dos solos e da água subterrânea na envolvente das Minas de Ervedosa (NE de Portugal). Produção de recursos didácticos. Tese de Mestrado. Universidade de Trás-os-Montes e Alto Douro, Vila Real.

24. FAOUN (1984) Physical and chemical methods of soil and water analysis. Food and Agriculture Organization of the United Nations. 2nd Ed. Soil Bulletin 10: 1–275.
25. ISO (2005) Soil quality — Determination of pH. ISO—The International Organization for Standardization, Geneva, Switzerland. ISO 10390:2005.
26. SPAC (2000) Handbook of Reference Methods. Soil and Plant Analysis Council.Boca Raton, Florida: CRC Press.
27. Houba VJG, Temminghoff EJM, Gaikhorst GA, van Vark W (2000) Soil analysis procedures using 0.01 M calcium chloride as extraction reagent. Commun Soil Sci Plant Anal 31: 1299–1396.
28. OECD (2006a) *Lemna* sp. Growth Inhibition Test. OECD - Guideline for Testing of Chemicals, Paris, France. 221.
29. OECD (1984) Terrestrial Plants, Growth Test. OECD - Guideline for Testing of Chemicals. Paris, France. 208.
30. Sims DA, Gamon JA (2002) Relationships between leaf pigment content and spectral reflectance across a wide range of species, leaf structures and developmental stages. Remote Sens Environ 81: 337–354.
31. Elkahoui E, Hernández JA, Abdelly C, Ghrir R, Limama F (2005) Effects of salt on lipid peroxidation and antioxidant enzyme activities of *Catharanthus roseus* suspension cells. Plant Sci 168: 607–613.
32. Khedr AHA, Abbas MA, Wahid AAA, Quick WP, Abogadallah GB (2003) Proline induces the expression of salt-stress-responsive proteins and may improve the adaptation of *Pancratium maritimum* L. to salt-stress. J Exp Bot 54(392): 2553–2562.
33. Lutts S, Kinet JM, Bouharmont J (1996) NaCl-induced senescence in leaves of rice (*Oryza sativa* L.) cultivars differing in salinity resistance. Ann Bot 78: 389–398.
34. Zar JH (1996) Biostatistical analysis. 3rd Edition. New Jersey: Prentice-Hall International Inc.
35. CEC - Commission of the European Communities (2006) Communication from the Commission to the Council, the European Parliament, the European Economic and Social Committee and the Committee of the regions. Thematic Strategy for Soil Protection. COM(2006)231final. Available: http://eur-lex.europa.eu/LexUriServ/LexUriServ.do?uri = COM:2006:0231:FIN:EN:PDF. Accessed 2012 Sep 25.
36. CEC - Commission of the European Communities (2006) Proposal for a Directive of the European Parliament and the Council establishing a framework for the protection of soil and amending Directive 2004/35/Ec (presented by the Commission) 2006/0086/COD. Available: http://eur-lex.europa.eu/LexUriServ/LexUriServ.do?uri = COM:2006:0232:FIN:EN:PDF. Accessed 2012 Sep 25.
37. Jensen J, Mesman M (2006) Ecological risk assessment of contaminated land. Decision support for site specific investigations. RIVM report number 711701047. The Netherlands: National Institute for Public Health and the Environment. 136 pp.
38. USEPA (2004) Framework for Inorganic Metals Risk Assessment. United States Environmental Protection Agency. Draft EPA/630/P-04/068B, 20460. Washington DC.
39. Jänsch S, Römbke J, Schallnaß HJ, Terytze K (2007) Derivation of soil values for the path 'soil – soil organisms' for metals and selected organic compounds using species sensitivity distributions. Environ Sci Pollut Res Int 14(5): 308–318.
40. Menzies NW, Donn MJ, Kopitteke PM (2007) Evaluation of extractants for estimation of the phytoavailable trace metals in soils. Environ Pollut 145: 121–130.
41. Clemens S (2006) Toxic metal accumulation, responses to exposure and mechanisms of tolerance in plants. Biochimie 88: 1707–1719.
42. Nagajyoti PC, Lee KD, Sreekanth TVM (2010) Heavy metals, occurrence and toxicity for plants: a review. Environ Chem Lett 8(3): 199–216.
43. Duester L, van der Geest HG, Moelleken S, Hirner AV, Kueppers K (2011) Comparative phytotoxicity of methylated and inorganic arsenic- and antimony species to *Lemna minor*, *Wolffia arrhiza* and *Selenenastrum capricornutum*. Microchem J 97: 30–37.
44. Teisseire H, Couderchet M, Vernet G (1998) Toxic responses and catalase activity of *Lemna minor* L. exposed to folpet, copper, and their combination. Ecotoxicol Environ Saf 40: 194–200.
45. Radić S, Stipaničev D, Cvjetko P, Mikelić IL, Rajčić MM, et al. (2010) Ecotoxicological assessment of industrial effluent using duckweed (*Lemna minor* L.) as a test organism. Ecotoxicology 19: 216–222.
46. Bedell J-P, Briant A, Delolme C, Lassabatère L, Perrodin Y (2006) Evaluation of the phytotoxicity of contaminated sediments deposited "on soil": II. Impact of water draining from deposits on the development and physiological status of neighbouring plants at growth stage. Chemosphere 62: 1311–1323.
47. Chapman EEV, Dave G, Murimboh JD (2010) Ecotoxicological risk assessment of undisturbed metal contaminated soil at two remote lighthouse sites. Ecotoxicol Environ Saf 73: 961–969.
48. Kranner I, Colville L (2011) Metals and seeds: biochemical and molecular implications and their significance for seed germination. Environ Exp Bot 72: 93–105.
49. Briat J-F, Lebrun M (1999) Plant responses to metal toxicity. C R Acad Sci III 322 (1): 43–54.
50. Halliwell B (1987) Oxidative damage, lipid peroxidation and antioxidant protection in chloroplasts. Chem Phys Lipids 44: 327–340.
51. Sun Y, Li Z, Guo B, Chu G, Wei C, et al. (2008) Arsenic mitigates cadmium toxicity in rice seedlings. Environ Exp Bot 64: 264–270.
52. Singh RP, Agrawal M (2010) Biochemical and physiological responses of rice (*Oryza sativa* L.) grown on different sewage sludge amendments rates. Bull Environ Contam Toxicol 84: 606–612.
53. Dinakar N, Nagajyothi PC, Suresh S, Udaykiran Y, Damodharam T (2008) Phytotoxicity of cadmium on protein, proline and antioxidant enzyme activities in growing *Arachis hypogaea* L. seedlings. J Environ Sci (China) 20: 199–206.
54. Vernay P, Gauthier-Moussard C, Jean L, Bordas F, Faure O, et al. (2008) Effect of chromium species on phytochemical and physiological parameters in *Datura innoxia*. Chemosphere 72: 763–771.
55. Jiang W, Liu D, Li H (2000) Effects of Cu^{2+} on root growth, cell division and nucleoulus of *Helianthus annus* L. Sci Total Environ 256: 59–65.
56. Li C-X, Feng S-L, Shao Y, Jiang L-N, Lu X-Y, et al. (2007) Effects of arsenic on seed germination and physiological activities of wheat seedlings. J Environ Sci 19: 725–732.
57. Björkman O, Demming B (1987) Photon yield of O_2 evolution and chlorophyll fluorescence characteristics among vascular plants of diverse origin. Planta 170: 489–504.
58. Küpper H, Küpper F, Spiller M (1996) Environmental relevance of heavy metal-substituted chlorophylls using the example of water plants. J Exp Bot 47: 259–266.
59. MacFarlane GR (2002) Leaf biochemical parameters in *Avicennia marina* (Forsk.) Vierh as potential biomarkers of heavy metal stress in estuarine ecosystems. Mar Pollut Bull 44: 244–256.
60. Fatima RA, Ahmad M (2005) Certain antioxidant enzymes of *Allium cepa* as biomarkers for the detection of toxic heavy metals in wastewater. Sci Total Environ 346: 256–273.
61. Corrêa AXR, Rörig LR, Verdinelli MA, Cotelle S, Férad JF, et al. (2006) Cadmium phytotoxicity : quantitative sensitivity relationships between classical endpoints and antioxidant enzyme biomarkers. Sci Total Environ 357 (1–3):120–127.
62. Sun B-Y, Kan S-H, Zhang Y-Z, Deng S-H, Wu J, et al. (2010) Certain antioxidant enzymes and lipid peroxidation of radish (*Raphanus sativus* L.) as early warning biomarkers of soil copper exposure. J Hazard Mater 183: 833–838.

20

Latitudinal Patterns in Phenotypic Plasticity and Fitness-Related Traits: Assessing the Climatic Variability Hypothesis (CVH) with an Invasive Plant Species

Marco A. Molina-Montenegro[1]*, Daniel E. Naya[2]

1 Centro de Estudios Avanzados en Zonas Áridas, Facultad de Ciencias del Mar, Universidad Católica del Norte, Coquimbo, Chile, 2 Departamento de Ecología y Evolución, Facultad de Ciencias and Centro Universitario de la Regional Este, Universidad de la República, Montevideo, Uruguay

Abstract

Phenotypic plasticity has been suggested as the main mechanism for species persistence under a global change scenario, and also as one of the main mechanisms that alien species use to tolerate and invade broad geographic areas. However, contrasting with this central role of phenotypic plasticity, standard models aimed to predict the effect of climatic change on species distributions do not allow for the inclusion of differences in plastic responses among populations. In this context, the climatic variability hypothesis (CVH), which states that higher thermal variability at higher latitudes should determine an increase in phenotypic plasticity with latitude, could be considered a timely and promising hypothesis. Accordingly, in this study we evaluated, for the first time in a plant species (*Taraxacum officinale*), the prediction of the CVH. Specifically, we measured plastic responses at different environmental temperatures (5 and 20°C), in several ecophysiological and fitness-related traits for five populations distributed along a broad latitudinal gradient. Overall, phenotypic plasticity increased with latitude for all six traits analyzed, and mean trait values increased with latitude at both experimental temperatures, the change was noticeably greater at 20° than at 5°C. Our results suggest that the positive relationship found between phenotypic plasticity and geographic latitude could have very deep implications on future species persistence and invasion processes under a scenario of climate change.

Editor: Frank Seebacher, University of Sydney, Australia

Funding: The authors have no support or funding to report.

Competing Interests: The authors have declared that no competing interests exist.

* E-mail: marco.molina@ceaza.cl

Introduction

Populations exposed to environmental changes may respond in four not mutually exclusive ways: they can become extinct, migrate to new areas, adapt via genetic change, or persist via phenotypic plasticity [1]. Although the last two alternatives avoid local extinction, current evidence suggests that for most populations coping with accelerated changing conditions, and thus local persistence, will be closely related to the amount of plasticity for fitness-related traits [2–4].

However, contrasting with the central role of phenotypic plasticity in population persistence, standard models aimed to predict the effect of climatic change on species distribution (i.e., the climate envelope models) do not allow for the inclusion of differences in plastic responses among populations. Obviously this is not a fanciful constraint; the quantification of plasticity for several traits in several populations is a difficult task, and thus available data in this regard are still very scarce (e.g., provenance tests). An encouraging pathway that would trade-off the large amount of data needed to include differences in plastic responses among populations (of each species to be modeled) with the fairly low predictive power of current climate envelope models [5], is the identification of global patterns in phenotypic plasticity, which could be easily incorporated into the models if they exist.

In this context, the climatic variability hypothesis (CVH) could be considered a timely and promising hypothesis, since it directly connects phenotypic plasticity with climatic and geographic variables at a global scale. Specifically, the CVH states that as the range of climatic fluctuation experienced by terrestrial animals increases with latitude, individuals at higher latitudes should have broader ranges of thermal tolerance and acclimation abilities that enable them to cope with the fluctuating environmental conditions [6; see also 7–8]. Given that tolerance ranges and acclimation responses are ultimately linked to mechanisms of morphological, physiological and/or behavioral flexibility, the central idea of the CVH has been recently extended to phenotypic plasticity in general [9–11]. Up to the present, empirical evidence supporting the CVH may be clustered in three major groups: (1) studies that directly evaluated the relationship between latitude and thermal tolerance range in ectothermic animals [e.g., 12–18]; (2) studies that directly evaluated the relationship between latitude and phenotypic flexibility for non-thermal traits in ecto- and endothermic animals [e.g., 19–22]; and (3) studies that analyzed different ecological patterns at the population and community levels that are expected to emerge from the CVH [e.g., 7, 23–28].

An important gap that still remains in our knowledge is the direct applicability of the CVH to plant species. On theoretical grounds there are two contrasting ideas about how plant species should change with geographic latitude to cope with the environment. On one hand, the limited vagility of plants may result in a great degree of adaptation to local conditions, resulting in a great population differentiation, which could preclude the existence of a latitudinal pattern in trait plasticity [29–31]. On the other hand, the limited vagility of plants may imply a reduced ability to avoid environmental influences, and thus the pattern predicted by the CVH could be more clearly observed than in animal species [10,32]. Although some studies have been conducted to assess the relationship between plasticity and climatic heterogeneity in plant species [33–35], to the best of our knowledge no empirical study has evaluated the validity of CVH in these organisms.

Accordingly, the aim of the present study was to analyze how plasticity for several ecophysiological and fitness-related traits changes with latitude in the invasive *Taraxacum officinale* (dandelion complex). Specifically, we evaluated plasticity –due to differences in environmental temperature (5 and 20°C)– in photosynthetic rate, water use efficiency, foliar angles, plant biomass, number of flowers and seed output in five populations that occur along a latitudinal gradient (from 0° to 54°S). We predicted that the reduction in the duration of the growing season observed at higher latitudes should determine that populations at higher latitudes will be more efficient at exploiting favorable thermal conditions (20°C) than populations from lower latitudes. Additionally, we predicted that higher thermal heterogeneity at higher latitudes should determine that populations at higher latitudes would be more plastic. Thus we expect to observe, in agreement with the CVH, a positive relationship between phenotypic plasticity and geographic latitude.

Methods

Populations, Traits and Experimental Environments

Taraxacum officinale is a member of the Asteraceae originally from Europe that has spread worldwide. It is one of the top invasive species around the world [36]. This plant can be found growing in sites with changing climatic characteristics, disturbance regimes, and along a wide range of altitudes and latitudes. Seeds of *T. officinale* were collected in five localities: Manta (Ecuador), Trujillo (Perú), La Serena (Chile), Valdivia (Chile) and Punta Arenas (Chile). The hemispheric latitudinal gradient covers from *ca.* 0° to *ca.* 54° S, including a notorious and significant thermal gradient (Fig. 1). All seeds were collected at sea level to reduce altitudinal effects (Fig. 1). Preliminarily, cytogenetic analysis showed that individuals from all localities sampled in this study have the same ploidy level (n = 24 chromosomes) and size of chromosomes, suggesting that all individuals sampled belong to the same species (data not shown).

A small number of seeds (four to five) per individual plant collected from a relatively large number of sampled plants (50–55) per population provided the initial pool of seeds. As *T. officinale* has apomictic reproduction [37], samples were taken from widely separated plants to avoid sampling the same genet twice. Seeds in each locality were collected from 3 populations separated by 1 km each. Finally, all seeds collected in the 3 populations of each locality were pooled and randomized before sorting them into experimental treatments. This was done because the aim of this study was to compare responses in a number of genotypes from different localities growing along the latitudinal gradient rather than to isolate genotypic effects from phenotypic effects [33,38]. No specific permits were required for seed collection in the localities sampled in this study and confirm that all populations are not privately-owned or protected in any way. Additionally, we confirm that the field studies did not involve endangered or protected species.

Seeds from all localities were germinated at 24±2°C on wet filter paper in Petri dishes and planted in 300-mL plastic pots filled with potting soil. First generation plants (F1) were generated from this initial seed pool and were grown in a greenhouse at the Universidad de Concepción (Concepción, Chile) under controlled conditions of light and temperature (1320 µmol m-2s-1±55 and 22±2°C, respectively). These plants were again put in 300-mL plastic pots filled with potting organic soil and irrigated every two days with 50 ml of water. After five months these plants produced the achenes that were used to obtain experimental plants (F_2). One week after of appearance of the first true leaf, F_2 seedlings were transferred to growth chambers (Forma Scientific Inc.) with a photon flux density (PFD) of 170 µmol m-2 s-1 and 16/8 h light/dark photoperiod. The temperature treatment consisted of transferring 20 individuals from each locality described above to a growth chamber set at 5 or 20°C for 90 days. These temperatures were chosen because they are close to the mean temperatures in each extreme of gradient. Plants were irrigated daily and supplemented with Phostrogen® (Solaris, NPK, 14:10:27) using 0.2 g L^{-1} once every 15 days. Plastic pot positions were randomized within the experimental plot every four days. Interplot distances were sufficient to prevent mutual shading. After 90 days we recorded three ecophysiological traits, net photosynthesis, water use efficiency (photosynthesis/foliar transpiration) and foliar angles, and three fitness-related traits (total dry biomass, flower production and seed output).

Climatic Data

For each sampled population we downloaded data from the WorldClim data base (http://www.worldclim.org/) on the following climatic variables: annual mean temperature (Tmed, in °C), minimum temperature of the coldest month (Tmin, in °C), maximum temperature of the warmest month (Tmax, in °C), temperature seasonality (TS: standard deviation of the mean monthly temperature, in °C), temperature annual range (TAR: difference between maximum temperature of warmest month and minimum temperature of the coldest month, in °C), accumulated annual precipitation (Rainfall, in mm), and rainfall seasonality (RS: standard deviation of the mean monthly rainfall, in mm) (Table 1).

Plasticity Estimations and Statistical Analyses

Phenotypic plasticity was considered as the environmentally-induced change in the expression of phenotypic traits at the end of the experimental period, whatever the mechanistic causes (e.g., ontogenetic, allometric) behind this differential expression [39]. Phenotypic plasticity was estimated for each trait and locality as the percentage of change in mean trait value from one environment to the other; that is, $P = [(X_{20} - X_5)/X_{20}] * 100$, where P is plasticity, X_{20} is the mean trait value at 20°C and X_5 is the mean trait value at 5°C [40]. In addition, a measurement of overall plasticity was estimated for each population as the arithmetic average of the percentage of change observed for all the traits. A bootstrapping procedure, with 1000 iterations was used in order to obtain error estimations of plasticity for each trait and population (i.e., a percentage of change was calculated from two randomly selected individuals, one in each temperature treatment, in each iteration). The error estimation for overall plasticity was calculated as the arithmetic mean of the errors obtained for all

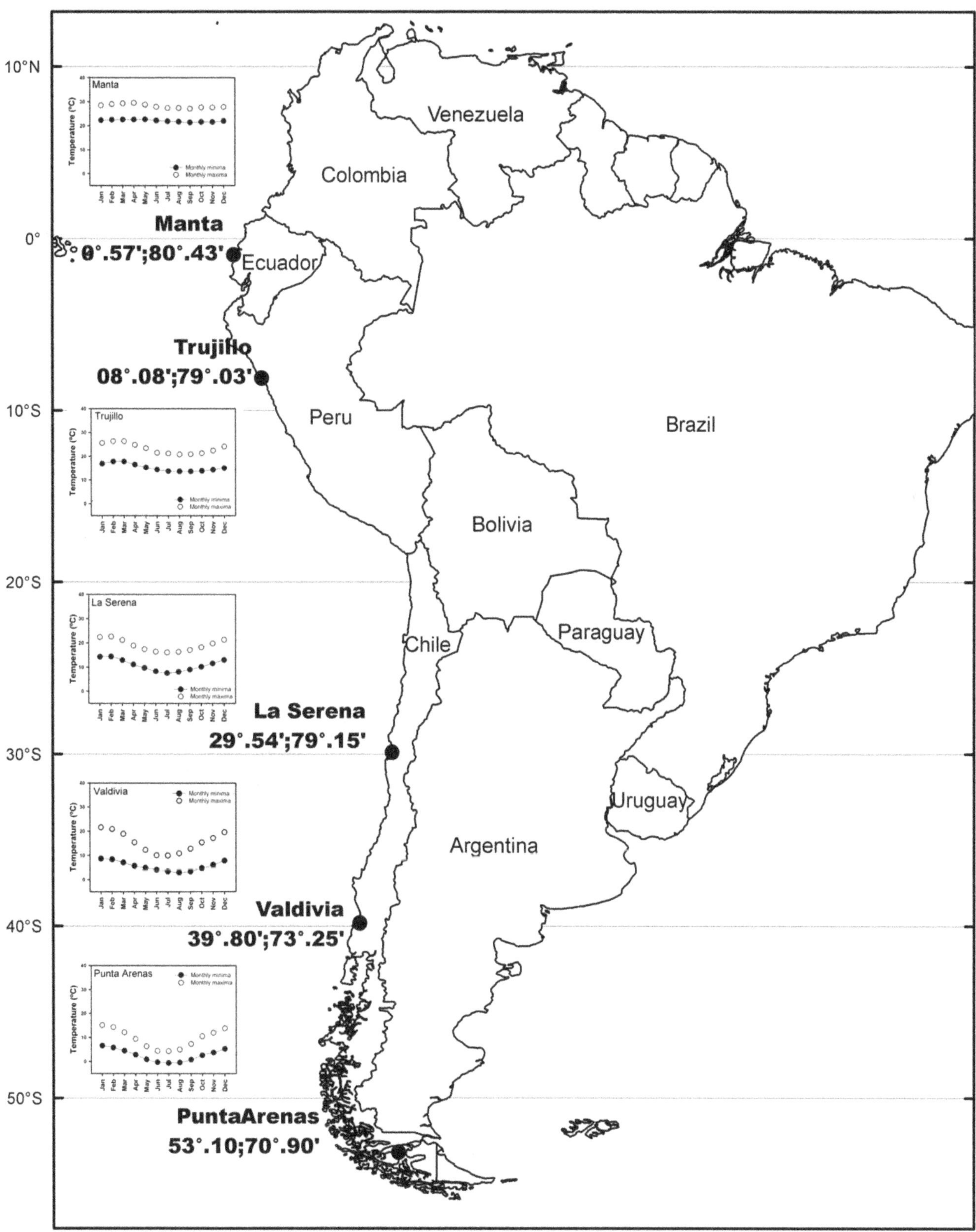

Figure 1. Map showing the five sampled localities along the latitudinal gradient. Monthly minimum and maximum temperatures are given for each locality.

the traits. The relationships between mean trait values at each temperature and trait plasticity with geographic latitude and climatic variables were evaluated separately using the Pearson product moment coefficient.

Table 1. Climatic variables for each locality (see Methods for abbreviations).

	Manta	Trujillo	La Serena	Valdivia	Punta Arenas
Tmed (°C)	25.1	19.2	14.9	10.5	6.1
Tmin (°C)	21.4	13.6	7.6	3.0	−0.6
Tmax (°C)	29.5	26.3	22.6	21.6	15.1
TS (°C)	62.7	187.2	241.3	309.3	327.1
TAR (°C)	8.1	12.7	15.0	18.6	15.7
Rainfall (mm)	1788	6	83	2211	431
RS (mm)	64.0	180.9	119.5	61.3	15.6

Results

Except for water use efficiency and foliar angle at 5°C, for which no latitudinal changes were observed, mean trait values increased with latitude at both experimental temperatures (Table 2). In addition, given that these increases in mean values were noticeably greater at 20°C than at 5°C, phenotypic plasticity also increased with latitude for all the six analyzed traits (Fig. 2). Regarding the relationship between phenotypic plasticity and climatic variables, it was observed that plasticity in photosynthetic rate was positively related to temperature annual range, plasticity in water use efficiency was negatively correlated with maximum temperature of the warmest month, and plasticity in foliar angle, flower production, seed output and overall plasticity were positively related to thermal seasonality and negatively related to environmental temperatures (Table 3).

Neither annual rainfall nor rainfall seasonality were correlated with plasticity for any trait (Table 3). In any case, the number of significant correlations between any single climatic variable and trait plasticity was always less than those observed for latitude, and moreover, latitude was by far a better predictor of overall plasticity than any climatic variable (Table 3, Fig. 3).

Discussion

The Earth is undergoing dramatic environmental changes (referred to as global change) which are mainly related to five different (but interacting) phenomena: climate change, land use change, resource overexploitation, pollution, and invasive species [41,42]. Two results obtained in the present study are relevant in this scenario of global change. First, in agreement with the climatic variability hypothesis (CVH), a clear association was found between latitude and plasticity for all the ecophysiological and fitness-related traits analyzed. Second, all the sampled populations of the highly invasive species considered here showed a great degree of plasticity for all the traits evaluated, regardless of their specific location along the latitudinal gradient. Thus, in what follows we discuss the implications of these results within the contexts of the CVH and species invasiveness ability.

Latitudinal Patterns in Phenotypic Plasticity

As mentioned above, the present study demonstrates for the first time in a plant species the occurrence of the latitudinal pattern of phenotypic plasticity predicted by the CVH. Moreover, correlation coefficients (ranging between 0.84 and 0.99) support the idea that latitudinal patterns in plasticity maybe stronger in plants than in animal species [10]. The positive relationship between plasticity and latitude was due to a slight increase (or no change) in trait values with latitude at 5°C but a strong increase at 20°C. This result is logical from a biological point of view, since: (1) Environmental temperatures of 5°C are close to monthly mean temperatures in high latitude localities such as Punta Arenas and Valdivia, but are far from monthly temperatures (even from monthly minima) in the other localities (Figure 1). Thus a selective advantage associated with an increase in trait values at this temperature should only be expected in the former localities. In line with this idea, it should be noted that the only traits that did not change with latitude were foliar angle and water use efficiency, i.e., traits for which a rise in mean values with latitude probably do not represent a selective advantage. (2) Environmental temperatures of 20°C are higher than monthly maxima in Punta Arenas, close to monthly maxima during four months in Valdivia and during seven months in La Serena, close to the monthly mean in Trujillo, and close to the monthly minima in Manta (Figure 1). Thus, a temperature of 20°C is probably associated with the short favourable season at middle and high latitudes, but only with normal or even unfavourable conditions at lower latitudes. Consequently, positive selection for higher trait values at this temperature is expected to be very strong in high latitude localities, but only weak (or null) in low latitude localities. It has been proposed that populations growing along latitudinal gradients in its poleward limit of distributions should show a trade-off between cold resistance and growth, thus any amelioration in thermal stress conditions will be positive for these populations, improving their ecophysiological performance compared with other populations distributed at lower latitudes (see [43]. Unfortunately, our experimental design did not contemplate environmental temperatures greater than 20°C, for which a strong positive selection for higher traits values may be expected for all the localities. Thus, it is possible that the inclusion of a higher temperature (e.g., 30°C) in the experimental setup could attenuate the strong latitudinal pattern for plasticity reported here.

On the other hand, reported data suggest that latitude is a better predictor of phenotypic plasticity than climatic variables. This result, usually reported in macrophysiological studies [e.g., 19, 44, 45], appears to be related to at least three different facts [21]. First, latitude is probably a better predictor of long-term regimes of climatic variables than current climate values provided by weather stations. Second, latitude is correlated with several other climatic, ecological, and historical factors that could affect phenotypic plasticity. Third, the smooth variation of climatic variables in space suggests that latitude may represent a weighted variable of climatic conditions acting over spatial scales more similar to those at which adaptation is expected to occur. In any case, recent studies evaluating the effect of latitude and climatic variability on phenotypic plasticity for areas where both sets of variables are not directly correlated have found that climatic variables, and not latitude, were the best predictors of phenotypic plasticity [20,23]. Thus, as stated by the CVH, climatic variability appears to be the proximal cause behind the latitudinal patterns in phenotypic plasticity. Thus, phenotypic plasticity can play an important role both in invasive and native plant species along gradients [42,46]. For example, Santamaría et al. (2003) [32] showed the positive role of phenotypic plasticity in aquatic plants across a latitudinal gradient. Additionally, Mou et al. 2012 [47], demonstrated that plasticity in morphological and physiological traits would improve the performance and resource acquisition in environmental variability conditions.

Finally, the results found in this study are in line with recent papers that also report an increase in phenotypic plasticity with latitude for different traits and taxonomic groups (see Introduction). Obviously, this does not mean that all the existing

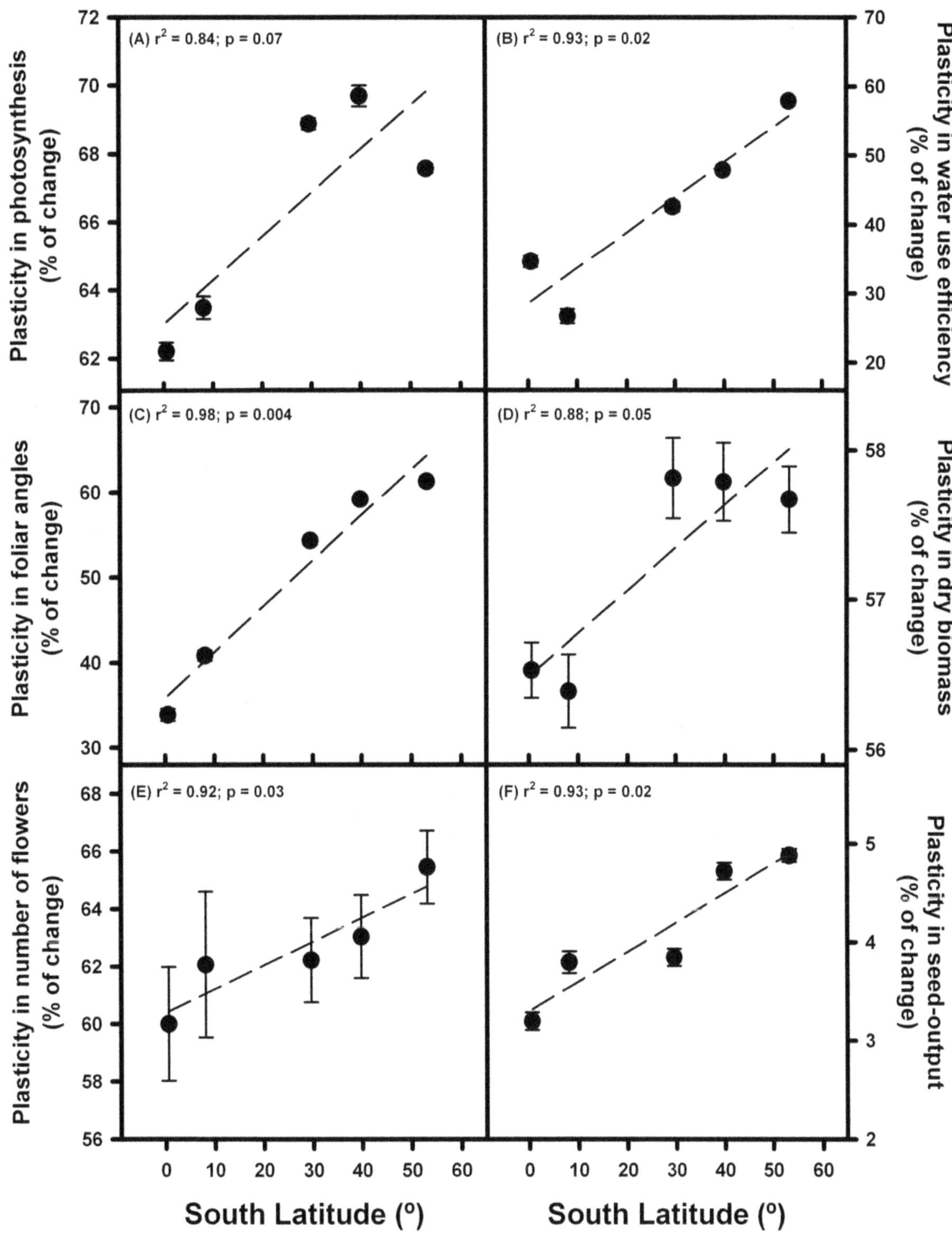

Figure 2. Relationship between trait plasticity (measured as percentage of change) and latitude: (A) photosynthetic efficiency, (B) water use efficiency, (C) foliar angle, (D) plant biomass, (E) number of flowers, (F) seed output.

evidence supports the prediction of the CVH. For instance, studies analyzing acclimation abilities in ectothermic animals do not show a clear pattern of latitudinal variation [12,48–50]. Although it is true that these studies had a more reduced taxonomic scope (e.g., one genus or one family) and geographic extension (e.g., one region or one continent) than most of the studies that support the CVH –suggesting that differences in results may be due in part to difference in the temporal and spatial scales considered– it could be expected that the prediction of the CVH does not hold for all phenotypic characters. As for phenotypic plasticity itself –although the lack of plasticity for one or more traits does not deny the adaptive

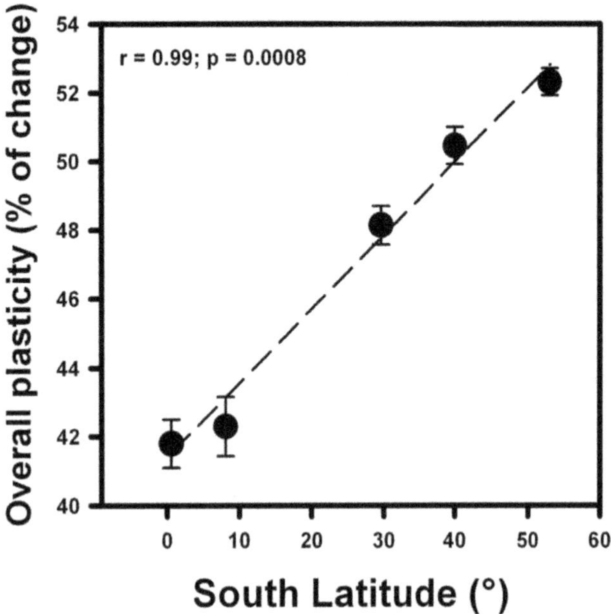

Figure 3. Relationship between overall plasticity (measured as the arithmetic mean of the percentage of change for all the six traits) and latitude.

value of plasticity in general– the lack of latitudinal patterns in plasticity for some traits does not deny the validity of the CVH in general terms. We believe that the evidence supporting the CVH is enough to warrant that it be seriously considered in models aimed at predicting the effect of global change (and particularly the effect of global warming) in future species distributions. A simple way to do this is to modify the current climate envelope models by increasing the potential niche of each population in parallel to latitude, but the specific details on how this could be done this exceed the aim of the present study.

The Relationship between Phenotypic Plasticity and Invasiveness

Although it has been reported that *T. officinale* may reduce the fitness of native herbs in many ways [35,50], little is known about the mechanisms associated with its wide distribution (but see [51].) Based on our results we suggest that *T. officinale* is capable of invading a broad latitudinal gradient mainly because, being an r-selected species, this species is also able to exhibit a great level of phenotypic plasticity. In fact, in a recent paper Richards et al. (2006) [52] suggested that biological invasions would be driven by phenotypic plasticity, playing an important role in successful plant invasions in wide clines. Thus plasticity in functional traits may enhance ecological niche breadth and therefore confer a fitness advantage [52]. Geographic gradients in abiotic conditions across a wide range could impose divergent selection pressures and promote genetically based differentiation among introduced populations. A classic manifestation of this is the evolution of geographic clines, often found in native species occurring across altitudinal gradients [53,54]. However, whether introduced plant populations rapidly evolve clines in response to environmental conditions across their introduced range has seldom been studied [54,55]. Furthermore, whether clines in traits among introduced plant populations broadly converge on those expressed among native conspecifics occurring over similar latitudinal or altitudinal gradients is unknown. Nevertheless, this may be the case for *T. officinale*, which shows not only high plasticity, but also local adaptation along a broad latitudinal gradient. Phenotypic plasticity and ecotypic differentiation are two complementary strategies to face environmental heterogeneity [29,55–57]. It has been shown that plasticity can initially allow exotic species to become naturalized across the non-native range [57] and, once naturalized, genetic recombination of heritable phenotypes may respond to local selection pressures giving rise to ecotypes with higher fitness [58].

The triploid condition of *T. officinale* individuals from all localities studied suggests they are apomictics, which usually is associated with low genetic variability and high plasticity as a successful strategy to cope with changing environments [35]. Nevertheless, unexpectedly high levels of genetic variation have

Table 2. Mean trait values measured at 5°C and 20°C for each locality, and the Pearson product moment correlation coefficient (together with the associated probability and the slope of the regression) for the correlation between mean trait values and latitude.

	Manta	Trujillo	La Serena	Valdivia	Pta. Arenas	r	P-value	Slope
A max (μmol m^{-2} s^{-1}) at 5°C	3.03 (0.06)	2.96 (0.08)	4.03 (0.08)	5.18 (0.19)	5.38 (0.06)	0.90	<0.000001	0.05
A max (μmol m^{-2} s^{-1}) at 20°C	8.03 (0.14)	8.11 (0.20)	12.94 (0.13)	17.09 (0.25)	16.59 (0.20)	0.94	<0.000001	0.19
WUE (μmol CO_2/mmol H_2O) at 5°C	1.22 (0.05)	1.28 (0.06)	1.19 (0.05)	1.23 (0.05)	1.23 (0.06)	0.04	0.71	0.0005
WUE (μmol CO_2/mmol H_2O) at 20°C	1.87 (0.05)	1.75 (0.05)	2.07 (0.04)	2.37 (0.04)	2.91 (0.04)	0.86	<0.000001	0.02
Foliar angle (°) at 5°C	12.33 (0.44)	11.20 (0.38)	12.07 (0.40)	11.27 (0.34)	11.67 (0.35)	−0.08	0.47	−0.007
Foliar angle (°) at 20°C	18.67 (0.42)	18.93 (0.43)	26.40 (0.55)	27.60 (0.54)	30.13 (0.46)	0.92	<0.000001	0.24
Plant biomass (mg) at 5°C	37.93 (0.55)	38.67 (0.81)	42.67 (0.91)	45.53 (0.92)	52.07 (0.92)	0.83	<0.000001	0.26
Plant biomass (mg) at 20°C	87.27 (0.78)	88.67 (0.61)	101.13 (1.47)	107.87 (1.46)	123.00 (1.36)	0.93	<0.000001	0.67
Number of flowers at 5°C	0.80 (0.14)	0.73 (0.18)	1.13 (0.17)	1.13 (0.17)	1.27 (0.18)	0.30	0.01	0.01
Number of flowers at 20°C	2.00 (0.17)	1.93 (0.18)	3.00 (0.17)	3.07 (0.18)	3.67 (0.16)	0.71	<0.000001	0.03
Seed-output at 5°C	0.901 (0.001)	0.901 (0.003)	0.915 (0.002)	0.922 (0.003)	0.926 (0.002)	0.80	<0.000001	0.0005
Seed-output at 20°C	0.931 (0.003)	0.937 (0.003)	0.952 (0.002)	0.968 (0.002)	0.973 (0.002)	0.86	<0.000001	0.001

A max = photosynthetic rate; WEU = water use efficiency.

Table 3. Pearson product moment correlation coefficient (and associated probability) for the correlations between ecophysiological and fitness-related trait plasticity (measured as percentage of change) and climatic variables.

	A max	WUE	Foliar angle	Plant biomass	Number of flowers	Seed output	Overall plasticity
Tmed (°C)	−0.81 (0.10)	−0.86 (0.06)	−0.97 (0.007)	−0.81 (0.10)	−0.96 (0.01)	−0.97 (0.007)	−0.96 (0.01)
Tmin (°C)	−0.86 (0.06)	−0.84 (0.08)	−0.98 (0.003)	−0.84 (0.08)	−0.93 (0.02)	−0.96 (0.01)	−0.96 (0.01)
Tmax (°C)	−0.71 (0.18)	−0.89 (0.04)	−0.92 (0.03)	−0.76 (0.13)	−0.98 (0.005)	−0.91 (0.03)	−0.94 (0.02)
TS (°C)	0.86 (0.06)	0.76 (0.13)	0.97 (0.006)	0.81 (0.10)	0.91 (0.03)	0.96 (0.01)	0.92 (0.03)
TAR (°C)	0.92 (0.03)	0.63 (0.26)	0.92 (0.03)	0. 81 (0.09)	0.73 (0.16)	0.88 (0.05)	0.84 (0.08)
Rainfall (mm)	0.05 (0.94)	0.14 (0.82)	−0.06 (0.93)	0.06 (0.92)	−0.28 (0.64)	0.04 (0.94)	0.02 (0.97)
RS (mm)	−0.31 (0.61)	−0.83 (0.08)	−0.46 (0.44)	−0.49 (0.40)	−0.43 (0.47)	−0.49 (0.40)	−0.62 (0.26)

A max = photosynthetic rate (μmol m^{-2} s^{-1}), WUE = water use efficiency (μmol CO_2/mmol H_2O).

been found in other apomictic species generated by subsexual reproduction [59], including genetic segregation and hybridization between sexual and apomictic individuals [60,61]. Exceptions to the general pattern of high cytogenetic variability in widely distributed alien species do exist. They tend to occur in species with very high levels of phenotypic plasticity (e.g., *Taraxacum*) which have a low level of cytogenetic variability [62].

The high invasive capacity of Eurasian species is considered to be a result of an evolutionary history on a large continental mass that suffered major upheavals during the glacial period and longer association between humans and plants than in the New World [63]; consequently, aliens of Eurasian origin, because of their higher competitiveness, vagility and plasticity are likely to invade areas even when interchange is fairly limited [64]. Finally, the fact that species that have been introduced into South America and can adapt to the wide range of environmental conditions found along fifty-five degrees of latitude could be surprising. Surely the high plasticity in the ecophysiological and fitness-related traits of *T. officinale* individuals from different populations suggests that some invaders could definitely adapt to changing environmental conditions found along broad gradients worldwide.

Conclusions

The present study shows that the prediction of the climatic variability hypothesis appears to hold for plant species, finding a strong positive relationship between phenotypic plasticity and geographic latitude. As discussed, this result could have very profound implications on future species persistence under a scenario of climate change. In addition, obtained data support the idea that the great invasiveness ability reported for *T. officinale* along broad gradients could be related with both, to being a weedy r-selected species and to having high plasticity levels for several ecophysiological and fitness-related traits.

Acknowledgments

This paper forms part of the research activities of the CYTED network ECONS (410RT0406).

Author Contributions

Conceived and designed the experiments: MAMM DEN. Performed the experiments: MAMM DEN. Analyzed the data: MAMM DEN. Contributed reagents/materials/analysis tools: MAMM DEN. Wrote the paper: MAMM DEN.

References

1. Fuller A, Dawson T, Helmuth B, Hetem RS, Mitchell D, et al. (2010) Physiological mechanisms in coping with climate change. Physiological and Biochemical Zoology 85: 713–720.
2. Deutsch CA, Tewksbury JJ, Huey RB, Sheldon KS, Ghalambor CK, et al. (2008) Impacts of climate warming on terrestrial ectotherms across latitude. Proceedings of the National Academy of Science USA 105: 6668–6672.
3. Teplitsky C, Mills JA, Alho JS, Yarrall JW, Merila J (2008) Bergmann's rule and climate change revisited: disentangling environmental and genetic responses in a wild bird population. Proceedings of the National Academy of Science USA 105: 13492–13496.
4. Hoffman AA, Sgro CM (2011) Climate change and evolutionary adaptation. Nature 470: 479–485.
5. Duncan RP, Cassey P, Blackburn TM (2009) Do climate envelope models transfer? A manipulative test using dung beetle introductions. Proceeding of the Royal Society B 276: 1449–1457.
6. Janzen DH (1967) Why mountain passes are higher in the tropics. American Naturalist 101: 233–249.
7. Stevens GC (1989) The latitudinal gradient in geographical range: how so many species coexist in the tropics. American Naturalist 133: 240–256.
8. Gaston KJ, Chown SL (1999) Why Rapoport's rule does not generalize. Oikos 84: 309–312.
9. Chown SL, Gaston KJ, Robinson D (2004) Macrophysiology: large-scale patterns in physiological traits and their ecological implications. Functional Ecology 18: 159–167.
10. Ghalambour CK, Huey RB, Martin PR, Tewksbury JJ, Wang G (2006) Are mountain passes higher in the tropics? Janzen's hypothesis revisited. Integrative and Comparative Biology 46: 5–17.
11. Kellerman V, van Heerwaarden B, Sgro CM, Hoffmann AA (2009) Fundamental evolutionary limits in ecological traits drive *Drosophila* species distributions. Science 325: 1244–1246.
12. Brattstrom BH (1968) Thermal acclimation in anuran amphibians as a function of latitude and altitude. Comparative Biochemistry and Physiology 24: 93–111.
13. Snyder GK, Weathers WW (1975) Temperature adaptations in amphibians. American Naturalist 109: 93–101.
14. van Berkum FH (1988) Latitudinal patterns of the thermal sensitivity of sprint speed in lizards. American Naturalist 132: 327–343.
15. Addo-Bediako A, Chown SL, Gaston KJ (2000) Thermal tolerance, climatic variability and latitude. Proceeding of the Royal Society of London B 267: 739–745.
16. Stillman JH, Somero GN (2000) A comparative analysis of the upper thermal tolerance limits of Eastern Pacific porcelain crabs, genus *Petrolisthes*: influences of latitude, vertical zonation, acclimation, and phylogeny. Physiological and Biochemical Zoology 73: 200–208.
17. Calosi P, Bilton DT, Spicer JI (2008) Thermal tolerance, acclimatory capacity and vulnerability to global climate change. Biological Letters 4: 99–102.
18. Sunday JM, Bates AE, Dulvy NK (2011) Global analysis of thermal tolerance and latitude in ectotherms. Prooceedings of the Royal Society B 278: 1823–1830.
19. Naya DE, Bozinovic F, Karasov WH (2008) Latitudinal trends in digestive flexibility: testing the climatic variability hypothesis with data on the intestinal length of rodents. American Naturalist 172: E122–E134.
20. Naya DE, Catalan C, Artacho P, Gaitán-Espitia JD, Nespolo RF (2011) Exploring the functional association between physiological flexibility, climatic variability and geographical latitude: lesson from land snails. Evolutionary Ecology Research 13: 647–659.

21. Naya DE, Spangenberg L, Naya H, Bozinovic F (2012) Latitudinal patterns in rodent metabolic flexibility. American Naturalist 179: E172–E179.
22. Maldonado K, Bozinovic F, Cavieres G, Fuentes C, Sabat P, et al. (2012) Phenotypic flexibility in basal metabolic rate is associated with rainfall variability among populations of Rufous-collared sparrow. Zoology 115: 128–133.
23. Maldonado K, Bozinovic F, Rojas JM, Sabat P (2011) Within-species digestive tract flexibility in roufous-collared sparrows and the climatic variability hypothesis. Physiological and Biochemical Zoology 84: 377–384.
24. Huey RB (1978) Latitudinal patterns of between-altitude faunal similarity: mountanins might be "higher" in the tropics. American Naturalist 112: 225–254.
25. Pagel MD, May RM, Collie AR (1991) Ecological aspects of the geographical distribution and diversity of mammalian species. American Naturalist 137: 791–815.
26. France R (1992) The North American latitudinal gradient in species richness and geographical range of freshwater crayfish and amphipods. American Naturalist 139: 342–354.
27. Letcher AJ, Harvey PH (1994) Variation in geographical range size among mammals of the Paleartic. American Naturalist 144: 30–42.
28. Hernández M, Vrba ES (2005) Rapoport effect and biomic specialization in African mammals: revisitiong the climatic variability hypothesis. Journal of Biogeography 32: 903–918.
29. Counts RL (1992) Phenotypic plasticity and genetic variability in annual *Zizania* spp. Along a latitudinal gradient. Canadian Journal of Botany 71: 145–154.
30. Joshi J, Schmid B, Caldeira MC, Dimitrakopoulus PG, Good J, et al. (2001). Local adaptation enhances performance of common plant species. Ecology Letters 4: 536–544.
31. Mooney HA, Cleland EE (2001) The evolutionary impact of invasive species. Proceedings of the National Academy of Sciences 98: 5446–5451.
32. Santamaría L, Figuerola J, Pilon JJ, Mjelde M, Geen AJ, et al. (2003) Plant performance across latitude: the role of plasticity and local adaptation in an aquatic plant. Ecology 84: 2454–2461.
33. Gianoli E (2004) Plasticity traits and correlations in two populations of *Convolvulus arvensis* (Convolvulaceae) differing in environmental heterogeneity. International Journal of Plant Sciences 165: 825–832.
34. Molina-Montenegro MA, Atala C, Gianoli E (2010) Phenotypic plasticity and performance of *Taraxacum officinale* (dandelion) in habitats of contrasting environmental heterogeneity. Biological Invasions 12: 2277–2284.
35. Molina-Montenegro MA, Peñuelas J, Munné-Bosch S, Sardans J (2012) Higher plasticity in ecophysiological traits enhances the performance and invasion success of *Taraxacum officinale* (dandelion) in alpine environments. Biological Invasions 14: 21–33.
36. Holm L, Doll L, Holm E, Pacheco J, Herberger J (1997) World Weeds. Natural Histories and Distributions. John Wiley & Sons, Inc. New York.
37. Vašut R (2003) *Taraxacum* sect. Erythrosperma in Moravia (Czech Republic): taxonomic notes and the distribution of previously described species. Preslia, Praha 75: 311–338.
38. Schlichting CD, Pigliucci M (1995) Lost in phenotypic space: environment-dependent morphology in *Phlox drummondii* (Polemoniaceae). International Journal of Plant Sciences 156: 542–546.
39. Gianoli E, Valladares F (2012) Studying phenotypic plasticity: the advantages of abroad approach. Biological Journal of the Linnean Society 105: 1–7.
40. Valladares F, Sanchez-Gomez D, Zavala MA (2006) Quantitative estimation of phenotypic plasticity: bridging the gap between the evolutionary concept and its ecological applications. Journal of Ecology 94: 1103–1116.
41. Gaston KJ, Spicer JI (2004) Biodiveristy. An introduction. Blackwell Publishing, Massachusetts.
42. Matesanz S, Gianoli E, Valladares F (2010) Global change and the evolution of phenotypic plasticity in plants. Annals of the New York Academy of Sciences 1206: 35–55.
43. Molina-Montenegro MA, Gallardo-Cerda J, Flores TSM, Atala C (2012) The trade-off between cold resistance and growth determined the *Nothofagus pumilio* treeline. Plant Ecology 213: 1333–142.
44. Speakman JR (2000) The cost of living: field metabolic rates of small mammals. Pages 178–294 *in* A. H. Fisher, and D. G. Raffaelli eds. Advances in ecological research. Academic Press, San Diego, California.
45. Rezende EL, Bozinovic F, Garland Jr T (2004) Climatic adaptation and the evolution of basal and maximum rates of metabolism in rodents. Evolution 58: 1361–1374.
46. Pichancourt JB, van Klinken RD (2012) Phenotypic plasticity influences the size, shape and dynamics of the geographic distribution of an invasive plant. PLoS-ONE 7: e32323. doi:10.1371/journal.pone.0032323.
47. Mou P, Jones RH, Tan Z, Bao Z, Chen (2012) Morphological and physiological plasticity of plant roots when nutrients are both spatially and temporally heterogeneous. Plant And Soil: doi: 10.1007/s11104-012-1336-y.
48. Brown JH, Feldmeth CR (1971) Evolution in constant and fluctuating environments: thermal tolerance of desert pupfish (*Cyprinodon*). Evolution 25: 390–398.
49. Mitchell KA, Sgro CM, Hoffmann AA (2011) Phenotypic plasticity in upper thermal limits is weakly related to *Drosophila* species distributions. Funcional Ecology 25: 661–670.
50. Overgaard J, Kristensen TN, Mitchell KA, Hoffmann AA (2011) Thermal tolerance in widespread and tropical *Drosophila* species: does phenotypic plasticity increase with latitude. American Naturalist 178: S80–S96.
51. Quiroz CL, Choler P, Bapist F, Molina-Montenegro MA, González-Teuber M, et al. (2009) Alpine dandelion originated in the native and introduced ranges differ in their responses to environmental constraints. Ecological Research 24: 175–183.
52. Richards CL, Bossdorf O, Muth NZ, Gurevitch J, Pigliucci M (2006) Jack of all trades, master of some? On the role of phenotypic in plant invasions. Ecology Letters 9: 981–993.
53. Jonas CS, Geber MA (1999) Variation among populations of *Clarkia unguiculata* (Onagraceae) along altitudinal and latitudinal gradients. American Journal of Botany 86: 333–343.
54. Neuffer B, Hurka H (1999) Colonization history and introduction dynamics of *Capsella bursa-pastoris* (Brasicaceae) in North America: isozymes and quantitative traits. Molecular Ecology 8: 1667–1681.
55. Maron JL, Vilà M, Bommarco R, Elmendorf S, Beardsley P (2004) Rapid evolution of an invasive plant. Ecological Monographs 74: 261–280.
56. Platenkamp GA (1990) Phenotypic plasticity and genetic differentiation in the demography of the grass *Anthoxanthum odoratum*. Journal of Ecology 78: 772–788.
57. Sexton JP, McKay JK, Sala A (2002) Plasticity and genetic diversity may allow saltcedar to invade cold climates in North America. Ecological Applications 12: 1652–1660.
58. Ellstrand NC, Schierenbeck KA (2000) Hybridization as a stimulus for the evolution of invasiveness in plants. Proceedings of the National Academy of Sciences 97: 7043–7050.
59. Darlington CD (1937) Recent advances in cytology. Sc. Edd. Churchill, London.
60. Van Baarlen P, van Dijk PJ, Hoekstra RF, de Jong HJ (2000) Meiotic recombination in sexual diploid and apomictic triploid dandelions (*Taraxacum officinale* L.). Genome 43: 827–835.
61. Van Dijk PJ, Tanja Bakx-Schotman JM (2004) Formation of unreduced megaspores (Diplospory) in apomictic dandelions (*Taraxacum officinale*, s.l) is controlled by a sex-specific dominant locus. Genetics 166: 483–492.
62. Novak SJ, Mack RN, Soltis DE (1991) Genetic variation in *Bromus tectorum* Poaceae: population differentiation in its North American range. American Journal of Botany 78: 1150–1161.
63. di Castri F (1989) History of biological invasions with special emphasis on the Old World. Pages 1–30 *in* J. R. Drake, H. A. Mooney, F. di Castri, R. Groves, F. J. Kruger, M. Rejmánek, and M. Williamson, eds. Biological invasions: a global perspective. John Wiley and Sons Press, Chichester.
64. Arroyo MTK, Marticorena C, Matthei O, Cavieres LA (2000) Plant invasions in Chile: present patterns and future predictions. Pages 385–421 *in* H. A. Mooney, and R. Hobbs, eds. Invasive species in a changing world. Island Press, Washington.

21

The Response of the Mediterranean Gorgonian *Eunicella singularis* to Thermal Stress Is Independent of Its Nutritional Regime

Leïla Ezzat[1,3]*, Pierre-Laurent Merle[2], Paola Furla[2], Alexandre Buttler[1,4,5], Christine Ferrier-Pagès[3]

1 School of Architecture, Civil and Environmental Engineering (ENAC), Ecole Polytechnique Fédérale de Lausanne (EPFL), Lausanne, Switzerland, **2** UMR 7138 UNS-UPMC-CNRS Equipe Symbiose Marine (SYMAR), Université Nice Sophia-Antipolis, Nice, France, **3** Ecophysiology Team, Centre Scientifique de Monaco, Monaco, Principality of Monaco, **4** Swiss Federal Institute for Forest, Snow and Landscape Research (WSL), Lausanne, Switzerland, **5** UMR CNRS 6249 Laboratoire de Chrono-Environnement, Université de Franche-Comté, Besançon, France

Abstract

Over the last few decades, sessile benthic organisms from the Mediterranean Sea have suffered from the global warming of the world's oceans, and several mass mortality events were observed during warm summers. It has been hypothesized that mortality could have been due to a nutrient (food) shortage following the stratification of the water column. However, the symbiotic gorgonian *Eunicella singularis* has also presented a locally exceptional mortality, despite its autotrophic capacities through the photosynthesis of its dinoflagellate symbionts. Thus, this study has experimentally investigated the response of *E. singularis* to a thermal stress (temperature increase from 18 to 26°C), with colonies maintained more than 2 months under four nutritional diets: autotrophy only (AO), autotrophy and inorganic nitrogen addition (AN), autotrophy and heterotrophy (AH), heterotrophy only (HO). At 18°C, and contrary to many other anthozoans, supplementation of autotrophy with either inorganic nitrogen or food (heterotrophy) had no effect on the rates of respiration, photosynthesis, as well as in the chlorophyll, lipid and protein content. In the dark, heterotrophy maintained the gorgonian's metabolism, except a bleaching (loss of pigments), which did not affect the rates of photosynthesis. At 24°C, rates of respiration, and photosynthesis significantly decreased in all treatments. At 26°C, in addition to a decrease in the lipid content of all treatments, a bleaching was observed after 1 week in the AO treatment, while the AH and AN treatments resisted three weeks before bleaching. These last results suggest that, temperatures above 24°C impair the energetic reserves of this species and might explain the mortality events in the Mediterranean.

Editor: Christian R. Voolstra, King Abdullah University of Science and Technology, Saudi Arabia

Funding: This study was funded by the Government of the Principality of Monaco through the Centre Scientifique de Monaco. The funders had no role in study design, data collection and analysis, decision to publish, or preparation of the manuscript.

Competing Interests: The authors have declared that no competing interests exist.

* E-mail: Leila.Ezzat@gmail.com

Introduction

Gorgonians are among the most emblematic and representative organisms of the Mediterranean sublittoral communities [1]. These ecosystem engineer species play important ecological roles, not only in the plankton-benthos coupling, but also because they provide shade and shelter to numerous other species, and therefore largely contribute to the biomass and diversity of the benthic community [2]. Thus, any environmental perturbation inducing significant changes in their abundance could affect the proper functioning and organization of the Mediterranean benthic ecosystem.

Over the last few decades, gorgonians and other sessile organisms have suffered from the rapid seawater warming observed throughout the world's oceans, and showed increased events of mass mortalities and/or diseases [3–8], as often documented for tropical corals [9–11]. Thereby, after particularly warm summers, when seawater temperatures increased to and above 24°C during several weeks, the four Mediterranean gorgonian species (*Paramuricea clavata*, *Eunicella singularis*, *Eunicella cavolinii* and *Corallium rubrum*), were impacted over kilometers in the North-Occidental Mediterranean, [3,4,6,12], attracting the attention of the scientific community. Several studies have therefore monitored *in situ* the growth, health and reproductive capacities of the gorgonian populations during and after a mass mortality event, to assess its impact on the structure and dynamics of the benthic community [6,12–14]. Fewer studies have considered the physiological response of gorgonians to a thermal stress [15–17]. The latter studies showed that deep populations of *E. singularis* were more resistant to a seawater temperature increase to 24–26°C than shallow populations, and even more resistant than other species. Another study [6] linked the mortality events to a particularly strong summer stratification of the water column, and a possible reduction in food resources.

The aim of the present work was therefore to improve our knowledge on factors affecting the thermal sensitivity of *E. singularis* (Esper, 1791), one of the most impacted species [4,12], which remained affected several years after the stress [18]. As suggested by Coma et al. [19], one important factor is the availability of food resources to sustain gorgonian metabolism during a thermal stress. In terms of energetic budget, *E. singularis* is an interesting model species because, as many other tropical scleratinian corals, it has a dual feeding mode, both through auto- and heterotrophy. Indeed, it is the only Mediterranean gorgonians to live in symbiosis with a dinoflagellate of the genus *Symbiodinium*, also commonly named zooxanthellae. The symbionts of tropical and temperate scleractinians are known to transfer most of the photosynthesized carbon (photosynthetates) to their host, which has therefore access to this autotrophic nutrition [20,21]. Besides, like any animal *E. singularis* feeds on dissolved and particulate organic matter (heterotrophy), composed mainly of algae and zooplankton [22–24]. In many tropical anthozoans, heterotrophy increases the general metabolism and can sustain the whole metabolism during stress events [25].

During thermal stress, *E. singularis* may lose both its autotrophic and heterotrophic feeding capacities. Indeed, under stressful conditions, heterotrophy is generally affected by polyp retraction, which decreases prey capture [26], and by the water column stratification, which prevents the upwelling of nutrients and the subsequent development of phyto-and zooplankton. In absence of heterotrophy, most nutrients have to be supplied by symbiont photosynthesis, which is however itself affected during thermal stress [26]. Indeed, in many anthozoans, elevation of seawater temperature generally induces bleaching, characterized by the loss of symbionts and/or their associated pigments [27], with a concomitant reduction in the rate of photosynthesis and autotrophic inputs. Nevertheless, these processes are still poorly known and there is a need to disentangle both nutrition modes and their role in the corals fitness and resistance to environmental changes

In order to assess the effect of climate change on Mediterranean populations of *E. singularis*, a better understanding of the auto- and heterotrophic energy acquisition in normal and stressful conditions is required. For this purpose, we exposed several colonies to 4 trophic conditions across a range of temperatures that might be experienced *in situ* (from 18 to 26°C): autotrophy only, autotrophy supplied with inorganic nitrogen, autotrophy and heterotrophy, heterotrophy only (organisms kept in the dark). The aims of the study were to: 1) evaluate the effect of auto-and heterotrophy, in combination or alone, on the protein, chlorophyll and lipid content, as well as on the rate of photosynthesis of *E. singularis* under non-stressful conditions, and 2) determine the response of *E. singularis* to a thermal stress, when maintained under the different feeding conditions. The results obtained will allow us to gain a better knowledge of the trophic functioning of *E. singularis* under laboratory conditions, to draw inferences about what might be happening in the field. The general hypothesis tested was that nutritional mode and temperature affect the performance of *E. singularis* and that these effects can help explain the observed mass mortality and disease events in the field. Several predictions can be made: 1) under non stressful conditions, heterotrophy, in combination with autotrophy, increases the tissue reserves and eventually the rate of photosynthesis, as previously observed in many, but not all, scleractinian tropical species [25]; 2) in addition, heterotrophy only sustains the basic metabolism of *E. singularis* in the dark, and addition of inorganic nitrogen promotes photosynthesis, as for plants; 3) under stress conditions, the loss of autotrophic energy, following thermally-induced bleaching, is compensated by heterotrophy, as observed in some tropical coral species [28,29]; 4) Alternatively, supply of inorganic nitrogen could also maintain the rates of photosynthesis. Finally, gorgonians maintained in the dark should suffer less from the thermal stress, since no reactive oxygen species, generally produced during photosynthesis [30], would induce oxidative stress in these gorgonians.

Materials and Methods

Biological Material

Experiments were performed on twelve mother colonies (named A to L) of the symbiotic gorgonian *Eunicella singularis* (Esper, 1791) which were randomly sampled by SCUBA diving off Sète, North West Mediterranean Sea (43°19′13.25″N, 3°59′24.48″E) in late January 2012. These mother colonies were located in shallow waters (~10–15m). Samples were maintained in aerated cool water boxes and directly transferred to the laboratory. They were first acclimated at their in situ temperature of 16°C for a week, until they recovered from sampling. About 32 gorgonian tips (5 to 7 cm long) were then cut from each mother colonies (384 tips), labeled, and distributed in eight experimental tanks placed under controlled conditions at 18°C.

Experimental Setup

The experimental setup included eight 20l tanks supplied with (oligotrophic) Mediterranean seawater pumped from 50 m depth at a flow rate of 20 l h^{-1}. Tanks were grouped by two in order to have nubbins from six colonies per aquarium, or the twelve mother colonies represented in two tanks. Acclimation to laboratory conditions lasted for one week with a constant temperature of 18 ± 0.5°C. Following the acclimation week, four nutritional regimes (kept during the whole experiment) were established and maintained at a constant temperature of 18±0.5°C during two months. The four regimes (2 tanks per regime) were: autotrophy only (AO), autotrophy and ammonium enrichment (AN), heterotrophy only (HO) and heterotrophy + autotrophy (HA). AO consisted in exposing gorgonian tips to an irradiance of 125–150 µmol photons m^{-2} s^{-1}, on a 12h light: 12h dark regime, to match as closely as possible the light intensity received by the gorgonians *in situ*. For AN diet, gorgonians were maintained in the same light level as in AO, but also received during half an hour, a daily pulse of 3 µM NH_4, sampled in a mother solution of NH_4Cl (80 mM). HO diet consisted in maintaining gorgonian tips in total darkness but feeding them, five times a week, with an equal quantity of krill, grinded frozen shrimps and mussels given at repletion. They were also fed twice a week with *Artemia salina* nauplii. For HA, gorgonian tips were exposed to an irradiance of 125–150 µmol photons m^{-2} s^{-1} and fed as for the HO diet.

After two months at 18°C, several physiological measurements (as described below) were performed and temperature was then increased in order to imitate a thermal stress event as monitored in the Mediterranean Sea [31]. It was first raised from 18°C to 22°C and kept constant during 10 to 12 days, then again from 22°C to 24°C and kept constant during 10–12 days. After this period, temperature was finally raised from 24°C to 26°C and kept constant during 3 weeks. Indeed, Rodolfo-Metalpa et al.[32] showed that during warm summers, the temperature of the surface layer increased from 20°C mid-June to 24°C mid-July, and then remained at elevated temperatures (>24°C) until August. Moreover, this thermal increase is similar to the one performed in a previous study [15] and allows comparison of the results obtained. Constant seawater temperature was maintained using temperature controllers (Toshniwal N6100, Toshcon®, West Instruments, Brighton, East Sussex, UK; ±0.1°C) and submersible resistance heaters (Visi-Therm® Deluxe, Aquarium Systems, Sarrebourg, France). Salinity values were constant at 38 PSU. All tanks were cleaned two times per week in order to prevent algal growth. Samples were collected first at the end of the 2 months period at 18°C (samples called T18), then at the end of the first week at 24°C and 26°C (samples called T24 and T26_1) and finally at the end of the third week at 26°C (samples called T26_3). Sample tips were used to assess the photosynthetic performances of the gorgonians, and then frozen at −20°C for the determination of chlorophyll, protein and lipid concentrations. At all sampling times, additional tips were directly frozen at −20°C to allow the determination of both the symbiont density and the two cellular stress marker levels.

Measurements

Chlorophyll concentration. Prior to the determination of chlorophyll content of the samples used in the experiment, two different protocols of chlorophyll extraction were tested on independent gorgonian tips to obtain the best fit between chlorophyll determination and solvent toxicity [33,34]. Tremblay et al. [34] method was based on acetone solvent extraction with a 4°C overnight incubation time, while Ritchie [33] protocol compared chlorophyll extraction according to three different solvents: acetone, methanol and ethanol with a 4°C overnight incubation time. The methanol and ethanol based protocols gave similar results while the acetone protocol was significantly less efficient in extracting chlorophyll pigments (ANOVA, $p < 0.05$, data not shown). We therefore chose ethanol solvent from Ritchie [33] protocol for the experiment. For this purpose, six gorgonian tips (representing six different colonies) that were first used for photosynthesis measurements, were each introduced in a glass tube containing pure ethanol. Pigments were extracted at −20°C during 24h, and this step was repeated twice to extract all pigments. The extracts were then centrifuged at 11, 000g for 10 min, and chl *a* and *c2* were measured according to Jeffrey and Humphrey [35]. Data were normalized to the tip surface area of gorgonians, which was measured using a caliper, taking into account the length and the width of the tip.

Protein content. The protein concentration was assessed by incubating gorgonian tips (3 tips per tank, representing six different colonies) in a water bath at 90°C for 30 min with a 1N NaOH solution. Samples were then placed overnight at 4°C. Protein standards were prepared using bovine serum albumin (BSA, Interchim) across a concentration range from 0 to 2000 µg ml^{-1} and, as for the gorgonian samples, incubated at 60°C for 30 min in 96-well microplates with a dye reagent (Uptima Reagents, Interchim). Samples were then homogenized for 30s at 400 r.p.m on a microplate shaker. Finally, absorbance was measured at 560nm and protein contents were normalized per surface area, measured as described above.

Lipid biomass. The lipid concentration was assessed by incubating gorgonian tips (3 tips per tank, representing six different colonies) in a water bath at 40°C during 1 hour with a solution of MeOH according to the method of Bligh and Dwyer [36]. Then, an equal volume of $CHCL_2$ and H_2O was added in each sample to obtain a bi-phasic medium. After centrifugation at 2000 rpm during 10 min, the lower phase was sampled and a second rinsing was necessary to retrieve all lipids. This lipid fraction was transferred in pre-weighted tubes, and each tube was re-weighted after 24 hours evaporation. The lipid contents were normalized to the surface area of the gorgonian.

Photosynthesis and respiration rates. These measurements were performed on 3 tips per tank (from 3 different colonies), and therefore 6 colonies per nutritional regime. Changes in oxygen production were monitored during 30 minutes at 0 µmol photons m^{-2} s^{-1} (respiration, R) and at the culture irradiance of 125–150 µmol photons m^{-2} s^{-1} (net photosynthesis, Pn). Measurements were repeated at the end of each temperature threshold (T18, T24, T26_1 and T26_3), at the given temperature. Tips were incubated in small glass chambers, filled with a known volume of 0.45µm filtered seawater (FSW) continuously stirred with a stirring bar, and equipped with an Unisense optode (oxygen-sensitive minisensor) connected to a computer with the Oxy-4 software (Chanel fiber-optic oxygen meter, PreSens, Regensburg, Germany). Optodes were calibrated before each experiment against nitrogen-saturated and air-saturated seawater for the 0% and 100% oxygen, respectively. Light was provided by a metal halide lamp (Philips, HPIT 400W, Distrilamp, Bossee, Indre et Loire, France). Pn and R rates were estimated by regressing oxygen data against time. At the end of the incubations, gorgonian tips were frozen for the subsequent determination of their surface area (cm^2) and chlorophyll concentration. These two parameters were used to normalize Pn, R and gross photosynthesis (Pg = Pn + R) measurements. Oxygen fluxes were converted to carbon equivalents based on molar weights according to [37]: $Pc = Pg \times 12/PQ$ and $Rc = R \times (12 \times RQ)$ where P_C is the amount of carbon acquired through photosynthesis; 12 is the molecular mass of C; PQ is the photosynthetic quotient, equal to 1.1 mol O_2: mol C [20]; R_C is the amount of carbon consumed by respiration and RQ is the respiratory quotient, equal to 0.8 mol C:mol O_2 [20] *Pc* and *Rc* rates were used to calculate P:R = (Pc ×12)/(Rc ×24), considering that photosynthesis was efficient during the 12 h light period while respiration ran for 24 h.When this ratio equals to or is above 1 on a 24 h basis, this shows the autotrophic capability of an organism to self-maintenance [38].

Chlorophyll *a* fluorescence of photosystem II (PS II). Measurements were performed on 3 gorgonian tips per tank, for a total of 6 different colonies per nutritional regime. A Pulse Amplitude Modulation (PAM) fluorometer [DIVING-PAM, Walz, Germany, [39]] was used to assess the maximal photosynthetic efficiency of photosystems II. The minimal (F_0) and maximal (F_m) fluorescence yields were measured, after putting the gorgonians 5 min in the dark, by applying a weak pulsed red light (max. intensity <1 mol photon m^{-2} s^{-1}, width 3µs, frequency 0.6kHz) and a saturating pulse of actinic light (max. intensity 5000 µmol photon m^{-2} s^{-1}, width 800ms) on gorgonian tips by mean of an optical fiber placed at a fixed distance from tips surface area. The following equation allows calculating the maximum photosynthetic efficiency: $F_v/F_m = (F_m - F_0)/F_m$, where F_v is the variable fluorescence. Rapid light curve (RLC) function of the PAM was also used to estimate at the end of each temperature threshold (T18, T24, T26_1, T26_3), the effective quantum yield ($\Delta F/F_{m'}$), after exposure for 10s to 9 different light intensities (from 0 to 2983 µmol photons m^{-2} s^{-1}). The last light level was followed by the 10min dark relaxation period where $\Delta F/F_{m'}$ was calculated after 30s, 1, 3, 5 and 10 mn.

Symbiont density and cellular stress markers. At all sampling times, 3 additional gorgonian tips per tank, for a total of 6 different colonies per nutritional regime, were directly frozen at −20°C, without any other manipulation, to avoid stress-on stress bias. Subsequently, each of these frozen fragments was weighed, measured in size, grinded in liquid nitrogen and powdered in a mortar. The symbiont density was first measured (using a modified Neubauer hemocytometer) and cytosoluble protein extracted by tissue homogenization using short sonication cycles following the protocols previously described [14]. On crude extracts, we then measured: the total oxyradical scavenging capacity (TOSC, using a spectro-fluorometer, SAFAS, Monaco) and the degree of protein ubiquitination (using dot blots, with a rabbit anti-ubiquitin antibody, DakoCytomation). TOSC and Ubiquitination results were normalized to the amount of extracted cytoplasmic proteins (BioRad Assay Kit, using Bovine Serum Albumin as standard).

Statistical Data Analysis

Statistical analyses were performed using Statistica 11 (Statsoft). Data were collected and tested at the end of each temperature threshold. All data were expressed as means ± standard error. Normality and homoscedasticity of the data residuals were tested using Kolmogorov-Smirnov (using Lilliefors corrections) and Levene tests, and data were log-transformed when required. A general linear model for parametric repeated measure analysis of variance (ANOVA) was performed on all response variables using the different temperature steps as dependent variables and nutritional diet and colony as categorical predictors. The assumption of sphericity (independency of the repeated measures) was tested. When not fulfilled, the hypotheses were tested using the multivariate approach (Wilks test) for repeated measurements. When there were significant differences, the analyses were followed by *a posteriori* testing (Tukey's test). P-values were considered for $p<\alpha$, $\alpha=0.05$.

Ethics Statement

Twelve mother colonies of *Eunicella singularis* (Cnidarian) were sampled off Sète, North West Mediterranean Sea (43°19′13.25″N, 3°59′24.48″E) in late January 2012 at 15m depth under collection permit of Sète Marine Station (OF4500055960), of the University of Montpellier II. As scientific organization, represented by Dr. François Bonhomme, Sète Marine Station is empowered to conduct sampling in nature.

Results

Chlorophyll, Protein and Lipid Extractions

There was a significant interaction between the nutritional diet and the temperature on the gorgonian chlorophyll content (Figure 1, Table 1). Indeed, HO samples showed a significant decrease of this variable as soon as temperature reached 24°C (Tukey HSD, $p<0.02$), and contained almost no chlorophyll in their tissue after three weeks at 26°C (Figure 1). Conversely, chlorophyll content was significantly decreased only after a week at 26°C for AO samples (Tukey HSD, $p=0.04$) and after three weeks at 26°C (T26_3) for HA and AN samples (Tukey HSD, $p<0.002$). The different regimes and the heat stress duration exerted similar effects on symbiont density, than those observed on the chlorophyll content (data not shown). Therefore, the amount of total chlorophyll content per symbiont remained stable at 4.7 ± 0.8 pg $cell^{-1}$ for all regimes and stress times (except for HO-T26_3, for which the calculation was not possible due to the lack of accurate algal density determination).

The nutritional diet induced no significant difference in protein contents (0.8–1.2 mg cm^{-2}) (Figure 2, Table 1). Indeed, heterotrophy allowed HO gorgonians to maintain their protein concentration at the same level than gorgonians exposed to light for the two first temperature thresholds. High temperatures had a significant effect on the protein concentration, which increased in treatments at T26_3 (Figure 2, Tukey HSD, $p < 0.001$) with a greater tendency for the heterotrophic diets (Tukey HSD, $p > 0.05$).

Regarding the lipid extraction, half of the 18°C samples (AO and AH) have been unfortunately lost (due to freezing problems). However, the two remaining samples were not significantly different from T24 (ANOVA, $p>0.05$, data not shown). We will therefore present only the results of the three following temperature steps (T24, T26_1 and T26_3). In the present case, only thermal stress had a significant effect (Table 2) by decreasing the lipid contents of all gorgonians between T24 and T26_1 (Figure 3, Tukey HSD, $p<0.04$).

Photosynthesis and Respiration Rates

In the following results, data are expressed per surface area of gorgonian colony, when not specified, but normalization per chlorophyll content is also discussed. Normalization per surface area allows comparison between treatments considering the total amount of carbon produced and respired by the gorgonian colony, while the normalization per chlorophyll content is rather an index of chlorophyll efficiency.

There was a significant interaction between the nutritional diet and the temperature for the response of R and Pn (Table 2) normalized to the surface area, but the interaction was mainly due to the HO treatment at 26°C that behaved differently than the others. Indeed, after two months at the control temperature of 18°C, no significant differences appeared between the four nutritional regimes for net photosynthesis, Pn, (Figure 4A, Tukey HSD, $p>0.05$) respiration rates, R (Figure 4B, Tukey HSD, $p>0.05$) and gross photosynthesis, Pg (Figure 4C, Tukey HSD, $p>0.05$). Gorgonians produced per surface area as much O_2 as they consumed (Pn: 0.45–0.65 μmol O_2 h^{-1} cm^{-2} and R: 0.5–0.8 μmol O_2 h^{-1} cm^{-2}). When normalized to the chlorophyll content, all treatments maintained in the light (AO, AN, AH) also presented equivalent rates of Pn (0.012 ± 0.002 μmol O_2 (μg chl a) $^{-1}$ h^{-1}), and R (-0.015 ± 0.005 μmol O_2 (μg chl a) $^{-1}$ h^{-1}). However, HO gorgonians had higher Pn (0.022 ± 0.003 μmol O_2 (μg chl a) $^{-1}$ h^{-1}), and R (-0.034 ± 0.005 μmol O_2 (μg chl a) $^{-1}$ h^{-1}), because their chlorophyll content was significantly lower. Finally, all treatments maintained at this control temperature showed a P:R ratio above 1, with a higher value for the AO treatment and a lower for the HO (Table 3).

After a week exposure at 24°C, Pn, R and Pg, either normalized to the surface area (Figure 4A, B, C) or to chlorophyll content, significantly and drastically decreased for all regimes compared to the control temperature of 18°C (Tukey HSD, $p<0.0001$), with no difference in R, Pn and Pg between nutritional treatments (Tukey HSD, $p>0.05$). Rates normalized to the chlorophyll content were equal to: -0.006 ± 0.005 μmol O_2 (μg chl a)$^{-1}$ h^{-1}) for Pn, and -0.009 ± 0.005 μmol O_2 (μg chl a) $^{-1}$ h^{-1}) for R. P:R ratios decreased to below 1 in all treatments, but more drastically in the gorgonians maintained under heterotrophy (HO and HA) than under autotrophy (AO, AN) (Table 3).

After one to three weeks at 26°C, Pn, R and thus Pg normalized to the surface area remained comparable to the measurements performed at 24°C (Tukey HSD, $p>0.05$ respectively for T26_1 and T26_3). However, Pn of gorgonians maintained in the dark (HO) were significantly lower than measurements performed at T24 (Tukey HSD, $p < 0.001$). After a week at 26°C, P:R ratios of AN and AO regimes remained above or equal to 1, allowing gorgonian nubbins to still meet their metabolic needs (Table 3). However, after three weeks at 26°C, all P:R ratio decreased to below 1, indicating that autotrophic contribution no longer allowed to meet metabolic needs (Table 3).

Photosynthetic Efficiency of Photosystem II (PSII)

Overall, temperature decreased the F_v/F_m in all treatments, from 0.65 ± 0.02 at T18 and T24, to 0.4 ± 0.15 at T26_3 (Figure 5A, Tukey HSD, $p<0.05$). Also, at each temperature step, maintaining the gorgonians in the dark (HO treatment) significantly lowered the F'_v/F'_m compared to the other treatments (Tukey HSD, $p<0.001$). This lower effective photosynthetic efficiency suggests the loss of some photoprotective pigments. This is explained by a higher increase in minimal fluorescence (F_0) for the HO treatment, compared to the others, suggesting a faster increase in inactive PSII reaction centers. There was also a significant interaction of the nutritional diet and the temperature

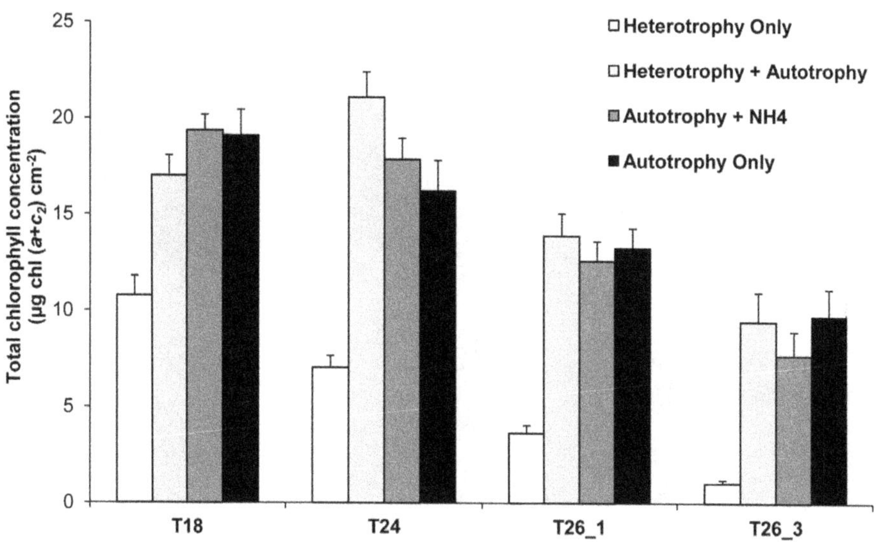

Figure 1. Effect of nutritional diets and thermal stress on total chlorophyll concentration of *E.singularis*. Total chlorophyll concentration (µg chl (*a+c2*) cm^{-2}) for the different nutritional treatments according to seawater temperature (18°C, 24°C, and 26°C after 1 and 3 weeks).

(Table 2) on both (F_v/F_m) and (F'_v/F'_m) responses. This interaction was mainly due to a significantly lower F'_v/F'_m at temperatures ≥ 24°C and lower F_v/F_m at T26_3 for gorgonians maintained in the dark than for those maintained in the light (Figure 5A, B, Tukey HSD, $p < 0.001$). These gorgonians did not recover their initial yield, even after more than 10 minutes recovery in the dark (Figure 6). Moreover, for gorgonians maintained in the light, HA treatment had a significantly lower F_v/F_m than AO treatment at T26_3 (Tukey HSD, $p = 0.036$) and AN treatment also showed a significant decrease in the F'_v/F'_m for temperatures ≥ 26°C (Tukey HSD, $p < 0.02$). HA and AO treatments maintained a similar F'_v/F'_m value between T18 until T26_3 (Figure 5B, Tukey HSD, $p < 0.01$).

Table 1. Results of the statistical analyses on the effects of the nutritional diet and temperature factors on the tissue parameters of gorgonians maintained in the different nutritional treatments according to seawater temperatures (18°C, 24°C, 26°C after 1 and 3 weeks).

	Degrees of freedom	P-value	F-Value
Chlorophyll content (µg chl (a+c2) cm^{-2})			
Temperature	3	**P<0.00001**	119.3119
Diet	3	**P<0.00001**	94.6450
Colony	5	**0.002**	3.5740
Temperature * Diet	9	**P<0.00001**	7.5778
Protein content (mg cm^{-2})			
Temperature	3	**P<0.0001**	8.8645
Diet	3	0.2315	1.5982
Colony	5	0.3408	1.2358
Temperature * Diet	9	0.2967	1.2380
Lipid content (mg cm^{-2})			
Temperature	2	**P<0.0001**	13.6454
Diet	3	0.1870	1.8185
Colony	5	0.5744	0.7877
Temperature * Diet	6	0.1803	1.6047
Ubiquitine content (arbitrary units)			
Temperature	3	**P<0.00001**	20.8791
Diet	3	0.7465	0.4124
Colony	5	**0.0051**	5.3431
Temperature * Diet	9	0.7720	4.0446

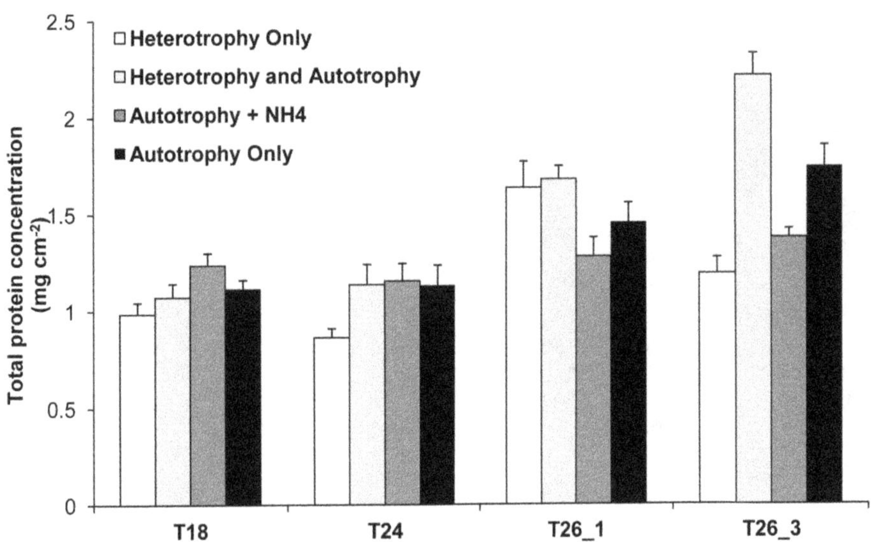

Figure 2. Effect of nutritional diets and thermal stress on total protein concentration of *E.singularis*. Total protein concentration (mg cm^{-2}) for the different nutritional treatments according to seawater temperature (18°C, 24°C, and 26°C after 1 and 3 weeks).

Stress Markers

No significant effect of the nutritional regimes and the heat stress duration was detected on the TOSC levels (ANOVA, $p>0.05$, data not shown), mainly due to high inter-individual variations (see the colony effects in Table 1 and 2). Measurements of ubiquitin conjugates showed a transient increase of this parameter at T26_1 (Figure 7, Table 1, Tukey HSD, $p<0.0001$), which returned to basal levels at the last sampling point (T26_3) (Tukey HSD, $p<0.001$).

Discussion

Nutritional Effects on *Eunicella singularis* Physiology

This paper presents for the first time the effect of autotrophy and/or heterotrophy on the metabolism and photosynthetic capacities of *E. singularis*, one of the most threatened gorgonians of the Mediterranean benthic ecosystems. This effect was evaluated on healthy and thermally-stress gorgonians. Results showed that *E. singularis* had a certain nutritional plasticity under non-stressful culture conditions. This species was indeed able to use either heterotrophy, or autotrophy only, to sustain its basic metabolism. The combination of the two nutritional regimes did not offer an additional advantage compared to the other conditions. The photosynthetic capacities of this gorgonian however rapidly collapsed as soon as temperature reached 24°C, independently of the trophic regime, and the supply of external nutrients was not used to sustain the rates of photosynthesis.

Under non-stressful conditions (18°), and in organisms maintained in the light, there was no significant difference in chlorophyll and protein contents, respiration, photosynthesis and photosynthetic efficiency, whether gorgonians were maintained in pure autotrophy, with a supply of inorganic nitrogen or with organic feeding. The lack of feeding effect on gorgonian metabolism is different from the effect generally observed on

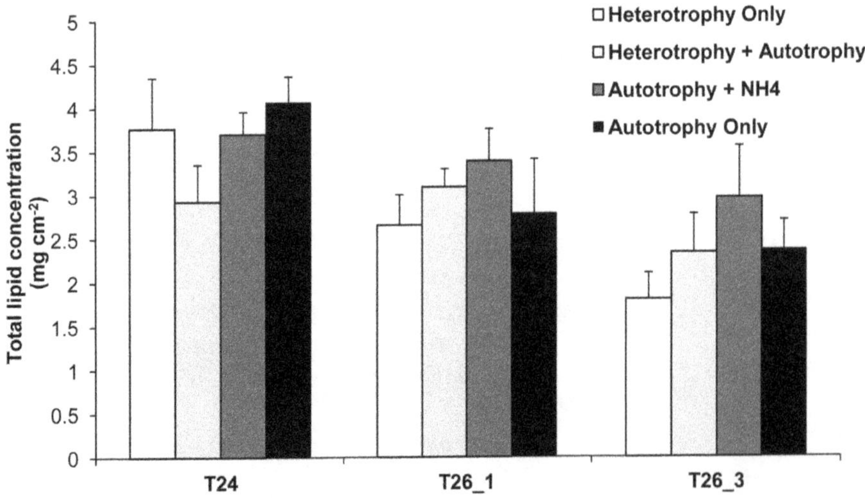

Figure 3. Effect of nutritional diets and thermal stress on total lipid concentration of *E.singularis*. Total lipid concentration (mg cm^{-2}) for the different nutritional treatments according to seawater temperature (24°C, and 26°C after 1 and 3 weeks).

Table 2. Results of the statistical analyses on the effects of the nutritional diet and temperature factors on the photosynthetic parameters of gorgonians maintained in the different nutritional treatments according to seawater temperatures (18°C, 24°C, 26°C after 1 and 3 weeks).

	Degrees of freedom	P-Value	F-Value
Maximum quantum yield (Fv/Fm)			
Temperature	3	**P<0.00001**	72.1850
Diet	3	**P<0.00001**	317.8420
Colony	5	**0.0245**	5.1110
Temperature * Diet	9	**P<0.0001**	6.1240
Effective quantum yield (Fv'/Fm')			
Temperature	3	**P<0.00001**	81.8900
Diet	3	**0.0157**	6.0000
Colony	5	0.8952	0.2000
Temperature * Diet	9	**P<0.00001**	8.8200
Respiration (μmol O2 h^{-1} cm^{-2})			
Temperature	3	**P<0.00001**	131.3409
Diet	3	0.9687	0.0821
Colony	5	0.8652	0.3640
Temperature * Diet	9	**0.0085**	2.9039
Net photosynthesis (μmol O2 h^{-1} cm^{-2})			
Temperature	3	**P<0.00001**	90.9526
Diet	3	**P<0.00001**	22.0642
Colony	5	0.2627	1.4524
Temperature * Diet	9	**0.0138**	2.8535
Gross photosynthesis (μmol O2 h^{-1} cm^{-2})			
Temperature	3	**P<0.00001**	96.3068
Diet	3	0.0932	2.6025
Colony	5	0.4530	0.9999
Temperature * Diet	9	0.9058	0.5283

other symbiotic anthozoan species, such as on tropical [25] and temperate scleractinian corals [21,40–42], for which feeding increases protein and chlorophyll concentrations, as well as symbiont densities, calcification and photosynthesis rates. It has however to be noticed that the effect of feeding on other temperate anthozoans is temperature and/or light dependent. Indeed, in the scleractinian coral *Cladocora caespitosa*, an enhancement of symbiont density by feeding was only observed under low temperatures (10–12°C, [42]) and increase in tissue biomass was significant only under high light intensity (250 μmol photons m^{-2} s^{-1}, [21]). In another temperate coral species, *Plesiastrea versipora*, the only significant effect of feeding also occurred under low temperature conditions [43]. In view of these observations, the effect of feeding on the metabolism of temperate species seems to be strongly correlated with or dependent on light and temperature. This might explain why, in our experimental conditions, where light and temperature were kept at a "medium" level, feeding had no significant effect on *E. singularis* metabolism. It can also be argued that *E. singularis* did not feed on the type of prey and dissolved material given as heterotrophic diet in this experiment. However two observations proved the contrary. Gorgonians maintained in the dark for more than two months did not decrease their respiration rates and tissue biomass compared to the other treatments, indicating that, in this condition, food served as a metabolic fuel to sustain metabolism. In addition, at 24°C, fed treatments (HO and HA) presented a P:R ratio less than 1 and lower compared to the unfed treatments. Thereby, gorgonians supplied with sufficient nutrients, decreased their production of symbiont photosynthates, suggesting a "down-regulation" phenomena with a shift from auto-to heterotrophy. Heterotrophy might therefore play an important role in gorgonians living in deep waters, where light levels are generally very low. Finally, *E. singularis* photosynthesis was not enhanced following an inorganic nitrogen enrichment. In tropical corals, such enrichment induced an increase in symbiont density [44,45], leading in some occasions to an enhancement in the rate of photosynthesis [46], but not always [45]. This lack of effect at least suggests that *Eunicella singularis* is not nutrient limited under our experimental conditions. Additional experiments using C:N measurements should be carried out to get a better understanding of the effect of heterotrophy on this species.

At 18°C, only gorgonians maintained fed in the dark, showed some physiological differences compared to the other treatments. As already discussed above, heterotrophy alone was sufficient to sustain the basic metabolism (tissue biomass and ,respiration rates), as already observed with the Mediterranean scleractinian coral *Cladocora caespitosa,* whose metabolism in the dark was sustained through heterotrophy for 2 months [21]. A bleaching however occurred in the gorgonians during this experiment, and has to be related to the occurrence, in situ, at 100 m depth, of aposymbiotic

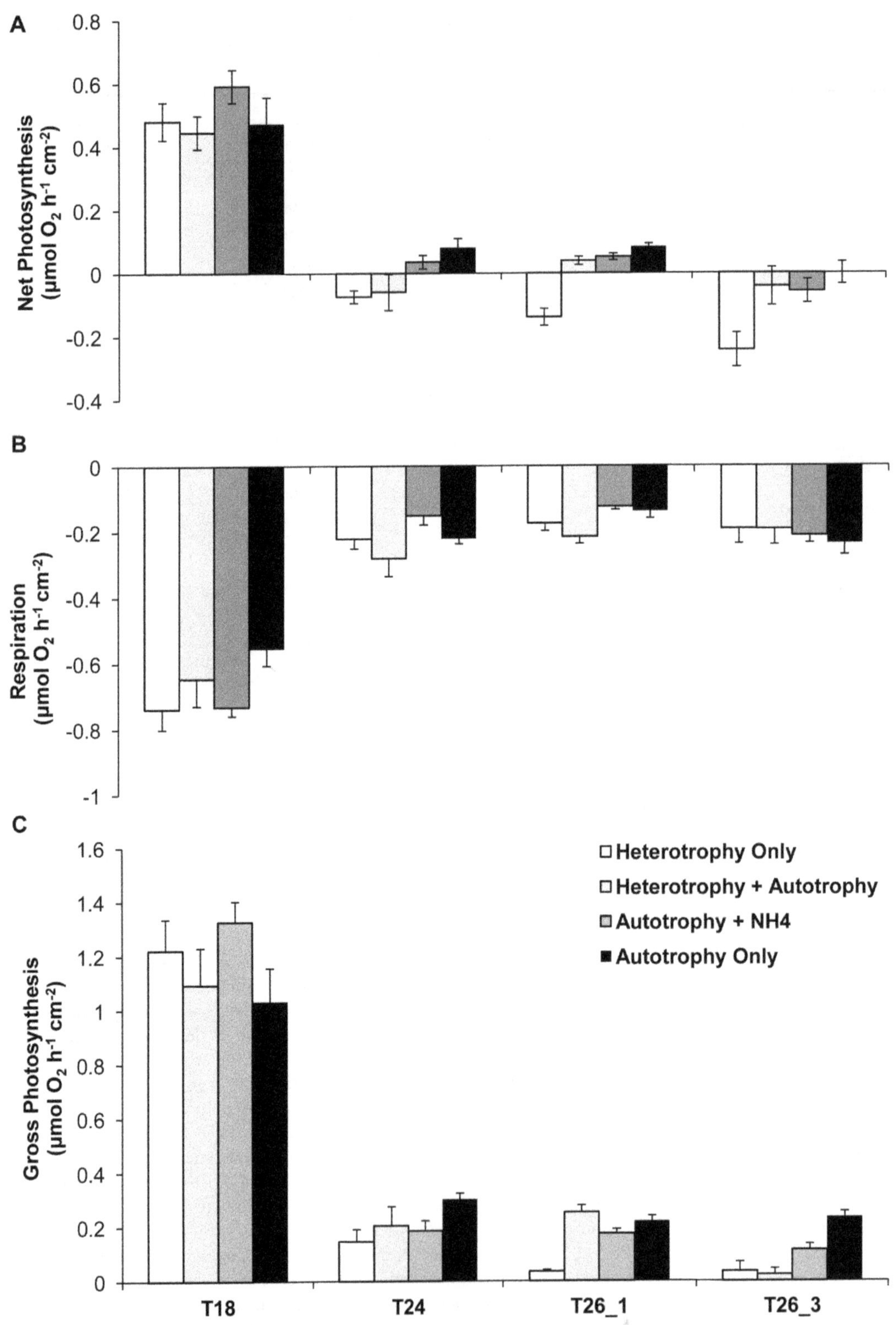

Figure 4. Effect of nutritional diets and thermal stress on photosynthesis and respirations rates of *E.singularis.* Net photosynthesis (Pn) (A), respiration (R) (B) and gross photosynthesis (Pg) (C) rates (µmol O_2 h^{-1} cm^{-2}) for the four nutritional treatments according to seawater temperature (18°C, 24°C, and 26°C after 1 and 3 weeks).

colonies of *E. singularis*. All together, these observations suggest that the loss of symbionts in the dark is a common process in this gorgonian species [47]. Indeed, such bleaching cannot be attributed to a degradation of the chlorophyll pigments in the dark, since the maximal photosynthetic efficiency (F_v/F_m), and the rates of photosynthesis were kept at the same level as in the light. Photosystems were rather inactivated than destructed. Symbiont loss seems thus to be due to elevated energetic costs for keeping a

Table 3. P:R (Photosynthesis-respiration ratio) of the gorgonians maintained in the different nutritional treatments according to seawater temperatures (18°C, 24°C, 26°C after 1 and 3 weeks).

Nutritional treatment	T18	T24	T26_1	T26_3
HO: Heterotrophy Only	1.13± 0.03	0.51± 0.04	0.19± 0.06	0.41± 0.20
HA: Heterotrophy + Autotrophy	1.18± 0.07	0.57± 0.11	0.82± 0.05	0.71± 0.07
AN: Autotrophy + NH_4	1.24± 0.03	0.95± 0.16	0.97± 0.06	0.38± 0.07
AO: Autotrophy Only	1.26± 0.07	0.96± 0.10	1.16± 0.09	0.80± 0.08

large symbiont population in the gorgonian tissue. This observation provides useful information about the ecology of this gorgonian species, which spends the winter months at very low levels of irradiance. It shows that temporary shading, although decreasing pigment and/or symbiont concentrations, does not have a serious impact on the autotrophic capacities of the gorgonian. Overall, symbionts of *E. singularis* behave like many phytoplankton species [48–50], but also like the symbionts of the Mediterranean scleractinian coral *Cladocora caespitosa* [21], which are also known to cope with a period of darkness without losing their photosynthetic capacities. For such Mediterranean symbiotic organisms, the maintenance of the photosynthetic apparatus is therefore vital for an efficient photosynthesis and an efficient input of autotrophic carbon, on return to favorable light levels.

Effects of Thermal Stress

This experiment has shown that as soon as the seawater temperature was increased to 24°C, many physiological processes in *E. singularis*, such as the rates of respiration and photosynthesis collapsed, independently of the trophic status of the gorgonians. A similar decrease in respiration and polyp activity above 20°C has been previously observed in *E. singularis* either in laboratory [17] or

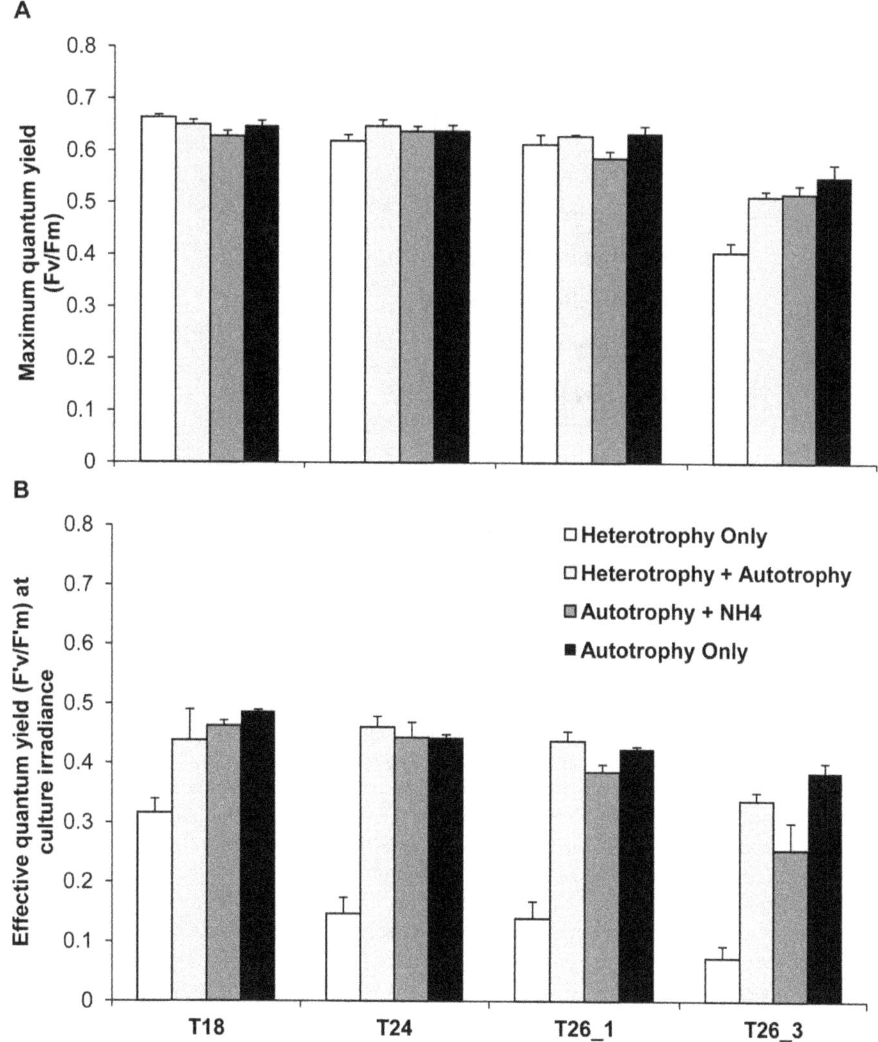

Figure 5. Effect of nutrional diets and thermal stress on maximum and effective quantum yield of *E.singularis*. Maximum quantum yield (F_v/F_m) (A), Effective quantum yield at culture irradiance (F'_v/F'_m) (B) for the different nutritional treatments according to sea seawater temperature (18°C, 24°C, and 26°C after 1 and 3 weeks).

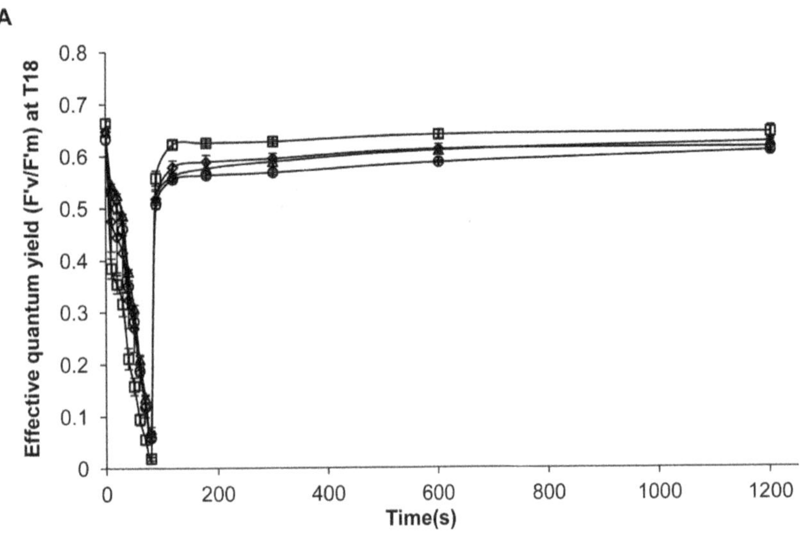

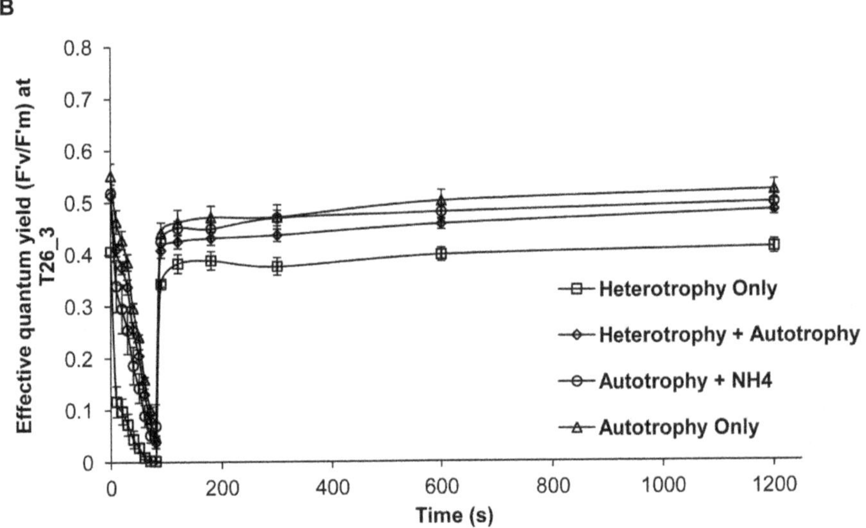

Figure 6. Effect of nutritional diets and thermal stress on effective quantum yield and recovery during the RLC (rapid light curve). Effective quantum yield (F'_v/F'_m) and recovery versus Time(s) at T18 (A) and T26_3 (B).

in situ, during the summer period when seawater temperature reached 22 to 24°C [51]. Several authors [51–53] related this decreased metabolic activity to the "summer dormancy", which corresponds to an adaptation to nutrient shortage conditions after the stratification of the upper water column. Gorgonians therefore tend to present a low polyp activity and a high consumption rate of internal lipid and protein reserves due to a lack of food in the surrounding environment [52]. However, results obtained in this experiment, in which half of the treatments were fed at repletion, suggest that the decrease in the general gorgonian metabolism is not only due to food shortage but rather to a real thermal stress. Indeed, both heterotrophically-fed and unfed gorgonians in this study presented the same decline in the rates of respiration, photosynthesis, and symbiont density. This lack of feeding effect on the physiology of *E.singularis* is therefore contrary to most of the previous observations made on other tropical and Mediterranean anthozoans, for which feeding generally enhances the metabolism and sustains it during thermal stress [25,26,29]. Another theory that has been put forward to explain the sensitivity of organisms to thermal stress is the "Oxygen- and Capacity-Limited Thermal Tolerance" [54]. It is how ambient oxygen levels shape and limit animal life. In the case of gorgonians, abnormal increase in seawater temperature may induce a stratification of the water column, reduce the amount of oxygen available, and thereby induce organisms' hypoxia. Although this process might occur in situ, it can not explain the decrease in metabolic activity observed in this study, as the aquaria were well-mixed and the seawater renewal rate was important.

The significant and drastic decrease in oxygen production, in all treatments at 24°C, suggests that the autotrophic capacity of the gorgonians was impacted at this temperature, and reached a critical state at 26°C. This photosynthetic shut-down was not due to a photoinhibition of the photosystems, or to a degradation of the D1 protein [55], since chlorophyll concentrations, and the maximal and effective quantum yields remained at a high level during the whole experiments, only decreasing at 26°C, as observed in a previous study [15]. Gorgonians also recovered a maximal yield after a dark recovery period, suggesting that the

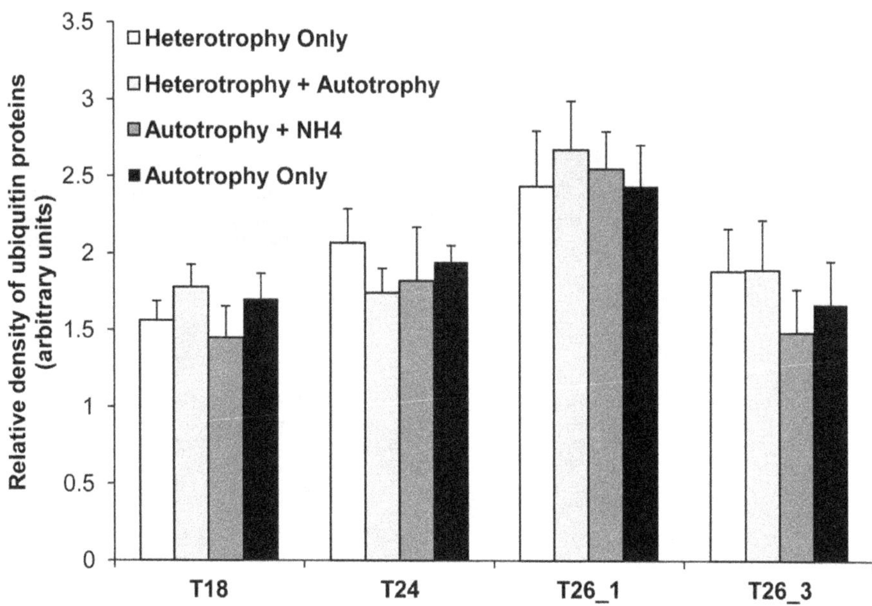

Figure 7. Effect of nutritional diets and thermal stress on ubiquitinylated proteins of *E.singularis.* Relative density of ubiquitin proteins (arbitrary units) for the different nutritional treatments according to seawater temperature (18°C, 24°C, and 26°C after 1 and 3 weeks).

symbionts were still photosynthetically competent. This response is also different than those observed in tropical corals, for which a decrease in oxygen production is often related to a loss in photosynthetic pigments and photosynthetic efficiency [56–60]. The uncoupling between chlorophyll fluorescence response and O_2 evolution has already been observed in phytoplankton, and might have several origins. Mitochondrial respiration can lead to an underestimation of the actual oxygen production [61]. Processes involved in chlororespiration, in the Mehler reaction or in plastoquinol oxidase (PTOX) can also explain this discrepancy [62]. Indeed, the PTOX pathway, for example, is used by prokaryotic and eukaryotic cells to generate ATP without producing reductant, leading to a dissipation of light energy with reduced O_2 production. The ATP generated is however used to support metabolic pathways, which can be the case here in the gorgonians exposed to a thermal stress. Further experiments are needed to draw the complete activation pathways involved in the heat-induced bleaching of this temperate symbiotic Cnidarian. However, since the amount of chlorophyll per cell did not change during the heat stress, the observed bleaching was associated with the degradation or the expulsion of the symbionts. Corroborating our previous observations with this same species, the level of ubiquitine conjugate remained stable at 24°C and only increased at the 26°C step, suggesting higher protein degradation processes at this stage [14]. Obviously, in the absence of proteasome inhibitors (in such circulating seawater systems), the amount of protein ubiquitine conjugates punctually reflects the very dynamic equilibriums between protein synthesis, stabilization and degradation. The fact that the ubiquitination levels returned to basal for a longer hyperthermia exposure (T26_3) is coherent with the fact that no tissue necrosis was observed throughout the experiment and is also reminiscent with the observations made in a previous experiment [15]. In this experiment [15], there was an effect of the thermal history on the gorgonian capacity to resist thermal stress. Indeed, calcification rates and photosynthetic efficiency were maintained at a higher level in gorgonian colonies collected in deep (35m) than in surface waters (ca. 15m). The main hypotheses were that either deep colonies had access to a larger quantity of zooplankton and particulate matter that supplied energy to resist to the stress, or that these deep colonies were exposed less frequently to high temperature levels and oxidative stress. It is obvious, from this study, that a lack of food cannot explain the lower resistance of surface populations to thermal stress, their sensitivity might therefore be due either to a higher occurrence of high temperature events, or to a combination of high temperature and high irradiance levels, inducing oxidative stress.

Conclusion

Under non-stressful conditions (18°C, 150 μmol photons m^{-2} s^{-1}), this study has shown that *E. singularis* was mainly autotroph, since heterotrophy, in our experimental conditions did not change any of the parameters tested. This is contrary to what was observed in many other scleractinian tropical species for which heterotrophy enhanced their metabolism, independently of the irradiance received. In the dark, however, feeding alone was able to maintain gorgonian's metabolism. This has important implications for understanding the ecology and physiology of these gorgonians *in situ*, since results suggest that autotrophy can supply most of the metabolic needs in summer, while heterotrophy can compensate for the lack of autotrophic input in winter. However, more studies, in which seasonal changes in light level are coupled with changes in seawater temperature, are needed to complement these first data. Also, the response of *E. singularis* to a thermal stress seems independent of the trophic conditions. Indeed, conversely to some other cnidarian species, heterotrophy could not prevent bleaching under thermal stress, nor compensate for the decrease in the rates of photosynthesis, which induced, on a long-term, the decrease in the lipid reserves. By decreasing the rates of respiration (i.e. the energetic expenses) *E. singularis* was however able to resist the stress for some weeks, without necrosis, as already observed in previous laboratory experiments [15–17]. This is contrary to other gorgonian species such as *P.clavata* and *C.rubrum* [17]. In the light of these results, the mortality events of *E. singularis* observed in nature [63], [4], after seawater temperatures had reached values above 24°C, are either due to a long-term impairment of the photosynthesis, or to a combination of several stresses, such as a combination of temperature increase with variation in hydrodynamic conditions, UV radiation increases or presence of pathogens or contaminants.

Acknowledgments

We thank S. Sikorski, C. Rottier and T. Zamoum for laboratory assistance, Rodolphe Schlaepfer (EPFL), P. Tremblay, J-O Irisson (Laboratoire d'Océanographie de Villefranche-sur-Mer), E.Béraud for statistical advices, and D. Allemand, Director of the Centre Scientifique de Monaco, for financial and scientific support. Many thanks to the undergraduate students: A. Degli-Albizi, S. Tahiri and L. Monticelli, who participated in soluble protein extraction and in algal density determination. Finally, we wish to thank two anonymous reviewers for providing constructive and helpful comments.

Author Contributions

Conceived and designed the experiments: LE CFP PF PLM. Performed the experiments: LE CFP PLM. Analyzed the data: LE AB. Contributed reagents/materials/analysis tools: CFP PF AB. Wrote the paper: LE CFP AB PLM.

References

1. Weinberg S, Weinberg F (1979) The life cycle of a Gorgonian: *Eunicella singularis* (Esper, 1794). Bijdr Dierk 48: 127–140. Bijdr Dierk 48: 127–140.
2. Wendt PH (1985) A comparative study of the invertebrate macrofauna associated with seven sponge and coral species collected from the South Atlantic Bight. J Elisha Mitchell Sci Soc 101: 187–203.
3. Cerrano C, Bavestrello G, Bianchi CN, Cattaneo-vietti R, Bava S, et al. (2000) A catastrophic mass-mortality episode of gorgonians and other organisms in the Ligurian Sea (North-western Mediterranean), summer 1999. Ecology Letters 3: 284–293.
4. Perez T, Garrabou J, Sartoretto S, Harmelin J-G, Francour P, et al. (2000) Mortalité massive d'invertébrés marins: un événement sans précédent en Méditerranée nord-occidentale. Comptes Rendus de l'Académie des Sciences - Series III - Sciences de la Vie 323: 853–865.
5. Garrabou J, Perez T, Sartoretto S, Harmelin J (2001) Mass mortality event in red coral Corallium rubrum populations in the Provence region (France, NW Mediterranean). 217.
6. Coma R, Linares C, Ribes M, Diaz D, Garrabou J, et al. (2006) Consequences of a mass mortality in populations of *Eunicella singularis* (Cnidaria: Octocorallia) in Menorca (NW Mediterranean). Marine Ecology Progress Series 327: 51–60.
7. Bally M, Garrabou J (2007) Thermodependent bacterial pathogens and mass mortalities in temperate benthic communities: a new case of emerging disease linked to climate change. Global Change Biology 13: 2078–2088.
8. Younes WAN, Bensoussan N, Romano J-C, Arlhac D, Lafont M-G (2003) Seasonal and interannual variations (1996–2000) of the coastal waters east of the Rhone river mouth as indicated by the SORCOM series. Oceanologica Acta 26: 311–321.
9. Harvell CD, Kim K, Burkholder JM, Colwell RR, Epstein PR, et al. (1999) Emerging Marine Diseases – Climate Links and Anthropogenic Factors. Science 285: 1505–1510.
10. Harvell D, Aronson R, Baron N, Connell J, Dobson A, et al. (2004) The rising tide of ocean diseases: unsolved problems and research priorities. Frontiers in Ecology and the Environment 2: 375–382.
11. Lesser MP (2007) Coral reef bleaching and global climate change: can corals survive the next century? Proceedings of the National Academy of Sciences of the United States of America 104: 5259–5260.
12. Garrabou J, Coma R, Bensoussan N, Bally M, ChevaldonnÉ P, et al. (2009) Mass mortality in Northwestern Mediterranean rocky benthic communities: effects of the 2003 heat wave. Global Change Biology 15: 1090–1103.
13. Linares C, Coma R, Diaz D, Zabala M, Hereu B, et al. (2005) Immediate and delayed effects of a mass mortality event on gorgonian population dynamics and benthic community structure in the NW Mediterranean Sea. Oldendorf, ALLEMAGNE: Inter-Research. 11 p.
14. Gori A, Linares C, Rossi S, Coma R, Gili J-M (2007) Spatial variability in reproductive cycle of the gorgonians *Paramuricea clavata* and *Eunicella singularis* (Anthozoa, Octocorallia) in the Western Mediterranean Sea. Marine Biology 151: 1571–1584.
15. Ferrier-Pagès C, Tambutté E, Zamoum T, Segonds N, Merle P-L, et al. (2009) Physiological response of the symbiotic gorgonian *Eunicella singularis* to a long-term temperature increase. Journal of Experimental Biology 212: 3007–3015.
16. Pey A, Zamoum T, Allemand D, Furla P, Merle P-L (2011) Depth-dependant thermotolerance of the symbiotic Mediterranean gorgonian *Eunicella singularis*: Evidence from cellular stress markers. Journal of Experimental Marine Biology and Ecology 404: 73–78.
17. Previati M, Scinto A, Cerrano C, Osinga R (2010) Oxygen consumption in Mediterranean octocorals under different temperatures. Journal of Experimental Marine Biology and Ecology 390: 39–48.
18. Linares C, Coma R, Garrabou J, Díaz D, Zabala M (2008) Size distribution, density and disturbance in two Mediterranean gorgonians: *Paramuricea clavata* and *Eunicella singularis*. Journal of Applied Ecology 45: 688–699.
19. Coma R, Ribes M (2003) Seasonal energetic constraints in Mediterranean benthic suspension feeders: effects at different levels of ecological organization. Oikos 101: 205–215.
20. Muscatine L, McCloskey LR, Marian RE (1981) Estimating the Daily Contribution of Carbon from Zooxanthellae to Coral Animal Respiration. Limnology and Oceanography 26: 601–611.
21. Hoogenboom M, Rodolfo-Metalpa R, Ferrier-Pagès C (2010) Co-variation between autotrophy and heterotrophy in the Mediterranean coral *Cladocora caespitosa*. The Journal of Experimental Biology 213: 2399–2409.
22. Coma R, Gili JM, Zabala M, Riera T (1994) Feeding and prey capture cycles in the aposymbiotic gorgonian *Paramuricea clavata*. 115: 257–270.
23. Wild C, Huettel M, Klueter A, Kremb SG, Rasheed MYM, et al. (2004) Coral mucus functions as an energy carrier and particle trap in the reef ecosystem. Nature 428: 66–70.
24. Cerrano C, Arillo A, Azzini F, Calcinai B, Castellano L, et al. (2005) Gorgonian population recovery after a mass mortality event. Aquatic Conservation: Marine and Freshwater Ecosystems 15: 147–157.
25. Houlbrèque F, Ferrier-Pagès C (2009) Heterotrophy in Tropical Scleractinian Corals. Biological Reviews 84: 1–17.
26. Ferrier-Pagès C, Rottier C, Beraud E, Levy O (2010) Experimental assessment of the feeding effort of three scleractinian coral species during a thermal stress: Effect on the rates of photosynthesis. Journal of Experimental Marine Biology and Ecology 390: 118–124.
27. Buddemeier RW, Fautin DG (1993) Coral bleaching as an adaptive mechanism. BioScience 43: 320–325.
28. Grottoli AG, Rodrigues LJ, Palardy JE (2006) Heterotrophic plasticity and resilience in bleached corals. Nature 440: 1186–1189.
29. Borell EM, Yuliantri AR, Bischof K, Richter C (2008) The effect of heterotrophy on photosynthesis and tissue composition of two scleractinian corals under elevated temperature. Journal of Experimental Marine Biology and Ecology 364: 116–123.
30. Suggett DJ, Warner ME, Smith DJ, Davey P, Hennige S, et al. (2008) Photosynthesis and production of hydrogen peroxide by symbiodinium (Pyrrhophyta) phylotypes with different thermal tolerances S1. Journal of Phycology 44: 948–956.
31. Rodolfo-Metalpa R, Richard C, Allemand D, Bianchi C, Morri C, et al. (2006) Response of zooxanthellae in symbiosis with the Mediterranean corals Cladocora caespitosa and Oculina patagonica to elevated temperatures. Marine Biology 150: 45–55.
32. Rodolfo-Metalpa R, Reynaud S, Allemand D, Ferrier-Pagès C (2008) Temporal and depth responses of two temperate corals, *Cladocora caespitosa* and *Oculina patagonica*, from the North Mediterranean Sea. Marine Ecology Progress Series 369: 103–114.
33. Ritchie R (2006) Consistent Sets of Spectrophotometric Chlorophyll Equations for Acetone, Methanol and Ethanol Solvents. Photosynthesis Research 89: 27–41.
34. Tremblay P, Grover R, Maguer JF, Legendre L, Ferrier-Pagès C (2012) Autotrophic carbon budget in coral tissue: a new 13C-based model of photosynthate translocation. The Journal of Experimental Biology 215: 1384–1393.
35. Jeffrey SW, Humphrey GF (1975) New spectrophotometric equations for determining chlorophylls a, b, c1, and c2 in higher plants, algae, and natural phytoplankton. Biochem Physiol Pflanz 167: 191–194.
36. Bligh EG, Dyer WJ (1959) A rapid method of total lipid extraction and purification. Canadian Journal of Biochemistry and Physiology 37: 911–917.
37. Anthony KRN, Fabricius KE (2000) Shifting roles of heterotrophy and autotrophy in coral energetics under varying turbidity. Journal of Experimental Marine Biology and Ecology 252: 221–253.
38. Odum HT, Odum EP (1955) Trophic Structure and Productivity of a Windward Coral Reef Community on Eniwetok Atoll. Ecological Monographs 25: 291–320.
39. Schreiber U, Schliwa U, Bilger W (1986) Continuous recording of photochemical and non-photochemical chlorophyll fluorescence quenching with a new type of modulation fluorometer. Photosynthesis Research 10: 51–62.
40. Miller M (1995) Growth of a temperate coral: effects of temperature, light, depth, and heterotrophy. Marine Ecology Progress Series 122: 217–225.
41. Rodolfo-Metalpa R, Huot Y, Ferrier-Pagès C (2008) Photosynthetic response of the Mediterranean zooxanthellate coral *Cladocora caespitosa* to the natural range of light and temperature. Journal of Experimental Biology 211: 1579–1586.
42. Rodolfo-Metalpa R, Peirano A, Houlbrèque F, Abbate M, Ferrier-Pagès C (2008) Effects of temperature, light and heterotrophy on the growth rate and budding of the temperate coral *Cladocora caespitosa*. Coral Reefs 27: 17–25.

43. Kevin KM, Hudson RCL (1979) The role of zooxanthellae in the hermatypic coral *Plesiastrea urvillei* (Milne Edwards and Haime) From cold waters. Journal of Experimental Marine Biology and Ecology 36: 157–170.
44. Marubini F, Davies PS (1996) Nitrate increases zooxanthellae population density and reduces skeletogenesis in corals. Marine Biology 127: 319–328.
45. Stambler N (1998) Effects of light intensity and ammonium enrichment on the hermatypic coral Stylophora pistillata and its zooxanthellae. Symbiosis 24: 127–146.
46. Titlyanov EA, Bil' K, Fomina I, Titlyanova T, Leletkin V, et al. (2000) Effects of dissolved ammonium addition and host feeding with *Artemia salina* on photoacclimation of the hermatypic coral *Stylophora pistillata*. Marine Biology 137: 463–472.
47. Gori A, Bramanti L, López-González P, Thoma J, Gili J-M, et al. (2012) Characterization of the zooxanthellate and azooxanthellate morphotypes of the Mediterranean gorgonian *Eunicella singularis*. Marine Biology 159: 1485–1496.
48. Deventer B, Heckman CW (1996) Effects of prolonged darkness on the relative pigment content of cultured diatoms and green algae. Aquatic Sciences – Research Across Boundaries 58: 241–252.
49. Peters E (1996) Prolonged darkness and diatom mortality: II. Marine temperate species. Journal of Experimental Marine Biology and Ecology 207: 43–58.
50. Peters E, Thomas DN (1996) Prolonged darkness and diatom mortality I: Marine Antarctic species. Journal of Experimental Marine Biology and Ecology 207: 25–41.
51. Coma R, Ribes M, Gili JM, Zabala M (2002) Seasonality of in situ respiration rate in three temperate benthic suspension feeders. Limnol Oceanogr 47: 324–331.
52. Rossi S, Ribes M, Coma R, Gili J-M (2004) Temporal variability in zooplankton prey capture rate of the passive suspension feeder *Leptogorgia sarmentosa* (Cnidaria: Octocorallia), a case study. Marine Biology 144: 89–99.
53. Rossi S, Gili J-M, Coma R, Linares C, Gori A, et al. (2006) Temporal variation in protein, carbohydratre, and lipid concentrations in *Paramuricea clavata* (Anthozoa, Octocorallia): evidence for summer-autumn feeding constraints. Marine Biology 149: 643–651.
54. Pörtner H-O (2010) Oxygen- and capacity-limitation of thermal tolerance: a matrix for integrating climate-related stressor effects in marine ecosystems. The Journal of Experimental Biology 213: 881–893.
55. Jones RJ, Hoegh-Guldberg O (2001) Diurnal changes in the photochemical efficiency of symbiotic dinoflagellates (Dinophyceae) of corals: photoprotection, photoinactivation and the relationship to coral bleaching. Plant, Cell and Environment 24: 89–99.
56. Brown BE, Suharsono S (1990) Damage and recovery of coral reefs affected by El Niño related seawater warming in the Thousand Islands, Indonesia. Coral Reefs 8: 163–170.
57. Coles SL, Fadlallah YH (1991) Reef coral survival and mortality at low temperatures in the Arabian Gulf: new species-specific lower temperature limits. Coral Reefs 9: 231–237.
58. Glynn PW (1993) Coral reef bleaching: ecological perspectives. Coral Reefs 12: 1–17.
59. Gates RD, Baghdasarian G, Muscatine L (1992) Temperature Stress Causes Host Cell Detachment in Symbiotic Cnidarians: Implications for Coral Bleaching. The Biological Bulletin 182: 324–332.
60. Williams E, Bunkley-Williams L (1990) The world wide coral reef bleaching cycle and related sources of coral mortality. Atoll Res Bull 335: 77.
61. Beardall J, Burger-Wiersma T, Rijkeboer M, Sukenik A, Lemoalle J, et al. (1994) Studies on enhanced post-illumination respiration in microalgae. Journal of Plankton Research 16: 1401–1410.
62. Zehr JP, Kudela RM (2009) Photosynthesis in the Open Ocean. Science 326: 945–946.
63. Weinberg S (1991) Faut-il protéger les gorgones de la Méditerranée ?; Posidonie G, editor. Marseille. 47–52 p.

PERMISSIONS

All chapters in this book were first published in PLOS ONE, by The Public Library of Science; hereby published with permission under the Creative Commons Attribution License or equivalent. Every chapter published in this book has been scrutinized by our experts. Their significance has been extensively debated. The topics covered herein carry significant findings which will fuel the growth of the discipline. They may even be implemented as practical applications or may be referred to as a beginning point for another development.

The contributors of this book come from diverse backgrounds, making this book a truly international effort. This book will bring forth new frontiers with its revolutionizing research information and detailed analysis of the nascent developments around the world.

We would like to thank all the contributing authors for lending their expertise to make the book truly unique. They have played a crucial role in the development of this book. Without their invaluable contributions this book wouldn't have been possible. They have made vital efforts to compile up to date information on the varied aspects of this subject to make this book a valuable addition to the collection of many professionals and students.

This book was conceptualized with the vision of imparting up-to-date information and advanced data in this field. To ensure the same, a matchless editorial board was set up. Every individual on the board went through rigorous rounds of assessment to prove their worth. After which they invested a large part of their time researching and compiling the most relevant data for our readers.

The editorial board has been involved in producing this book since its inception. They have spent rigorous hours researching and exploring the diverse topics which have resulted in the successful publishing of this book. They have passed on their knowledge of decades through this book. To expedite this challenging task, the publisher supported the team at every step. A small team of assistant editors was also appointed to further simplify the editing procedure and attain best results for the readers.

Apart from the editorial board, the designing team has also invested a significant amount of their time in understanding the subject and creating the most relevant covers. They scrutinized every image to scout for the most suitable representation of the subject and create an appropriate cover for the book.

The publishing team has been an ardent support to the editorial, designing and production team. Their endless efforts to recruit the best for this project, has resulted in the accomplishment of this book. They are a veteran in the field of academics and their pool of knowledge is as vast as their experience in printing. Their expertise and guidance has proved useful at every step. Their uncompromising quality standards have made this book an exceptional effort. Their encouragement from time to time has been an inspiration for everyone.

The publisher and the editorial board hope that this book will prove to be a valuable piece of knowledge for researchers, students, practitioners and scholars across the globe.

LIST OF CONTRIBUTORS

Chao Wang
Key Laboratory of Experimental Marine Biology, Institute of Oceanology, Chinese Academy of Sciences, Qingdao, China
Graduate University of the Chinese Academy of Sciences, Beijing, China

Xiaolei Fan
Qingdao Institute of Bioenergy and Bioprocess Technology, Chinese Academy of Sciences, Qingdao, China

Guangce Wang, Jianfeng Niu and Baicheng Zhou
Key Laboratory of Experimental Marine Biology, Institute of Oceanology, Chinese Academy of Sciences, Qingdao, China

Eleonora Sforza and Alberto Bertucco
Dipartimento di Ingegneria Industriale DII, Università di Padova, Padova, Italy

Diana Simionato, Giorgio Mario Giacometti and Tomas Morosinotto
Dipartimento di Biologia, Università di Padova, Padova, Italy

Shuxia Zheng, Haiyan Ren, Wenhuai Li and Zhichun Lan
State Key Laboratory of Vegetation and Environmental Change, Institute of Botany, Chinese Academy of Sciences, Beijing, China

Alexander Punnoose
Instituto de Física Teórica, Universidade Estadual Paulista, São Paulo, Brazil
Department of Physics, City College of the City University of New York, New York, United States of America

Liza McConnell, Wei Liu, Andrew C. Mutter and Ronald Koder
Instituto de Física Teórica, Universidade Estadual Paulista, São Paulo, Brazil
Department of Physics, City College of the City University of New York, New York, United States of America

Junmin Li
College of Life and Environmental Sciences, Hangzhou Normal University, Hangzhou, China
Institute of Ecology, Taizhou University, Linhai, China
State Key Laboratory of Vegetation and Environmental Change, Institute of Botany, Chinese Academy of Sciences, Beijing, China

Tao Xiao and Qiong Zhang
College of Life and Environmental Sciences, Hangzhou Normal University, Hangzhou, China

Ming Dong
College of Life and Environmental Sciences, Hangzhou Normal University, Hangzhou, China
State Key Laboratory of Vegetation and Environmental Change, Institute of Botany, Chinese Academy of Sciences, Beijing, China

Guangrong Hu, Yong Fan, Lei Zhang, Cheng Yuan and Fuli Li
Shandong Provincial Key Laboratory of Energy Genetics, Qingdao Institute of Bioenergy and Bioprocess Technology, Chinese Academy of Sciences, Qingdao, PR China

Jufang Wang and Wenjian Li
Institute of Modern Physics, Chinese Academy of Sciences, Lanzhou, PR China

Qiang Hu
Laboratory for Algae Research and Biotechnology (LARB), College of Technology and Innovation, Arizona State University, Mesa, Arizona, United States of America

Charles P. Deblois, Axelle Marchand and Philippe Juneau
Department of Biological Sciences, TOXEN, Ecotoxicology of Aquatic Microorganisms Laboratory, Université du Québec à Montréal, Montréal, Québec, Canada

Daniel Manzano
Instituto Carlos I de Fisica Teorica y Computacional, University of Granada, Granada, Spain,
Institute for Theoretical Physics, University of Innsbruck, Innsbruck, Austria

Maria Teresa Giardi, Giuseppina Rea, Maya D. Lambreva, Amina Antonacci and Sandro Pastorelli
Institute of Crystallography, National Research Council of Italy, CNR, Rome, Italy

Ivo Bertalan and Udo Johanningmeier
Institute of Plant Physiology, Martin-Luther University Halle-Wittenberg, Halle (Saale), Germany

Autar K. Mattoo
The Henry A. Wallace Beltsville Agricultural Research Center, United States Department of Agriculture, Agricultural Research Service, Sustainable Agricultural Systems Laboratory, Beltsville, Maryland, United States of America

Yong Li
College of Resources and Environmental Sciences, Nanjing Agricultural University, Nanjing, Jiangsu, China
National Key Laboratory of Crop Genetic Improvement, MOA Key Laboratory of Crop Ecophysiology and Farming System in the Middle Reaches of the Yangtze River, College of Plant Science and Technology, Huazhong Agricultural University, Wuhan, Hubei, China

Binbin Ren, Lei Ding, Qirong Shen and Shiwei Guo
College of Resources and Environmental Sciences, Nanjing Agricultural University, Nanjing, Jiangsu, China

Shaobing Peng
National Key Laboratory of Crop Genetic Improvement, MOA Key Laboratory of Crop Ecophysiology and Farming System in the Middle Reaches of the Yangtze River, College of Plant Science and Technology, Huazhong Agricultural University, Wuhan, Hubei, China

Virginia Matzek
Department of Environmental Studies and Sciences, Santa Clara University, Santa Clara, California, United States of America

Jacob A. Nelson and Bruce Bugbee
Crop Physiology Laboratory, Department of Plant Soils and Climate, Utah State University, Logan, Utah, United States of America

Wenhua You, Shufeng Fan, Dan Yu, Dong Xie and Chunhua Liu
The National Field Station of Lake Ecosystem of Liangzi Lake, College of Life Science, Wuhan University, Wuhan, P.R. China

Catherine J. Collier, Cecilia Villacorta-Rath and Miwa Takahashi
School of Marine and Tropical Biology, James Cook University, Townsville, Australia

Kor-jent van Dijk
School of Marine and Tropical Biology, James Cook University, Townsville, Australia
School of Earth and Environmental Science, Australian Centre for Evolutionary Biology and Biodiversity, University of Adelaide, Adelaide, Australia

Michelle Waycott
School of Earth and Environmental Science, Australian Centre for Evolutionary Biology and Biodiversity, University of Adelaide, Adelaide, Australia

Fong-Lee Ng and Siew-Moi Phang
Institute of Ocean and Earth Sciences, University of Malaya, Kuala Lumpur, Malaysia
Institute of Biological Sciences, Faculty of Science, University of Malaya, Kuala Lumpur, Malaysia

Vengadesh Periasamy
Low Dimensional Materials Research Centre, Department of Physics, Faculty of Science, University of Malaya, Kuala Lumpur, Malaysia

Kamran Yunus and Adrian C. Fisher
Centre of Research for Electrochemical, Science and Technology (CREST), Department of Chemical Engineering and Biotechnology, University of Cambridge, Cambridge, United Kingdom

Jennifer Firn
School of Earth, Environment and Biological Sciences, Queensland University of Technology, Brisbane, Queensland, Australia

Suzanne M. Prober
Ecosystem Sciences, CSIRO, Wembley, Western Australia, Australia

Yvonne M. Buckley
School of Biological Sciences, The University of Queensland, St. Lucia, Queensland, Australia
Ecosystem Sciences CSIRO, Dutton Park, Queensland, Australia

Antonietta Quigg
Department of Marine Biology, Texas A&M University at Galveston, Galveston, Texas, United States of America
Department of Oceanography, Texas A&M University, College Station, Texas, United States of America

Eva Kotabová, Jana Jarešová, Radek Kaňa and Ondřej Prášil
Institute of Microbiology, Academy of Sciences of the Czech Republic, Třeboň , Czech Republic
Faculty of Sciences, University of South Bohemia, ČeskéBudě jovice, Czech Republic

Jiří Šetlík, Barbora Šedivá and Ondřej Komárek
Institute of Microbiology, Academy of Sciences of the Czech Republic, Třeboň , Czech Republic

Xia Li, Anzhi Ren, Rong Han, Lijia Yin, Maoying Wei and Yubao Gao
College of Life Sciences, Nankai University, Tianjin, P. R. China

Ana Gavina, Glória Pinto, Maria Teresa Claro, Conceição Santos and Fernando Gonçalves
Departamento de Biologia, Universidade de Aveiro, Campus de Santiago, Aveiro, Portugal
CESAM - Centro de Estudos do Ambiente e do Mar, Universidade de Aveiro, Campus de Santiago, Aveiro, Portugal

Sara C. Antunes and Ruth Pereira
CESAM - Centro de Estudos do Ambiente e do Mar, Universidade de Aveiro, Campus de Santiago, Aveiro, Portugal
Departamento de Biologia da Faculdade de Ciências, Universidade do Porto, Rua do Campo Alegre, Porto, Portugal

Marco A. Molina-Montenegro
Centro de Estudios Avanzados en Zonas Áridas, Facultad de Ciencias del Mar, Universidad Católica del Norte, Coquimbo, Chile

Daniel E. Naya
Departamento de Ecología y Evolución, Facultad de Ciencias and Centro Universitario de la Regional Este, Universidad de la República, Montevideo, Uruguay

Leïla Ezzat
School of Architecture, Civil and Environmental Engineering (ENAC), Ecole Polytechnique Fédérale de Lausanne (EPFL), Lausanne, Switzerland
Ecophysiology Team, Centre Scientifique de Monaco, Monaco, Principality of Monaco

Pierre-Laurent Merle and Paola Furla
UMR 7138 UNS-UPMCCNRS Equipe Symbiose Marine (SYMAR), Université Nice Sophia-Antipolis, Nice, France

Alexandre Buttler
School of Architecture, Civil and Environmental Engineering (ENAC), Ecole Polytechnique Fédérale de Lausanne (EPFL), Lausanne, Switzerland
Swiss Federal Institute for Forest, Snow and Landscape Research (WSL), Lausanne, Switzerland
UMR CNRS 6249 Laboratoire de Chrono-Environnement, Universite´ de Franche-Comté, Besançon, France

Christine Ferrier-Pagès
Ecophysiology Team, Centre Scientifique de Monaco, Monaco, Principality of Monaco

Index

P

Q

R

S

T

U

W

www.ingramcontent.com/pod-product-compliance
Lightning Source LLC
Chambersburg PA
CBHW080528200326
41458CB00012B/4363
* 9 7 8 1 6 8 2 8 6 5 8 5 9 *